Green Energy and Technology

Climate change, environmental impact and the limited natural resources urge scientific research and novel technical solutions. The monograph series Green Energy and Technology serves as a publishing platform for scientific and technological approaches to "green"—i.e. environmentally friendly and sustainable—technologies. While a focus lies on energy and power supply, it also covers "green" solutions in industrial engineering and engineering design. Green Energy and Technology addresses researchers, advanced students, technical consultants as well as decision makers in industries and politics. Hence, the level of presentation spans from instructional to highly technical.

Indexed in Scopus.

Indexed in Ei Compendex.

Mardiana Idayu Ahmad • Abdul Khalil H. P. S.

Environmental and Energy Technology

Micro to Nano Bio-Based Materials

Springer

Mardiana Idayu Ahmad
School of Industrial Technology
Universiti Sains Malaysia
Penang, Malaysia

Abdul Khalil H. P. S.
Faculty of Business, Hospitality and Technology
Universiti Islam Melaka
Kuala Sungai Baru, Melaka, Malaysia

ISSN 1865-3529 ISSN 1865-3537 (electronic)
Green Energy and Technology
ISBN 978-3-032-10705-3 ISBN 978-3-032-10706-0 (eBook)
https://doi.org/10.1007/978-3-032-10706-0

This Springer imprint is published by the registered company Springer Nature Switzerland AG
The registered company address is: Gewerbestrasse 11, 6330 Cham, Switzerland

Preface

The book *Environmental and Energy Technology: Micro to Nano Bio-Based Materials* presents a comprehensive and timely exploration into the growing field of green and sustainable technologies. In an era where environmental degradation and rising energy demands threaten global well-being, this book offers an integrated perspective on how bio-based materials, spanning the micro to nano scale, can be strategically leveraged for advanced environmental and energy applications.

This book bridges the domains of environmental science, materials engineering, and renewable energy technologies. It provides readers whether from academia, industry, or policy with a grounded understanding of the transformative potential of micro- and nano-scale bio-based materials in addressing pressing sustainability challenges. This book examines the fundamentals, applications, and future trajectories of emerging green materials and technologies. It offers a multidisciplinary platform, merging insights from green chemistry, nanotechnology, energy systems, and the circular bioeconomy. Through this, it illuminates how technological innovation, when aligned with Sustainable Development Goals (SDGs), can lead to impactful, real-world solutions.

This book consists of ten carefully structured chapters. Chapter 1 introduces the foundational context of environmental and energy challenges, setting the stage for how bio-based materials offer sustainable solutions to current ecological pressures. Chapter 2 explores the origin, synthesis, and essential characteristics of bio-based materials, especially those derived from renewable sources such as plants, microbes, and biomass waste. Chapter 3 explores the intersection of nanotechnology with bio-based materials, demonstrating how nanostructuring enhances performance in energy, packaging, and environmental applications. Chapter 4 focuses on the role of these materials in pollution control, detailing their capabilities in filtration, adsorption, and catalytic degradation across air, water, and soil environments. Chapter 5 presents their integration into renewable energy systems, from hydrogen storage to solar cells and biofuels. Chapter 6 highlights the unique advantages of nanostructured bio-aerogels, emphasizing their low-density, high-surface-area properties for energy storage and thermal management. Chapter 7 investigates the production of biodiesel from waste frying oils, underscoring the benefits of nanocatalysis in waste-to-energy conversion. Chapter 8 examines how nanomaterial-based insulation contributes to enhanced energy efficiency in buildings, aligning green construction with climate goals. Chapter 9 provides insight into air purification using bio-aerogels, including applications for indoor air quality improvement and industrial emission control. Finally, Chapter 10 reflects on current limitations and future directions for bio-based nanotechnologies, including scalability, safety, and integration with emerging technologies such as artificial intelligence and digital monitoring.

We believe that this book will serve as a valuable reference for a wide range of audiences, including researchers and graduate students in materials science, environmental engineering, and energy technology who are seeking to deepen their understanding of sustainable innovations. It also aims to benefit industry professionals and innovators committed to advancing green and renewable solutions, as well as policymakers and institutions involved in shaping the future of green infrastructure and the responsible use of natural resources. We extend our deepest gratitude to all contributors for their scholarly insights and to the editorial and production teams whose efforts made this publication possible. It is our sincere hope that this work will inspire continued research, foster innovation, and encourage meaningful collaboration to drive forward the global transition toward clean energy and a sustainable future.

Penang, Malaysia — Mardiana Idayu Ahmad
Kuala Sungai Baru, Melaka, Malaysia — Abdul Khalil H. P. S.
August 2025

Contents

About the Authors

Mardiana Idayu Ahmad received her PhD in Sustainable Energy Technologies University of Nottingham, United Kingdom, and is currently an Associate Professor in Environmental Technology at Universiti Sains Malaysia. She is also a Fellow of the World Society on Sustainable Energy Technologies (WSSET). Her research expertise spans environmental technology, environmental management, and engineering science, with a particular focus on sustainable energy technologies and bio-based micro-nano materials for environmental applications. Over her career, she has served as principal investigator and co-investigator in numerous high-impact projects, and has published over 200 scholarly works, including journal articles, research books, book chapters, and conference papers. An active member of the WSSET, Malaysian Board of Technologists and the Institution of Engineering and Technology (IET), she has engaged extensively with global experts, and contributed as a committee member for multiple regional and international conferences. Beyond academia, she has been actively involved in consultancy projects in sustainable energy and environmental technology, collaborating with industry and client organizations, effectively bridging research with practical applications to address complex real-world challenges.

Abdul Khalil H. P. S. is an internationally renowned expert in biopolymers and biocomposites, particularly biopolymer-based aerogels. He currently serves as the Vice Chancellor of Universiti Islam Melaka (UNIMEL), Malaysia. He earned his Doctoral Degree in Biocomposites from the University of Wales, United Kingdom, and has been affiliated with the School of Industrial Technology, Universiti Sains Malaysia (USM) since 1992. Over more than 32 years, he has built an extensive career in teaching and research, focusing on sustainable materials for diverse applications in engineering, environmental science, biomaterial, biomedical, and other industrial sectors. Abdul Khalil H. P. S. is a Fellow of the International Association of Advanced Materials (FIAAM), Sweden, and the Institute of Materials, Minerals and Mining (FIMMM), United Kingdom. He has received numerous prestigious accolades, including the World Class Professor title (2018–2022) from the

Indonesian Ministry of Education and Culture, the Malaysia Rising Star Award, multiple Malaysia Research Star Awards, and the Top Research Scientist Malaysia designation by the Academy of Sciences Malaysia. He has published over 400 indexed papers, garnered more than 20,000 citations, and holds an H-index of 70 (Scopus). In addition, he has authored seven books, contributed to 50 book chapters, and serves on the editorial boards of six ISI/Scopus-indexed journals, while having reviewed over 300 reputable research articles. His research has catalyzed numerous international collaborations, and his expertise is recognized through adjunct professorships at Universiti Pendidikan Sultan Idris (UPSI), Malaysia; Universitas Sumatera Utara (USU), Indonesia; and Universitas Syiah Kuala (USK), Indonesia. In recognition of his exceptional contributions to global research, scholarly publications, and academic leadership, Abdul Khalil H. P. S. has been awarded the Doctor Honoris Causa (D.HC.) by the Ministry of Education, Culture, Research, and Technology (MoECRT), Indonesia, through Universitas Syiah Kuala.

Chapter 1
Introduction to Environmental and Energy Technologies

1.1 Overview of Environmental and Energy Challenges

The global environment and energy landscape are experiencing unprecedented challenges driven by rapid industrialisation, urbanisation, and population growth. Human activities have intensified climate change, air and water pollution, resource depletion, and biodiversity loss, all of which pose significant threats to ecological balance and human well-being. The continued reliance on fossil fuels remains a major driver of greenhouse gas emissions, exacerbating global warming and disrupting natural systems. Furthermore, unsustainable consumption patterns and inefficient energy use have placed immense pressure on natural resources, making the transition to renewable energy sources an urgent necessity. These challenges are intricately interconnected, requiring a multidimensional approach that integrates scientific innovation, policy interventions, and sustainable practices. A comprehensive understanding of environmental and energy complexities is essential for developing effective strategies to mitigate negative impacts, enhance resilience, and ensure a sustainable future for both the planet and future generations.

Environmental Challenges The environmental challenges in the twenty-first century are becoming more severe at an alarming rate, which is being driven by rapid industrialisation, urban development, and consumption of resources that are not consistent with sustainable practices. Because of the intricate connections between these concerns, a cascading effect is being created on ecosystems, public health, and the climate of the entire planet. The adverse effects of human activities highlight the urgent need for new solutions and sustainable practices. These consequences include the growth in greenhouse gas emissions as well as the widespread loss of biodiversity caused by human activities. This section explores the causes, impacts, and necessity of transformative actions for five main environmental challenges.

M. I. Ahmad, A. K. H. P. S., *Environmental and Energy Technology*, Green Energy and Technology, https://doi.org/10.1007/978-3-032-10706-0_1

(i) **Climate change and global warming**
Anthropogenic greenhouse gas (GHG) emissions, primarily from the combustion of fossil fuels, have significantly contributed to rising global temperatures, intensifying the frequency and severity of extreme weather events, accelerating sea-level rise, and causing widespread ecological disruptions. According to the Intergovernmental Panel on Climate Change [11], without immediate and substantial mitigation efforts, global temperatures could exceed 2 °C above pre-industrial levels by the end of the century, leading to catastrophic consequences for both natural ecosystems and human societies. One of the most pressing concerns is ocean acidification, a direct result of elevated atmospheric CO_2, which threatens marine biodiversity, particularly coral reefs and shellfish population which are the key components of coastal ecosystems and economies. While proactive interventions may still mitigate some of these impacts, the consequences of climate change are already unfolding, manifesting as severe wildfires, intensifying storms, and other extreme weather events. These escalating crises highlight the urgent need for global action, as climate change is no longer a distant concern but a present reality with far-reaching implications for both the environment and human well-being.

(ii) **Air pollution**
Air pollution refers to the presence of one or more contaminants or pollutant substances in the atmosphere at concentrations that pose risks to human health, ecosystems, and climate stability. Technically, it encompasses the emission, dispersion, chemical transformation, and deposition of gases, particulate matter (PM), and biological molecules from both anthropogenic and natural sources. Common pollutants include nitrogen oxides (NO_x), sulfur dioxide (SO_2), carbon monoxide (CO), ozone (O_3), volatile organic compounds (VOCs), and fine particulate matter ($PM_{2.5}$ and PM_{10}) [18]. These pollutants interact through complex photochemical and thermodynamic processes, resulting in secondary pollutants such as tropospheric ozone and secondary organic aerosols (SOAs), which further exacerbate environmental degradation. Advanced studies utilize atmospheric dispersion modeling, source apportionment, and satellite remote sensing to monitor and analyze air quality dynamics. Air pollution is not merely an environmental issue but also a transboundary and multi-sectoral challenge interlinked with industrial processes, urbanization, energy systems, and transport infrastructure.

(iii) **Water scarcity and pollution**
Water scarcity and pollution represent interconnected global challenges that critically impact environmental sustainability, economic development, and public health. Water scarcity is defined as the physical or economic unavailability of freshwater resources to meet the demands of a region, often quantified by thresholds such as below 1000 m^3 per capita annually (Falkenmark index). It arises from population growth, climate change, over-abstraction, and inefficient water use, leading to stress on hydrological cycles and reduced water security. Concurrently, water pollution involves the degradation of water quality due to the discharge of contaminants such as nutrients, heavy metals,

pharmaceuticals, and microplastics into surface and groundwater bodies. These pollutants originate from domestic, industrial, and agricultural sources and are exacerbated by inadequate wastewater treatment infrastructure. Technically, pollution alters the physicochemical and biological characteristics of water, impairing ecosystem functions and rendering water unsuitable for human consumption, agriculture, or industrial use. The compounding effects of scarcity and pollution create a feedback loop that diminishes the resilience of water systems and undermines progress toward achieving Sustainable Development Goal 6 (clean water and sanitation). According to the United Nations World Water Development Report [23], over two billion people currently live in water-stressed regions, while nearly 80% of global wastewater is discharged untreated into the environment.

(iv) **Biodiversity loss**

Biodiversity loss refers to the decline in the variety, variability, and abundance of living organisms across genetic, species, and ecosystem levels, resulting in diminished ecological integrity and resilience. It is driven by a combination of anthropogenic pressures including habitat destruction, climate change, overexploitation, invasive species, and pollution. Technically, biodiversity encompasses ecosystem diversity (e.g., forests, wetlands), species richness (taxonomic diversity), and genetic heterogeneity within populations, all of which underpin essential ecosystem services such as pollination, nutrient cycling, climate regulation, and disease control. The disruption of these components compromises ecological stability and reduces adaptive capacity in the face of environmental change. Recent global assessments, such as the Intergovernmental Science-Policy Platform on Biodiversity and Ecosystem Services [10], estimate that approximately one million species are at risk of extinction within decades, indicating an unprecedented rate of biodiversity decline primarily due to land-use change and unsustainable resource consumption. Biodiversity loss also undermines human well-being, particularly in communities that depend directly on natural ecosystems for livelihoods, food security, and cultural identity. Addressing this issue requires systemic conservation strategies, ecological restoration, sustainable land management, and integration of biodiversity values into economic and policy frameworks.

(v) **Soil pollution**

Soil pollution is a significant environmental challenge, arising from the accumulation of harmful substances, including industrial chemicals, heavy metals, pesticides, and untreated waste, which degrade soil quality and threaten ecosystems. The primary sources of soil pollution include industrial waste disposal, agricultural runoff laden with synthetic fertilizers and pesticides, mining activities, and improper management of municipal and hazardous waste [20]. Contaminated soil not only reduces agricultural productivity by disrupting the delicate balance of soil nutrients and microorganisms but also poses serious health risks to humans and animals through the food chain. For instance, crops grown in polluted soil can absorb toxic substances, such as cadmium or arsenic, which, when consumed, contribute to chronic health issues like cancer, developmental disorders, and neurological damage. Moreover, soil pollution exacerbates climate change by releasing stored carbon dioxide and methane into the atmosphere, further intensifying global warming. Efforts to address soil pollution must focus on sustainable land management practices, such as organic farming, effective waste management, and bioremediation techniques that utilize plants and microorganisms to detoxify contaminated soils. By mitigating soil pollution, we can safeguard food security, protect biodiversity, and ensure the sustainability of natural resources for future generations.

(vi) **Ineffective waste management**

Growing concerns about environmental sustainability have been exacerbated by inefficient waste management practices, leading to severe environmental and health risks. Industrial waste, including solid and hazardous waste, is a major contributor to pollution, as many industries discharge untreated or partially treated waste containing organic and inorganic contaminants. While industries are vital to economic growth, their role in pollution cannot be overlooked. Industrial pollution degrades soil and water quality, posing significant threats to ecosystems and human health. Among various pollution sources, industrial waste is particularly concerning due to the widespread use of non-biodegradable, poorly biodegradable, or low-cost chemicals to maximize production efficiency. However, the long-term environmental and health effects of these chemicals are often disregarded. Additionally, improper disposal of solid and hazardous waste intensifies contamination, further endangering public health and ecosystems [15]. Addressing these challenges requires robust regulatory frameworks and sustainable waste management strategies to mitigate industrial pollution and safeguard the environment.

Energy Challenges The world's energy systems are facing unprecedented challenges driven by growing demand, finite resources, and the urgent need to mitigate environmental impacts. The energy sector accounts for a significant share of global greenhouse gas emissions, making its transformation essential for addressing climate change. Moreover, disparities in energy access, inefficiencies in energy use, and the integration of renewable energy systems further complicate the global energy landscape. This section examines the energy challenges, including dependence on fossil fuels, energy access and inequality, inefficiencies, renewable energy integration, and the interdependence between energy and water resources.

(i) **Dependence on fossil fuels**

The global energy system remains overwhelmingly dependent on fossil fuels in terms of coal, oil, and natural gas which collectively supply over 80% of global energy demand [9]. This persistent reliance presents a twofold crisis: not only are fossil fuels the dominant source of anthropogenic greenhouse gas emissions, intensifying global warming and environmental degradation, but they are also finite resources, thus threatening long-term energy security. The environmental consequences are extensive. Coal combustion contributes significantly to carbon dioxide and particulate emissions, deteriorating air quality and increasing mortality from respiratory diseases. Oil exploitation, transport, and accidental spills lead to marine pollution and biodiversity loss, while natural gas, although often labeled a "bridge fuel" emits methane during extraction and distribution, a greenhouse gas far more potent than CO_2 in the short term. In response, transitioning to renewable energy sources such as solar, wind, and hydropower has become central to achieving global climate goals and sustainable development. These alternatives are not only low-carbon and abundant but are increasingly cost-competitive due to advancements in technology and supportive policies. Nonetheless, the transition is fraught with challenges, including grid integration complexities, high upfront capital costs, and institutional resistance from entrenched fossil fuel stakeholders. To overcome these barriers, policy mechanisms such as carbon pricing, fossil fuel subsidy reform, and investments in innovation ecosystems are vital. Addressing fossil fuel dependency is no longer optional but it is a strategic imperative for environmental sustainability, economic resilience, and global equity in the face of a changing climate.

(ii) **Energy access and inequality**

Energy access remains a critical global challenge, with hundreds of millions of people still living without electricity, highlighting the widespread issue of energy poverty. This lack of access limits economic growth by reducing productivity and hindering industrial development, while also impacting essential services such as education and healthcare. Students struggle to study after dark, and healthcare facilities often lack reliable power for lighting, vaccine storage, and life-saving equipment. Energy poverty further reinforces social inequalities, particularly affecting women and children who often collect traditional fuels that are inefficient and harmful to health. Addressing this issue is vital for achieving Sustainable Development Goal 7, which calls for universal access to affordable, reliable, and modern energy. Expanding energy access through inclusive and sustainable solutions such as decentralized renewable energy systems like solar microgrids and wind turbines can improve livelihoods, enhance infrastructure, foster innovation, and promote equitable development across all communities.

(iii) **Energy inefficiency**

Energy inefficiency continues to be a critical challenge in the global energy sector, with significant losses occurring during the stages of generation, transmission, and consumption. These inefficiencies are not only economically costly but also environmentally damaging, as they lead to excessive energy production to meet demand, thereby increasing greenhouse gas emissions. Key to overcoming energy inefficiency is the adoption of advanced technologies and innovative solutions. Smart grids, for example, can enhance the efficiency of energy distribution by balancing supply and demand in real-time, reducing waste and ensuring a more reliable energy supply. Similarly, energy-efficient appliances, such as LED lighting and modern HVAC systems, contribute to substantial energy savings at the consumer level. Sustainable building designs that incorporate high-performance insulation, renewable energy integration, and passive cooling techniques further optimize energy use in the residential and commercial sectors. Investments in these innovations not only reduce energy inefficiency but also promote economic and social benefits. Lower energy costs enable industries to operate more competitively, while consumers experience reduced utility bills. Additionally, optimizing energy use contributes to energy security by reducing reliance on fossil fuels and minimizing the risks associated with energy supply disruptions. A concerted global effort to improve energy efficiency is vital to achieving a sustainable energy future, ensuring that the world's growing energy needs are met without compromising environmental integrity or resource availability.

(iv) **Renewable energy integration**

Integrating renewable energy sources, such as solar and wind, into existing energy systems remains a critical yet complex endeavor. The inherent intermittency of these sources, driven by natural fluctuations in sunlight and wind patterns, poses significant challenges to maintaining grid stability. This variability can lead to periods of surplus energy generation during peak conditions and shortages when conditions are unfavorable, disrupting the balance of energy supply and demand. Addressing these challenges is essential to achieving a sustainable and reliable energy transition. Investments in advanced energy storage technologies are fundamental to overcoming the intermittency issue and ensuring the seamless integration of renewables. Innovations such as lithium-ion batteries, hydrogen storage systems, and pumped hydro storage provide viable solutions for storing excess energy generated during high-production periods and deploying it during low-production intervals. These technologies not only enhance grid stability but also support the scalability of renewable energy by accommodating fluctuations in energy output. Modernizing energy grids is equally important to support renewable energy integration. Smart grids equipped with advanced digital communication and control systems enable real-time monitoring and optimization of energy flow, enhancing grid resilience and reliability. Decentralized energy systems, including microgrids and community-based renewable energy projects, further contribute to the diversification and resilience of energy distribution networks. These systems reduce dependence on centralized power plants and minimize transmission losses, particularly in remote and underserved areas. By addressing the challenges of renewable energy integration through strategic investments and technological advancements, the global energy sector can move closer to achieving a sustainable and resilient energy future. Such efforts not only reduce reliance on fossil fuels but also contribute to environmental preservation, energy security, and the achievement of international climate goals.

(v) **Nexus of energy and water**

Energy production is intrinsically tied to water resources, with significant reliance on water for processes such as cooling, extraction, and fuel processing. This dependency creates acute vulnerabilities in regions already experiencing water scarcity, where competition for limited water supplies exacerbates challenges across both energy and agricultural sectors. Thermal power plants, for example, require vast quantities of water for cooling, while fossil fuel extraction and processing further amplify water demands, making the energy-water nexus a critical area of concern. Conversely, water supply systems themselves are heavily energy-intensive. Activities such as pumping, treatment, and desalination of water often involve substantial energy consumption, especially in areas with limited freshwater availability or high salinity levels. For instance, desalination plants, while instrumental in meeting water needs in arid regions, are known for their high energy requirements, which increase operational costs and environmental impacts. This reciprocal relationship underscores the intricate interdependence between energy and water systems, making efficiency improvements in one sector vital for the sustainability of the other. To address this dual challenge, developing water-efficient energy technologies and integrating water-saving strategies in power generation have become imperatives. Innovations such as dry cooling systems, which significantly reduce water usage in thermal power plants, and the adoption of closed-loop water systems are proving to be effective. Furthermore, renewable energy technologies like solar and wind, which require minimal water for operation, present a promising solution to decouple energy production from water dependence. Similarly, integrating energy-efficient practices into water systems such as upgrading pumping technologies and incorporating renewable energy into treatment facilities can drastically lower energy consumption in water supply processes.

1.2 Role of Technology in Addressing Global Sustainability Issues

The urgent need to mitigate climate change, reduce environmental degradation, and ensure sustainable resource management has placed technology at the forefront of global sustainability efforts. As the world grapples with rising energy demands and increasing pollution, technological advancements in energy production, environmental protection, and material science offer promising solutions. The development of renewable energy, smart energy systems, green technology, and sustainable materials is critical in addressing these challenges. By integrating these innovations, societies can transition towards a low-carbon economy while preserving natural resources for future generations. A key strategy in reducing carbon emissions and mitigating climate change is the shift from fossil fuels to renewable energy sources. Solar, wind, hydro, and bioenergy have gained global attention due to their potential to generate clean energy while minimizing environmental impact. Innovations such as high-efficiency photovoltaic (PV) cells and floating solar farms have improved the accessibility and effectiveness of solar energy. Similarly, wind energy technologies, particularly offshore wind farms with enhanced turbine designs, contribute to reliable and large-scale renewable electricity production. However, the intermittent nature of renewable energy sources necessitates advancements in energy storage systems, such as lithium-ion and solid-state batteries, which enhance grid stability and ensure a consistent power supply.

In parallel, smart energy systems are revolutionizing the way energy is managed and consumed. The integration of artificial intelligence (AI) and the Internet of Things (IoT) in energy grids allows for real-time monitoring, predictive maintenance, and optimization of energy usage. Smart grids enhance electricity distribution efficiency, reducing energy losses and improving demand-response mechanisms. Additionally, IoT-enabled smart meters and automated home energy systems empower consumers to reduce electricity waste, contributing to energy conservation efforts. These advancements in energy production and management pave the way for a more resilient and adaptive energy infrastructure, aligning with global sustainability goals. While renewable energy addresses energy production challenges, green technology and sustainable materials play a vital role in reducing environmental impact across industries. One of the most significant technological innovations in this area is carbon capture and storage (CCS), which enables industries to capture and store CO_2 emissions, thereby preventing them from entering the atmosphere. Additionally, green hydrogen production, powered by renewable energy, is emerging as a key solution for decarbonizing transportation and heavy industries.

Alongside green technology, the development of sustainable materials is transforming industries such as construction, manufacturing, and packaging. The adoption of biodegradable polymers, recycled composites, and plant-based materials reduces dependency on non-renewable resources and minimizes waste generation [15]. In the construction sector, eco-friendly materials such as geopolymer concrete and bio-based insulation improve energy efficiency and reduce carbon footprints [8]. Additionally, nanotechnology-based materials are being explored for energy-efficient insulation, enhancing building sustainability and reducing heating and cooling demands [1]. These innovations demonstrate how material science and green technology work together to create more sustainable production and consumption patterns. Beyond energy and materials, environmental protection technologies are crucial in addressing pollution, waste management, and ecosystem restoration. Advanced water purification systems, including reverse osmosis, membrane filtration, and AI-powered water quality monitoring, enhance access to clean water while reducing contamination risks. Similarly, AI and robotics are improving waste management through automated sorting systems, which increase recycling efficiency and resource recovery.

One of the most promising approaches to environmental sustainability is the circular economy, which emphasizes reducing waste, reusing materials, and recycling resources. Waste-to-energy technologies, such as anaerobic digestion and gasification, convert organic waste into bioenergy, reducing landfill dependency while generating renewable power. Bioremediation techniques, which use microorganisms to break down pollutants, are being employed to clean up oil spills, plastic waste, and heavy metal contamination [17]. These interconnected efforts demonstrate how technology supports the transition from a linear "take-make-dispose" model to a sustainable, regenerative economic system. The integration of technological advancements in renewable energy, smart energy systems, green materials, and environmental protection is pivotal in addressing global sustainability challenges. By harnessing these innovations, societies can transition towards a low-carbon economy, mitigate climate change, and reduce environmental degradation. However, achieving a truly sustainable future requires not only technological progress but also strong policy support, financial investments, and international collaboration. Governments, industries, and researchers must work together to accelerate the adoption of these technologies and ensure that sustainability remains a priority in global development. As the world moves towards a more sustainable future, technology will continue to be a driving force in creating resilient, efficient, and environmentally friendly solutions.

1.3 Fundamentals of Environmental Technologies

1.3.1 Definition and Classification of Environmental Technologies

Environmental technologies, also known as green technologies or clean technologies, encompass a wide range of innovations designed to minimize environmental impact, enhance sustainability, and promote the efficient use of natural resources. These technologies aim to mitigate pollution, optimize energy and water use, and support ecological balance while enabling industrial and economic development The increasing global focus on climate change and environmental degradation has led to the rapid advancement and adoption of environmental technologies across various sectors, including energy production, waste management, and industrial processes.

The classification of environmental technologies can be broadly divided into preventive, control, and remediation technologies. Table 1.1 shows the classification of environmental technologies. Preventive technologies focus on reducing environmental harm at the source by using cleaner production methods, energy-efficient processes, and sustainable materials.

1.3.2 Pollution Control, Waste Management, and Resource Efficiency

Pollution control is a foundational component of environmental protection strategies, involving the deployment of engineering solutions, regulatory mechanisms, and behavioral changes to mitigate the release of harmful substances into the environment. This domain spans the management of air, water, and soil contaminants through both end-of-pipe technologies such as electrostatic precipitators and

Table 1.1 Classification of environmental technologies

Classification	Definition	Examples
Preventive technologies	Focus on reducing environmental harm at the source through cleaner production, energy efficiency, and sustainable materials	Green chemistry, eco-friendly industrial processes, closed-loop manufacturing systems
Control technologies	Aim to capture and reduce pollutants before they are released into the environment	Carbon capture and storage (CCS), electrostatic precipitators, wastewater treatment facilities
Remediation technologies	Involve restoring polluted environments through biological or chemical means	Bioremediation, phytoremediation, chemical neutralization of contaminants in soil and water

activated carbon filters and process-integrated interventions that promote cleaner production. Contemporary shifts in pollution control have emphasized proactive, source-reduction strategies, particularly in light of the emergence of persistent pollutants like microplastics and per- and polyfluoroalkyl substances (PFAS), which conventional technologies struggle to remove [4]. These developments underscore the necessity for dynamic and adaptive pollution control frameworks that are integrated with real-time monitoring systems and supported by robust environmental governance.

Building upon pollution control, waste management addresses the lifecycle of residual materials, providing a structured approach to collection, treatment, and final disposal (Fig. 1.1). Modern waste management systems are guided by the waste hierarchy, which prioritizes strategies such as prevention, reuse, recycling, and energy recovery over conventional landfilling or incineration [15]. The technological landscape has evolved to include advanced approaches such as mechanical-biological treatment, anaerobic digestion, pyrolysis, and automated material recovery facilities (MRFs), which enable transformation of waste into value-added products and energy [22]. However, technological efficiency must be coupled with institutional coordination, public engagement, and financial viability to ensure sustained performance. Moreover, improper waste disposal remains a major contributor to environmental pollution, reinforcing the inherent linkage between waste management and pollution control [3]. Therefore, a systemic and integrated framework is essential to address both waste streams and pollutant pathways.

Resource efficiency extends these efforts by promoting the sustainable and optimized use of materials and energy across entire value chains. It focuses on minimizing environmental impacts per unit of output through strategies such as dematerialization, process redesign, and product eco-design. Analytical tools such as material flow analysis (MFA), life cycle assessment (LCA), and input-output modeling are commonly employed to quantify resource intensity and inform strategic interventions. Importantly, resource efficiency addresses the root causes of waste and pollution by enhancing upstream decision-making. For instance, modular

Fig. 1.1 Waste management for pollution control

product design and material substitution in electronics manufacturing have shown significant reductions in both resource demand and end-of-life waste. As such, resource efficiency acts as a linchpin in unifying industrial productivity with environmental sustainability, reinforcing and complementing the principles of pollution control and waste management.

In synthesis, the interdependence of pollution control, waste management, and resource efficiency represents a cohesive triad essential for advancing global sustainability agendas, particularly under the UN Sustainable Development Goals (SDGs). These domains are increasingly integrated through circular economy models and industrial symbiosis frameworks, which emphasize closed-loop systems and multi-stakeholder collaboration. Synergies are evident in initiatives such as waste-to-energy (WtE) systems, which simultaneously reduce landfill volumes, recover energy, and mitigate emissions. Furthermore, policy mechanisms like extended producer responsibility (EPR) encourage upstream accountability, incentivizing sustainable design and improved material recovery. Promoting coherence across these three areas is vital not only for ecological preservation but also for fostering economic resilience and reducing systemic vulnerabilities in a resource-constrained world.

1.4 Evolution of Energy Technologies and Innovations

The evolution of energy technologies reflects a dynamic interplay between technological innovation, socio-economic demand, and environmental constraints. Historically, energy systems were dominated by biomass and manual labor, which gave way to coal during the Industrial Revolution, marking the onset of centralized fossil-based power generation. The twentieth century saw a rapid expansion of oil and natural gas infrastructure, along with the introduction of nuclear power. However, this fossil-fuel-centric trajectory has increasingly come under scrutiny due to concerns over greenhouse gas emissions, energy insecurity, and resource depletion. In response, a global transition is underway, steering toward renewable and low-carbon energy systems driven by policy frameworks, decarbonization targets, and technological advancements.

Technological innovations in renewable energy systems such as solar photovoltaics, wind turbines, and bioenergy have substantially altered the global energy landscape. Advances in material science and manufacturing techniques have driven down the levelized cost of electricity (LCOE) for renewables, making them competitive with or even cheaper than fossil fuels in many regions Additionally, the integration of digital technologies, including the Internet of Things (IoT), artificial intelligence (AI), and blockchain, is reshaping energy management through smart grids, predictive maintenance, and decentralized generation. These technologies enhance energy efficiency, grid resilience, and user autonomy, promoting a more participatory and responsive energy ecosystem. Moreover, innovations in hydrogen production, carbon capture and storage (CCS), and next-generation nuclear reactors are expanding the frontier of clean energy possibilities, albeit with ongoing challenges in scalability, cost, and regulatory acceptance.

The shift towards cleaner energy technologies is deeply interlinked with broader sustainability efforts, particularly in how energy systems intersect with resource efficiency and waste minimization. The integration of circular economy principles into the energy sector encourages the reuse and recycling of components such as solar panels, wind turbine blades, and battery cells thereby reducing raw material dependence and lifecycle emissions. For instance, the deployment of second-life batteries from electric vehicles for stationary energy storage exemplifies a resource-efficient innovation that bridges energy and waste management sectors. The rise of industrial symbiosis allows waste heat and by-products from one industry to serve as input for another, maximizing energy recovery and minimizing environmental burden. These practices underline a systemic approach to innovation, where energy transitions are pursued in harmony with material and ecological flows.

Looking forward, the evolution of energy technologies must navigate complex trade-offs between scalability, affordability, social equity, and ecological boundaries. While technological progress has been remarkable, the pace and scope of deployment remain uneven across regions due to infrastructural, financial, and institutional barriers. In low- and middle-income countries, energy access and affordability remain critical issues, highlighting the need for inclusive innovation pathways that address development needs while enabling decarbonization. Simultaneously, the ethical and environmental implications of resource extraction for clean energy technologies such as lithium, cobalt, and rare earth elements necessitate more sustainable and transparent supply chains. Future innovations must therefore emphasize not only technological efficacy but also system-wide sustainability, reinforcing the imperative for integrated policies, interdisciplinary research, and collaborative governance mechanisms.

1.5 Interconnection Between Environmental and Energy Technologies

The relationship between environmental and energy technologies is a crucial aspect of sustainable development. As global energy consumption continues to rise, so does its environmental impact, contributing to climate change, pollution, and resource. The interconnected nature of these two domains necessitates an integrated approach where technological innovations in energy production align with environmental conservation strategies. Sustainable energy solutions must not only meet the increasing global demand for power but also mitigate ecological damage and promote long-term environmental resilience.

Historically, the industrial revolution marked a significant increase in energy consumption, primarily relying on fossil fuels. The reliance on coal, oil, and natural gas led to severe environmental consequences, including air and water pollution, biodiversity loss, and greenhouse gas (GHG) emissions. The continued use of conventional energy sources has accelerated global warming, necessitating a shift toward renewable energy, energy efficiency, and carbon mitigation strategies. In response, modern energy technologies have been developed to balance energy demands with sustainability goals, creating a strong link between the environmental and energy sectors.

The transition towards sustainable energy is not just about replacing fossil fuels with renewables. It involves systemic changes in energy infrastructure, resource management, industrial processes, and economic policies. Technologies such as smart grids, energy storage, and carbon capture play an essential role in mitigating the negative environmental impacts of energy systems while ensuring economic viability. Additionally, circular economy models have been introduced to optimize material use and minimize waste, strengthening the synergy between environmental conservation and energy efficiency.

1.5.1 Energy Production and Environmental Impact

Energy production is a key contributor to environmental degradation, primarily due to emissions from fossil fuel combustion. Carbon dioxide (CO_2), methane (CH_4), and nitrous oxide (N_2O) emissions from energy-intensive industries, transportation, and power generation are responsible for climate change and severe air pollution [12]. The extraction, processing, and transportation of fossil fuels also have significant ecological consequences, including habitat destruction, groundwater contamination, and oil spills.

Beyond air pollution, the water footprint of energy production is another environmental concern. Traditional power plants, including coal and nuclear facilities, require large amounts of water for cooling, leading to thermal pollution and water scarcity in certain regions. Hydroelectric projects, despite being a renewable source of energy, can also disrupt aquatic ecosystems by altering water flow, affecting fish migration, and contributing to habitat loss. The environmental impacts of energy production emphasize the need for cleaner energy alternatives and improved efficiency in energy systems.

The shift to renewable energy sources has been widely recognized as a solution to mitigate these environmental impacts. Solar, wind, geothermal, and bioenergy technologies provide sustainable alternatives that significantly reduce emissions and reliance on finite resources. However, even renewable energy systems have environmental implications. The production of solar panels requires mining for rare-earth elements, while wind turbines demand vast land areas, which can lead to ecosystem disruption. Therefore, developing sustainable lifecycle strategies for these technologies is crucial in balancing energy generation with environmental responsibility.

1.5.2 Circular Economy and Resource Recovery Systems

The circular economy is a model that seeks to eliminate waste and ensure the continuous use of materials through reuse, recycling, and recovery. Unlike the traditional linear economy, where products are made, used, and disposed of, a circular approach integrates sustainable resource management into energy systems. Circularity in energy production focuses on maximizing efficiency, minimizing waste, and recovering valuable resources from end-of-life energy technologies. One application of circular economy principles is the reuse and recycling of solar panels and wind turbine components. Solar PV panels have a lifespan of around 25–30 years, after which they generate large amounts of electronic waste [6]. Advanced recycling technologies are now being developed to recover materials such as silicon, silver, and aluminum, reducing the need for new resource extraction. Similarly, wind turbine blades, traditionally made from composite materials, are being repurposed for construction materials and energy recovery projects [7].

Another critical area of circularity in energy systems is waste-to-energy (WTE) technologies. These processes convert municipal, agricultural, and industrial waste into usable energy, reducing landfill dependency and mitigating methane emissions from organic waste decomposition. Technologies such as anaerobic digestion, pyrolysis, and gasification are increasingly being integrated into urban sustainability plans to enhance waste recovery efficiency. The implementation of industrial symbiosis, where waste from one industry serves as input for another, further reinforces circular economy principles. For example, excess heat from industrial processes can be captured and utilized in district heating systems, improving energy efficiency while reducing emissions. These interconnected systems demonstrate how circularity can transform energy infrastructure, making it more resilient and environmentally responsible.

1.5.3 Integrated Approaches for Sustainable Development

Achieving sustainable development requires a holistic approach that incorporates environmental, social, and economic considerations into energy policies and technological innovations. The United Nations Sustainable Development Goals (SDGs) emphasize the importance of integrating clean energy solutions with climate action, economic growth, and societal well-being. Sustainable development frameworks must align with technological advancements, policy reforms, and global cooperation to ensure long-term benefits for both the environment and the economy. One such integrated approach is the nexus framework, which interlinks energy, water, and food security. This model recognizes that energy production impacts water availability and agricultural productivity, necessitating cross-sectoral strategies to enhance efficiency and sustainability. For example, solar-powered irrigation systems are now being deployed in water-scarce regions to support sustainable agriculture while reducing reliance on fossil fuels.

The integration of digitalization and artificial intelligence (AI) is also playing a transformative role in sustainable energy management. Smart grids, AI-driven energy forecasting, and automated demand-response systems optimize energy distribution, reduce waste, and enhance grid stability [5]. These advancements contribute to a more adaptive and resilient energy infrastructure capable of responding to climate variability and energy demand fluctuations. Furthermore, green financing and policy incentives are critical drivers of sustainable energy adoption. Governments worldwide are implementing carbon pricing, renewable energy subsidies, and emission trading schemes to encourage industries and consumers to shift toward cleaner energy solutions. These regulatory mechanisms create economic incentives for innovation while ensuring that environmental externalities are accounted for in energy production and consumption.

The integration of environmental and energy technologies is fundamental to achieving a low-carbon future. By aligning circular economy principles, technological innovation, and sustainable development strategies, societies can transition toward a cleaner, more resilient global energy system. Future advancements in hydrogen energy, nuclear fusion, and energy-efficient materials will further shape the next phase of sustainable energy evolution, making it imperative for policymakers, researchers, and industry stakeholders to collaborate in achieving long-term environmental and energy security.

1.5.4 The Importance of Environmental Technologies and Energy Technologies

Environmental and energy technologies have emerged as critical pillars in the pursuit of sustainable development, particularly in mitigating pollution, conserving finite natural resources, and restoring ecological integrity. These technologies encompass a wide array of innovations, including cleaner production processes, pollution control systems, renewable energy solutions, and circular economy models aimed at reducing the environmental footprint of human activities. Environmental technologies, such as green manufacturing, resource-efficient production, and waste minimization strategies, are instrumental in curbing industrial emissions and reducing the burden on ecosystems.

Simultaneously, energy technologies particularly those based on renewables like solar, wind, and bioenergy are transforming global energy systems toward low-carbon pathways. These technologies not only reduce greenhouse gas emissions but also enhance energy security and resilience, especially in underserved and climate-vulnerable regions. Moreover, integrated environmental-energy technologies, such as carbon capture and storage (CCS) and smart grid systems, demonstrate the synergistic potential of addressing environmental degradation and meeting rising energy demands simultaneously. With climate change, biodiversity loss, and resource scarcity intensifying, the advancement and deployment of these technologies are no longer optional but essential. Effective policy frameworks, green financing mechanisms, and international cooperation are needed to scale up their implementation across sectors and regions. Ultimately, the strategic adoption of environmental and energy technologies is foundational to achieving the targets of the United Nations Sustainable Development Goals (SDGs), particularly SDG 7 (Affordable and Clean Energy) and SDG 13 (Climate Action), while paving the way for a more resilient, equitable, and sustainable future. Table 1.2 shows the emerging environmental and energy technologies with recent innovations.

Table 1.2 The environmental and energy technologies with recent innovations

Category	Technologies	Applications	Recent innovations
Environmental technologies			
Air pollution control technologies	Electrostatic precipitators, scrubbers, catalytic converters, CCUS, direct air capture (DAC)	To reduce harmful emissions (e.g., PM, SO_2, NO_x, CO_2) from industrial, transportation, and energy sources	Carbon capture and storage (CCUS), direct air capture systems
Water and wastewater treatment technologies	Reverse osmosis (RO), ultrafiltration (UF), advanced oxidation processes (AOPs), nanotechnology, AI-integrated monitoring	To remove contaminants and ensure clean water for human, agricultural, and industrial use	Graphene oxide membranes, AI for real-time water quality monitoring
Soil remediation techniques	Bioremediation, phytoremediation, electrokinetic remediation	To restore polluted land by removing or neutralizing contaminants in soil	Electrokinetic remediation, use of genetically engineered microbes
Waste management and recycling innovations	Automated sorting systems, waste-to-energy (WTE), anaerobic digestion, gasification, biodegradable materials	To reduce landfill dependency, increase recycling rates, and recover energy or materials from waste	AI-driven waste sorting, bio-based plastics from renewable feedstocks
Energy technologies			
Solar energy systems	Photovoltaics (PV), concentrated solar power (CSP), solar thermal, perovskite solar cells, solar-battery integration	Electricity generation, heating, and grid integration in residential, commercial, and industrial sectors	High-efficiency perovskite solar cells, AI-based solar tracking systems
Wind and hydropower technologies	Offshore and onshore wind turbines, floating wind turbines, run-of-river hydro, pumped storage hydro, small-scale hydro	Utility-scale renewable energy generation and grid stability	Floating wind farms, fish-friendly turbines for small hydro
Bioenergy and biomass conversion technologies	Second-generation biofuels, biogas production, biomass gasification, algae-based bioenergy	Renewable fuels for transport, electricity generation, and industrial processes	Algae-derived biofuels, integrated waste-to-energy systems
Advanced battery and energy storage systems	Solid-state batteries, lithium-sulfur batteries, hydrogen fuel cells	Stabilizing renewable energy supply, electric vehicles, and backup power systems	Longer-lasting solid-state batteries, reversible hydrogen storage
Emerging energy technologies	Green hydrogen production, electrolysis, nuclear fusion	Decarbonization of hard-to-abate sectors, long-term clean energy supply	Low-cost electrolysis technologies, fusion plasma control systems

1.6 Micro to Nano Scales Bio-based Materials as Green Technology

The transition from micro- to nano-scaled bio-based materials has emerged as a transformative approach in advancing sustainable technologies. These materials, derived from renewable biological resources such as plant biomass, algae, or agricultural waste, offer a green alternative to petroleum-based and non-renewable materials. When engineered at the micro- and nanoscale, these bio-based materials exhibit enhanced physicochemical properties including increased surface area, mechanical strength, reactivity, and functionality, making them attractive for applications in packaging, biomedical devices, energy storage, and environmental remediation [14]. he utilization of such materials aligns with the principles of green chemistry and circular economy, emphasizing biodegradability, low toxicity, resource renewability, and reduced environmental impact throughout their life cycle.

1.6.1 Fundamentals and Definitions

The development of micro- to nano-scale bio-based materials represents a significant advancement in the pursuit of sustainable and green technologies. By integrating concepts from materials science, nanotechnology, and environmental engineering, these materials offer innovative solutions that are not only functionally superior but also environmentally responsible. Engineered from renewable biological resources such as plant fibers, biopolymers, or microbial biomass, these materials are increasingly replacing conventional, fossil-based materials across various applications.

Micro-scale structures are defined as those with dimensions ranging from 1 micrometer (µm) to 1000 µm, while nano-scale materials range from 1 nanometer (nm) to 100 nm. The size of a material plays a crucial role in determining its physical, chemical, and biological behavior. As shown in Fig. 1.2 (adapted from Mejía-Rosales et al. [19]), materials span a wide spectrum of length scales from the atomic-level pico-scale, through the nano- and micro-scales, and up to the macro-scale, highlighting the dimensional diversity relevant in materials science and engineering.

Microtechnology refers to the design, fabrication, and application of devices and structures at the micrometer scale. Techniques such as photolithography, soft lithography, micro-molding, and laser ablation are commonly used to develop functional microstructures, including channels, pores, and patterned surfaces. Microtechnology underpins key innovations in fields such as microelectromechanical systems (MEMS), biosensors, microneedles, and microfluidics, all of which benefit from miniaturization, high sensitivity, and material efficiency. In parallel, nanotechnology involves the manipulation of matter at dimensions less than 100 nm, where unique phenomena emerge due to quantum effects and a high surface-area-to-volume ratio. Nanostructured materials demonstrate enhanced mechanical strength, chemical reactivity, thermal conductivity, and bioavailability compared to their bulk

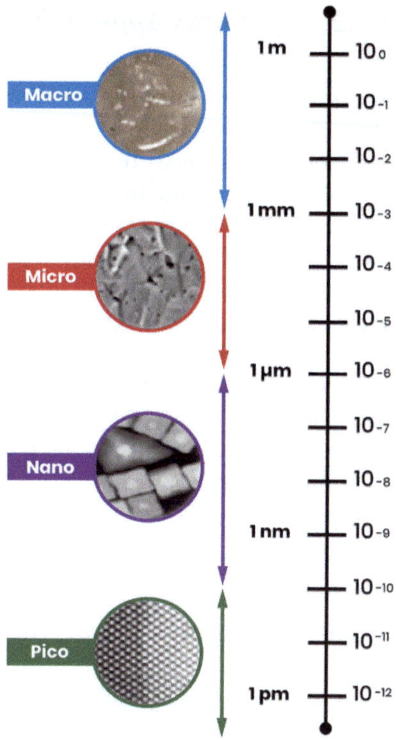

Fig. 1.2 Schematic illustration of material length scales, ranging from the picometer to the meter scale. (Adapted from Mejía-Rosales et al. [19])

counterparts [13]. hese features enable the creation of next-generation materials for drug delivery, water purification, energy storage, and smart packaging.

The integration of bio-based materials into both micro- and nanotechnologies enhances their environmental value. These materials such as nanocellulose, chitosan, polylactic acid (PLA), and starch nanoparticles are renewable, biodegradable, and often derived from agricultural or marine biomass. When processed at the micro- or nano-scale, bio-based materials can be engineered to exhibit superior performance while maintaining ecological compatibility. For example, nanocellulose exhibits high tensile strength, barrier properties, and biocompatibility, making it ideal for applications in biomedical scaffolds, coatings, and environmental filters [2]. y advancing the development of micro- to nano-scale bio-based materials, researchers and industries are contributing to global sustainability goals. These materials support several United Nations Sustainable Development Goals (SDGs), particularly SDG 9 (Industry, Innovation and Infrastructure), SDG 12 (Responsible Consumption and Production), and SDG 13 (Climate Action). Their use enables the creation of low-impact, high-performance materials that reduce dependency on fossil-based resources, minimize waste, and offer biodegradable or recyclable end-of-life options. In summary, micro- and nanotechnologies when combined with bio-based material innovation offer a powerful platform for developing advanced green technologies. This convergence represents not only scientific and technological progress but also a vital strategy for addressing pressing environmental challenges.

1.6.2 Synthesis Approaches and Material Types

The synthesis of micro to nano scale bio-based materials often involves green chemistry approaches that prioritize low energy inputs, minimal waste generation, and benign solvents. Techniques such as high-pressure homogenization, acid hydrolysis, enzymatic treatment, and electrospinning are widely used for producing nanocellulose and chitosan nanofibers, while pyrolysis and ball-milling methods are applied in the preparation of biochar nanoparticles [16]. These methods are frequently enhanced with ultrasonication or microwave irradiation to improve yield and material performance. Materials commonly developed at these scales include nanocellulose (from plant fibers or bacterial fermentation), biochar nanoparticles (from agricultural residues), lignin-derived carbon nanostructures, and starch or chitosan-based nanocomposites. The ability to tune particle morphology, surface chemistry, and porosity offers extensive control over functionality in subsequent applications. Table 1.3 shows the green synthesis approaches for bio-based materials.

1.6.3 Functional Properties and Environmental Significance

Bio-based materials at micro and nano dimensions demonstrate multifunctional characteristics that make them integral to the advancement of green technologies. Nanocellulose, for example, exhibits high tensile strength, thermal stability, and

Table 1.3 Green synthesis approaches for bio-based nanomaterials

Material type	Primary source	Synthesis techniques	Functional tunability
Nanocellulose	Plant fibers, bacterial cellulose	High-pressure homogenization, acid hydrolysis, enzymatic treatment, ultrasonication	Morphology, crystallinity, mechanical strength
Chitosan nanofibers	Crustacean shells	Electrospinning, acidic solubilization, microwave irradiation	Surface area, antimicrobial properties, film-forming
Biochar nanoparticles	Agricultural residues	Pyrolysis, ball-milling, ultrasonication	Porosity, surface chemistry, adsorption efficiency
Lignin-derived carbon nanostructures	Lignocellulosic biomass	Carbonization, microwave-assisted activation	Electrical conductivity, pore structure
Starch-based nanocomposites	Starch-rich plants (e.g., corn, cassava)	Nanoprecipitation, ultrasonication	Flexibility, barrier properties, biodegradability

barrier properties, making it suitable for biodegradable packaging and as reinforcement in composite materials. Biochar nanoparticles possess exceptional adsorption capabilities, attributed to their porous structure and surface chemistry, and are therefore utilized in water purification, heavy metal remediation, and soil amendment [21]. Lignin-based carbon materials show electrochemical activity and conductivity, enabling their application in electrodes for batteries and supercapacitors [24]. The environmental footprint of these materials is considerably lower than that of synthetic nanomaterials, as they are renewable, biodegradable, and often derived from agro-waste or industrial by-products, aligning well with circular economy goals.

1.6.4 Applications Across Sectors

The practical applications of micro to nano scale bio-based materials span several domains. In environmental technology, they are used for pollutant adsorption, catalysis, and membrane filtration in air and water treatment systems. For instance, functionalized nanocellulose membranes offer selective removal of contaminants and pathogens, while biochar enhances nutrient retention and carbon sequestration in soils. In the energy sector, lignin-derived carbon and chitosan nanocomposites are explored for electrodes, separators, and electrolytes in flexible, biodegradable energy storage systems. Furthermore, in the biomedical field, chitosan nanoparticles are applied in drug delivery systems, tissue engineering scaffolds, and wound dressings due to their biocompatibility and antimicrobial activity. The packaging industry benefits from starch and cellulose-based nanomaterials that replace petroleum-based plastics with compostable alternatives, contributing to plastic pollution reduction.

Despite their potential, the commercialization and large-scale implementation of micro to nano scale bio-based materials are constrained by several technical and socio-economic challenges. These include inconsistent material quality, limitations in scalability, high production costs, and the absence of standardized metrics for performance and safety. Moreover, while these materials are biodegradable, their behavior in complex environmental matrices remains underexplored, necessitating comprehensive life cycle assessment (LCA) and ecotoxicological studies. Regulatory frameworks also lag behind technological advancements, creating uncertainty for industries seeking to adopt these innovations. Nevertheless, with growing global emphasis on green manufacturing, decarbonization, and waste valorization, future research should focus on process optimization, hybrid material development, and system-level integration. Strengthening cross-disciplinary collaboration and policy support will be essential to accelerate their transition from lab-scale prototypes to market-ready solutions. Table 1.4 shows classification, properties, and applications of micro to nano bio-based materials.

Table 1.4 Classification, properties, and applications of micro to nano bio-based materials

Material type	Scale	Source	Properties	Primary applications
Nanocellulose	Nano	Plant fibers, bacteria	High tensile strength, barrier function	Packaging, composites, water filtration
Biochar nanoparticles	Nano	Agricultural waste	High surface area, adsorptive capacity	Water/soil remediation, carbon sequestration
Lignin-based carbon	Nano/ micro	Lignocellulosic biomass	Electrical conductivity, renewability	Electrodes in batteries/supercapacitors
Chitosan nanoparticles	Nano	Crustacean shells	Biocompatibility, antimicrobial action	Drug delivery, wound healing, food packaging
Starch nanoparticles	Nano	Corn, cassava	Biodegradability, flexibility	Bioplastics, biodegradable packaging

1.7 Sustainability and Environmental Impact

Sustainability is a fundamental principle in modern technological advancements, ensuring that current resource use does not compromise the needs of future generations. The rapid expansion of industrialization and urbanization has led to increasing environmental challenges, including climate change, resource depletion, and pollution. To counter these effects, sustainability-focused technologies and policies have been developed to reduce carbon footprints, optimize resource efficiency, and support circular economy models. Environmental impact assessments are essential in measuring the sustainability of technologies. Life Cycle Assessment (LCA), carbon footprint analysis, and climate mitigation strategies help determine the environmental costs associated with energy production, material extraction, and waste disposal. The integration of sustainability indicators in technological advancements ensures that innovations contribute positively to environmental goals while maintaining economic viability.

1.7.1 Life Cycle Assessment of Technologies

Life Cycle Assessment (LCA) is a systematic approach used to evaluate the environmental impact of a product, process, or technology throughout its entire lifecycle. LCA considers all stages of production, from raw material extraction and manufacturing to transportation, usage, and disposal. This methodology provides a holistic view of the environmental burden associated with energy and environmental technologies, allowing for informed decision-making in sustainability practices. LCA is particularly important in renewable energy technologies, where material use, energy efficiency, and end-of-life disposal must be assessed. For example, while solar panels and wind turbines offer low operational emissions, their manufacturing involves significant resource extraction, such as rare earth metals and silicon. Studies have shown that recycling solar panels and repurposing wind turbine blades can significantly reduce their environmental footprint by minimizing material waste and improving circular economy applications.

In the field of battery technology, LCA is used to evaluate the sustainability of lithium-ion batteries, solid-state batteries, and emerging energy storage solutions. The mining of lithium, cobalt, and nickel poses ecological and ethical challenges, making the development of sustainable battery recycling techniques and alternative materials a key focus of research. LCA models also assess the long-term benefits of hydrogen energy systems, identifying green hydrogen as a more sustainable alternative to fossil fuels due to its zero-emission potential when produced using renewable energy.

The importance of LCA extends to construction materials, transportation technologies, and industrial processes. Green building materials, such as bio-based insulation, recycled concrete, and sustainable polymers, undergo LCA assessments to determine their overall environmental impact compared to conventional materials. Similarly, electric vehicles (EVs) are evaluated based on battery lifecycle emissions, electricity sourcing, and recyclability, ensuring they offer a genuine advantage over internal combustion engines. By integrating LCA into sustainability frameworks, policymakers, researchers, and industries can identify the most effective strategies for reducing environmental harm, prioritizing innovations that maximize efficiency while minimizing ecological footprints.

1.7.2 Carbon Footprint and Climate Mitigation Strategies

A carbon footprint represents the total greenhouse gas (GHG) emissions generated by a specific activity, process, or technology, serving as a critical indicator of environmental impact. As carbon emissions are the leading cause of global warming, reducing them is essential for effective climate change mitigation and long-term planetary sustainability. The energy sector, particularly fossil fuel-based power generation, remains the largest contributor to global carbon emissions. Addressing this challenge requires a multifaceted approach that includes transitioning to renewable energy sources, enhancing energy efficiency, promoting carbon capture and storage technologies, adopting sustainable transportation systems, and encouraging behavioral changes across industries and communities. Climate mitigation strategies include:

- Carbon pricing and emission trading systems (ETS)—Implementing carbon taxes and cap-and-trade programs incentivizes industries to reduce emissions by placing financial costs on carbon output (OECD,
- Carbon capture, utilization, and storage (CCUS)—Capturing CO_2 emissions from power plants and industrial facilities prevents GHG accumulation in the atmosphere. Advanced direct air capture (DAC) technologies are now emerging as viable solutions for large-scale carbon removal.

- Reforestation and afforestation—Natural climate solutions, such as tree planting and ecosystem restoration, act as carbon sinks, absorbing atmospheric CO_2 and improving biodiversity.
- Energy efficiency in industry and buildings—Smart grid technologies, AI-driven energy management, and high-efficiency appliances reduce energy consumption, minimizing indirect carbon emissions.
- Sustainable urban planning—Designing low-carbon cities with integrated public transport, green spaces, and renewable energy systems significantly lowers urban emissions.

1.7.3 Role of Environmental and Energy Technologies in Achieving Sustainable Development Goals (SDGs

The United Nations Sustainable Development Goals (SDGs) outline a global blueprint for addressing sustainability challenges by 2030. Environmental and energy technologies play a transformative role in accelerating progress toward the United Nations Sustainable Development Goals (SDGs), particularly in addressing interconnected challenges related to climate change, resource efficiency, clean energy access, and environmental protection. Technologies that improve energy efficiency, enable the shift to renewable sources, reduce pollution, and support circular economy practices are central to realizing goals such as SDG 7 (Affordable and Clean Energy), SDG 13 (Climate Action), SDG 6 (Clean Water and Sanitation), and SDG 12 (Responsible Consumption and Production).

Environmental technologies including cleaner production systems, carbon capture, air and water purification, and sustainable waste management, reduce the environmental footprint of industries while supporting economic development. These technologies are essential in ensuring that economic growth aligns with ecological sustainability, thereby contributing to SDG 9 (Industry, Innovation and Infrastructure) and SDG 11 (Sustainable Cities and Communities. Concurrently, renewable energy technologies such as solar photovoltaics, wind turbines, and bioenergy systems are facilitating the decarbonization of energy systems worldwide, helping countries reduce greenhouse gas emissions and meet their climate commitments under the Paris Agreement. Moreover, advancements in energy storage, smart grid systems, and decentralized microgrids are enabling more inclusive energy access, particularly in remote and energy-poor regions, thereby addressing inequalities (SDG 10) and promoting quality education and healthcare (SDGs 3 and 4) through electrified services. However, achieving these synergies requires coordinated policy support, investments in clean technology innovation, and strong institutional frameworks. In summary, environmental and energy technologies serve as cross-cutting enablers of multiple SDGs, bridging economic, environmental, and social objectives. Their strategic integration into national development plans is essential to drive systemic transformation toward sustainability and resilience in a rapidly changing world. Figure 1.3 below shows the role of environmental and energy technologies in achieving sustainable development goals.

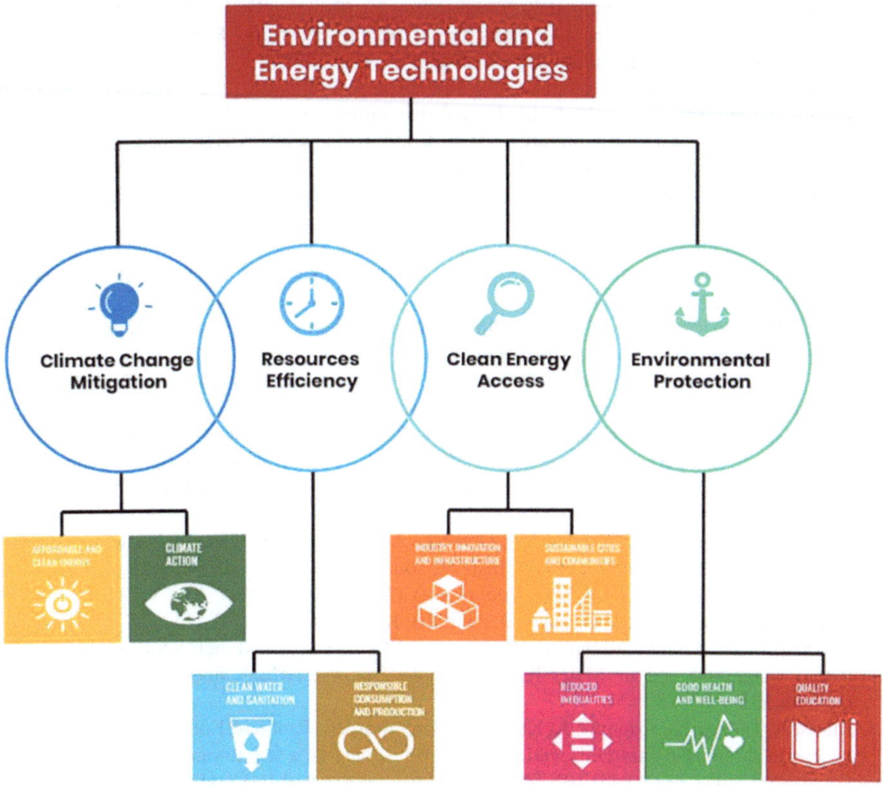

Fig. 1.3 Role of environmental and energy technologies in achieving sustainable development goals

References

1. Ali N, Sadek S, Abdin A (2023) A review on innovative nanomaterials for enhancing energy performance of the building envelope. Curr Nanomat 9(4):287–302. https://doi.org/10.2174/0124054615248038231020054831
2. Ansari MM, Heo Y, Do K, Ghosh M, Son YO (2024) Nanocellulose derived from agricultural biowaste by-products-sustainable synthesis, biocompatibility, biomedical applications, and future perspectives: a review. Carbohydr Polym Technol Appl 8(June):100529. https://doi.org/10.1016/j.carpta.2024.100529
3. Awasthi AK, Cheela VRS, D'Adamo I, Iacovidou E, Islam MR, Johnson M, Miller TR, Parajuly K, Parchomenko A, Radhakrishan L, Zhao M, Zhang C, & Li J. (2021). Zero waste approach towards a sustainable waste management. Resources, Environment and Sustainability, 3, 100014. https://doi.org/10.1016/j.resenv.2021.100014
4. Awewomom J, Dzeble F, Takyi YD, Ashie WB, Ettey ENYO, Afua PE, Sackey LNA, Opoku F, Akoto O (2024) Addressing global environmental pollution using environmental control techniques: a focus on environmental policy and preventive environmental management. Dis Environ 2(1). https://doi.org/10.1007/s44274-024-00033-5
5. Beck T (2024) The role of artificial intelligence in sustainable energy management. Innov Energy Res 13(4). https://doi.org/10.4172/2576-1463.1000412
6. Chowdhury MS, Rahman KS, Chowdhury T, Nuthammachot N, Techato K, Akhtaruzzaman M, Tiong SK, Sopian K, Amin N (2020) An overview of solar photovoltaic panels' end-of-life material recycling. Energ Strat Rev 27:100431. https://doi.org/10.1016/j.esr.2019.100431

7. Deeney P, Leahy PG, Campbell K, Ducourtieux C, Mullally G, Dunphy NP (2025) End-of-life wind turbine blades and paths to a circular economy. Renew Sust Energ Rev 212(January):115418. https://doi.org/10.1016/j.rser.2025.115418
8. Firoozi AA, Firoozi AA, Oyejobi DO, Avudaiappan S, Flores ES (2024) Emerging trends in sustainable building materials: technological innovations, enhanced performance, and future directions. Results Eng 24(September):103521. https://doi.org/10.1016/j.rineng.2024.103521
9. IEA (2023) World energy outlook 2023 | enhanced reader. World Energy Outlook. pp 23–28. https://www.iea.org/news/the-energy-world-is-set-to-change-significantly-by-2030-based-on-today-s-policy-settings-alone
10. IPBES (2019) Global assessment report on biodiversity and ecosystem services of the intergovernmental science-policy platform on biodiversity and ecosystem services. IPBES. https://zenodo.org/records/6417333
11. IPCC (2023) AR6 synthesis report: climate change 2023 – IPCC. United Nations. https://www.ipcc.ch/report/sixth-assessment-report-cycle/
12. IPCC (2022) Climate Change 2022: Impacts, Adaptation, and Vulnerability. Contribution of Working Group II to the Sixth Assessment Report of the Intergovernmental Panel on Climate Change [H.-O. Pörtner, D.C. Roberts, M. Tignor, E.S. Poloczanska, K. Mintenbeck, A. Alegría, M. Craig, S. Langsdorf, S. Löschke, V. Möller, A. Okem, B. Rama (eds.)]. Cambridge University Press. Cambridge University Press, Cambridge, UK and New York, NY, USA, 3056 pp. https://doi.org/10.1017/9781009325844
13. Jeevanandam J, Barhoum A, Chan YS, Dufresne A, Danquah MK (2018) Review on nanoparticles and nanostructured materials: history, sources, toxicity and regulations. Beilstein J Nanotechnol 9(1):1050–1074. https://doi.org/10.3762/bjnano.9.98
14. Khedulkar AP, Bobade RG, Doong R-a, Pandit B, Ky NM, Ambare RC, Hoang T-D, Kumar KJ (2025) Bio-based nanomaterials as effective, friendly solutions and their applications for protecting water, soil, and air. Mater Today Chem 46(January):102688. https://doi.org/10.1016/j.mtchem.2025.102688
15. Kumar A, Thakur AK, Gaurav GK, Klemeš JJ, Sandhwar VK, Pant KK, Kumar R (2023) A critical review on sustainable hazardous waste management strategies: a step towards a circular economy. Environ Sci Pollut Res 30(48):105030–105055. https://doi.org/10.1007/s11356-023-29511-8
16. Kumar A, AL-Jumaili A, Bazaka O, Ivanova EP, Levchenko I, Bazaka K, Jacob MV (2021) Functional nanomaterials, synergisms, and biomimicry for environmentally benign marine antifouling technology. Materials Horizons 8(12):3201–3238. https://doi.org/10.1039/D1MH01103K
17. Kuppan N, Padman M, Mahadeva M, Srinivasan S, Devarajan R (2024) A comprehensive review of sustainable bioremediation techniques: eco friendly solutions for waste and pollution management. Waste Manag Bull 2(3):154–171. https://doi.org/10.1016/j.wmb.2024.07.005
18. Manisalidis I, Stavropoulou E, Stavropoulos A, Bezirtzoglou E (2020) Environmental and health impacts of air pollution: a review. Front Public Health 8(February):1–13. https://doi.org/10.3389/fpubh.2020.00014
19. Mejía-Rosales S, Ponce A, José-Yacamán M (2013) Experimental techniques for structural characterization. In: Nanoalloys: from fundamentals to emergent applications (Issue July 2018). Elsevier Inc. https://doi.org/10.1016/B978-0-12-394401-6.00004-7
20. Münzel T, Hahad O, Daiber A, Landrigan PJ (2023) Soil and water pollution and human health: what should cardiologists worry about? Cardiovasc Res 119(2):440–449. https://doi.org/10.1093/cvr/cvac082
21. Nosratabad N, Yan Q, Cai Z, Wan C (2024) Exploring nanomaterial-modified biochar for environmental remediation applications. Heliyon 10(18):e37123. https://doi.org/10.1016/j.heliyon.2024.e37123
22. Singh R, Singh A, Sheoran P, Fagodiya RK, Rai AK, Chandra P, et al. (2022) Energy efficiency and carbon footprints of the rice-wheat system under long-term tillage and residue management practices in western Indo-Gangetic Plains in India. Energy 244:122655. https://doi.org/10.1016/j.energy.2021.122655
23. UNESCO (2023) The United Nations world water development report 2023: partnerships and cooperation for water – UNESCO digital library. UNESCO. https://unesdoc.unesco.org/ark:/48223/pf0000384655
24. Wang H, Fu F, Huang M, Feng Y, Han D, Xi Y, Xiong W, Yang D, Niu L (2023) Lignin-based materials for electrochemical energy storage devices. Nano Mat Sci 5(2):141–160. https://doi.org/10.1016/j.nanoms.2022.01.002

Chapter 2
Bio-based Materials from Biomass: A Foundation for Green Innovation

2.1 Overview of Biomass

In recent decades, rapid economic and technological advancements have significantly improved global living standards. However, this progress has also accelerated environmental degradation, primarily due to the heavy reliance on fossil fuels. This dependence has led to rising levels of air pollution, global warming, and public health deterioration, prompting urgent calls for a shift toward cleaner, more sustainable energy alternatives. In response, governments and international organizations have introduced comprehensive policy frameworks that support the adoption of renewable energy technologies, aiming to mitigate carbon emissions and reduce fossil fuel dependency. Among the various renewable resources, biomass has emerged as a particularly versatile and abundant option. It contributes not only to energy security but also to economic development especially in rural areas where access to conventional energy infrastructure may be limited.

Modern technologies now enable the efficient conversion of biomass into various energy products, including biofuels, electricity, and thermal energy. Feedstocks such as agricultural residues, forestry by-products, and organic municipal waste are increasingly utilized in anaerobic digestion, gasification, and fermentation systems to maximize energy recovery. Simultaneously, many countries are aligning biomass utilization with broader socio-economic goals integrating it into agricultural, land use, and rural development strategies. These efforts support income diversification, reduce rural-to-urban migration, and promote environmental conservation. The global urgency to combat climate change has further accelerated interest in biomass. Organizations such as the Intergovernmental Panel on Climate Change (IPCC) have emphasized its role in achieving net-zero emissions targets, highlighting the potential of biomass as more than just an energy source, but a foundational element in comprehensive sustainable development strategies. Recent innovations in biorefinery systems have also demonstrated that biomass can yield not only

M. I. Ahmad, A. K. H. P. S., *Environmental and Energy Technology*, Green Energy and Technology, https://doi.org/10.1007/978-3-032-10706-0_2

electricity and heat but also next-generation fuels and bio-based chemicals, offering cost-effective, low-carbon alternatives to fossil-based systems.

Despite its biological origin, biomass is often mistakenly used interchangeably with bio-based materials. While both stem from organic matter, they serve distinct purposes. Biomass refers to any biologically sourced organic material from plants, animals, algae, and microorganisms that can be used to generate energy. Common sources include agricultural residues (e.g., corn stover, rice husks), forestry waste (e.g., sawdust, wood chips), and organic municipal waste. These materials undergo thermochemical or biochemical conversion to produce energy carriers like bioethanol, biodiesel, biogas, and electricity [42]. Bio-based materials, on the other hand, are derived from biomass but are engineered into many products such as biodegradable plastics, composites, textiles, adhesives, and construction materials. For example, polylactic acid (PLA) sourced from corn starch is widely used for compostable packaging, while natural fibers like hemp and flax are now integrated into automotive panels and building insulation. These materials substitute conventional fossil-derived inputs, helping to decarbonize manufacturing sectors. From a carbon accounting perspective, biomass is considered carbon-neutral due to its short-cycle CO_2 loop. Plants absorb atmospheric CO_2 during growth, and this is re-released during combustion or decomposition. This biogenic carbon cycle ensures that no new carbon is added to the atmosphere, unlike fossil fuels that release carbon sequestered over millions of years.

2.1.1 Chemical Composition of Biomass

Biomass exhibits a highly heterogeneous composition depending on its origin. Plant-based residues predominantly consist of structural biopolymers such as cellulose, hemicellulose, and lignin (Fig. 2.1) although their relative proportions vary significantly across species and processing methods [27]. In contrast, animal-derived biomass, such as cattle manure, tends to be richer in proteins, while agricultural residues like cereal grains are primarily composed of starch. These compositional variations directly influence the physicochemical behavior and conversion efficiency of biomass during thermochemical and biochemical processing. Table 2.1 shows the compositional elements of cellulose, hemicellulose and lignin.

(i) **Cellulose**
Cellulose is a naturally abundant linear polysaccharide composed of D-glucose units linked through β-1,4-glycosidic bonds, with the general molecular formula $(C_6H_{12}O_6)_n$, where *n* represents the degree of polymerization. The fundamental repeating unit of cellulose is cellobiose, which consists of two glucose molecules (4-O-β-D-glucopyranosyl-D-glucopyranose). As a principal structural component of plant cell walls, cellulose accounts for approximately 90% of cotton and around 50% of wood composition, highlighting its prevalence in terrestrial biomass. The structural configuration and reactivity of

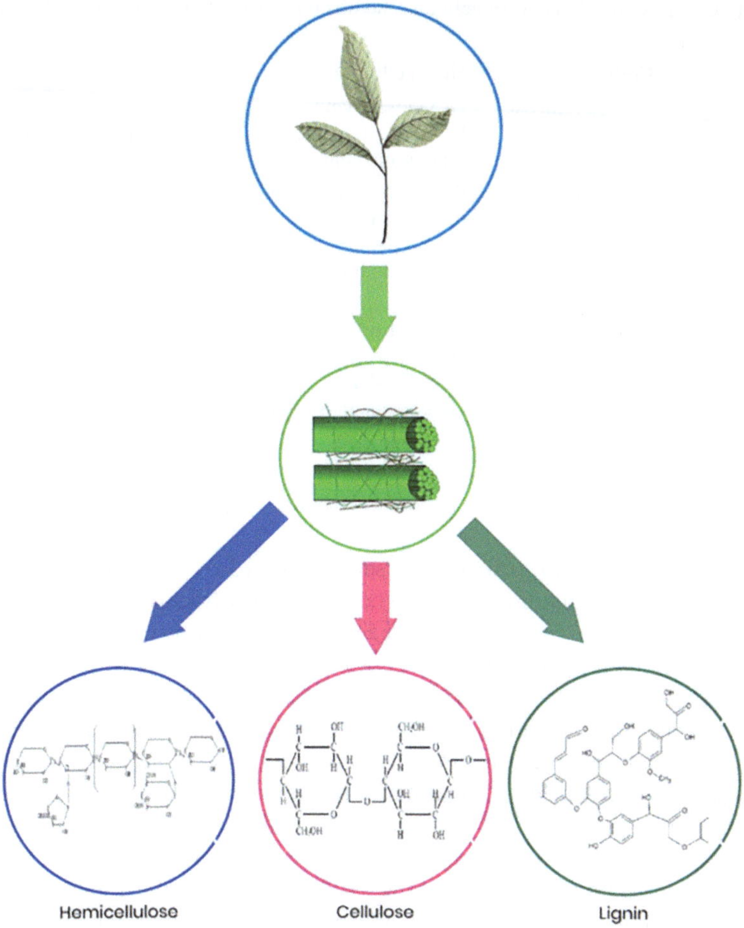

Fig. 2.1 Plant cellulose, hemicellulose, and lignin

cellulose are significantly influenced by its extensive hydrogen bonding network. In particular, intermolecular hydrogen bonds between hydroxyl groups on C-3 and adjacent glycosidic oxygen atoms stabilize the polymer chains and enhance rigidity. These interactions promote the formation of both crystalline and amorphous domains within the cellulose structure. Crystalline regions, where molecules are closely packed, contribute to the polymer's strength and thermal stability, whereas amorphous regions are more disordered and accessible to chemical modification or enzymatic attack.

X-ray diffraction analyses reveal that cellulose crystallinity typically ranges from 30% to 80%, with a density of approximately 1.588 g/cm^3 in crystalline regions and 1.5 g/cm^3 in amorphous areas [72]. The alternating presence of crystalline and amorphous regions directly affects cellulose's reactivity, particularly regarding its hydroxyl groups. Primary hydroxyl groups (located on

Table 2.1 Composition elements of cellulose, hemicellulose and lignin in plant cell walls

Compositional elements	Cellulose	Hemicellulose	Lignin
Subunits	D-Pyranglucose units	D-Xylose, mannose, L-arabinose, galactose, glucuronic acid	Guaiacylpropane (G), syringylpropane (S), p-hydroxyphenylpropane (H)
Subunits bonds	β-1,4-Glycosidic bonds	β-1,4-Glycosidic bonds in main chains; β-1.2-, β −1.3-, β-1.6-glycosidic bonds in side chains	Various ether bonds (mainly β-O-4); carbon-carbon bond
Composition	β-Glucan	Polyxylose Galactoglucomannan (Gal-Glu-Nan) Glucomannan (Glu-Man)	G lignin, GS lignin, GSH lignin
Polymer	Several hundred to tens of thousands	Less than 200	4000

C-6) display greater reactivity than secondary hydroxyls (on C-2 and C-3) due to reduced steric hindrance [17]. The amorphous fraction of cellulose, being more exposed and hydrophilic, is readily hydrolyzed by cellulolytic enzymes. Enzymatic hydrolysis initially breaks down these regions to yield cellobiose and short-chain oligosaccharides, eventually resulting in glucose monomers. In contrast, the crystalline fraction resists enzymatic degradation due to its compact hydrophobic structure. To improve its accessibility, cellulose is often subjected to chemical pretreatments like mercerization, which enhance its hydrophilicity and reduce crystallinity. Beyond its structural role in plants, cellulose has become a valuable industrial material. Its applications span multiple sectors, including paper and textile production, biomedical engineering, and environmental remediation. Particularly in wastewater treatment, cellulose-derived adsorbents have been successfully employed to remove pollutants such as heavy metals and hydrocarbons, demonstrating high adsorption efficiency and environmental compatibility.

(ii) **Hemicellulose**

Hemicellulose is a crucial structural component of plant cell walls, composed of heterogeneous, branched polysaccharides that form physical bonds with cellulose microfibrils, contributing to the wall's integrity and flexibility. Unlike the linear structure of cellulose, hemicelluloses are amorphous, less crystalline, and display greater solubility and reactivity due to their side-chain substitutions. The composition and architecture of hemicellulose vary significantly among plant species, typically comprising both pentoses (e.g., xylose, arabinose) and hexoses (e.g., glucose, galactose, mannose, rhamnose), with an average molecular weight of less than 30,000 Da [91, 92]. Hemicelluloses are thermally less stable than cellulose, with decomposition temperatures ranging from 180 °C to 350 °C, producing volatile compounds such as acids, ketones, aldehydes, furans, and non-condensable gases. This thermal behavior makes them promising feedstocks for bio-oil and syngas production in thermochemical conversion technologies. The hydrophilic and adhesive nature of hemicellulose also supports its functional role in maintaining moisture content and plasticity within cell walls. Structurally, hemicellulose is composed of several subtypes including xylans, mannans, galactans, and arabinogalactans, depending on the predominant sugar residues and branching patterns. These macromolecules often exhibit side groups such as acetyl, methyl, and uronic acid substitutions, which influence their solubility and reactivity in both natural and industrial applications.

(iii) **Lignin**

Lignin is an essential structural biopolymer in plant cell walls that provides rigidity, hydrophobicity, and resistance to microbial attack. It acts as a natural adhesive, embedding and reinforcing cellulose and hemicellulose fibers to strengthen plant tissues and support vertical growth. Due to its encrusting and cementing function, lignin forms a protective barrier around polysaccharide components, making its removal a crucial step in the isolation of cellulosic fibers for industrial applications. The lignin content varies significantly depending on the type and age of the plant ranging from approximately 10–12% in annual crops, to 25–30% in softwoods, and up to 50% in dense hardwoods like ebony. Its elemental composition typically includes 60–65% carbon, 5–7% hydrogen, and the remainder is oxygen, making it one of the most carbon-rich and energy-dense natural biopolymers [91, 92].

Structurally, lignin is a highly branched, amorphous aromatic polymer built from phenylpropanoid units, primarily synthesized through the oxidative radical polymerization of three major monolignols: p-coumaryl alcohol (H-unit), coniferyl alcohol (G-unit), and sinapyl alcohol (S-unit). These monomers differ in their methoxylation patterns, which influence the physicochemical properties of lignin and vary across plant species. The resulting polymer consists of complex C-C and C-O-C linkages, forming an irregular, cross-linked matrix. Lignin's macromolecular structure contains numerous hydroxyl and polar functional groups, enabling extensive intra- and intermolecular hydrogen bonding [76]. This network confers low solubility in most solvents, particularly under neutral pH conditions. However, lignin becomes soluble under alkaline treatment, which facilitates its depolymerization into soluble and insoluble fractions during biomass pretreatment processes. The condensation reactions that occur under such conditions lead to the formation of insoluble lignin precipitates, which can be valorized for use in biochemicals, carbon materials, and composites.

2.2 Definition and Classification of Bio-based Materials

Bio-based materials are defined as products predominantly composed of substances derived from living organisms or biomass such as plants, animals, algae, or microorganisms and may occur naturally, be chemically synthesized, or produced through biomass-based processes. In the construction sector, these renewable materials are increasingly being used for structural components, furniture, interior finishes, and decorative elements. By leveraging locally available biomass, the use of bio-based materials not only supports sustainability but also reduces reliance on imported and non-renewable resources. Zambon et al. [87] emphasize that the twenty-first century will undergo a fundamental transformation in material sourcing from the extractive approaches of the industrial era to regenerative practices involving cultivation, breeding, and farming of future materials. This shift reflects a broader reorientation toward biological production systems that prioritize renewable, bio-derived inputs over finite mineral resources. The vision is to "grow" building materials instead of mining them, aligning technological progress with ecological responsibility.

Bio-based materials offer a wide range of applications across multiple industries including construction, automotive, biomedicine, textiles, and packaging. Their ecological advantages position them as sustainable alternatives to fossil fuel- and mineral-based products. For example, conventional plastics can be replaced by bio-based polymers (bioplastics), while lightweight bio-composites can substitute for metal parts in vehicles. Furthermore, biofuels produced from biomass provide renewable alternatives to fossil-derived energy. Classified by their composition and function, bio-based materials can be categorized into polymers, composites, chemicals, fuels, hydrogels, and structural products. Bio-based polymers may be naturally occurring (e.g., cellulose), chemically modified (e.g., cellulose acetate), or synthetically produced from biological feedstocks (e.g., bio-based polyethylene) [48]. Bio-based composites combine renewable matrices with reinforcing agents such as natural fibers to enhance strength and functionality. Meanwhile, bio-based chemicals and fuels serve as replacements for petroleum-based compounds and energy sources, supporting cleaner industrial practices. Hydrogels derived from biological sources are applied in the medical field, particularly in wound care. Structural bio-based materials are increasingly used in packaging, furniture, and even building construction, offering renewable, biodegradable, and low-impact alternatives. Together, these innovations in bio-based materials signal a paradigm shift in material science one that aligns economic progress with environmental stewardship and supports global efforts toward sustainability.

2.2.1 Plant-Derived Materials

Plant-derived materials refer to substances obtained from renewable parts of plants such as roots, stems, leaves, and seeds that are increasingly utilized as sustainable alternatives to synthetic and fossil-based materials. These resources have been used since early human history, initially for medicinal and utilitarian purposes. Technological advancements have since expanded their applications to modern industries, ranging from construction and textiles to bio-plastics and packaging. Their abundance, renewability, and lower environmental impact make them a promising solution to pressing global challenges such as resource depletion, pollution, and climate change. The increasing environmental concerns associated with synthetic materials have driven a global search for more eco-friendly options. The shift toward plant-derived materials aligns with sustainability goals due to their biodegradability, reduced carbon footprint, and potential to support circular economic models. These materials also offer favorable mechanical properties such as flexibility, strength, and durability making them suitable for diverse applications in industry. Unlike petroleum-based plastics, plant-based materials decompose naturally, thereby helping to reduce persistent waste accumulation and supporting waste minimization strategies in global production systems.

One of the most widely used plant-derived substances is cellulose, a natural polymer found in plant cell walls. It serves as a key raw material in paper manufacturing and is also used in textiles and biodegradable plastics. As highlighted by Kaushal and Singh [31], cellulose derivatives such as cellulose acetate are now employed in eco-friendly coatings, films, and fibers used in the medical and fashion industries. Cotton, another cellulose-based fiber, remains a staple in the textile sector due to its softness, breathability, and durability. Lignin, a complex polymer that strengthens plant cell walls, was once considered a waste by-product in the pulp and biofuel sectors. However, recent innovations have enabled its conversion into value-added products such as bio-based adhesives, carbon fibers, resins, and energy storage components.

In the energy and chemical sectors, plant-derived materials serve as crucial feedstocks for biofuels and bio-based chemicals. Biodiesel, for instance, is produced from oils extracted from soybeans, palm, coconut, and sunflower plants and offers a renewable substitute for petroleum diesel, significantly reducing greenhouse gas emissions. Similarly, starch derived from crops like corn, potatoes, and wheat is used to produce bioethanol, which is often blended with gasoline to form a cleaner-burning fuel. Additionally, compounds such as terpenes, flavonoids, and alkaloids sourced from various plants have gained prominence in the food, cosmetic, and pharmaceutical industries. Essential oils extracted from plants like eucalyptus, tea tree, and lavender are valued for their antimicrobial and anti-inflammatory properties and are widely used in skin and hair care formulations. Beyond their role in textiles and energy, plant-derived materials are gaining prominence in construction, packaging, and industrial applications. Bamboo, known for its rapid growth and mechanical strength, is being increasingly used as a sustainable alternative to conventional hardwoods in flooring, furniture, and structural components. Hemp and flax fibers are incorporated into bio-composites for automotive parts, insulation, and even high-performance sports equipment. As noted by Atiwesh et al. [4], the packaging industry is also adopting materials like polylactic acid (PLA), a bio-based plastic made from corn starch or sugarcane as replacements for petroleum-based plastics. Other innovations include molded pulp from recycled fibers, biodegradable films made from starch and algae, and natural resins used in food-safe containers. With rising environmental awareness among consumers and increasing regulatory support from governments and sustainability-focused organizations, the demand for renewable materials is accelerating. As innovation and research progress,

Table 2.2 Renewable resources of plant-derived materials

Material	Source plant/part	Applications	Properties	References
Cellulose	Plant cell walls (wood, cotton, flax)	Paper, textiles, biodegradable plastics, coatings, medical fibers	Biodegradable, strong, abundant	[31]
Cotton	Cotton fiber	Apparel, home textiles, medical fabrics	Soft, breathable, durable	[31]
Lignin	Woody plants, pulp industry residue	Bio-resins, adhesives, carbon fibers, energy storage	Functional, renewable polymer	[31]
Starch	Corn, potatoes, wheat	Bioethanol, bioplastics, biodegradable films	Biodegradable, renewable	[29]
PLA	Corn starch, sugarcane	Packaging, disposable items, 3D printing	Compostable, clear finish	[4]
Bamboo	Bamboo culms	Flooring, furniture, construction materials	Strong, fast-growing, sustainable	[4]
Hemp fiber	Hemp stalks	Bio-composites, automotive parts, textiles	Lightweight, durable	[70]
Durian peel bio-aerogel	Durian fruit peel	Bio-adsorbents, filtration, packaging	Porous, lightweight, high adsorption	[2, 75]

plant-derived materials are expected to play an increasingly vital role in building a sustainable, low-impact, and resource-efficient future. Table 2.2 summarizes the renewable resources of plant-derived materials.

2.2.2 *Agricultural Waste and Byproduct*

Agricultural waste represents a significant reservoir of biomass, comprising organic residues generated during agricultural activities. These wastes originate from diverse plant parts, including switchgrass, hemp, miscanthus, wheat straw, corn, poplar, willow, and sugarcane tops. Agricultural residues are typically biodegradable and combustible, offering potential as feedstock for bioenergy applications. Their importance is largely influenced by their molecular composition and physicochemical characteristics. As noted by [63]. these by-products vary depending on the specific agricultural processes involved and the final products obtained. Notably, they often remain underutilized and are disposed of via open burning, dumping, or unmanaged landfilling practices that exacerbate greenhouse gas emissions such as CO_2 and CH_4, thereby posing serious threats to ecosystems. Agricultural wastes can be broadly categorized into field residues and processed residues, as depicted in Fig. 2.2. Field residues, which include stems, stalks, leaves, and seed pods, are left behind after harvesting.

Even after the primary crop is processed, these parts including husks, straw, stubble, shells, peels, pulp, bagasse, roots, seeds, and molasses remain available. They serve as lignocellulosic substrates for both first- and second-generation biofuels,

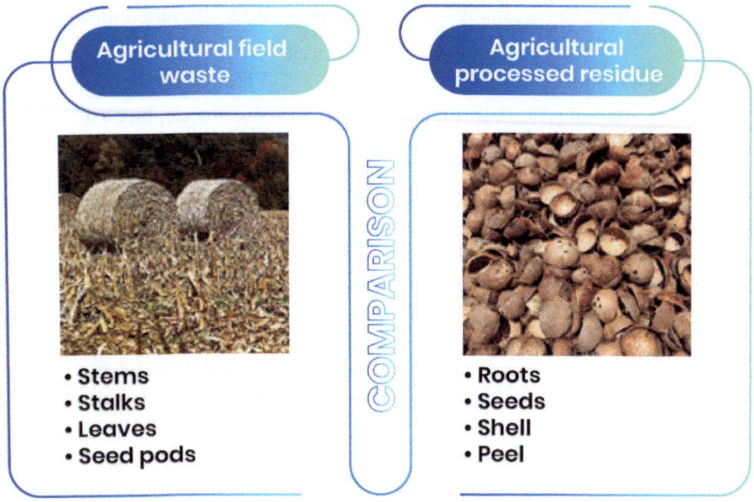

Fig. 2.2 The two primary groups of agricultural waste

offering viable alternatives to conventional energy sources. Food processing industries also contribute significantly to agricultural waste through the separation of valuable products from non-edible by-products. This waste includes spoiled or excess produce, fruit and vegetable peels, tea leaves, coffee grounds, poultry remains, jam residues, and bakery scraps. Both liquid and solid waste forms are produced, with solids often originating from unusable portions of raw materials. Waste generated from foodservice sectors canteens, hotels, and restaurants includes leftovers of rice, vegetables, noodles, eggs, oil, and meats, varying by origin. These residues are rich in organic compounds such as cellulose, carbohydrates, lipids, and proteins. Additionally, materials like olive waste and sugarcane bagasse contain high lignin content, limiting their use in animal feed despite being repurposed occasionally as dried pellets.

Beyond plant-based sources, animal-derived residues such as manure and carcasses are also integral components of agricultural waste. These by-products, along with hazardous residues like pesticides and insecticides, are non-food outputs of the agricultural sector. In regions specializing in either crop or livestock production, waste accumulation can pose environmental and economic challenges including bacterial contamination and elevated greenhouse gases (GHG) emissions. However, as Donner et al. [20] and Duque-Acevedo et al. [21] highlighted, with appropriate conversion technologies, these wastes can be transformed into valuable products like bioenergy, bio-fertilizers, biomaterials, and biomolecules. Valorizing these residues supports resource decoupling by reducing pressure on primary materials and land, thereby safeguarding biodiversity and food security. Yet, effective waste valorization requires a holistic understanding of technological, economic, and societal factors. While studies on anaerobic digestion, biorefineries, and enzymatic catalysis have focused on technological solutions, the socioeconomic dimensions remain

Table 2.3 Classification of agricultural waste and by-products as bio-based materials

Agricultural waste	Examples	Bio-based material type	Potential applications	References
Crop residues	Straw, husks, stalks, seed pods	Lignocellulosic biomass	Bioethanol, biochar, biogas, composite fillers	[18]
Fruit and vegetable waste	Peels, pulp, seeds, skins	Cellulose/pectin-rich biomass	Bioplastics, dietary fiber, bio-adsorbents	[41]
Tea and coffee processing waste	Spent coffee grounds, tea leaves	Polyphenol-rich biomass	Antioxidants, biochemicals, biochar, biogas	[66]
Rice husk (cereal by-product)	Rice husk	Silica, cellulose nanomaterials	Construction fillers, nanocellulose, feed enhancers	[85]
Animal-derived waste	Manure, bones, skin, blood residues	Protein-rich biomass	Biofertilizer, biopolymers, nutrient recovery	[73]
Mixed organic waste	Household food scraps, market waste	Mixed organic biomass	Compost, anaerobic digestion substrate	[62]

underexplored. For instance, long-standing research in the United States, India, and China emphasizes nutrient recycling and residue reintegration into agricultural systems [14, 20].

The adoption of circular economy principles and bioeconomy strategies has intensified research efforts globally. One effective valorization pathway is industrial symbiosis, where industries co-locate to share waste and by-products as inputs, enhancing overall resource efficiency. This model is widely applied in eco-industrial parks although predominantly within petrochemical and mixed industrial sectors. To tap the full potential of agricultural biomass without compromising food production or biodiversity, an integrated approach combining research and development (R&D), policy frameworks, and consumer engagement is necessary. Acceptance of waste-based products must grow, alongside the development of business models and marketing strategies that align with sustainability goals [14, 20]. Table 2.3 shows the classification of agricultural waste and by-products as bio-based materials.

2.2.3 *Microbial Biomass and Algal Sources*

Microbial biomass is highly dynamics and essential for the production of soil organic matter and nutrient distribution. Around 5% of soil organic C and N is made up of microbial biomass carbon and microbial biomass nitrogen. Therefore, it is crucial for the construction of structures, soil fertility, and the cycling of all biophilic elements. Because of exoenzymes, microorganisms in soil play a significant role in the release of carbon and nutrients from organic matter and plant litter.

Microbial biomass has a higher turnover rate and a shorter mean residence time that ranges from a few hours to weeks. Some agricultural soils, such as paddy soil, have a special management system that regulates redox-driven processes that impact the turnover and transformation of soil organic matter and nutrient cycling, which sets them apart from upland soils. The biogeochemistry of the soils is significantly influenced by redox and pH dynamics, as well as the solubility and biodegradation of organic matter. Due to the diffusion of dissolved oxygen, flooded paddy soil has a unique oxic surface and is not entirely anoxic from the water on top to the soil pores at the bottom. Furthermore, within an anoxic backdrop, rice roots produce distinct oxic microbial communities in the rhizosphere and radiate oxygen into the soil. Bacteria, fungi, and archaea are significantly impacted by the vertical and horizontal redox gradients that develop in paddy soils. For a variety of causes, the microbial biomass content of paddy soils is significantly different from that of upland soils. The microbially mediated turnover of soil organic matter and chemical conversions are impacted by redox dynamics in paddy soils. The concentration of water-soluble intermediate metabolites (such acetate, formate, propionate, and lactate) in paddy soils rises with extended anaerobic conditions. The root properties of crops grown in upland and paddy soils are different. Compared to upland soils, paddy soils integrate more plant-derived C into microbial biomass carbon. The contribution of microbial residues (microbial necromass, exoenzymes, and other extracellular polymeric compounds) to soil organic matter in a paddy must eventually increase as microbial biomass increases. Lastly, the redox conditions have a major role in the precipitation of dissolved organic matter by clay minerals and iron (oxyhydr)oxides, as well as the stabilization of dissolved organic matter in relation to the impacts of microbial degradation [83]. The primary dissolved organic matter sorbents in oxic environments are iron oxyhydroxides, which can also dissolve in reducing circumstances. The primary distinction between paddy and upland soils in terms of dissolved organic matter concentrations and fluxes is caused by this reduction in sorptive retention. Additionally, the C and N cycles in paddy-upland rotation soils are impacted by repeated drying-wetting cycles, which results in seasonal variations in chemical speciation and biological nutrient use. Analyzing the variables influencing microbial biomass concentrations is challenging due to the increased diversity of microorganisms caused by frequent anaerobic-aerobic cycles.

Algal biomass has the potential to produce a wide range of products, including proteins, lipids, carbohydrates, polysaccharides, polymers, and other bioproducts. Algae are an effective instrument for producing a variety of bioproducts due to their low cost and sustainable nature. Algae are now commonly used as a biomass feedstock due to the bioresource industry quick development. There are around 72,500 strains of algae that have been identified, among them only 36,250 have been researched for offering a variety of potential uses [30]. The algae utility as a feedstock for various processes is determined by the amounts of proteins, lipids, carbs, and fats it contains. Table 2.4 shows analysis of constituents of some algae strains. Recent advancements have been made to combine the generation of bioproducts from algal biomass with downstream processing. Biopolymers are widely used in medicines, cosmetics, medicine, and food packaging. They are used as a possible

Table 2.4 Contents of protein, carbohydrates, and lipids for some strains of algal biomass

Algae strain	Protein (wt %)	Carbohydrate (wt %)	Lipid (wt %)	References
Ecklonia radiata	7.60	65.40	3.60	[22, 23, 86, 90]
Undaria pinnatifida	12.50	53.90	2.20	
Carpophyllum flexuosum	5.90	69.00	2.60	
Carpophyllum plumosum	7.20	66.80	1.70	
Nannochloropsis sp.	31.68	8.99	15.04	
Spirulina	70.30	23.90	5.80	
Chlorella vulgaris	54.70	19.20	25.70	

food additive because of their great nutritional content. Emerging biopolymers like chitosan significantly improve processes like tissue regeneration and medication delivery. Biopolymers and their composites are also widely used in contemporary technologies like 3D printing. There have been several advancements in the utilization of algae to create various biopolymers in recent years. The cellulose and starch polymers found in algae are naturally occurring and can be removed using a variety of methods. Algae are a valuable source of feedstock for the synthesis of biopolymers due to their large number of photosynthetic factories. Several polymeric classes, including polyesters, polyamides, and polyolefins, can be produced by fermenting the biomass of algae. Natural polymers derived from algal biomass have potent antimicrobial, anticancer, antioxidant, and immunomodulatory properties and are less harmful. Additionally, the algal biomass can be utilized to create biodegradable composites with high mechanical qualities including tensile strength and stickiness.

2.3 Bio-based Material Synthesis

The process of converting the renewable resources into useful products that can take the place of traditional fossil-based products is known as bio-based materials synthesis. These components come from biomass, which includes lignocellulosic feedstocks, plant oils, microbiological sources, and agricultural waste. The creation of biodegradable polymers, biofuels, and high-performance biomaterials is made possible by the sustainable synthesis pathways, which lessen reliance on fossil fuels and advance environmental sustainability [63]. There are three interconnected routes to synthesize bio-based materials which are biochemical route, chemical route, and thermochemical route as illustrated in Fig. 2.3.

2.3.1 Biochemical Routes

Biochemical routes use biological factors such as microorganisms or enzymes to convert biomass into bio-based materials. Biomass are primarily converted into energy and other useful substances using biochemical methods. Biomass is made up

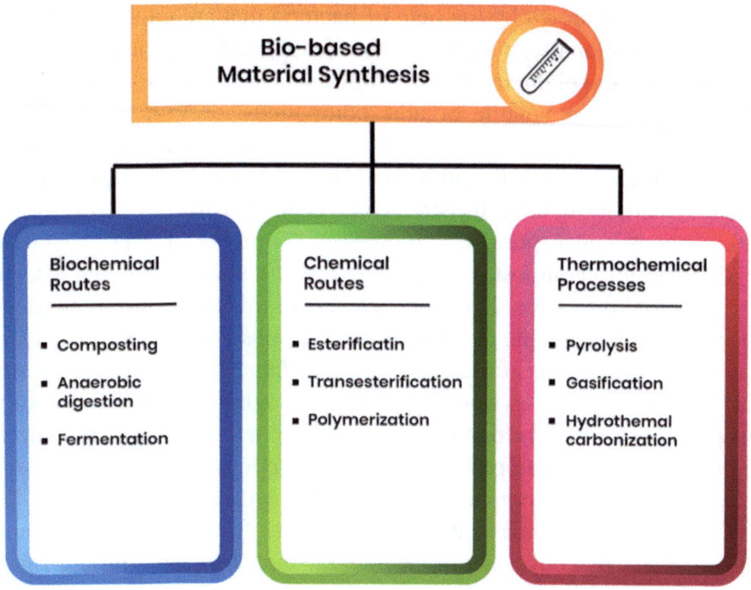

Fig. 2.3 The three routs of bio-based materials synthesis

of different organic materials as well as other beneficial substances. For anaerobic digestion, composting, fermentation, and other biological processes, bio-waste provides the perfect substrate. According to several past research, anaerobic digestion would be used to produce the biogas, which may be used as a natural gas alternative energy source [71]. Composting and fermentation are useful conversion techniques for bio-waste and its byproducts in addition to anaerobic digestion.

(i) **Composting**

Composting turns organic matter into nutrients-rich fertilizer in the presence of heat, moisture, and microorganisms. Composting can be done with almost anything that grows in the ground, including fruits, vegetables, seeds, grass, plants, and trees. Three key components are needed for this process which are water, green materials, and brown. In addition, it needs areas that are shaded and dry. For compost to form, precise ratios of water, green, and brown material are necessary. Alternating layers of varying particle sizes make up the composting ingredients. Water supplies moisture to aid in the breakdown of organic matter, green materials supply nitrogen, and brown materials supply carbon for the compost. Actinomycetes, fungus, and bacteria break down biomass during composting to produce organic acids, sugars, and starches. Methane and carbon dioxide are released by any organic biomass that ends up in a landfill, which contributes to global warming. In addition to lowering carbon footprints and harmful gas emissions, composting has other benefits. The use of compost as fertilizer lessens the need for potentially environmentally harmful chemical solutions. Humus is a nutrient-rich substance that is produced when organic matter is broken down by the help of bacteria that are encouraged to grow by compost. Additionally, it improves soil quality, aids in moisture retention, and inhibits pests and plant diseases. Compost has the ability to absorb and remove 99.6% of airborne volatile organic compounds, or VOCs, according to the EPA. These gasses and vapors can have negative health effects, such as throat discomfort and nausea. Compost reduces the amount of carbon in our atmosphere, where it can have disastrous effects on the earth, by acting as a "carbon sink" once it is incorporated into the soil. A number of small businesses turn solid waste into compost. Composting has the benefit of being a technically straightforward process. However, interest in this procedure is limited by the poor economic revenue. A straightforward substitute for making vermicompost with earthworms, *Eisenia fetida*, and *Lumbricus rubellis* is vermicomposting, often known as worm farming. Through the natural action of worms, this process involves the decomposition of kitchen waste, such as fruit peelings and vegetables, and can provide high-quality compost and a source of protein. Exposure to airborne organic waste results in an aerobic state.

Composting is a safe and environmentally friendly method of managing organic waste by converting biodegradable materials into nutrient-rich biofertilizers. Conventional waste management methods, such as landfilling, incineration, and open dumping, contribute to pollution, greenhouse gas emissions, and soil and water contamination. Unlike these methods, composting promotes environmental sustainability by enhancing soil structure, preventing erosion, and supporting biodiversity through beneficial microorganisms, insects, and fungi. Additionally, composting aids in bioremediation, helping to restore polluted soils while suppressing plant diseases, leading to healthier crops. In contrast to conventional methods that release harmful gases and odors, composting significantly reduces methane emissions, making it a more climate-friendly alternative. Although composting cannot break down recalcitrant materials like plastics and polythene bags, it effectively reduces overall waste volume, ensuring organic materials remain in a controlled, beneficial cycle rather than accumulating in landfills or polluting the environment [5]. Moreover, composting contributes to sustainable agriculture by improving soil fertility and water retention, reducing the need for chemical fertilizers. By integrating composting into waste management strategies, communities can lower their ecological footprint, reduce landfill dependency, and mitigate climate change. While composting requires proper conditions such as aeration, moisture, and microbial activity, its benefits far outweigh the limitations. In conclusion, composting offers a practical and sustainable alternative to conventional waste disposal methods, transforming organic waste into a valuable resource for soil health and environmental conservation. Table 2.5 outlines the advantages of composting over traditional waste management approaches.

(ii) **Anaerobic digestion**

The breakdown of intricate natural tissues into more specialized synthetic components without oxygen is known as anaerobic processing. This method works very well for handling large amounts of organic waste, such as food or

Table 2.5 The differences between composting as an effective method of waste management and conventional approaches of waste management

Composting as an effective waste management	Conventional methods of waste management
Wastes are kept in a controlled environment	Wastes contributes to the pollution of soil, water, and air
Composting reduces the waste volume	Wastes are piled dramatically
It reduces the greenhouses gas emissions	They release huge amounts of greenhouses gases
It surpasses pathogens and diseases	They spread pathogens and diseases causing impacts on human, animals, and plants
Composting promotes sustainability since it recycles biomass into useful form of bio-based materials	They are unsustainable practices since the track of biomass is on one track
Some wastes like plastic and polythene cannot be composted	They treat the stubborn materials like plastic and polyethylene

Table 2.6 Anaerobic digestion processes for some bio-based materials

Feed	Time (days)	Product	Quantity	References
Municipal food waste with cow slurry co-substrate	17.5	Methane	444.7 mL/g	[45]
Food waste with wheat straw co-substrate	Unstable	Methane	369 mL/g	[69]
Food waste	20–30	Methane	446 L/kg	[25]
Kitchen waste with cow manure	45	Methane	179.8 mL/g	[89]
Dairy and chicken manures	30	Methane	234.7	[78]

agricultural waste. The products of anaerobic digestion are digestate and biogas as shown in Table 2.6. These residues have the potential to be a great source of income. The majority of the time, biogas can be enhanced by eliminating carbon dioxide and water vapors before being utilized to generate heat and power [55]. In addition, the digestate can be used as compost or manure.

Anaerobic treatment is commonly used to treat the majority of food or agricultural processing wastes. An effective technique that produces methane as an energy vector and drastically reduces pollutant content is anaerobic digestion. The anaerobic approach is being utilized more as a substitute for energy vector production through pretreatment of high-strength organic wastewater as energy costs rise. In the anaerobic process, the digester can be made to execute a variety of tasks, such as reducing volume, producing biogas, eliminating pathogens, controlling odors, and mixing. Biogas, a byproduct of anaerobic digestion, is produced by these digesters. Methane makes up around 60% of biogas. Straw is utilized as an alternate method for biogas production. However, because it is difficult for anaerobic bacteria to break down agricultural straw in order to produce biogas, there are many limitations in the fermentation process of biogas generation. Therefore, many pretreatment techniques such as physical, chemical, biological, and combined are employed to boost the production of biogas from agricultural straw. Four processes of hydrolysis, acidogenesis, acetogenesis, and methanogenesis are interrelated in anaerobic digestion [63].

Hydrolysis is considered the primary stage of anaerobic digestion where the fermentation-causing bacteria convert complex insoluble organic materials like cellulose into soluble molecules like sugars and amino acids. In order to hydrolyze the complex substance into sugars, alcohols, peptides, and amino acids, the fermentative bacteria produce a variety of hydrolytic enzymes, including lipases, proteases, amylases, and cellulases. Following enzymatic cleavage, the hydrolysis yields are capable of being dispersed by acidogenic bacteria' cell membranes. Hydrolysis can be a rate-determining process. However, some studies depicted that methanogenesis may be a rate-deciding advance dependents upon hydrolytic to methanogenic microorganisms ratio. The anaerobic digestion process greatly depends on the hydrolysis stage. This has led to an intense focus on improving hydrolysis in anaerobic digesters [40]. During acidogenesis, simple sugars, fatty acids, and amino acids decompose into volatile fatty acids (VFAs) and alcohols. These and other VFAs are produced by acidogenic bacteria. The digester type affects the concentration of the compounds produced during acidogenesis. With a regeneration time of less than 36 h for acidogenic bacteria, this AD stage is quicker than the other process step. This phase involves a variety of microorganisms, including methanogens, acetogens, acidogens, and homoacetogens [26].

Alcohols and volatile fatty acids undergo acetogenesis to produce hydrogen, carbon dioxide, and acetic acid. Among the microorganisms involved in this stage for product creation are *Syntrophobacter wolinii, Sytrophomonos wolfei, Clostridium* spp., *Peptococcus anerobus, Lactobacillus*, and *Actinomyces*. An inevitable byproduct, hydrogen influences the components of the products made from organic matter and is crucial in maintaining the equilibrium of the redox reaction of this stage [55]. Following the formation of hydrogen and acetic acid, methanogenic bacteria enable methanogenesis, which transforms these substances into carbon dioxide and methane either by reducing carbon dioxide with hydrogen and methanogens such as *Methanobacterium, Methanobacillus, Methanococcus, and Methanosarcina*, or by breaking down acetic acid into carbon dioxide and methane. Lipid-rich waste from slaughterhouses, food processing facilities is hydrolyzed into glycerol and free long-chain fatty acids in the anaerobic process. While acetogenesis converts the long-chain fatty acids to acetate, acidogenesis converts glycerol to acetate. Moreover, methane is produced from the acetate. Similarly, methane from animal manure ranges from 25 to 50 m3 biogas/m^3 [63].

(iii) **Fermentation**

Fermentation is one of the oldest and most widely used biotechnological processes, allowing microorganisms to transform raw materials into valuable products. This technique has been utilized for centuries in food, beverage, and pharmaceutical industries, among others. During fermentation, microbes break down complex organic substances in the presence or absence of oxygen, depending on the specific process being employed. Under anaerobic and aerobic conditions, bacteria and other microorganisms consume nutrients from the substrate, generating secondary metabolites and performing various physiological functions essential to the process. Different fermentation techniques have been developed to obtain specific products based on industrial applications. These methods enable the efficient production of enzymes, biofuels, organic acids, antibiotics, and other value-added compounds from a variety of raw materials. One of the main advantages of fermentation-based production is that it does not require a complete redesign of industrial processes. Instead, it serves as an innovative and adaptable alternative that integrates easily into existing systems [55, 63]. Moreover, fermentation processes facilitate the large-scale treatment of organic waste without the need for extensive preprocessing. This makes them environmentally sustainable and cost-effective solutions for waste management and resource recovery. Depending on the type of substrate used, fermentation can be classified into two main types which are submerged fermentation and solid-state fermentation.

Solid-state fermentation is utilizing microorganisms on a solid substrate devoid of liquid. To achieve the largest and most nutritional production, a variety of microorganisms were employed, including yeast, bacteria, fungi, and others. Depending on the culture type, solid-state fermentation is further divided into pure and mixed cultures. While microorganisms are employed for fermentation in mixed cultures, specific strains are used for fermentation in pure cultures. Additionally, the type of substrate that is utilized to support solid surfaces and the source of nutrients during fermentation determines the solid-state fermentation. Green walnut husks, pomegranate and lemon peels, grape marcs, groundnut oil cake, pine apple waste, grape seeds, sweet potato waste, wheat and paddy straw, rice and wheat bran, black-eyed pea, cereals, peanut press cake, pulses, and *Citrullus vulgaris* seeds were among the many substrate types used for solid-state fermentation.

The procedure known as submerged fermentation uses a liquid fermentation substrate that contains nutrients to improve the growth of microorganisms. Common substrates used by submerged fermentation are free-flowing liquids such fruit and vegetable juices, molasses, soluble sugars, and liquid media. The optimum bacteria for this fermentation process are those that need a lot of moisture. Submerged fermentation is mostly used to extract the secondary metabolites. The choice of strain is one of the key components of an effective industrial process. Additionally, the appropriateness and properties of fermentation, such as viscosity or recoverability, vary throughout bacteria. To cultivate the microorganisms in the medium, submerged fermentation uses three distinct methods: fed-batch culture, batch culture, and continuous culture. In fed-batch culture, nutrients are continually given to the batch culture, while in batch culture, a fixed volume of medium is used. On the other hand, in continuous culture, the medium containing the product is extracted and new medium

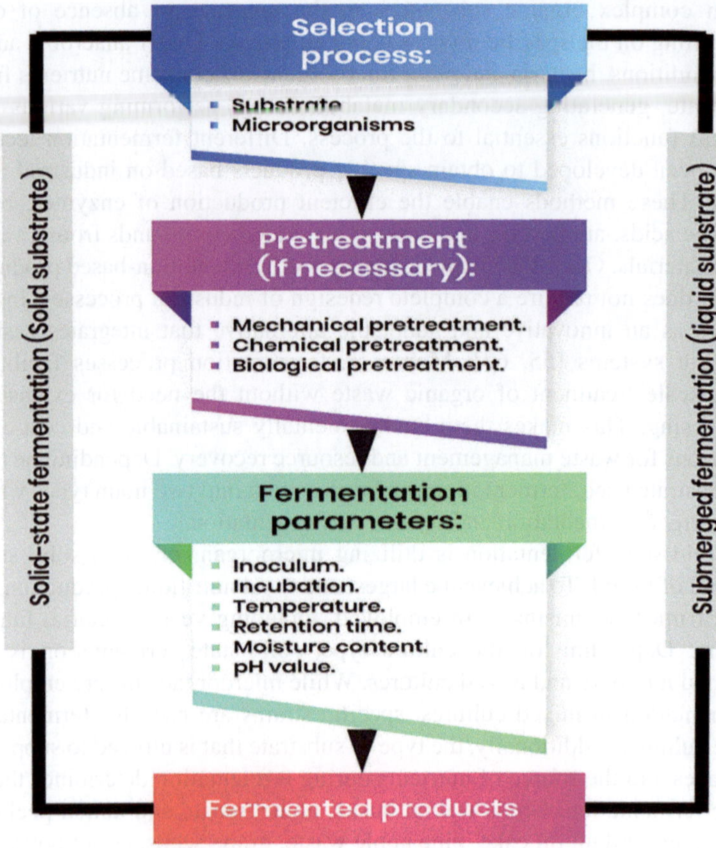

Fig. 2.4 Sequences of procedures and parameters for fermentation process

is continuously supplied as needed. Beginning with substrate selection and concluding with the fermented product, Fig. 2.4 depicts a number of processes and parameters for the fermentation process.

2.3.2 *Chemical Routes*

Chemical routes for the synthesis of bio-based materials involve converting renewable biomass into useful materials via a variety of chemical reactions. These approaches alter or improve the characteristics of bio-derived substances via solvent-based, thermochemical, and catalytic processes, allowing for their use in a variety of sectors. Common processes that transform basic biomass components, such as lignocellulose, lipids, and carbohydrates, into biopolymers, biofuels, and functionalized compounds include esterification, transesterification, and polymerization. These pathways seek to increase productivity, decrease waste, and lessen dependency on fossil fuels by utilizing the concepts of green chemistry, which will support the development of sustainable materials.

(i) **Esterification**

Esterification is one of the most important processes in chemical synthesis. Esters are ubiquitous organic molecules, both natural and manufactured. Biofuels like biodiesel, solvents like ethyl and methyl acetate, paints and varnishes, medications, polymers and coatings, and some insecticides and herbicides are the main examples of esterification products. Esters are widely used as preservatives and flavorings in food goods, as significant perfume ingredients, as perfumes in soap and cosmetics, and in the creation of personal care products. In a typical esterification reaction as shown in Fig. 2.5, a catalyst is used to accelerate a combination of carboxylic acids and an excess of matching alcohols. After a while, the reaction reaches equilibrium. To move the equilibrium forward, either an excess of one reactant, typically alcohol, must be added, or water must be continuously removed. Due to the development of thermodynamic equilibrium, the reaction is constrained by a low overall conversion and a slow pace of reaction. The esterification process has yet to get past these obstacles in a way that is both economical and ecologically friendly, despite its commercial significance. According to reports, employing a range of alcohols and acids under typical reflux circumstances, ester yields between 58% and 75% after 1–4 h of reaction [32].

A range of acid molecules, including carboxylic acids, acid chlorides, acid anhydrides, and lower esters, are used with various alcohols in the synthesis of esters. Even though acid chlorides are the most widely used acylating reagents, their high reactivity, low commercial availability, and sensitivity to moisture can occasionally result in a complex variety of products from the acylation reaction. Because of their comparable chemical characteristics, acid anhydrides and acyl chlorides are not usually the optimal reagents to use when creating esters. Alcohols and carboxylic acids react as a condensation reaction, which necessitates constant dehydration and stoichiometric quantities of the condensation reagents [32]. Appropriate organic or inorganic catalysts are used to help speed up the reaction and produce a wide range of ester compounds in order to compensate for the low reactivity of carboxylic acids and enable the reaction to proceed. This reaction capacity to run in neutral environments while maintaining various acid- or base-sensitive capabilities emphasizes its appeal and benefits over the other two reaction pathways requiring acid chlorides and acid anhydrides.

Fig. 2.5 Typical esterification reaction

Unavoidably, a sufficient catalyst is needed. Therefore, the rate of reaction is greatly decreased in the absence of a catalyst. High catalytic activity, improved kinetics, handling capabilities, low waste generation, separability from the end product combination, and recyclability are all characteristics of an ideal catalyst. By acting as a proton donor, the acid catalyst renders the carboxylic acid labile, allowing alcohol to quickly attack it nucleophilically. By altering the reaction conditions, such as temperature, employing excessive amounts of alcohol, or mixing the catalyst with an activator or promoter, the activity of a catalyst can be further increased. Traditional homogeneous catalysts like sulfuric and hydrochloric acids, as well as heterogeneous catalysts like ion exchange resins, are the most commonly used catalysts for esterification. The most widely used commercial esterification method is homogeneous acid catalysis. The most commonly used catalysts are p-toluene sulfonic acid, HI, HCl, and H_2SO_4. Mineral acids are examples of homogeneous catalysts that offer high conversion at little cost. However, as was previously mentioned, they are hazardous to the environment, challenging to manage, and complicate downstream separation by producing byproducts during neutralization. Solid acid catalysts were investigated as a potential replacement for liquid acid catalysts in order to address these issues. In order to create catalysts with accessible active sites for quick reactivity, solid acid catalysts are typically made by physically or chemically connecting acid functional groups to solid supports. Zeolites, heteropolyacids, ion-exchange resins, and supported chlorides are examples of heterogeneous acid catalysts that are being extensively studied as an environmentally acceptable way to make the esterification process green. However, since these catalysts are not renewable, there is a growing movement to produce solid acid catalysts that are carbon-based and renewable. Carbon supports for solid acid catalysts come from renewable resources. After that, acid functional groups are joined to carbon atoms through covalent bonds. Beside homogeneous and heterogeneous catalyst, enzymes employed to catalyze esterification. However, enzymes work in reversible reactions like all other catalysts affect the rate of reaction in both directions, the direction taken. So, it is determined by thermodynamics. [32]. Table 2.7 explores esterification processes using homogenous, heterogenous, and enzymatic catalysts.

(ii) **Transesterification**

Transesterification is the process of an oil or fat reacting with alcohol to form glycerol and ester when a catalyst is present as illustrated in Fig. 2.6. A high yield of esters requires the addition of extra alcohol, which helps to propel the reaction forward because the process is reversible. Two alcohols that are frequently utilized in the transesterification reaction which are methanol and ethanol. Nonetheless, methanol is favored because of its inexpensive cost and physical and chemical benefits, including a low boiling point, high reactivity because of its short carbon chain, and the ease of recovery provided by its inability to form azeotropes. Using either a homogeneous or heterogeneous base catalyst, the catalytic transesterification of oils with methanol has been thoroughly investigated. Alkali catalysts are most frequently used for trans-

Table 2.7 Homogeneous, heterogenous, and enzymatic catalyzed-esterification processes

Esterification type	Reactant	Alcohol	Catalyst	Catalyst concentration (wt%)	Time (min)	Resultant	Resultant percentage	References
Homogeneous catalyzed-esterification	Tall oil	Methanol	H_2SO_4	0.5	60	FAMEs	96.76	[35]
	Citric acid	Ethanol	Methanesulfonic acid	2	20	Triethyl citrate	91	[11]
	FFA	Methanol	BF_3	5.33	15	FAMEs	100	[33]
Heterogeneous catalyzed-esterification	Oleic acid	Methanol	Sulphonated carbonbased waste ginger straw	7	210	Methyl oleate	93.20	[84]
	FFA	Methanol	Acid-modified montmorillonite (k10) clay	6	120	FAMEs	80.8	[47]
	Acetic acid	n-hexanol	Sulfonic acidfunctionalized MIL-101	0.05	300	Hexyl acetate	65.1	[88]
Enzymatic catalyzed-esterification	Formic acid	Hexanol	Candida antarctica lipase (Novozym 435)	15	90	Hexyl formate	95	[65]
	Stigmasterol	Organic solvent	Candida rugosa lipase	8	960	Stigmasteryl oleate	97.33	[15]
	Octanoic acid	Hexanol	Candida antarctica lipase (Novozym 435)	–	60	Hexyl octanoate	90	[38]

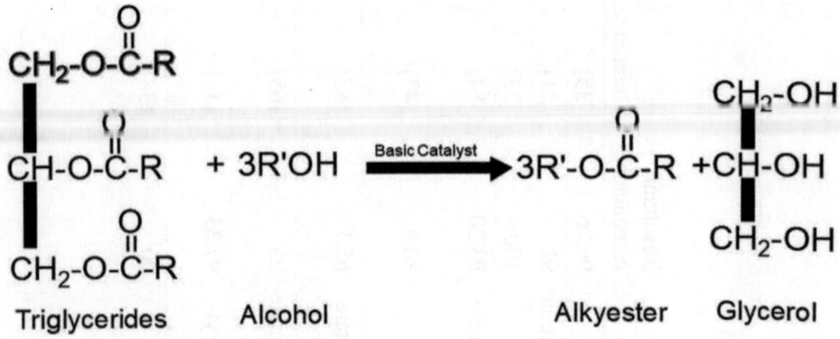

Fig. 2.6 Typical transesterification reaction

esterification because, in contrast to acid catalysts, they provide quick reactivity with tiny catalyst concentrations at low temperatures in a comparatively short amount of time. The sensitivity of base catalysts to the presence of water or high FFA concentration in oil, which can result in soap formation with subsequent emulsification, is a typical issue with their utilization [49].

Homogeneous catalysts play a crucial role in the transesterification process, particularly in biodiesel production. Among these, homogeneous base catalysts such as sodium hydroxide, potassium hydroxide, and sodium methoxide are widely used due to their high catalytic efficiency. These alkaline liquid catalysts promote rapid reactions, making them suitable for large-scale biodiesel production. In addition to homogeneous catalysts, various heterogeneous catalysts have been explored, including alkaline-earth metal compounds, titanium silicates, anion exchange resins, and immobilized enzymes, as outlined in Table 2.8. Heterogeneous catalysts have gained increasing attention due to their potential for reusability and lower environmental impact [49]. However, despite these advantages, homogeneous catalysts remain dominant in industrial applications. Their widespread use is attributed to their high catalytic activity, cost-effectiveness, and minimal mass transfer limitations. While heterogeneous catalysts offer promising benefits, further research is required to enhance their efficiency and scalability for commercial biodiesel production.

While there are many benefits to transesterification for the production of fatty acid alkyl esters, there are also some disadvantages, such as the lengthy processing time needed to transesterify vegetable oils and the post-treatment steps necessary for product purification, such as heating, neutralization, separation, and water washing when using homogeneous acid/base catalysts. In contrast to their homogeneous counterparts, heterogeneous base catalysts require a comparatively high temperature to stimulate reaction, although being more affordable. Environmental concerns have also been raised by the issue of potential catalyst leaching. High free fatty acid (FFA) level in the feedstock causes saponification, which consumes catalyst and results in an incomplete reaction with emulsion formation, making glycerol extraction challenging. Preventing soap production with a two-step transesterification process that

Table 2.8 Homogeneous and heterogeneous catalyzed-transesterification reaction

Transesterification type	Oil source	Catalyst	Parameters				Biodiesel yield (%)	References
			Molar ratio	Catalyst ratio (wt%)	Temperature (°C)	Time (min)		
Homogeneous catalyzed-transesterification	Sunflower	$NaOCH_3$	1:3.75	0.5	60	60	100	[34]
	Cottonseed	KOH	1:6	0.6	55	60	96	[54]
	Neem	KOH	1:6	1	65	40–50	96	[1]
	Jatropha curcas	NaOH	1:9	0.8	45	30	96.29	[53]
	Karanja	NaOH	1:9	0.5	–	30	89.5	[68]
Heterogeneous catalyzed-transesterification	Palm	Rice husk-derived sodium silicate	1:12	2.5	65	30	97	[61]
	Soybean	Cs-Na_2ZrO_3	1:30	1	65	15	98.8	[74]
	Soybean	Micro-structured CaO of eggshells	1:12	5	65	180	85.83	[74]
	Jatropha	KOH/calcined waste animal bones	1:9	6	70	180	96.10	[50]
	Castor	Iron (II) doped ZnO nanocatalyst	1:12	14	55	50	91	[7]

involves treating high FFA concentration with an acid catalyst beforehand can address this issue, but it also entails significant operating and instrumentation costs. Lipases are the most often researched enzyme catalysts for transesterification. The use of an enzyme catalyst can save all the separation and purification steps, but there are drawbacks to its occasional use, including the inhibition of the enzyme by high alcohol concentrations, a longer reaction time, high cost, and commercialization.

(iii) **Polymerization**

Polymerization is a basic chemical process that creates long-chain macromolecules called polymers by the chemical bonding of tiny molecules called monomers. Strength, flexibility, and durability are frequently improved by this transformation, which also drastically changes the material chemical and physical characteristics. There are various methods that polymerization can happen, but it is generally divided into two categories which addition polymerization and condensation polymerization. Furthermore, monomers having double or triple bonds react during polymerization without losing any atoms. Polyethylene (PE), polypropylene (PP), and polystyrene (PS) are common examples and are used extensively in consumer goods, construction, and packaging [16]. When monomers react in condensation polymerization, covalent bonds are formed and tiny molecules like water, methanol, or hydrochloric acid are released. Polyesters, polyamides (like nylon), and polycarbonates, all of which are utilized in textiles, engineering plastics, and biomedical applications, are frequently formed using this process. Depending on where they come from, polymers can be categorized as either natural or synthetic. Proteins (such silk, collagen, and keratin), polysaccharides (like cellulose, starch, and chitin), and nucleic acids (like DNA and RNA) are examples of natural polymers that are necessary for biological processes and the structural integrity of living things. However, because of their specific qualities, synthetic polymers such as polyethylene, polypropylene, and polytetrafluoroethylene (PTFE) or Teflon are manufactured and utilized extensively in a variety of sectors. Because of its versatility, polymerization is a fundamental component of bio-based materials, facilitating the creation of a wide range of applications, including enhanced biomedical materials, high-performance composites, and biodegradable plastics.

Bio-based polymerization is essential to the synthesis of bio-based materials in order to convert renewable monomers into sustainable polymers with a variety of uses. Environmentally acceptable substitutes for petroleum-based plastics are produced using bio-based polymerization, which frequently uses monomers made from plant oils, lignocellulosic biomass, or microbial fermentation. Polylactic acid (PLA), which is made via ring-opening polymerization of lactic acid, and polyhydroxyalkanoates (PHAs), which are naturally produced by bacteria, are examples of common bio-based polymers. These sustainable materials have enormous potential for use in packaging, biomedical disciplines, and as biodegradable substitutes for traditional plastics as long as bio-based polymerization continues to progress. Bio-based polymers are an environmentally favorable substitute for traditional plastics since they are made in part from sustainable and renewable source materials. Their biodegradability, environmental effect, and general degrading behavior under natural conditions are all greatly influenced by their production pathway, which includes the raw ingredients and synthesis techniques utilized. Three primary groups of bio-based polymers are depicted in Fig. 2.6. They include those produced biologically, those generated from biomass, and those synthesized from the action of microorganisms. The first group, known as bio-based polymers obtained from biomass, comprises natural polymers sourced from plants, animals, and lipids, such as wax, gluten, soy, gelatin, and collagen [52]. The second group, which is made up of pullulan, xanthan, and curdlan, is derived from microbial activity. Bio-based polymers, such as polyesters, bionolle, and polylactate, fall under the third group, which is biologically synthesized. These bio-based polymers are produced either directly by living things or via bio-based monomer polymerization. This classification emphasizes the promise of bio-based polymers for sustainable material development by highlighting their varied origins and paths. Figure 2.7 shows the sources of bio-based polymers.

2.3.3 *Thermochemical Processes*

Thermochemical processes are essential to the synthesis of bio-based materials by transforming biomass into useful chemicals through heat-driven reactions. Pyrolysis, gasification, hydrothermal carbonization are important techniques that yield charcoal, biofuels, and bio-based compounds.

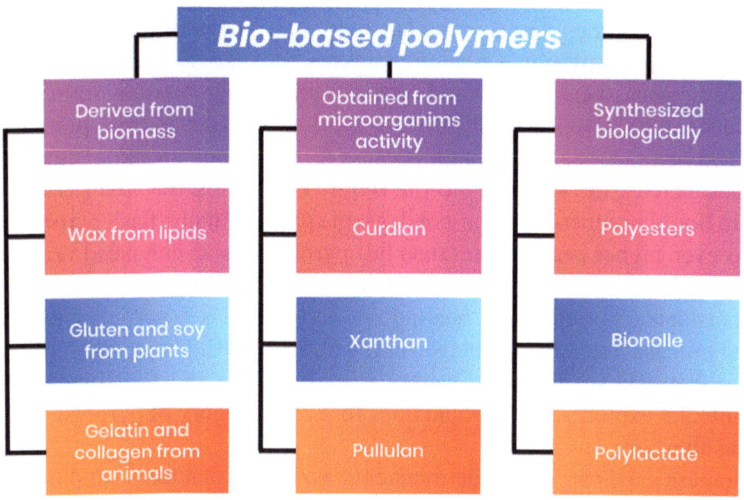

Fig. 2.7 Sources of bio-based polymers

(i) **Pyrolysis**

Pyrolysis is the thermochemical conversion process of any carbon-based product ess known. In the past, it was employed to produce charcoal. The word "pyrolysis" comes from Greek, where "pyro" means "fire" and "lysis" means "breakdown". Thus, it can be described as the decomposition of lignocellulosic derivatives in conditions of high temperature and low oxygen content. Charcoal, biochar, and liquid and non-condensable gases (H2, CH4, CO, CO_2, and N_2) are produced when lignocellulosic derivatives undergo thermal breakdown. The biomass is heated from room temperature to a predetermined temperature during pyrolysis. The biomass is moved via a screw conveyor at a predetermined speed while still inside the pyrolysis unit before the process is finished. The pyrolysis process yields three different forms of carbonaceous material which are biochar, char, and charcoal [3]. Char is any carbonaceous residue left over after pyrolysis, while charcoal is made by pyrolyzing vegetable or animal material in kilns for heating or cooking. In addition, biochar is a carbonaceous substance that is specifically made to be applied to soil as a component of environmental or agronomic management.

Based on the employed temperature and time, pyrolyses are categorized into slow pyrolysis, intermediate pyrolysis, fast pyrolysis, and flash pyrolysis. Biochar is the primary result of the slow pyrolysis, which is often a batch operation. Bio-oil and syngas are by-products. Moderate temperatures (300–700 °C), a ramping rate (<1 °C s^{-1}), and a longer residence/holding duration (>5 min) are the primary conditions for slow pyrolysis. intermediate pyrolysis under the subsequent circumstances. It has become a stand-alone synthesis method with a residence time of 2–4 s, a process temperature of up to 500 °C, and a process time of 30–1500 s. The primary product is the liquid fraction (40–60%), which is followed by solid biochar (16–25%) and gases (20–30%). The increasing desire to produce liquid products from biomass is the reason for the interest in the fast pyrolysis. The preferred operating parameters for quick pyrolysis are temperatures between 450 and 800 °C and brief holding times between 0.5 and 10 s. By preventing volatiles from taking part in secondary reactions and preventing primary products from participating in cracking reactions, these high process temperatures and brief holding durations increase the yields of bio-oil. Before char forms, the biomass is converted into liquid products and non-condensable gases. Finally, flash pyrolysis, which has even higher heating rates than fast pyrolysis, is an enhanced version of fast pyrolysis. Flash pyrolysis frequently uses the following operational parameters of residence periods <2 s, heating rates >1000 °C s^{-1}, particle size <0.1 mm, and process temperatures between 800 and 1200 °C. Due to the rapid ramping rate involved in flash pyrolysis, the particle size must be tiny (<0.1 mm). Table 2.9 explores the pros and cons for each category of pyrolysis [77].

(ii) **Gasification**

Biomass gasification is a thermochemical process that turns solid biomass into useful syngas and a solid residue known as char by treating it with a small amount of gasifying medium. Typically, methane, carbon dioxide, carbon

Table 2.9 Advantages and disadvantages of pyrolyses types

Pyrolysis	Advantages	Disadvantages
Slow pyrolysis	Standard technique and basic equipment Substantial biochar outcome Able to accommodate a broad range of particle sizes	More treatment is required Potential need for repeating
Intermediate pyrolysis	Able to handle feedstock with a high moisture content A variety of feedstocks can be treated, and the operating temperature is still low Bio-oil has less tar and can be used straight away to generate heat	The development of operational parameters is currently ongoing Lack of investigations conducted at the pilot or industrial size in this category
Fast pyrolysis	Appropriate technique for producing bio-oil Easy and quick procedures Economically, scaling up is possible Effective conversion of energy It is possible to improve bio-oil to "drop-in" fuel	Demands careful consideration of both functioning and design Little experience in business The bio-oil that is produced is not very stable It is necessary to use biomass feed in fine particles (1–2 mm) Choose biomass with moisture less 10%
Flash pyrolysis	It is possible to achieve >75% bio-oil outcome Remarkably low impurity levels in bio-oil Manufacturing of more energy-dense fuel	Application on an industrial scale may be hampered by operating parameters Oil has poor chemical stability, corrosion resistance, and thermal stability Oil solid phase and gradual increase in viscosity Alkali from biochar dissolves in bio-oil

monoxide, and hydrogen make up syngas. The gases can be further transformed into liquid fuels or into heat and energy for power plants. A gas such as oxygen, supercritical water, or air could serve as the gasifying medium. The production of tar, which is challenging to clean and reduces the hydrogen yield, is the main challenge encountered during the gasification process. In general, a high content synthesis gas can be produced by a high ratio of cellulose or hemicellulose and lignin. In order to produce synthesis gas (CO and H_2), which can then be converted into more valuable chemicals and biofuels, biomass gasification needs a gasifying media. Air, steam, or oxygen can all be used as the gasifying medium. Due to its affordability and ease of availability, air is the most often utilized gasifying medium. The resulting gas has a higher H/C ratio, though, if steam serves as the gasifying medium [44].

The machine that turns biomass into syngas is called a gasifier. Gasifier is separated into four distinct zones which are drying, pyrolysis, combustion, and reduction zones. The drying zone is where biomass gets dehydrated. Moisture content varies among biomass contents. The quality of the final product is mostly dictated by the biomass that is used for gasification. For syngas with a high heating value, biomass with a moisture content of 10–20% is usually advised. Before being used for gasification, biomass with a high moisture content is typically dried in the drying zone. However, a high moisture level degrades the product quality and results in energy loss. Biochar, liquid products, and gaseous molecules are produced in the pyrolysis zone when a number of big molecular groups are broken down into smaller hydrocarbons. The least stable substance, hemicellulose, breaks down to produce tar, char, and vapors. In theory, the simplest and most straightforward method for processing biomass degradation is combustion. Exothermic chemical processes raise the temperature inside the combustion zone. Additionally, the amount of gasifying agent is often regulated to prevent it from reaching the ash slagging temperature, which could interfere with its functionality. The final products produced in this zone are CO, CO_2, H_2, and H_2O [44]. The heat released is employed in the pyrolysis process and to partially dry the components. Because it raises the temperature of the tar particles in the generated gas, the reduction zone gets its name. The largest issue produced by biomass gasification is tar. An excessive amount of tar in the fuel gas will lower biomass overall efficiency and raise the plant total separation costs. Tar can block filters and potentially polymerize into complicated compounds if it is not treated. Gasification processes on bio-based materials are explored in Table 2.10.

(iii) **Hydrothermal carbonization**

Hydrothermal carbonization is a thermochemical process used to transform organic materials in presence of water into hydrochar which a substance rich in carbon. This process dramatically speeds up the natural carbonization that would otherwise take hundreds to millions of years to turn biomass into fossil fuels like coal. It takes place between 180 and 350 °C of temperature and 2 and 10 MPa of pressure [63]. As an effective and environmentally friendly substitute for biomass conversion, hydrothermal carbonization can accomplish this change in a few hours or days. Because hydrothermal carbonization does not require energy-intensive drying, it is especially beneficial for processing wet biomass, including sewage sludge, food waste, and agricultural leftovers. In contrast to traditional thermochemical processes like pyrolysis or combustion,

Table 2.10 Explorations of gasification processes for bio-based materials

Biomass source	Temperature (°C)	Products	References
Sewage sludge	700	Hydrogen	[24]
Algae slurry	605	syngas	[59]
Biowaste	>1000	Combined gases (H_2, CO, CO_2, H_2O, and CH_4)	[43]
Sugarcane bagasse	700	Hydrogen	[64]
Empty fruit bunch	900	Hydrogen-rich gas	[51]

Table 2.11 Hydrothermal carbonization of bio-based materials

Biomass source	Parameters	Product	References
Banana stalk	Temperature: 160 to 200 °C Time: 1–3 h	Hydrochar	[28]
Eucommia ulmoides	Temperature: 180 to 320 °C with 20 °C interval Time: 30 min	Hydrochar	[81]
Banana peel	Temperature: 190, 210, and 230 °C Time: 30 and 60 min	Hydrochar	[58]
Olive trimmings	Temperature: 180 to 250 °C Time: 0, 0.5, 1, 3, 6, and 8 h	Hydrochar	[39]
Grape marc	Temperature: 180, 220, and 250 °C Time: 1, 3, 8 h	Hydrochar	[8]
Wastes of olive mill, canned artichoke, and orange	Temperature: 200 to 250 °C Time: 2,4,8 and 24 h	Useful bioenergy	[10]
Coconut fiber and eucalyptus leaves	Temperature: 150 to 375 °C Time: 30 min	Hydrochar	[36]
Coconut fiber and dead eucalyptus leaves	Temperature: 150 to 375 °C Time: 30 min	Solid biochar	[37]

which require dry biomass for efficient processing, this makes it an extremely appropriate technology for handling high-moisture feedstocks. The hydrothermal carbonization process also resembles the natural geological processes that over millions of years produced coal, natural gas, and petroleum from decomposing plants and animals. This process, which turns wate into biofuels, soil amendments, and other value-added products, is well known for being resource-efficient and environmentally benign. It also supports the circular economy. Hydrothermal reactions are still being optimized by researchers for a number of uses, such as the synthesis of actated carbon, platform chemicals, and bio-crude oil. Two categories of thermochemical procedures are employed, contingent on the biomass's water content: hydrothermal carbonization for high-moisture feedstocks and dry processes (pyrolysis, combustion) for low-moisture feedstocks. Table 2.11 explores hydrothermal carbonization for some bio-based materials.

2.4 Characterization of Bio-Based Materials

Characteristics of bio-based materials must fulfil the required standards. The encouragement to transform into green manufacturing comes from the efficient, effective, clean, and recyclable alternatives provided by the bio-based materials. In this context, this subsection discusses the properties of bio-based materials.

2.4.1 Physical and Chemical Properties

Gas and water vapor permeability, sealing ability, thermoforming properties, transparency, antifogging ability, printability, availability, UV resistance, water resistance, and resistance to acid, alkali, and grease are some of the characteristics that should be taken into consideration when using food bio-based materials. Therefore, excellent mechanical, thermal, and gas barrier properties, water vapor permeability, and disposability are necessary for bio-based materials in relation to their final uses in the food and other industries. Besides, the production cost is one of the key characteristics of bio-based materials [57]. The majority of bio-based materials are hydrophilic, and because humidity affects gas barrier qualities, hydrophobic bio-based materials may become significantly more gas permeable when humidity rises. Water vapor barrier and gas barrier properties are necessary in many bio-based food packaging applications. In order for future bio-based materials to be able to replicate the water vapor barriers of the conventional materials, innovative chemical modifications and treatments are being investigated to address this issue.

Composites made of bio-based materials are primarily used as structural elements in textiles, packaging, and disposables. They can be made stronger for use in the transportation and construction industries to reduce weight and save costs without sacrificing load-bearing capacity. The development of technologies to innovate chemical and physical modifications for optimal composite properties pertaining to air, water permeability, water resistance/hydrophobicity, surface adhesion and frictional resistance, mechanical strength, fracture resistance, fire retardance, sound barrier, thermal insulation, optical and conductive properties, and more are all part of the current research on the properties of bio-based composite materials. The stiffness and modulus of the mechanical properties are fairly similar to those of traditional polymers. There are bio-based polymers that resemble polystyrene with their relative stiffness at moderate service temperatures or polyethylene with their relative flexibility at the same temperature. The biochar material, which was produced by treating waste wood at 1000 °C, was found to be extremely conductive, with a resistance of $3.1 \pm 1\ \Omega$/cm, according to resistivity measurements. PVA films packed with biochar particles have appropriate electrical conductivity and piezoresistive response, making them potential materials for piezoresistive sensors [57]. Additional study on charge transport in biopolymers would expand applicability to sensors, medical devices, and flexible biodegradable electronics for regulated implant and healing use.

The properties of bio-based materials for thermal insulation are amazing. Because of their low heat conductivity, rigid polyurethane and polyisocyanurate foam materials are widely employed as heat-insulating materials in the civil engineering and refrigeration industries. Other polyurethane compositions are used in coatings and adhesives. Through condensation with isocyanates, renewable resources were added to the polyurethane polymer matrix in the form of polyol, a hydroxyl derivative of vegetable and other natural oils, to create the stiff polyurethane product. The industrial standard for thermal insulation materials is

22 mW/m.K., which falls within the range of materials developed from vegetable polyols. Therefore, bio-based polyols and nanoparticles can be combined to create sustainable flame retardants in polyurethane-based materials, replacing chemicals sourced from petroleum. In recent years, bio-based insulating materials have become more widely used, usually in permeable wall constructions. The insulating material might affect the indoor air quality when the wall is constructed with air permeability. Compared to traditional insulation materials, bio-based materials can provide enough thermal conductivity with a lower environmental impact. Certain bio-based materials have high sound insulation in addition to thermal insulation. In order to comply with health and safety regulations and lessen the impact on the environment, noise reduction and control are crucial factors in contemporary urban building. There is a lot of research being done on sustainable alternatives to asbestos-based and synthetic fiber-based sound-absorbing materials with better applications. Compared to solid glass fibers, natural fibers can attenuate sound energy through a variety of ways because of their lightweight, porous interior structure and network of air spaces in their hollow cell lumens. These routes account for the fiber structure's superior sound-absorbing qualities. When sound waves strike a natural fiber, they travel through several paths into the fiber lumen, where friction causes energy loss and conversion to heat. Fiber vibration is caused by the air in the bulk and lumen vibrating. As the number of bamboo fibers grows for the same cross-sectional area, the sound absorption coefficient rises as the diameter of the fibers decreases. Activated carbon fiber composites provide for increased air capacity to absorb sound since the non-woven structure of activated carbon fiber is integrated with the fibers [57]. Inorganic particles, such as precipitated calcium carbonate, were added to the composite to improve its rigidity and sound absorption. Another way to accomplish the desired qualities of sound absorption with shade, insulation, moisture barrier, and mechanical strength is by using nano-fillers in composite mixes. Table 2.12 shows characteristics of some bio-based materials.

2.4.2 Analytical Techniques

A wide variety of analytical techniques are used in the evaluation of bio-based materials in order to assess their mechanical strength, moisture-related characteristics, thermal behavior, and chemical composition. These techniques allow researchers to probe material behavior under various environmental and operational conditions.

(i) Spectroscopic methods, such as Fourier Transform Infrared Spectroscopy (FTIR) and Raman Spectroscopy, are widely used for identifying functional groups, detecting molecular vibrations, and assessing chemical modifications in lignocellulosic structures and other natural polymers. These insights are crucial for tracking structural changes during processing or modification.

Table 2.12 Properties of some bio-based materials

Bio-based material	Density (g/cm^3)	Porosity (%)	Capacity thermal mass (J/kg.K)	Thermal conductivity (W/m.K)	Moisture buffer value (g/m^2.%HR)	Water vapor permeability (kg/m.s.pa)	Factor of resistance to dissemination of steam
Wooden concrete	1.090–1.294	N/A	952–1035	0.31–0.42	1.96–2.55	2.19×10^{-12}	36
Cork concrete	0.908–1.937	N/A	N/A	0.5–0.194	0.39–1.2	(2.7×10^{-12})–(21.4×10^{-12})	9.4–71.9
Concrete date palm	0.68–0.968	27–64	1500–2341	0.091–0.201	1.79–4.05	(2.22×10^{-11})–(3.59×10^{-11})	3.16–8.87
Straw concrete	0.4–0.56	75.1	1168	0.0371–0.094	2.9–2.59	(1.89×10^{-11})–(2.56×10^{-11})	7.87–10.58
Hemp concrete	0.2–0.96	76	1250–2900	0.06–0.6	2.02	10^{-11}–10^{-10}	5–25
Linen concrete	0.598	71	1144	0.168	2.27	2.34×10^{-11}	N/A
Wood lime panels	0.875–0.984	1.33–1.4	N/A	0.0742–0.1078	N/A	N/A	N/A
Wood fiber panels	0.15–0.32	89–97	1700–2000	0.04–0.053	2–3.5	N/A	2.6–16.8
Bamboo panels	0.538	63.17	1760	0.088	3.9	1.77×10^{-11}	13.1
Solid wood	0.368–0.58	67	1280–1500	0.080–0.192	N/A	N/A	17–160.9
Expanded cork	0.065–0.24	19–86	1600–2100	0.037–0.044	N/A	N/A	5–54.61
Wood wool	0.47	N/A	1330	0.041	N/A	N/A	3.6
Hemp wool	0.35	95	1200	0.04	N/A	N/A	3.7
Wood fiber	0.5–0.27	N/A	1900–2100	0.038–0.050	N/A	N/A	1–5

Adapted from [12]

(ii) X-ray Diffraction (XRD) is applied to determine the crystalline versus amorphous content in materials like cellulose, lignin, and starch. This information directly relates to mechanical strength and chemical reactivity, influencing applications in bio-composites and packaging.
(iii) Thermal analysis techniques include Thermogravimetric Analysis (TGA) and Differential Scanning Calorimetry (DSC), which provide critical data on decomposition temperatures, thermal transitions (e.g., melting point, glass transition temperature), and thermal stability. Such insights are vital for ensuring performance under thermal stress in end-use environments.
(iv) Surface morphology and microstructural features are examined using Scanning Electron Microscopy (SEM) and Atomic Force Microscopy (AFM). These techniques reveal fiber-matrix interfaces, porosity, and surface roughness, which affect composite performance, water absorption, and durability.
(v) Mechanical testing, including tensile, compressive, and flexural strength tests, quantifies the structural integrity and durability of bio-composites and films.
(vi) Moisture-related evaluations, such as water vapor permeability (WVP) and moisture buffer value (MBV), are conducted to assess barrier properties and hygroscopic behavior, crucial for applications like bio-packaging and insulation.
(vii) Chromatographic techniques, such as high-performance liquid chromatography (HPLC) and gas chromatography-mass spectrometry (GC-MS), are essential for the separation, identification, and quantification of bioactive compounds, additives, or degradation products.

2.5 Micro to Nano-scale Advancements in Bio-based Materials

Advancements in bio-based material opened up new opportunities across a range of industries. Micro-scale forms of bio-based materials, including cellulose fibers, chitosan, and starch, were initially used in a variety of industries, including packaging, textiles, and biomedicine. But thanks to the development of nanotechnology, scientists can now improve these materials on a nanoscale. Nanotechnology can improve the qualities and reduce the cost of biomaterials by changing the manufacturing and fabrication processes. Recently, bio-nanocomposites, which include chitosan, proteins, cellulose, and starch, have been observed to perform better in food packaging. Plastisizers are typically added to protein and polysaccharide-based films to improve performance and improve mechanical and barrier qualities. It has also been acknowledged that metal oxides enhance the polymer matrix barrier qualities. The most widely employed nanomaterials to improve the mechanical, thermal, barrier, and antibacterial qualities of nanocomposites are clay and silver nanoparticles. Additionally, by adding antioxidants, oxygen scavengers, antibacterial activity, scent, color, and other physiologically active substances, bio-nanocomposites may provide better biochemical packing and active functionalities.

Bio-based nanocomposites demonstrated effective outcomes in a range of applications. Natural bio-nanocomposites, including proteins, cellulose, chitosan, and starch, as well as materials like zein and nano-clay, have been recognized for use in environmentally friendly packaging. As shown in Table 2.10, these polymers have been identified for a number of uses in food industry. Since the usage of plastic has resulted in significant pollution and health risks, bioplastics have drawn the attention of numerous academics in recent years. Thus, biodegradable or renewable polymers became popular as a substitute for petroleum-based plastic. Microorganisms have the ability to break down bioplastics and biopolymers into organic chemicals, carbon dioxide, hydrogen, and water. Numerous researches have demonstrated that nanocomposites made of bio-based materials provide a number of benefits along with increased product safety. Table 2.13 examines polylactic acid-based synthetic bioplastic.

Table 2.13 Natural and synthetic bio-nanocomposites derived from bio-based materials and their applications

Bio-nanocomposite type	Polymeric matrix	Nanomaterial	Application	References
Natural nanocomposite	Chitosan from marine source	Sodium montmorillonite (Np-Clay), NP-ZnO	Mechanical, thermal, antibacterial enhancement in food packaging	[60]
	Starch from potato	Starch nanocrystals	Enhancement in mechanical and thermal properties for food industry	[67]
	Cellulose from mango leaf	Nanofibrillated cellulose films with mango leaf extract (NFC-MLE)	UV-light protection and antioxidant properties	[9]
Synthetic bio-nanocomposite	PLA (Polylactic acid)	Graphite and silica-based nanoparticle	Water vapor and oxygen barrier, storage modulus, tensile strength, crystallinity, glass transition temperature (Tg) value, antimicrobial	[46]
	Poly (3-hydroxybutyrate-co-3-hydroxyhexanoate) (PHBHHx)	PHBHHx/graphene oxide (GO) nanocomposites	Thermal stability, gas and vapor barrier, biocide action	[19]
	Carboxymethyl chitosan (CMCS)/carboxymethyl cellulose sodium (CMC)/PLA	Nano zinc oxide (ZnO) (CMCS/CMC+ PLA/ZnO NPs nanocomposite)	Water vapor barrier, oil resistance, antibacterial	[82]

Food quality can be negatively impacted by a variety of germs, poisons, and diseases, which is why they must be identified. Toxins, infections, and microorganisms have all been observed using fluorescent nanoparticles. A bioreceptor in a nano-biosensor that binds to analytes (microbes or related toxins) is used to detect toxins produced by different food deteriorating bacteria. Through the analysis of gas emissions or changes in food color brought on by spoiling, a number of gas sensors have been created to identify microorganisms. Gas sensors based on metal oxides have been employed because of their great stability and sensitivity. An overview of different nano-biosensors for the detection of toxins and foodborne pathogens is provided in Table 2.11. The most frequent pollutants identified in food products are various heavy metals, medications, pesticides, veterinary antibiotics, and allergens that may have a negative impact on human health. Several platforms have been used to detect these contaminants which are: (i) optical and electrochemical sensors (using nanomaterials like Au and Ag NPs with good conductivity and surface platform resonance characteristics); (ii) magnetic nanoparticles (Fe_2O_3 NPs) for better analyst separation and enrichment; and (iii) electrochemical sensors based on carbon nanotubes and graphene (for better electrical conductivity). The most widely used nano-biosensors using nanomaterials and their purposes of toxins and heavy metals detections are mentioned in Table 2.14.

Table 2.14 Bio-based materials nano-biosensors and their applications for toxins and heavy metals detections

Nano-biosensor type	Sensor	Nanomaterial	Function	References
Toxins and pathogens detector	Aptamer-based biosensor	carbon dots on gold nanoparticles	Toxin detection (aflatoxin B_1)	[79]
	Graphene-interfaced electrical biosensor	Graphene NPs	Microbe detection (*E. coli* O157:H7)	[56]
	Electrochemical immunobiosensor	(Cadmium sulfide-quantum dots) Cd QDs encapsulated in ZIF-8 metal organic framework	Electrochemicalsignal generation	[93]
Heavy metals and drug detector	Electrochemical	$GSH@Fe_3O_4$ (glutathione coated magnetic nanoparticles)	Preconcentration (Heavy meats (Pb^{2+}, Cd^{2+}))	[6]
	Colorimetric sensing	CeO_2 NPs	Catalytic activity (Mycotoxin-Ochratoxin)	[13]
	Fluorescence sensor	Nitrogen-doped carbon dots	Fluorescent response (Hg^{2+}, Cu^{2+}, Fe^{3+})	[80]

2.6 Summary

Bio-based materials emerged as sustainable alternatives to conventional petroleum-based materials. They offer eco-friendly solutions across various industries. These materials are derived from renewable biological sources and can be classified into three main groups which are plant-derived materials such as cellulose and lignin, agricultural waste and byproducts including husks and shells residues, and microbial and algal sources that can produce proteins, carbohydrates, and lipids. Utilizing these resources reduces reliance on fossil fuels and supports a circular economy. The synthesis of bio-based materials occurs through biochemical, chemical, and thermochemical routes. Biochemical processes involve microbial fermentation and enzymatic transformations to produce biopolymers and bio-based chemicals. Chemical synthesis modifies bio-derived monomers through polymerization and functionalization to enhance material properties. Thermochemical methods, such as pyrolysis, gasification, and hydrothermal carbonization convert biomass into biofuels, biochars, and bio-based resins. Bio-based materials have diverse applications across multiple sectors. In the food industry, they provide biodegradable alternatives to plastic films and coatings. In biomedicine, bio-based polymers are used in tissue engineering, wound dressings, and drug delivery systems. Additionally, in construction and automotive industries, bio-composites enhance sustainability by replacing synthetic materials with biofibers and biopolymeric matrices. Recent advancements in bio-based materials at the micro and nanoscale have significantly improved their properties to open up new opportunities across a range of industries. Nanotechnology can improve the qualities and reduce the cost of biomaterials by changing the manufacturing and fabrication processes. These innovations are driving the next generation of bio-based materials toward competition with conventional materials while maintaining environmental benefits.

References

1. Abba EC, Nwakuba NR, Obasi SN, Enem JI (2017) Effect of reaction time on the yield of biodiesel from neem seed oil. Am J Energy Sci 4(2):5–9
2. Adunphatcharaphon S, Petchkongkaew A, Greco D, D'Ascanio V, Visessanguan W, Avantaggiato G (2020) The E ectiveness of durian peel as a multi-mycotoxin adsorbent. Toxins (Basel) 12(2):1–17
3. Akhil D, Lakshmi D, Kartik A, Vo D-VN, Arun J, Gopinath KP (2021) Production, characterization, activation and environmental applications of engineered biochar: a review. Environ Chem Lett 19(3):2261–2297. https://doi.org/10.1007/s10311-020-01167-7
4. Atiwesh G, Mikhael A, Parrish CC, Banoub J, Le TT (2021) Heliyon environmental impact of bioplastic use: a review. HLY 7(9):e07918. https://doi.org/10.1016/j.heliyon.2021.e07918
5. Ayilara M, Olanrewaju O, Babalola O, Odeyemi O (2020) Waste management through composting: challenges and potentials. Sustainability 12(11):4456. https://doi.org/10.3390/su12114456
6. Baghayeri M, Amiri A, Maleki B, Alizadeh Z, Reiser O (2018) A simple approach for simultaneous detection of cadmium(II) and lead(II) based on glutathione coated magnetic nanopar-

ticles as a highly selective electrochemical probe. Sensors Actuators B Chem 273:1442–1450. https://doi.org/10.1016/j.snb.2018.07.063
7. Baskar G, Soumiya S (2016) Production of biodiesel from castor oil using iron (II) doped zinc oxide nanocatalyst. Renew Energy 98:101–107. https://doi.org/10.1016/j.renene.2016.02.068
8. Basso D, Patuzzi F, Castello D, Baratieri M, Rada EC, Weiss-Hortala E, Fiori L (2016) Agro-industrial waste to solid biofuel through hydrothermal carbonization. Waste Manag 47:114–121. https://doi.org/10.1016/j.wasman.2015.05.013
9. Bastante CC, Silva NHCS, Cardoso LC, Serrano CM, Martínez de la Ossa EJ, Freire CSR, Vilela C (2021) Biobased films of nanocellulose and mango leaf extract for active food packaging: supercritical impregnation versus solvent casting. Food Hydrocoll 117:106709. https://doi.org/10.1016/j.foodhyd.2021.106709
10. Benavente V, Calabuig E, Fullana A (2015) Upgrading of moist agro-industrial wastes by hydrothermal carbonization. J Anal Appl Pyrolysis 113:89–98. https://doi.org/10.1016/j.jaap.2014.11.004
11. Bohórquez WF, Osorio-Pascuas OM, Santaella MA, Orjuela A (2020) Homogeneous and heterogeneous catalytic kinetics in the production of Triethyl citrate. Ind Eng Chem Res 59(43):19203–19211. https://doi.org/10.1021/acs.iecr.0c03690
12. Bourbia S, Kazeoui H, Belarbi R (2023) A review on recent research on bio-based building materials and their applications. Mat Renew Sustain Energy 12(2):117–139. https://doi.org/10.1007/s40243-023-00234-7
13. Bülbül G, Hayat A, Andreescu S (2016) ssDNA-functionalized Nanoceria: a redox-active Aptaswitch for biomolecular recognition. Adv Healthc Mater 5(7):822–828. https://doi.org/10.1002/adhm.201500705
14. Camacho-Otero J, Boks C, Pettersen IN (2018) Consumption in the circular economy: a literature review. Sustainability 10(8):2758. https://doi.org/10.3390/su10082758
15. Chang M, Zhang T, Feng W, Wang T, Liu R, Jin Q, Wang X (2020) Preparation of highly pure stigmasteryl oleate by enzymatic esterification of stigmasterol enriched from soybean phytosterols. LWT 128:109464. https://doi.org/10.1016/j.lwt.2020.109464
16. Cornel EJ, Jiang J, Chen S, Du J (2021) Principles and characteristics of polymerization-induced self-assembly with various polymerization techniques. CCS Chem 3(4):2104–2125. https://doi.org/10.31635/ccschem.020.202000470
17. de Freitas FA, de Sá Barros S, Saron C, Lobo WV, dos Santos RIL, Las-Casas B, Yupanqui-Mendoza SL, de Souza LKC (2024) Lignocellulose characterization and exploitation. In: M. R. B. T.-E. of R. E (ed) Rahimpour sustainability and the environment, First edn. Elsevier, pp 565–576. https://doi.org/10.1016/B978-0-323-93940-9.00022-0
18. Devi A, Bajar S, Kour H, Kothari R, Pant D, Singh A (2022) Lignocellulosic biomass valorization for bioethanol production: a circular bioeconomy approach. Bioenergy Res 15(4):1820–1841. https://doi.org/10.1007/s12155-022-10401-9
19. Díez-Pascual AM (2021) Effect of graphene oxide on the properties of poly(3-hydroxybutyrate-co-3-hydroxyhexanoate). Polymers 13(14):2233. https://doi.org/10.3390/polym13142233
20. Donner M, Verniquet A, Broeze J, Kayser K, De Vries H (2021) Critical success and risk factors for circular business models valorising agricultural waste and by-products. Resour Conserv Recycl 165:105236. https://doi.org/10.1016/j.resconrec.2020.105236
21. Duque-Acevedo M, Belmonte-Ureña LJ, Cortés-García FJ, Camacho-Ferre F (2020) Agricultural waste: review of the evolution, approaches and perspectives on alternative uses. Glob Ecol Conserv 22:e00902. https://doi.org/10.1016/j.gecco.2020.e00902
22. Feng H, Zhang B, He Z, Wang S, Salih O, Wang Q (2018) Study on co-liquefaction of spirulina and spartina alterniflora in ethanol-water co-solvent for bio-oil. Energy 155:1093–1101. https://doi.org/10.1016/j.energy.2018.02.146
23. Fithriani D, Ambarwaty D, Nurhayati. (2020) Identification of bioactive compounds from Nannochloropsis sp. IOP Conf Ser Earth Environ Sci 404(1):012064. https://doi.org/10.1088/1755-1315/404/1/012064

24. Gasafi E, Reinecke M-Y, Kruse A, Schebek L (2008) Economic analysis of sewage sludge gasification in supercritical water for hydrogen production. Biomass Bioenergy 32(12):1085–1096. https://doi.org/10.1016/j.biombioe.2008.02.021
25. Grimberg SJ, Hilderbrandt D, Kinnunen M, Rogers S (2015) Anaerobic digestion of food waste through the operation of a mesophilic two-phase pilot scale digester – assessment of variable loadings on system performance. Bioresour Technol 178:226–229. https://doi.org/10.1016/j.biortech.2014.09.001
26. Habagil M, Keucken A, Sárvári Horváth I (2020) Biogas production from food residues—the role of trace metals and co-digestion with primary sludge. Environments 7(6):42. https://doi.org/10.3390/environments7060042
27. Ingle AP, Saxena S, Moharil MP, Rivaldi JD, Ramos L (2025) Production of biomaterials and biochemicals from lignocellulosic biomass through sustainable approaches: current scenario and future perspectives. Biotechnol Sustain Mat. https://doi.org/10.1186/s44316-025-00025-2
28. Islam MA, Akber MA, Limon SH, Akbor MA, Islam MA (2019) Characterization of solid biofuel produced from banana stalk via hydrothermal carbonization. Biomass Convers Biorefinery 9(4):651–658. https://doi.org/10.1007/s13399-019-00405-5
29. Karnwal A, Rauf A, Jassim AY, Selvaraj M, Al-Tawaha ARMS, Kashyap P, Kumar D, Malik T (2025) Advanced starch-based films for food packaging: innovations in sustainability and functional properties. Food Chem X 29(February):102662. https://doi.org/10.1016/j.fochx.2025.102662
30. Kartik A, Akhil D, Lakshmi D, Panchamoorthy Gopinath K, Arun J, Sivaramakrishnan R, Pugazhendhi A (2021) A critical review on production of biopolymers from algae biomass and their applications. Bioresour Technol 329:124868. https://doi.org/10.1016/j.biortech.2021.124868
31. Kaushal N, Singh AK (2023) International journal of biological macromolecules advancement in utilization of bio-based materials including cellulose, lignin, chitosan for bio-inspired surface coatings with special wetting behavior: a review on fabrication and applications. Int J Biol Macromol 246(March):125709. https://doi.org/10.1016/j.ijbiomac.2023.125709
32. Khan Z, Javed F, Shamair Z, Hafeez A, Fazal T, Aslam A, Zimmerman WB, Rehman F (2021) Current developments in esterification reaction: a review on process and parameters. J Ind Eng Chem 103:80–101. https://doi.org/10.1016/j.jiec.2021.07.018
33. Kolet M, Zerbib D, Molina K, Nakonechny F, Nisnevitch M (2020) Biodiesel production using Lewis catalysts under ultrasonic activation. Israel J Chem 60(5–6):644–651. https://doi.org/10.1002/ijch.202000034
34. KoohiKamali S, Tan CP, Ling TC (2012) Optimization of sunflower oil transesterification process using sodium methoxide. Sci World J 2012:1–8. https://doi.org/10.1100/2012/475027
35. Lawer-Yolar G, Dawson-Andoh B, Atta-Obeng E (2021) Synthesis of biodiesel from tall oil fatty acids by homogeneous and heterogeneous catalysis. Sustain Chem 2(1):206–221. https://doi.org/10.3390/suschem2010012
36. Liu Z, Balasubramanian R (2012) Hydrothermal carbonization of waste biomass for energy generation. Procedia Environ Sci 16:159–166. https://doi.org/10.1016/j.proenv.2012.10.022
37. Liu Z, Quek A, Kent Hoekman S, Balasubramanian R (2013) Production of solid biochar fuel from waste biomass by hydrothermal carbonization. Fuel 103:943–949. https://doi.org/10.1016/j.fuel.2012.07.069
38. Lopresto CG, Calabrò V, Woodley JM, Tufvesson P (2014) Kinetic study on the enzymatic esterification of octanoic acid and hexanol by immobilized Candida Antarctica lipase B. J Mol Catal B Enzym 110:64–71. https://doi.org/10.1016/j.molcatb.2014.09.011
39. Lucian M, Volpe M, Fiori L (2019) Hydrothermal carbonization kinetics of lignocellulosic agro-wastes: experimental data and modeling. Energies 12(3):516. https://doi.org/10.3390/en12030516
40. Meegoda JN, Li B, Patel K, Wang LB (2018) A review of the processes, parameters, and optimization of anaerobic digestion. Int J Environ Res Public Health 15(10):2224. https://doi.org/10.3390/ijerph15102224

41. Merino D, Quilez-Molina AI, Perotto G, Bassani A, Spigno G, Athanassiou A (2022) A second life for fruit and vegetable waste: a review on bioplastic films and coatings for potential food protection applications. Green Chem 24(12):4703–4727. https://doi.org/10.1039/d1gc03904k
42. Mignogna D, Ceci P (2024) Biomass energy and biofuels: perspective, potentials, and challenges in the energy transition. pp 1–33
43. Mirzaei M, Ahmadi MH, Ghazvini M, Sobhani E, Lorenzini G, Khalilpoor N (2018) Energetic study of gasification system for bio-waste as renewable energy resource: case study. Defect Diffus Forum 388:44–60. https://doi.org/10.4028/www.scientific.net/DDF.388.44
44. Mishra S, Upadhyay RK (2021) Review on biomass gasification: gasifiers, gasifying mediums, and operational parameters. Mat Sci Energy Technol 4:329–340. https://doi.org/10.1016/j.mset.2021.08.009
45. Morken J, Gjetmundsen M, Fjørtoft K (2018) Determination of kinetic constants from the co-digestion of dairy cow slurry and municipal food waste at increasing organic loading rates. Renew Energy 117:46–51. https://doi.org/10.1016/j.renene.2017.09.081
46. Mulla MZ, Rahman MRT, Marcos B, Tiwari B, Pathania S (2021) Poly lactic acid (PLA) nanocomposites: effect of inorganic nanoparticles reinforcement on its performance and food packaging applications. Molecules 26(7):1967. https://doi.org/10.3390/molecules26071967
47. Naik BD, Udayakumar M (2021) Optimization studies on esterification of waste cooking oil using sulfated montmorillonite clay acidic catalyst. Mater Today Proc 46:9855–9861. https://doi.org/10.1016/j.matpr.2020.11.419
48. Nanda S, Patra BR, Patel R, Bakos J, Dalai AK (2022) Innovations in applications and prospects of bioplastics and biopolymers: a review. Environ Chem Lett 20(1):379–395. https://doi.org/10.1007/s10311-021-01334-4
49. Naseef HH, Tulaimat RH (2025) Transesterification and esterification for biodiesel production: a comprehensive review of catalysts and palm oil feedstocks. Energy Conv Manag X 26:100931. https://doi.org/10.1016/j.ecmx.2025.100931
50. Nisar J, Razaq R, Farooq M, Iqbal M, Khan RA, Sayed M, Shah A, Rahman I, ur. (2017) Enhanced biodiesel production from Jatropha oil using calcined waste animal bones as catalyst. Renew Energy 101:111–119. https://doi.org/10.1016/j.renene.2016.08.048
51. Ogi T, Nakanishi M, Fukuda Y, Matsumoto K (2013) Gasification of oil palm residues (empty fruit bunch) in an entrained-flow gasifier. Fuel 104:28–35. https://doi.org/10.1016/j.fuel.2010.08.028
52. Okolie O, Kumar A, Edwards C, Lawton LA, Oke A, McDonald S, Thakur VK, Njuguna J (2023) Bio-based sustainable polymers and materials: from processing to biodegradation. J Comp Sci 7(6):213. https://doi.org/10.3390/jcs7060213
53. Om Tapanes NC, Gomes Aranda DA, de Mesquita Carneiro JW, Ceva Antunes OA (2008) Transesterification of Jatropha curcas oil glycerides: theoretical and experimental studies of biodiesel reaction. Fuel 87(10–11):2286–2295. https://doi.org/10.1016/j.fuel.2007.12.006
54. Onukwuli DO, Emembolu LN, Ude CN, Aliozo SO, Menkiti MC (2017) Optimization of biodiesel production from refined cotton seed oil and its characterization. Egypt J Pet 26(1):103–110. https://doi.org/10.1016/j.ejpe.2016.02.001
55. Pan S-Y, Tsai C-Y, Liu C-W, Wang S-W, Kim H, Fan C (2021) Anaerobic co-digestion of agricultural wastes toward circular bioeconomy. IScience 24(7):102704. https://doi.org/10.1016/j.isci.2021.102704
56. Pandey A, Gurbuz Y, Ozguz V, Niazi JH, Qureshi A (2017) Graphene-interfaced electrical biosensor for label-free and sensitive detection of foodborne pathogenic E. coli O157:H7. Biosens Bioelectron 91:225–231. https://doi.org/10.1016/j.bios.2016.12.041
57. Pandit P, Nadathur GT, Maiti S, Regubalan B (2018) Functionality and properties of bio-based materials. In: Bio-based materials for food packaging. Springer, Singapore, pp 81–103. https://doi.org/10.1007/978-981-13-1909-9_4
58. Putra HE, Damanhuri E, Dewi K, Pasek AD (2018) Hydrothermal carbonization of biomass waste under low temperature condition. MATEC Web Conf 154:01025. https://doi.org/10.1051/matecconf/201815401025

59. Rahbari A, Shirazi A, Venkataraman MB, Pye J (2019) A solar fuel plant via supercritical water gasification integrated with Fischer-Tropsch synthesis: steady-state modelling and techno-economic assessment. Energy Convers Manag 184:636–648. https://doi.org/10.1016/j.enconman.2019.01.033
60. Rodrigues C, de Mello JMM, Dalcanton F, Macuvele DLP, Padoin N, Fiori MA, Soares C, Riella HG (2020) Mechanical, thermal and antimicrobial properties of chitosan-based-nanocomposite with potential applications for food packaging. J Polym Environ 28(4):1216–1236. https://doi.org/10.1007/s10924-020-01678-y
61. Roschat W, Siritanon T, Yoosuk B, Promarak V (2016) Rice husk-derived sodium silicate as a highly efficient and low-cost basic heterogeneous catalyst for biodiesel production. Energy Convers Manag 119:453–462. https://doi.org/10.1016/j.enconman.2016.04.071
62. Roy P, Mohanty AK, Dick P, Misra M (2023) A review on the challenges and choices for food waste valorization: environmental and economic impacts. ACS Environ Au 3(2):58–75. https://doi.org/10.1021/acsenvironau.2c00050
63. Sadh PK, Chawla P, Kumar S, Das A, Kumar R, Bains A, Sridhar K, Duhan JS, Sharma M (2023) Recovery of agricultural waste biomass: a path for circular bioeconomy. Sci Total Environ 870(January):161904. https://doi.org/10.1016/j.scitotenv.2023.161904
64. Sahoo A, Ram DK (2015) Gasifier performance and energy analysis for fluidized bed gasification of sugarcane bagasse. Energy 90:1420–1425. https://doi.org/10.1016/j.energy.2015.06.096
65. Seo J, Shin M, Lee J, Lee T, Oh J-M, Park C (2021) Novel and highly efficient lipase-catalyzed esterification of formic acid with hexanol for waste gas reutilization. J Ind Eng Chem 93:430–435. https://doi.org/10.1016/j.jiec.2020.10.022
66. Sermyagina E, Mendoza Martinez CL, Nikku M, Vakkilainen E (2021) Spent coffee grounds and tea leaf residues: characterization, evaluation of thermal reactivity and recovery of high-value compounds. Biomass Bioenergy 150(May):106141. https://doi.org/10.1016/j.biombioe.2021.106141
67. Sessini V, Arrieta MP, Kenny JM, Peponi L (2016) Processing of edible films based on nanoreinforced gelatinized starch. Polym Degrad Stab 132:157–168. https://doi.org/10.1016/j.polymdegradstab.2016.02.026
68. Sharma YC, Singh B (2008) Development of biodiesel from karanja, a tree found in rural India. Fuel 87(8–9):1740–1742. https://doi.org/10.1016/j.fuel.2007.08.001
69. Shi X, Guo X, Zuo J, Wang Y, Zhang M (2018) A comparative study of thermophilic and mesophilic anaerobic co-digestion of food waste and wheat straw: process stability and microbial community structure shifts. Waste Manag 75:261–269. https://doi.org/10.1016/j.wasman.2018.02.004
70. Singh S, Deepak D, Aggarwal L, Gupta VK (2014) Tensile and flexural behavior of hemp Fiber reinforced virgin-recycled HDPE matrix composites. Proc Mater Sci 6(Icmpc):1696–1702. https://doi.org/10.1016/j.mspro.2014.07.155
71. Srivastava RK, Shetti NP, Reddy KR, Aminabhavi TM (2020) Biofuels, biodiesel and biohydrogen production using bioprocesses. A review. Environ Chem Lett 18(4):1049–1072. https://doi.org/10.1007/s10311-020-00999-7
72. Stanciu M, Tanas F, Teac C (2025) Crystallinity changes in modified cellulose substrates evidenced by spectral and X-ray diffraction data. Polysaccharides 6(2):30
73. Tarafdar A, Gaur VK, Rawat N, Wankhade PR, Gaur GK, Awasthi MK, Sagar NA, Sirohi R (2021) Advances in biomaterial production from animal derived waste. Bioengineered 12(1):8247–8258. https://doi.org/10.1080/21655979.2021.1982321
74. Torres-Rodríguez DA, Romero-Ibarra IC, Ibarra IA, Pfeiffer H (2016) Biodiesel production from soybean and Jatropha oils using cesium impregnated sodium zirconate as a heterogeneous base catalyst. Renew Energy 93:323–331. https://doi.org/10.1016/j.renene.2016.02.061
75. Trong NT, Tan PHL, Ngoc DN, Le Huy B, Thanh DT, Van NT (2024) Optimizing the synthesis conditions of aerogels based on cellulose fiber extracted from rambutan peel using response surface methodology. AIMS Environ Sci 11(4):576–592. https://doi.org/10.3934/environsci.2024028
76. Vasile C, Baican M (2023) Lignins as promising renewable biopolymers and bioactive compounds for high-performance materials. Polymers (Basel) 15(15):1–52

77. Vuppaladadiyam AK, Vuppaladadiyam SSV, Awasthi A, Sahoo A, Rehman S, Pant KK, Murugavelh S, Huang Q, Anthony E, Fennel P, Bhattacharya S, Leu S-Y (2022) Biomass pyrolysis: a review on recent advancements and green hydrogen production. Bioresour Technol 364:128087. https://doi.org/10.1016/j.biortech.2022.128087
78. Wang X, Yang G, Feng Y, Ren G, Han X (2012) Optimizing feeding composition and carbon-nitrogen ratios for improved methane yield during anaerobic co-digestion of dairy, chicken manure and wheat straw. Bioresour Technol 120:78–83. https://doi.org/10.1016/j.biortech.2012.06.058
79. Wang B, Chen Y, Wu Y, Weng B, Liu Y, Lu Z, Li CM, Yu C (2016) Aptamer induced assembly of fluorescent nitrogen-doped carbon dots on gold nanoparticles for sensitive detection of AFB1. Biosens Bioelectron 78:23–30. https://doi.org/10.1016/j.bios.2015.11.015
80. Wang Z, Xu C, Lu Y, Chen X, Yuan H, Wei G, Ye G, Chen J (2017) Fluorescence sensor array based on amino acid derived carbon dots for pattern-based detection of toxic metal ions. Sensors Actuators B Chem 241:1324–1330. https://doi.org/10.1016/j.snb.2016.09.186
81. Wang Y, Qiu L, Zhu M, Sun G, Zhang T, Kang K (2019) Comparative evaluation of hydrothermal carbonization and low temperature pyrolysis of Eucommia ulmoides Oliver for the production of solid biofuel. Sci Rep 9(1):5535. https://doi.org/10.1038/s41598-019-38849-4
82. Wang F, Wang L, Zhang X, Ma S, Zhao Z (2022) Study on the barrier properties and antibacterial properties of cellulose-based multilayer coated paperboard used for fast food packaging. Food Biosci 46:101398. https://doi.org/10.1016/j.fbio.2021.101398
83. Wei L, Ge T, Zhu Z, Ye R, Peñuelas J, Li Y, Lynn TM, Jones DL, Wu J, Kuzyakov Y (2022) Paddy soils have a much higher microbial biomass content than upland soils: a review of the origin, mechanisms, and drivers. Agric Ecosyst Environ 326:107798. https://doi.org/10.1016/j.agee.2021.107798
84. Yu H, Cao Y, Li H, Zhao G, Zhang X, Cheng S, Wei W (2021) An efficient heterogeneous acid catalyst derived from waste ginger straw for biodiesel production. Renew Energy 176:533–542. https://doi.org/10.1016/j.renene.2021.05.098
85. Yuan S, Hou Y, Liu S, Ma Y (2024) A comparative study on Rice husk, as agricultural waste, in the production of silica nanoparticles via different methods. Materials 17(6):1–11. https://doi.org/10.3390/ma17061271
86. Zainan NH, Srivatsa SC, Li F, Bhattacharya S (2018) Quality of bio-oil from catalytic pyrolysis of microalgae Chlorella vulgaris. Fuel 223:12–19. https://doi.org/10.1016/j.fuel.2018.02.166
87. Zambon I, Cecchini M, Egidi G, Saporito MG, Colantoni A (2019) Revolution 4.0: industry vs. agriculture in a future development for SMEs. Processes 7(1):36. https://doi.org/10.3390/pr7010036
88. Zang Y, Shi J, Zhang F, Zhong Y, Zhu W (2013) Sulfonic acid-functionalized MIL-101 as a highly recyclable catalyst for esterification. Cat Sci Technol 3(8):2044. https://doi.org/10.1039/c3cy00044c
89. Zhai N, Zhang T, Yin D, Yang G, Wang X, Ren G, Feng Y (2015) Effect of initial pH on anaerobic co-digestion of kitchen waste and cow manure. Waste Manag 38:126–131. https://doi.org/10.1016/j.wasman.2014.12.027
90. Zhang R, Yuen AKL, de Nys R, Masters AF, Maschmeyer T (2020) Step by step extraction of bio-actives from the brown seaweeds, Carpophyllum flexuosum, Carpophyllum plumosum, Ecklonia radiata and Undaria pinnatifida. Algal Res 52:102092. https://doi.org/10.1016/j.algal.2020.102092
91. Zhang L, Larsson A, Moldin A, Edlund U (2022a) Industrial crops & products comparison of lignin distribution, structure, and morphology in wheat straw and wood. Indus Crop Prod 187(PB):115432. https://doi.org/10.1016/j.indcrop.2022.115432
92. Zhang X, Li L, Xu F (2022b) Chemical characteristics of wood cell wall with an emphasis on ultrastructure: a mini-review. Forests 13(3):439
93. Zhong M, Yang L, Yang H, Cheng C, Deng W, Tan Y, Xie Q, Yao S (2019) An electrochemical immunobiosensor for ultrasensitive detection of Escherichia coli O157:H7 using CdS quantum dots-encapsulated metal-organic frameworks as signal-amplifying tags. Biosens Bioelectron 126:493–500. https://doi.org/10.1016/j.bios.2018.11.001

77. Vuppaladadiyam AK, Vuppaladadiyam SSV, Awasthi A, Sahoo A, Rehman S, Pant KK, Murugavelh S, Huang Q, Anthony E, Fennel P, Bhattacharya S, Leu S-Y (2022) Biomass pyrolysis: A review on recent advancements and green hydrogen production. Bioresour Technol 364:128087. https://doi.org/10.1016/j.biortech.2022.128087
78. Wang X, Yang G, Feng Y, Ren G, Han X (2012) Optimizing feeding composition and carbon-nitrogen ratios for improved methane yield during anaerobic co-digestion of dairy, chicken manure and wheat straw. Bioresour Technol 120:78–83. https://doi.org/10.1016/j.biortech.2012.06.058
79. Wang G, Chen Y, Wu Y, Wang B, Liu Y, Lu Z, Li CM, Yu C (2016) Aptamer-induced assembly of fluorescent nitrogen-doped carbon dots on gold nanoparticles for sensitive detection of AFB1. Biosens Bioelectron 78:23–30. https://doi.org/10.1016/j.bios.2015.11.015
80. Wang Z, Xu C, Lu Y, Chen X, Yuan H, Wei G, Ye G, Chen J (2017) Fluorescence sensor array based on amino acid derived carbon dots for pattern-based detection of toxic metal ions. Sensors Actuators B Chem 241:1324–1330. https://doi.org/10.1016/j.snb.2016.09.186
81. Wang Y, Qiu L, Zhu M, Sun G, Zhang T, Kang K (2019) Comparative evaluation of hydrothermal carbonization and low temperature pyrolysis of Eucommia ulmoides Oliver for the production of solid biofuel. Sci Rep 9(1):5535. https://doi.org/10.1038/s41598-019-38849-4
82. Wang F, Wang L, Zhang X, Ma S, Zhao Z (2022) Study on the barrier properties and antibacterial properties of cellulose-based multilayer coated paperboard used for fast food packaging. Food Biosci 46:101398. https://doi.org/10.1016/j.fbio.2021.101398
83. Wei L, Ge T, Zhu Z, Ye R, Peñuelas J, Li Y, Lynn TM, Jones DL, Wu J, Kuzyakov Y (2022) Paddy soils have a much higher microbial biomass content than upland soils: a review of the origin, mechanisms, and drivers. Agric Ecosyst Environ 326:107798. https://doi.org/10.1016/j.agee.2021.107798
84. Yu H, Cao Y, Li H, Zhao G, Zhang X, Cheng S, Wei W (2021) An efficient heterogeneous acid catalyst derived from waste ginger straw for biodiesel production. Renew Energy 176:533–542. https://doi.org/10.1016/j.renene.2021.05.098
85. Yuan S, Hou Y, Liu S, Ma Y (2024) A comparative study on Rice husk, as agricultural waste, in the production of silica nanoparticles via different methods. Materials 17(6):1271. https://doi.org/10.3390/ma17061271
86. Zainan NH, Srivatsa SC, Li F, Bhattacharya S (2018) Quality of bio-oil from catalytic pyrolysis of microalgae Chlorella vulgaris. Fuel 223:12–19. https://doi.org/10.1016/j.fuel.2018.02.166
87. Zambon I, Colosimo F, [illegible] (2019) Revolution 4.0: industry vs. agriculture in a future development for SMEs. Processes 7:36. https://doi.org/10.3390/pr7010036
88. Zhang C, Shi L, Zhang B, Zhang Y, Zhu X (2016) Sulfonic acid functionalized [illegible] highly recyclable catalyst for esterification. Catal Sci Technol [illegible]. https://doi.org/10.1039/c5cy00848c
89. Zhai N, Zhang T, Yin D, Yang G, Wang X, Ren G, Feng Y (2015) Effect of initial pH on anaerobic co-digestion of kitchen waste and cow manure. Waste Manag 38:126–131. https://doi.org/10.1016/j.wasman.2014.12.027
90. Zhang K, [illegible] (2020) Step by step extraction of bioactive [illegible] Comparative [illegible] and [illegible]. https://doi.org/10.1016/j.[illegible]
91. Zheng L, Gustafsson A, Moldin A, Lund [illegible] (2022) [illegible] lignin distribution, structure, and morphology in wheat straw and wood. Ind Crops Prod 187(PB):115427. https://doi.org/10.1016/j.indcrop.2022.115427
92. Zhou X, [illegible] (2022) Chemical characteristics of [illegible] wheat straw [illegible] ultrastructure: a mini review. [illegible]
93. Zhong M, Yang L, Yang H, Cheng C, Deng W, Tan Y, Xie Q, Yao S (2019) An electrochemical immunobiosensor for ultrasensitive detection of Escherichia coli O157:H7 using CdS quantum dots-encapsulated metal-organic frameworks as signal-amplifying tags. Biosens Bioelectron 126:493–500. https://doi.org/10.1016/j.bios.2018.11.001

Chapter 3
Nanotechnology in Bio-based Materials

3.1 Introduction to Nanotechnology in Bio-based Materials

Nanotechnology has emerged as a transformative force in the field of material science, offering remarkable potential to enhance the properties and functionalities of bio-based materials. At the nanoscale, materials exhibit unique physicochemical characteristics such as increased surface area, enhanced mechanical strength, improved thermal stability, and tailored barrier properties, which are highly desirable for various advanced applications. Integrating nanotechnology into bio-based materials not only enables the development of sustainable and high-performance materials but also addresses key environmental challenges by promoting the use of renewable resources and reducing dependency on petroleum-based products. This section provides an overview of the fundamental concepts of nanotechnology as applied to bio-based materials, discusses the types of nanomaterials commonly used, and highlights the innovative strategies for incorporating nanoscale components into bio-polymers, composites, and functional materials. Furthermore, it sets the stage for exploring specific applications in packaging, biomedical engineering, environmental remediation, energy conversion and storage, and other industrial sectors in the subsequent sections of this chapter.

3.1.1 Concept and Definition of Nanotechnology, Nanomaterials, Nanoparticles and Nano Structure

Nanotechnology is the multidisciplinary field of science and engineering that involves the design, production, and application of materials and devices by manipulating matter at the nanoscale, typically between 1 and 100 nanometers (nm). At this scale, materials exhibit novel physical, chemical, and biological properties that

M. I. Ahmad, A. K. H. P. S., *Environmental and Energy Technology*, Green Energy and Technology, https://doi.org/10.1007/978-3-032-10706-0_3

differ significantly from their bulk counterparts due to quantum effects and increased surface-to-volume ratios. To put this into perspective, a nanometer is one-billionth of a meter roughly 100,000 times smaller than the width of a human hair. At this scale, the behavior of materials changes dramatically, enabling the development of highly functional and miniaturized systems with enhanced performance. These unique characteristics enable breakthroughs in various fields such as medicine, electronics, energy, environmental science, and agriculture.

At the core of nanotechnology are nanomaterials, which refer to materials with at least one external dimension in the nanoscale range. Nanomaterials may exist in various forms including particles, tubes, films, or composites and are engineered to exploit novel behaviors arising at the nanoscale. Their enhanced electrical, thermal, optical, and mechanical properties are central to the development of high-performance, miniaturized devices and systems. According to Grand View Research [26], the global nanomaterials market was valued at USD 12.42 billion in 2023 and is projected to reach USD 32.77 billion by 2030, growing at a compound annual growth rate (CAGR) of 15.0%.

A fundamental type of nanomaterial is the nanoparticle a discrete entity with dimensions typically less than 100 nm in all directions. Nanoparticles can be composed of metals (e.g., gold, silver), metal oxides (e.g., TiO_2, ZnO), carbon-based structures (e.g., fullerenes, graphene), or polymers. Due to their small size, nanoparticles exhibit exceptional properties such as high reactivity, tunable surface chemistry, and the ability to cross biological barriers, making them highly valuable for targeted drug delivery, catalysis, imaging, and sensor technologies. Closely related to nanoparticles are nanostructures, which refer more broadly to material architectures that incorporate nanoscale features in one, two, or three dimensions. Examples include nanowires (1D), thin films (2D), and porous nanocomposites (3D). These nanostructures serve as the scaffolds or frameworks within which nanomaterials function, offering structural integrity and often playing a crucial role in determining the material's overall performance. For instance, a nanostructured surface can improve the adhesion and interaction of biomolecules in biosensors or enhance light absorption in photovoltaic cells.

3.1.2 *Synthesis of Nanomaterials*

The synthesis of nanomaterials is a foundational step in nanotechnology that determines the size, shape, surface characteristics, and overall functionality of the resulting nanoparticles [2]. These parameters are crucial because they directly influence the physical, chemical, optical, magnetic, and biological properties of nanomaterials, which in turn govern their effectiveness in specific applications such as catalysis, drug delivery, environmental remediation, and energy storage Hossain et al. [31].

Nanomaterial synthesis methods are broadly categorized into two main approaches:

- Top-down approach
- This method involves the physical breakdown of bulk materials into nanoscale structures using mechanical or lithographic techniques. Common techniques include mechanical milling, laser ablation, and electron beam lithography. The top-down approach allows precise control over the shape and size of nanoparticles but may introduce structural defects or surface irregularities. It is particularly useful for producing nanostructured films and patterns for microelectronics and sensor technologies.
- Bottom-up approach
  The bottom-up strategy assembles nanoparticles from atomic or molecular precursors through chemical or biological reactions. This method mimics natural growth processes and often yields materials with more uniform morphology and fewer defects. Examples include sol-gel processing, co-precipitation, hydrothermal synthesis, and biological synthesis. The bottom-up approach is highly versatile and is widely used for fabricating metal nanoparticles, oxides, quantum dots, and polymer-based nanocomposites.

Based on the two classification approaches, it can be divided into physical, chemical, and biological methods.

- Physical methods—These involve processes such as evaporation-condensation, laser ablation, sputtering, and arc discharge. Physical techniques offer high purity products but may require high energy input and specialized equipment.
- Chemical methods—Chemical synthesis is the most widely used route due to its scalability, reproducibility, and control over particle morphology. Techniques include chemical vapor deposition (CVD), microemulsion, sol-gel, precipitation, and reduction reactions. Parameters such as pH, temperature, and precursor concentration can be tuned to tailor nanoparticle characteristics.
- Biological (green) methods—Driven by sustainability and safety concerns, green synthesis methods employ biological agents like plant extracts, bacteria, fungi, or enzymes to reduce metal salts into nanoparticles. These methods are environmentally friendly and avoid the use of hazardous chemicals, but may require longer reaction times and present challenges in standardization. Bio-reducing agents provide natural capping and stabilizing functionalities, enhancing nanoparticle biocompatibility.

Several factors must be considered during nanomaterial synthesis to ensure reproducibility, efficiency, and functionality include particle size and distribution; shape and crystallinity; surface area and charge; stability and agglomeration tendency and; functionalization for targeted applications. Advancements in synthesis techniques have enabled the fine-tuning of nanomaterial characteristics to suit specific applications. For instance, in catalysis, high surface area and active sites are essential, while in biomedical fields, biocompatibility and controlled release properties are critical. Emerging trends in nanomaterial synthesis include scalable green manufacturing, template-assisted synthesis, and multi-functional hybrid nanostructures. Integrating synthesis techniques with real-time monitoring and AI-guided optimization is expected to accelerate innovation. However, challenges remain in terms of cost-effectiveness, environmental impact, and regulatory compliance, particularly for large-scale production.

3.1.3 Role of Nanotechnology in Enhancing Material Properties

Materials in nanoparticles forms exhibited enhanced physical and chemical properties. Nanoparticles were first used in the electronic sector because of their distinct electrical and thermal characteristics. However, carbon-based nanomaterials and other modifications offer possibilities for novel applications in water treatment because of their large surface area, ease of modification, and excellent capacity for microbial disinfection, which made them an effective method for the removal of organic and inorganic contaminants. Graphene, titanium dioxide, and other chemical-based nanomaterials, on the other hand, are also employed in many applications. However, some metal-based nanoparticles are said to be persistent and difficult to degrade, which can result in a number of environmental problems. Some disadvantages of the chemical processes used to create nanomaterials include the potential for contamination from precursor chemicals and the production of dangerous byproducts. Thus, there is a great demand for highly effective, economical, and ecologically friendly techniques for the synthesis of nanoscale materials.

Finding reducing agents with a biological origin (bio-based nanomaterials), which can be derived from bacteria, viruses, fungus, plants, and algae, is necessary to eliminate environmental toxicity. People refer to them as "bio-nano factories." Because they are macroscopic, they exhibit high efficiency and work best in applications involving heavy metal remediation. Different microorganisms, such as plants and their extracts, agricultural and agro-industrial waste, biomolecules and enzymes, and algae, such as diatoms, phytoplankton, zooplankton, and micro-seaweeds, can be used to create bio-based nanomaterials. Materials with biological origins are known as bio-based nanomaterials. Both organic and inorganic nanomaterials can be endogenously synthesized by living things. The interaction of biological species with inorganic materials resulted in the creation of nanomaterials through a variety of intracellular and extracellular mechanisms. A series of nanomaterials/biological interfaces that depend on colloidal forces and dynamic bio-physiochemical interactions are created when nanoparticles interact with tiny organic ligands, therapeutic compounds, proteins, DNA, and cell membranes. These interactions provide the nanomaterials unique physical characteristics while also altering the bio-nanoconjugates' organic reactions [13, 92, 94].

The monomeric building blocks of biomaterials, known as biopolymers, undergo polymerization to create bio-based products. Adding live organisms and naturally degradable material, which is made from natural organic resources like bacteria and plants, to polymers is what is meant by the name "bio". A variety of microorganisms are essential to the production of various biopolymers, such as polyester, polyamides, and polysaccharides. The functional process of creating biopolymers from microorganisms is frequently thought of as a storage material or a byproduct of its own defense mechanism. It is commonly recognized that these types of compounds seem to be broken down by enzyme activity and microbes in order to eventually be reabsorbed into the environment. Their constitution, design, and molecular weight all influence their physical attributes. The production of nanomaterials by microbe-based synthesis is relatively slower than that of plant extracts, which may be a significant disadvantage [68]. Additionally, there are benefits to employing plants as a source of raw materials for the synthesis of nanomaterials, including their availability, high medicinal value, safe handling, and the presence of metabolites that may aid in the reduction process.

The role of nanomaterials in enhancing the properties can be noticed by comparing the significant differences of properties between the nanomaterials and its bult materials as exhibited in Table 3.1. Bulk materials have dimensions above 100 nanometers (nm) in all directions, making them visible to the naked eye, whereas nanomaterials have at least one dimension within the 1–100 nm range and require advanced microscopes for observation. Due to their larger size, bulk materials have a low surface area and contain a smaller percentage of available molecules and atoms, resulting in lower reactivity. In contrast, nanomaterials possess a high surface area and a greater number of exposed molecules, which enhances their reactivity and functionality. Additionally, the properties of bulk materials are generally fixed and cannot be tuned, whereas nanomaterials offer the possibility of tuning their properties for specific applications. This also influences their absorption and adsorption behaviors, with bulk materials exhibiting slow and low rates, while nanomaterials demonstrate rapid and high absorption and adsorption. These

Table 3.1 Comparing the general properties of bulk materials and nanomaterials

Property	Bulk materials	Nanomaterials
Size	Above 100 nm in all dimensions	At least one dimension in scale of 1–100 nm.
Visibility	Can be seen by naked eye	Can be seen by advanced microscope
Surface area	Low surface area.	Large surface area.
Molecules availability	Little percentage of molecules and atoms	High percentage of molecules and atoms
Reactivity	Lower reactivity	Higher reactivity
Functionality	Offer limited functions to products	Offer a variety of functions to products
Tuning	Their properties cannot be tuned	There is a possibility of tuning properties
Absorption and adsorption	Low and slow absorption and adsorption	High and fast absorption and adsorption
Examples	Nanocellulose and nanosilica	Salt and sand

distinctions make nanomaterials more versatile, offering a wider range of applications compared to bulk materials.

3.1.4 The Convergence of Nanotechnology and Bio-based Materials

The integration of nanotechnology with bio-based materials offers a sustainable and innovative platform for addressing challenges in sectors such as agriculture, healthcare, environmental remediation, and packaging. Nanotechnology, which involves the manipulation of materials at the nanoscale (1–100 nm), enables the precise engineering of physical and chemical properties such as surface reactivity, mechanical strength, and bioavailability. As noted by Khan et al. [41], nanoscale materials exhibit significantly enhanced functional performance due to their large surface-to-volume ratios and quantum confinement effects. Bio-based materials, on the other hand, are derived from renewable biological sources such as plants, microbes, algae, and agricultural waste. These materials are inherently biodegradable, non-toxic, and eco-friendly [48] making them ideal candidates for integration with nanotechnology to create high-value, sustainable nanomaterials. This convergence gives rise to bio-based nanomaterials, which combine the performance advantages of nanotechnology with the environmental benefits of biologically sourced feedstocks.

Natural polymers such as cellulose, lignin, starch, and chitosan have been successfully processed into nanostructures including nanocrystals, nanofibers, and nanogels. Nanocellulose derived from plant biomass exhibits superior mechanical properties and thermal stability, making it suitable for use in active food packaging and biomedical applications. Additionally, these biopolymers serve as effective carriers for agrochemicals and drugs due to their functional surface chemistry and biodegradability. Another significant area of advancement is the green synthesis of nanoparticles using biological agents such as plant extracts, enzymes, and microbial cultures. This method provides a safer and more sustainable alternative to conventional chemical synthesis, which often involves toxic reducing agents and high energy consumption. Bio-based nanocomposites synthesized from these agricultural residues show promise in water purification, biodegradable packaging, and filtration applications, enhancing environmental sustainability. Furthermore, Kargarzadeh et al. [39] elaborated on the structural features and functional applications of cellulose nanomaterials derived from agro-wastes, demonstrating their role in resource-efficient green technologies.

3.2 Types of Bio-based Nanomaterials

There are various followed criteria to classify nanomaterials. Among them is categorizing nanomaterials based on their origin source, composition, and functionality. In this base, nanomaterials can be classified into five types of nanocellulose, nanoclay, metal, carbon-based nanomaterials, and nanocatalyst as depicted in Fig. 3.1.

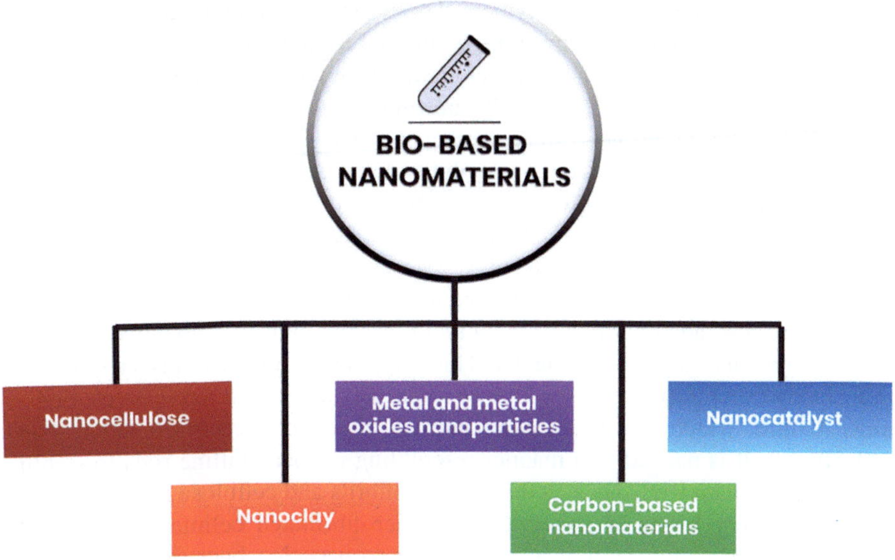

Fig. 3.1 The five types of bio-based nanomaterials

3.2.1 Nanocellulose

Similar to cellulose, nanocellulose exhibits high surface area, hydrophilicity, and crystalline structure. To extract nanocellulose from cellulose, top-down techniques are used which include mechanical (crushing and cryocrushing), chemical (alkaline and acidic hydrolysis, ionic treatments, etc.), and physical (microwave, gamma ray irradiation ultrasonification). There are two primary phases to the isolation. Purified cellulosic components (microfibrilar and/or crystalline components) are separated in the second step after the source material has been purified and homogenized in the first. The transition from cellulose to nanocellulose-based materials was made possible by the creation of nanofibrils from cellulose fibers, which has drawn attention from researchers in recent years. For the purpose of purifying water, cellulose-based nanomaterials have been used extensively. Several mechanical (such as cryocrushing and refining) and chemical (such as sulfonation, esterification, oxidation, etc.) processes are first used to prepare the cellulose-containing feedstock for pretreatments [69, 87]. The purpose of the pretreatment is to separate the cellulose from the other substances. Heavy metal sorption from water has been accomplished using cellulose nanomateials including carboxylated, succinylated, organosilylated, thiolated, and acetylated nanocelluloses, among others. Additionally, carboxylated nanocellulose, amino-functionalized nanocellulose, and cationic nanocellulose are used to separate or remove organic pollutants from water.

Aerogels are another method made from nanocellulose. These are novel classes of nanoscale material that has a number of benefits over inorganic aerogels, including high porosity, high surface area, and low density. Aerogel is a kind of material that has a three-dimensional porous network with gas as a dispersion medium. With many of the characteristics of cellulose, nanocellulose aerogels are an excellent alternative to petroleum-based polymers. These can be used in a number of applications, such as adsorption, energy storage, thermal insulation. They can make up for the disadvantages of traditional aerogel materials in terms of expense, ease of degradation, and environmentally friendly substitute. Compared to other types of aerogels, cellulose-based aerogels offer a number of benefits. Their use is environmentally friendly due to their high degree of flexibility, synthesis from renewable resources, and lack of hazardous ingredients or solvents. An effective class of nanomaterials for a variety of environmental remediation procedures is nanocellulose. However, there are still several obstacles to overcome before nanocellulose can be widely used in water treatment and environmental remediation [15, 47]. These obstacles include low cost, adsorption, adsorbent recovery, and multifunctionality.

A significant organic component and biopolyymer, cellulose holds a unique position among other polymers. 36 cellulose molecules are joined to create elementary units (5 nm), which are then further packed to create microfibrillated cellulose (20–50 nm). It is packed in a manner resembling nanocrystalline rods to stabilize the cellulose chain [69]. Thus, cellulose microfibrils and cellulose nanocrystals are the two primary forms of nanocellulose. Microfibrillated cellulose, plant fibers, microcrystalline cellulose, tunicate cellulose nanoparticles, bacterial cellulose particles, algal cellulose nanocrystals, and wood fiber are the nine different forms of cellulose-based particles. Table 3.2 provides a quick overview of some cellulose-based nanomaterials and their sources, and isolation technique.

3.2.2 *Nanoclays*

Nanoclays are naturally occurring or synthetically modified clay minerals with at least one dimension in the nanometer range (1–100 nm). Among the most widely used types are montmorillonite, a layered silicate, and allophane, a non-crystalline

Table 3.2 Some nanocellulose composites and their sources

Nanocellulose	Size (nm)	Source	Isolation technique	References
Cellulose nanocrystals	Length: 150–300 nm Diameter: 5–10 nm	Wood, cotton, hemp	Acid hydrolysis	[11]
Cellulose nanocrystals	3–10 nm	wood fiber, plant fibers, microcrystalline cellulose, microfibrillated cellulos or nanofiber cellulose	Acid hydrolysis	[9]
Nanofibrillated cellulose/nanofibers	<100	Cellulose fiber	Mechanical	[52]
Cellulose nanofibres	5–50 nm	Wood, rice flax, husk	Mechanical disintegration	[37]
Bacterial nanocellulose	5–70	Sugar, alcohol	–	[87]

aluminosilicate derived from volcanic ash. These materials possess unique physicochemical properties, including high specific surface area, ion exchange capacity, chemical stability, layered morphology, and low toxicity. These characteristics make nanoclays exceptionally attractive for a range of applications across environmental, energy, biomedical, and industrial sectors. Table 3.3 shows the summary of different nanoclays, their characteristics, benefits, and drawbacks adapted from Uddin et al. [86].

Table 3.3 Properties, advantages, and disadvantages of various nanoclays

Nanoclay	Classification	Main properties	Advantages	Limitations
Montmorillonite	Cationic, layered silicate	Exhibits excellent swelling in water, high cation exchange, and a large specific surface area	Enhances composite strength; highly effective in pollutant adsorption	Requires chemical surface modification for compatibility with non-polar systems
Kaolinite	Neutral, layered silicate	Non-swelling, chemically stable, relatively low surface area	Cost-effective, abundant, and thermally stable	Limited interlayer spacing and lower adsorption capacity
Hectorite	Cationic, trioctahedral	Contains lithium; has high swelling capacity and thermal resistance	Improves viscosity and thermal stability in formulations	Less naturally available and more costly
Laponite	Synthetic, cationic clay	Transparent, nanoscale platelets with uniform size and excellent colloidal stability	Ideal for biomedical and cosmetic applications due to high purity	Synthetic origin leads to higher production costs
Smectite	Cationic, layered silicate	High water-swelling ability, strong cation exchange, and excellent adsorption performance	Used extensively for pollutant removal; enhances mechanical strength in polymers	Excessive swelling can hinder uniform dispersion
Attapulgite	Neutral, fibrous clay	Rod-like morphology, high thermal resistance, and strong adsorption capacity	Effective in nanocomposites and as an adsorbent	Lower ion exchange capability compared to layered clays
Palygorskite	Neutral, fibrous clay	High surface area and thermal durability; fibrous structure aids dispersion	Useful in environmental remediation and drug delivery	Abrasiveness and polymer incompatibility in some formulations
Sepiolite	Neutral, fibrous clay	Porous and thermally stable; non-swelling	Suitable for high-temperature applications and gas adsorption	Reduced interaction with some polymers

(continued)

Table 3.3 (continued)

Nanoclay	Classification	Main properties	Advantages	Limitations
Halloysite	Neutral, tubular clay	Nanotubular structure with moderate surface area and biocompatibility	Well-suited for controlled release systems and biomedical uses	Lower surface area compared to layered nanoclays
Zeolite	Anionic, microporous clay	Crystalline aluminosilicate with uniform pores and high thermal/ chemical resistance	Excellent for catalysis, selective ion exchange, and molecular separation	Expensive synthesis and limited natural sources

Adapted from Uddin et al. [86])

In environmental applications, nanoclays are widely used to eliminate contaminants such as heavy metals, pesticides, pharmaceuticals, and dyes from water. Their effectiveness is due to combined mechanisms of surface adsorption, interlayer trapping, and ion exchange, which vary depending on the pollutant and the clay's structure [10]. For example, montmorillonite has shown remarkable efficiency in removing lead, cadmium, and arsenic ions, while allophane has proven effective against phosphate and phenolic compounds [85]. Further enhancementssuch as pillared or organically modified clays have demonstrated increased surface areas and stronger affinities for polar molecules, significantly boosting pollutant adsorption performance [65].

Nanoclays are also gaining attention in air purification and gas separation technologies. Their layered morphology and customizable surface chemistry enable them to adsorb acidic gases such as CO_2, SO_2, NO_x, and volatile organic compounds (VOCs) directly from industrial emissions. Pillared nanoclays, in particular, have been reported to have high adsorption capacities for these gases, making them suitable for integration into filtration systems [63]. Additionally, when formulated into mixed matrix membranes (MMMs), nanoclays have been shown to improve CO_2/CH_4 and CO_2/H_2 separation efficiency across multiple pressures, which not only helps reduce greenhouse gas emissions but also enhances energy efficiency during gas processing [33]. In energy sectors, nanoclays are being utilized to create advanced materials for fuel cells, batteries, thermal insulation, and enhanced oil recovery (EOR). For example, nanoclays have been shown to stabilize CO_2 foams used in EOR; when combined with surfactants, they improve foam durability and structure, enhancing oil displacement in porous rocks [27]. In solid-state lithium batteries, adding nanoclays like montmorillonite and halloysite into polymer electrolytes boosts ionic conductivity, enhances mechanical stability, and helps suppress dendrite growth which significantly extends battery lifespan and safety [95].

Nanoclays have emerged as multifunctional materials in a variety of energy and environmental applications. In the field of photocatalysis, nanoclays serve as supporting matrices for nanoparticles such as TiO_2 and ZnO. Their layered structure enhances the dispersion of photocatalysts, prevents agglomeration, and improves overall light-harvesting efficiency and surface activity, which are critical for efficient degradation of pollutants under UV and visible light. In the building and construction sector, nanoclay-based composites have been integrated into thermal insulation coatings due to their low thermal conductivity, contributing to passive energy savings and enhancing building sustainability. Structurally, nanoclays consist of stacked tetrahedral (SiO_4) and octahedral (Al, Mg, Fe) sheets, typically 1–2 nm thick with lateral dimensions extending from tens of nanometers to several micrometers. This unique morphology yields high aspect ratios and enables intercalation of polymers or functional molecules, which is pivotal in fabricating polymer/nanoclay composites. These composites are now widely used in food packaging for enhancing barrier properties against oxygen, moisture, and gases. For instance, ethylene-vinyl alcohol (EVOH)/montmorillonite nanocomposites significantly improve gas impermeability, thus extending the shelf life of perishable goods [64]. In biomedical applications, nanoclays are utilized in drug delivery systems, biosensing platforms, and tissue engineering scaffolds due to their biocompatibility and controllable degradation profiles. Specifically, polymer/nanoclay scaffolds have demonstrated excellent support for cell adhesion, proliferation, and differentiation, making them suitable for wound healing and bone regeneration. Furthermore, they are being employed in the development of electrochemical biosensors for real-time monitoring of physiological markers such as glucose [23].

3.2.3 *Metal and Metal Oxide Nanoparticles*

Nanomaterials based on metals have been widely employed in many different applications. They can be divided into many categories, such as metal nanoparticles, metal-oxide nanoparticles, trimetallic, and bimetallic nanoparticles. Because of their smaller size, increased surface area, and reactivity, certain metal nanoparticles, including gold (Au) and silver (Ag), have exceptional optical, electrical, and chemical capabilities when compared to bulk materials. The synthesis process typically determines the size of metal nanoparticles. Interestingly, metal nanoparticles have properties that vary with size and form, opening up a wide range of intriguing possibilities. For instance, it has been reported that an Au nanoparticle solution will appear bright red when the particle size is less than 100 nm and filthy yellowish when the particle size is bigger [77]. Likewise, a color difference between spherical and rod-shaped Au nanoparticles is seen. Au nanoparticles have found extensive use in optical sensors and imaging applications due to their optical characteristics. Applications for Ag nanoparticles are numerous and include biomedical imaging and medicinal equipment. They have special antibacterial qualities. To combat the problem of rising antibiotic resistance, the majority of metal nanoparticles are in fact being studied as a substitute antibiotic agent. Numerous different kinds of metallic nanoparticles, besides Au and Ag, have been produced and documented for a range of uses. Usually, the appropriate salts are chemically reduced to create metal nanoparticles.

Another kind of nanomaterial that is frequently used in several research fields is metal oxide nanoparticles. Metal oxide nanoparticles such as Fe_2O_3, Al_2O_3, TiO_2, SiO_2, CuO, SnO_2, WO_3, and others are the most commonly used. Their small size and high density of corner or edge surface sites are what give them their distinct physicochemical characteristics. Applications for metal oxide nanoparticles in biomedicine, energy storage, sensing, catalysis, and optoelectronics are particularly intriguing. They are frequently created using a hydrothermal or sol-gel method [67]. Controlling the properties of metal oxide nanoparticles, which are dependent on nucleation, development, and aging, requires a thorough understanding of the synthesis process. Iron oxide nanoparticles are frequently used in separation procedures, especially magnetite nanoparticles. Nanoparticles of semiconducting metal oxide are widely employed in gas sensors. It is true that certain metal oxide nanoparticles can have a variety of electronic structures and structural geometries, making them insulators, semiconductors, and metallic materials. Furthermore, a variety of additional materials, including polymers, small molecules, and organics, can be added to metal oxide nanoparticles to improve their characteristics and target particular uses.

Compared to monometallic nanoparticles, bimetallic nanoparticles, which are composed of two metals, have attracted greater scientific and technological attention. Their properties are determined by their constituent metals and nanometric size. Different metallic nanoparticle structures are combined to create these. Size-dependent optical, electrical, thermal, and catalytic properties are a few examples of these traits. Various techniques have been devised for their thorough characterization and preparation. In contrast to monometallic catalysts, bimetallization can enhance the catalytic properties of nanoparticles to a greater extent. Charge transport in bimetallic catalysts is explained by the electronic effect. The structure of bimetallic nanoparticles can be altered by alloying the constituent elements [74]. The structure and miscibility of bimetallic nanoparticles are influenced by the synthesis circumstances. Bimetallic nanoparticles are typically created by concurrently reducing two metal ions using an appropriate stabilizing technique. The core-shell and homogeneous alloy particle structures are determined by the reduction conditions.

Three metals combine to form trimetallic nanoparticles. Bimetallic catalysts were not as effective as trimetallic catalysts. Because their catalytic properties can be better tuned than those of monometallic catalysts, trimetallic and bimetallic nanoparticles are more interesting from a technological and scientific standpoint than monometallic ones. Trimetallic nanoparticles have been created via hydrothermal, microwave, coprecipitation, microemulsion, and selective catalytic reduction methods [73]. Trimetallic nanoparticles catalytic activity is decreased by their unstable surface area and ease of precipitation from liquids. Trimetallic nanoparticle stabilizers include dendrimers, organic ligands, block copolymers, and surfactants. Compared to monometallic or bimetallic nanoparticles, these nanoparticles exhibit higher electrochemical catalytic activity.

3.2.4 Carbon-Based Nanomaterials

Carbon-based nanomaterials have found extensive use in a wide range of applications because of their superior qualities. Graphene, carbon quantum dots, and carbon nanotubes are the most well-known materials in this group. The interior diameters of carbon nanotubes are in the nanometer range, while their lengths fall into the micrometer range. Carbon nanotubes are divided into two categories which are single-walled and multi-walled. While multi-walled carbon nanotubes are made up of two or more cylindrically wrapped concentric graphene sheets, single-walled carbon nanotubes are made up of a single graphene sheet that has been rolled up into a tubular shape. The adjacent sheets are separated by 0.34 nm. Van der Waals force was the binding force that kept the sheets together. Three shapes are produced when graphene sheets are rolled into carbon nanotubes: chiral, armchair, and zigzag. The diameter and helicity of graphitic rings determine whether single-walled carbon nanotubes behave in a metallic or semiconducting manner. Additionally, single-walled carbon nanotubes have electrical conductivities (102–106 S cm^{-1}, multi-walled carbon nanotubes = 102–106 S cm^{-1}). Their exceptional thermal conductivities (multi-walled carbon nanotubes ~2000 W m^{-1} K^{-1}, single-walled carbon nanotubes ~6000 W m^{-1} K^{-1}) are another feature [62, 67]. They are employed as reinforcing agents in various composites and products because of their extremely high tensile strength. Because only the outer surface is impacted by chemical treatment, multi-walled carbon nanotubes can maintain their properties while single-walled carbon nanotubes may change due to flaws caused by the breaking of carbon-carbon bonds. Due to their complex structure, multi-walled carbon nanotubes have a lower specific surface area and pore volume than single-walled carbon nanotubes, but they also have many benefits, including inexpensive production costs and catalyst-free synthesis. Numerous techniques, including laser ablation, flame synthesis, chemical vapor deposition, electrolysis, and others, can be used to create carbon nanotubes.

Graphene is a one-atom-thick sheet of sp^2 hybridized carbon atoms packed in a honeycomb-like crystal lattice. Since its initial synthesis using mechanical exfoliation in 2004, it has garnered a great deal of attention. Graphene is the strongest and thinnest material in the universe, in addition to having superior physical and chemical qualities. Excellent electrical conductivity (2000 S m^{-1}), enormous theoretical surface area (2630 m^2 g^{-1}), outstanding carrier mobility (200,000 cm^2 V^{-1} s^{-1}), and very high thermal conductivity (5000 W m^{-1} K^{-1}) are all features of graphene [19, 84]. It has great chemical and mechanical strength, as well as an amazing quantum hall effect. The cost of manufacture is modest. Exfoliation and cleavage, thermal or plasma-based chemical vapor deposition, chemical techniques, thermal breakdown of SiC and other substrates, and unzipping of carbon nanotubes are some of the ways used to create graphene. Both graphene oxide and reduced graphene oxide are graphene-related compounds. One of the most widely used materials is graphene oxide, which is made from graphite powder. Chemically modified graphene is created by starting with graphene oxide. Due to its abundance of oxygen-containing functional groups, graphene oxide is easily functionalized with various materials and moieties. In certain instances, the inclusion of functional groups based on oxygen lowers graphene oxide's performance. It is possible to decrease such functional groups via thermal, chemical, or electrochemical methods. Reduced graphene oxide is the term for the final phase of graphene. The degree and efficiency of graphene oxide reduction will determine how much reduced graphene oxide resembles pure graphene. In addition to having graphene-like characteristics, reduced graphene oxide has comparatively high conductivity. It is simply made from inexpensive graphene oxide in the required quantities. Excellent performance is demonstrated by reduced graphene oxide in electrochemical and other applications.

There are three types of carbon dots which are graphene quantum dots, carbon nanodots, and carbon quantum dots. Carbon nanodots are amorphous carbon nanoparticles that do not exhibit quantum confinement. Both carbon and graphene quantum dots exhibit crystal structure and quantum effects. When fluorescence contributes from both the structure and excited chemical states, this classification becomes difficult. To keep things simple, carbon dots can be used to refer to any quasi-spherical carbon nanoparticle. Zero-dimensional carbon particles smaller than 10 nm are known as carbon quantum dots. They were unintentionally created while single-walled carbon nanotubes were being separated electrophoretically. They are also referred to as "carbon nanolights" because of their fluorescence characteristics. Graphene quantum dots are graphene sheets with a lateral size of fewer than 100 nm. Because of their oxygen moieties, carbon quantum dots are less harmful, biocompatible, and soluble in water [25, 43]. It is possible to adjust their photoluminescence characteristics. They exhibit unique properties associated with quantum confinement effects. Because they have functional groups that contain oxygen, they are easily functionalized. Because of their special optical and photoluminescence qualities, they have been employed to create a variety of sensors. Numerous methods, such as electrochemical synthesis, microwave/ultrasonic-based synthesis, combustion/thermal/hydrothermal/acidic oxidation, arc discharge, laser ablation, plasma treatment, and others, can be used to create carbon nanodots.

3.2.5 *Nanocatalyst*

Catalysis is one of the widely used applications of nanoparticles. Over the past few decades, a variety of bio-based materials have been used as nanoscale catalysts. In any event, it has not yet been fully understood how its enormous synergist conduct manifests itself through nanoparticles. A nanoparticle large surface area directly improves reaction rate and could also provide a logical explanation for the movement of its reactants. Any substance structure and shape-dependent characteristics at the nanoscale can also affect how a material reacts with its surroundings. More notable selectivity has been attained through the calibration of the nanocatalyst in terms of synthesis, shape, and size. In this sense, the question is how the physical characteristics of nanoparticles affect the characteristics of their reactants and how manufacturing boundaries can affect those characteristics. A deeper understanding of them will enable researchers to create nanocatalysts that are incredibly resistant, dynamic, and deeply unique. In order to counteract the ecological impact caused by our reliance on the synthesis process, all of these points of interest will enable modern synthetic reactions to become more resource efficient, use less energy, and generate less waste. Nanoparticles are thought to be the most important contemporary catalyst and have a wider range of uses, including energy conversion, storage, and chemical synthesis [78]. Nanocatalysts differ from homogeneous and heterogeneous in various characteristics that impact their efficiency and application as shown in Table 3.3. Homogeneous catalysts have high activity and selectivity but suffer from low stability and difficult recoverability, making them costly despite requiring minimal preparation. In contrast, heterogeneous catalysts offer high stability and easy recoverability, but their lower activity and selectivity result in longer reaction times. Nanocatalysts, a specialized form of heterogeneous catalysts, provide the best balance by combining high activity, high selectivity, and fast reaction times, while also being cost-effective due to their easy recoverability and enhanced performance. However, their preparation requires precise control, making them more complex to synthesize compared to conventional heterogeneous catalysts. Table 3.4 shows the distinctive characteristics of homogeneous, heterogeneous, nanocatalysts.

A material or a group of atoms performance differs from that of a confined particle mostly due to the interatomic contact. The properties of a nanocatalyst can be adjusted in relation to those of its bulk equivalent by changing the relative amount of the under-composed surface atoms. Therefore, in chemical and physical exhibits, the contribution of the under-composed atoms and the participation of the interatomic interface might be the first step in resolving any problems between isolated atoms and a bulk material. In terms of bond unwinding and its effects on bond vitality, the atomic coordination decrease has a significant impact and consistently combines the display of a surface, a nanocatalyst, and an amorphous state. As aspects of atomic coordination diminish and its subordinates (size dependence) depend on the atomic catching potential, crystal binding intensities, and electron-phonon coupling, the irregular conduct of a surface and a nano-catalyst has been consistently

Table 3.4 Distinctive characteristics of homogeneous, heterogeneous, nanocatalysts

Characteristic	Homogeneous catalyst	Heterogeneous catalyst	Nanocatalyst
Preparation	No-preparation	Complicated preparation	Preparation is needed
Activity	High activity	Less activity	High activity
Stability	Less stability	High stability	High stability
Selectivity	High selectivity	Less selectivity	High selectivity
Reaction time	Moderate duration	Long duration	Fast duration
Recoverability	Difficult recoverability	Easy recoverability	Easy recoverability
Economic feasibility	Costly	Less costly	Cost-effective

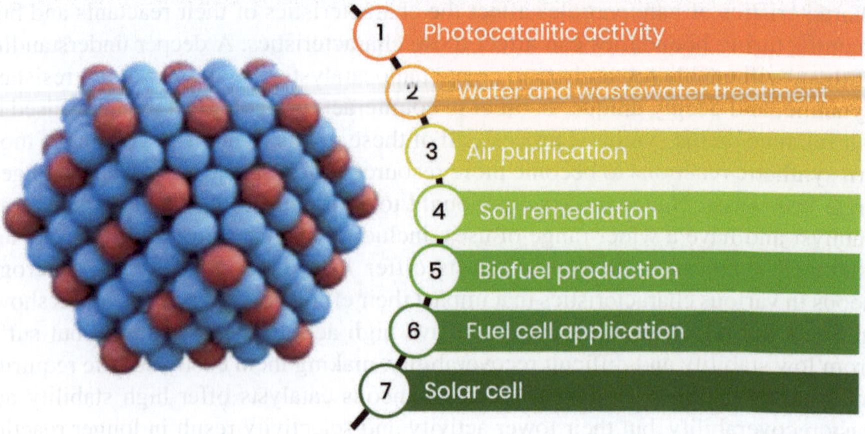

Fig. 3.2 Different applications of nanocatalyst

understood and effectively detailed. It may be widely used in catalysis with newer features and activity by precisely managing the size, shape, spatial distribution, surface compositions and electronic structure, as well as the temperature and synthetic dependability of the individual nano segments [78]. Due to the numerous potential benefits that can accrue from their application, nanoscale catalysts have recently attracted a lot of attention from both academic and industry researchers. Nanocatalysts have a wide range of applications due to their high surface area, enhanced reactivity, and selectivity. As shown in Fig. 3.2, they play a crucial role in photocatalytic activity, where they facilitate chemical reactions under light exposure, making them useful for solar cells and fuel cell applications. Additionally, nanocatalysts contribute to biofuel production by improving reaction efficiency in biomass conversion processes. They are also widely used in air purification and soil remediation, helping to break down pollutants and improve environmental quality. Furthermore, nanocatalysts are employed in water and wastewater treatment, where they aid in the removal of contaminants, ensuring cleaner water sources. These diverse applications highlight the importance of nanocatalysts in energy production, environmental protection, and sustainable industrial processes.

3.3 Synthesis Techniques of Nano-Bio-based Materials

Nanomaterials can be synthesized using the primary methods of top-down and bottom-up approaches. The top-down approach involves breaking down bulk materials into nanoscale structures through physical or mechanical techniques such as milling, lithography, and sputtering. In contrast, the bottom-up approach assembles nanomaterials from atomic or molecular components through chemical or biological processes like sol-gel synthesis and chemical vapor deposition. Additionally, green synthesis methods have emerged, utilizing environmentally friendly techniques involving plant extracts, microorganisms, and biopolymers, ensuring the sustainable production of bio-based nanomaterials with reduced environmental impact and improved biocompatibility. Thes methods were discussed in details in the following subsections.

3.3.1 Top-Down Approaches

In top-down approaches, bulk materials are systematically broken down into smaller nanoscale structures using various physical and mechanical techniques. As illustrated in Fig. 3.3, these methods include mechanical milling, electrospinning, lithography, sputtering, arc discharge method, and laser ablation.

Mechanical milling is one economical technique for creating materials at the nanoscale level from bulk materials. An efficient technique for creating phase blends and a useful tool for creating nanocomposites is mechanical milling. The ball milling method basic idea. Aluminum alloys enhanced by oxide and carbide,

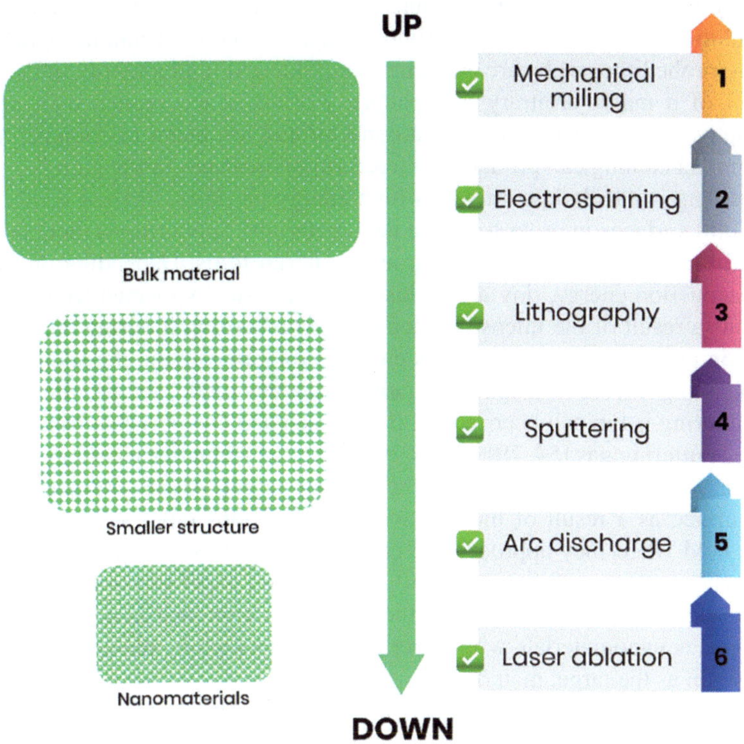

Fig. 3.3 Top-down techniques for the synthesis of bio-based nanomaterials

wear-resistant spray coatings, aluminum/nickel/magnesium/copper-based nanoalloys, and numerous other nanocomposite materials are all made by mechanical milling [44]. Ball-milled carbon nanoparticles are regarded as a novel class of nanomaterials that provide the potential to meet the needs of energy conversion, energy storage, and environmental cleanup.

Electrospinning is one of the most straightforward top-down techniques for creating nanostructured materials. Generally speaking, it is employed to create nanofibers from a broad range of materials, most commonly polymers. Coaxial electrospinning was one of the significant advances in electrospinning. Two coaxial capillaries make up the spinneret in coaxial electrospinning. Core-shell nanoarchitectures can be created in these capillaries using two viscous liquids, or a viscous liquid as the shell and a non-viscous liquid as the core, under an electric field [8]. A straightforward and efficient top-down method for producing core-shell ultrathin fibers on a large scale is coaxial electrospinning. These very thin nanomaterials can reach lengths of many centimeters. Core-shell and hollow polymer, inorganic, organic, and hybrid materials have all been developed using this technique.

Lithography is a practical technique for creating nanoarchitectures using a concentrated beam of light or electrons. Mask lithography and maskless lithography are the two primary categories of lithography. In masked nanolithography, a particular mask or template is used to transfer nanopatterns across a sizable surface area. Photolithography, soft lithography, and nanoimprint lithography are all forms of masked lithography. Electron beam lithography, focused ion beam lithography, and scanning probe lithography are all forms of maskless lithography [49, 89]. Without the need of a mask, arbitrary nanopattern writing is accomplished in maskless lithography. Ion implantation using a concentrated ion beam in conjunction with wet chemical etching can produce 3D freeform micro-nano-fabrication.

Sputtering is a technique that uses high-energy particles, like gas or plasma, to attack solid surfaces in order to create nanomaterials. Sputtering is thought to be a successful technique for creating thin nanomaterial films. Depending on the incident gaseous-ion energy, tiny atom clusters are physically ejected from the target surface as a result of the intense gaseous ions bombarding it during the sputtering deposition process. There are various methods for carrying out the sputtering process, including the use of magnetron, radio-frequency diode, and DC diode sputtering. Sputtering is typically carried out in an evacuated chamber that has been filled with the sputtering gas [54, 79]. Gas ions are created when free electrons clash with the gas at a high voltage applied to the cathode target. Atoms are ejected off the target surface as a result of the positively charged ions rapid acceleration in the electric field when they approach the cathode target. WSe_2-layered nanofilms are created via magnetron sputtering on SiO_2 and carbon paper substrates. When compared to electron-beam lithography, the sputtering approach is more economical and produces nanomaterials with fewer contaminants while maintaining the same composition as the target material.

The arc discharge technique helps to produce a variety of nanostructured materials. Fullerenes, carbon nanohorns, carbon nanotubes, few-layer graphene, and amorphous spherical carbon nanoparticles are among the carbon-based materials for which it is best known. The production of fullerene nanomaterials greatly benefits from the use of the arc discharge process. Two graphite rods are adjusted in a chamber with a certain helium pressure maintained during the formation process. Since fullerene synthesis is inhibited by the presence of oxygen or moisture, it is crucial to fill the chamber with pure helium. Arc discharge between the graphite rod ends propels the vaporization of carbon rods [8]. The circumstances surrounding arc discharge are crucial in producing novel nanomaterial structures. Because distinct carbon-based nanomaterials have diverse development methods, they are collected from various locations using the arc discharge method. Anode or cathode deposits, or deposits at both electrodes, can provide high-purity polyhedral graphite particles, pyrolytic graphite, multi-wall carbon nanohorns, and nano-graphite particles. In addition to the electrodes, the inner chamber can also yield carbon-based nanomaterials. Under various atmospheric conditions, single-wall carbon nanohorns can take on a variety of morphologies. For instance, single-wall carbon nanohorns that resemble dahlias are created in ambient air, while those that resemble buds are created in CO and CO_2 atmospheres. It is possible to effectively create graphene nanostructures using the arc discharge technique. The conditions under which graphene is synthesized can have an impact on its characteristics. Comparing graphene sheets made by hydrogen arc discharge exfoliation to those made by argon arc discharge, it is discovered that the former have better electrical conductivity and better thermal stability [93].

In laser ablation synthesis, a strong laser beam strikes the target material to create nanoparticles. Because of the high energy of the laser irradiation, the source material or precursor vaporizes during the laser ablation process, forming nanoparticles. Since no stabilizing agents or other chemicals are required, using laser ablation to create noble metal nanoparticles can be regarded as a green process [20]. This method can be used to create a variety of nanomaterials, including ceramics, oxide composites, metal nanoparticles, and carbon nanomaterials. An intriguing method for creating monodisperse colloidal nanoparticle solutions without the need of ligands or surfactants is pulsed laser ablation in liquids. The fluence, wavelength, and laser salt addition can all be changed to modify the nanoparticle's characteristics, including its average size and distribution [34].

3.3.2 *Bottom-Up Approaches*

In bottom-up approaches, tiny molecules combine to form nanomaterials. As illustrated in Fig. 3.4, these methods include chemical vapor deposition, solvothermal and hydrothermal methods, the sol-gelmethod, soft and hard templating methods, and reverse micelle methods.

The production of carbon-based nanomaterials greatly benefits from the use of chemical vapor deposition techniques. In chemical vapor deposition, vapor-phase precursors undergo a chemical reaction to create a thin coating on the substrate surface. If a precursor has sufficient volatility, high chemical purity, good stability

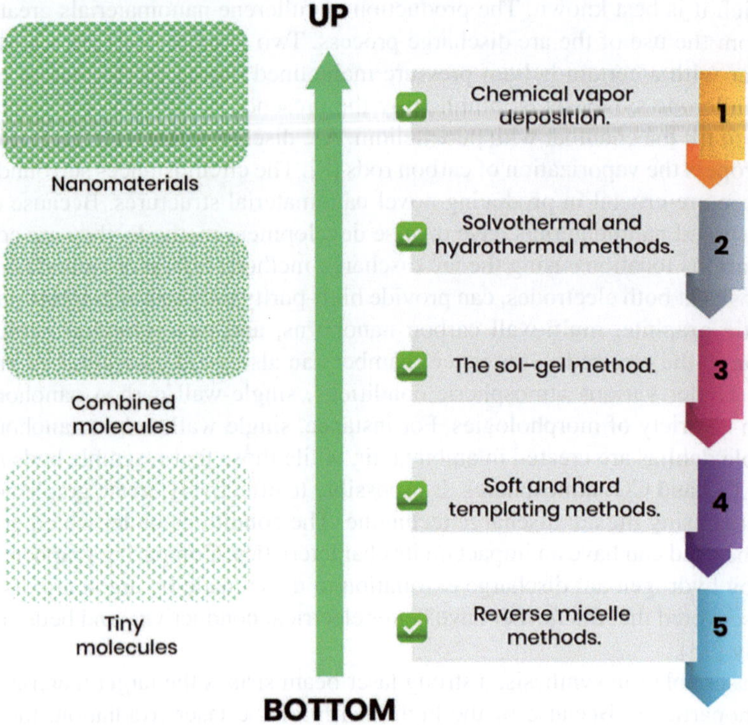

Fig. 3.4 Bottom-up techniques for the synthesis of bio-based nanomaterials

during evaporation, low cost, non-hazardous nature, and a long shelf-life, it is deemed acceptable for chemical vapor deposition [45]. Furthermore, there should not be any leftover contaminants after its breakdown. For example, a substrate is heated to high temperatures in an oven for creating carbon nanotubes using chemical vapor deposition. A carbon-containing gas, such hydrocarbons, is then gradually added to the system as a precursor. When the gas breaks down at high temperatures, carbon atoms are released. These atoms then recombine to create carbon nanotubes on the substrate. However, the nature and morphology of the resulting nanomaterial are greatly influenced by the catalyst selection. Ni and Co catalysts produce multi-layer graphene in the chemical vapor deposition-based graphene manufacturing process, while a Cu catalyst produces monolayer graphene. Chemical vapor deposition is a well-known process for creating two-dimensional nanomaterials and is generally a great way to create high-quality nanomaterials [70].

One of the most popular and widely utilized techniques for creating nanostructured materials is the hydrothermal process. Through a heterogeneous reaction conducted in an aqueous medium at high pressure and temperature near the critical point in a sealed vessel, the hydrothermal process produces nanostructured materials. The hydrothermal method and the solvothermal method are similar. The fact that it is conducted in a non-aqueous liquid is the only distinction [12]. Typically, closed systems are used for hydrothermal and solvothermal processes. Combining the advantages of hydrothermal and microwave techniques, the microwave-assisted hydrothermal approach has attracted a lot of interest lately for the engineering of nanomaterials. Exciting and practical techniques for creating different nano-geometries of materials, including nanowires, nanorods, nanosheets, and nano-spheres, are hydrothermal and solvothermal procedures [18].

One wet-chemical process that is widely utilized for creating nanomaterials is the sol-gel method. Numerous types of superior metal-oxide-based nanomaterials are developed using this technique. This process is known as the sol-gelmethod because, when creating the metal-oxide nanoparticles, the liquid precursor is changed into a sol, which is then changed into a network structure known as a gel. Metal alkoxides are the typical precursors used in the sol-gel process to create nano-materials. There are multiple steps involved in the sol-gel method of creating nanoparticles. In order to create a sol, the metal oxide is first hydrolyzed in water or with the aid of alcohol. Condensation occurs next, increasing the solvent's viscosity to create porous structures that are then allowed to mature. Hydroxo- (M-OH-M) or oxo- (M-O-M) bridges develop during the condensation or polycondensation process, forming metal-hydroxo- or metal-oxo-polymers in solution. Polycondensation persists as the material ages, resulting in modifications to its structure, characteristics, and porosity [16]. The porosity diminishes and the separation between the colloidal particles widens with age. Following the aging process, the gel is dried, which eliminates organic solvents and water. Last but not least, calcination is used to produce nanoparticles. The nature of the precursor, the rate of hydrolysis, the aging period, the pH, and the molar ratio of the precursor to H_2O all influence the end product that is produced using the sol-gel process. In addition to being cost-effective, the sol-gel process has other benefits, including the production of uniform material, a low processing temperature, and an easy way to create complex nanostructures and composites [58].

Nanoporous materials are widely produced using both soft and hard template techniques. A straightforward traditional technique for creating nanostructured materials is the soft template method. The production of materials with a variety of morphologies, easy implementation, and very moderate experimental conditions have all been cited as benefits of the soft template method. Many soft templates, including block copolymers, flexible organic molecules, and anionic, cationic, and non-ionic surfactants, are used in the soft templating process to create nanoporous materials [8]. Van der Waals forces, electrostatic forces, and hydrogen bonds are the main ways that the soft templates and the precursors interact. 3D ordered mesoporous structures are created using soft templates of precisely structured liquid crystalline micelles. The production of mesoporous solids utilizing alkyltrimethylammonium surfactant, such as lamellar (MCM-50), cubic (MCM-48), and hexagonal (MCM-41) ordered mesoporous silicas, is one of the classic examples. Two procedures known as cooperative self-assembly and "true" liquid-crystal templating are typically used for the soft templating approach of creating ordered mesoporous materials [61].

Another helpful way for creating nanomaterials with the appropriate sizes and shapes is the reverse micelle method. Normal micelles are produced by an oil-in-water emulsion, whose hydrophobic tails point toward a core that contains trapped oil droplets. However, in the situation of a water-in-oil emulsion, where the hydrophilic heads are pointing towards a water-containing core, reverse micelles are created. The center of the reverse micelles works as a nanoreactor for the creation of nanoparticles. It serves as a reservoir for the creation of nanomaterials [8]. By altering the water-to-surfactant ratio, the size of these nanoreactors may be regulated, which in turn influences the size of the nanoparticles produced using this technique. Reduced water concentration causes smaller water droplets, which in turn leads to the creation of smaller nanoparticles. As a result, the reverse micelle approach offers a simple way to create homogeneous nanoparticles with exact size control. Reverse micelle-produced nanoparticles are remarkably fine and uniformly distributed [75].

3.3.3 Green Synthesis Methods

Green synthesis of nanomaterials is an eco-friendly approach that utilizes biological systems such as plants, microorganisms, enzymes, and biopolymers to fabricate nanoparticles. This method minimizes environmental hazards by avoiding toxic chemicals, reducing energy consumption, and utilizing sustainable resources. Unlike traditional chemical and physical synthesis techniques, which often involve hazardous reagents and high-energy processes, green synthesis offers a safer and more sustainable alternative. Plant extracts, for example, contain phytochemicals like flavonoids, polyphenols, and alkaloids that act as natural reducing and stabilizing agents in nanoparticle production. Similarly, microorganisms such as bacteria, fungi, and algae mediate the synthesis of metal and metal oxide nanoparticles through biological processes that mimic natural biomineralization [32]. Enzyme-assisted synthesis provides another green route, utilizing biocatalysts to facilitate controlled nanoparticle formation. Additionally, biopolymers such as chitosan, alginate, and cellulose serve as templates or stabilizers in nanomaterial fabrication, enhancing their biocompatibility. These techniques further contribute to sustainable nanotechnology by reducing reaction time, energy consumption, and waste production. The efficacy of green synthesis methods is comparable to conventional approaches, yielding nanoparticles with desirable physicochemical properties for applications in medicine, environmental remediation, and energy storage. As the demand for sustainable nanotechnology grows, green synthesis methods will play a crucial role in advancing eco-conscious material science while ensuring minimal harm to human health and the environment. Table 3.5 shows examples of nanomaterials synthesized by green techniques.

Table 3.5 Green synthesis techniques for nanomaterials

Source	Nanoparticles	Active molecules	Shape	Size (nm)	References
Plant	Ag	Flavonoids, quercetin, glucosides	Spherical	12.5	[91]
Plant	Se	Polysaccharides, phenols, flavonoids, proteins	Spherical	50	[21]
Algae	Au	Phytochemicals with hydroxyl, carboxyl and amino functional groups	Spatial array of self assembled structures	2–10	[4]
Algae	Pd	Sulfated polysaccharides	Octahedral	5–10	[53]
Bacteria	PbS	Cysteine desulfhydrase	Spherical	10.5	[7]
Fungi	CdSe	Superoxidedisumutase	Spherical	9–15	[90]
Yeast	Sb_2O_3	Oxidoreductase and quinone	Spherical	2–10	[36]

3.4 Applications of Nanotechnology in Bio-based Materials

Nanotechnology enhances the performance of bio-based materials by improving their mechanical, thermal, and barrier properties, making them more durable and functional for various applications. In this chapter, the discussion includes the application of bio-based materials nanotechnology in packaging, biomedical, energy storage, and construction.

3.4.1 *Packaging Industry*

Packaging is crucial when it comes to handling and maintaining quality. Conventional packaging serves four primary purposes which are communication, containment, convenience, and protection or preservation. These fundamental features are enhanced to provide better, more intelligent, and active packaging. In order to provide creative, dynamic, and responsive food packaging with enhanced functionality, active or innovative packaging is replacing passive or standard packaging these days. When it comes to better food product preservation and quality maintenance, edible coatings and nanomaterials enhanced with nanoparticles are superior to traditional packaging materials. By enhancing the strength, durability, flexibility, barrier, and reuse qualities of packaging polymers, nanoparticles can change their mechanical and physical characteristics.

To enhance the functional qualities of packaging materials, inorganic (metal and metal oxide), organic (natural goods including protein, carbohydrate, and fat), and a combination of these two (nano clay) nanoparticles are typically added to a polymer matrix. Some methods that nanotechnology can enhance the quality and safety of foods through packaging include the use of nanofillers to increase vapor and gas barrier properties, the antibacterial qualities of nanocomposite films, and smart packaging based on nano-sensors. Although nanotechnology has many advantages, there are also some risks that should be taken into account before using it on a regular basis [5]. When compared to their bulk counterparts, nanoparticles exhibit distinct features. Potential health risks could result from reduced size and increased surface area per unit volume, migration of nanomaterials into food from a package, and consumer exposure to nanoparticles.

The demand for bio-based packaging grew globally and was predicted to reach 884,000 tons by 2020, with a compound annual growth rate of 18.3%. Additionally, a "Credence Research" analysis projects that the market for biodegradable food packaging would double in volume from 2014 to 2020–2021 and reach USD 7058.8 million by the end of 2023, with a compound annual growth rate of 11% from 2016 to 2023 [14]. A variety of nanomaterials from bio-based materials used in the packaging sector are examined in Table 3.6.

Table 3.6 Bio-based nanomaterials for packaging

Nanomaterials	Polymer matrix	Enhancement	References
Nanocellulose	Cellulose nanocrystal-reinforced polyvinyl alcohol/chitosan films	Tensile strength and thermal stability both improved. There was good antifungal and antibacterial action	[60]
Nanocellulose	Blended films of poly (3-hydroxybutyrate) and poly(ε-caprolactone) with cellulose nanocrystals	For mechanical, wettability, optical, and thermal qualities, the blend containing 3% weight cellulose nanocrystals performed best	[24]
Nanostarch	Starch nanocrystals loaded with β-Carotene integrated into chitosan film	The film solubility, moisture content, swelling index, and degradation all declined. The improved film's radical scavenging activity was 91.5%	[29]
Nanostarch	Cellulose and starch nanocrystals are used to strengthen mango kernel films	Tensile strength, modulus, and a water vapor barrier all improved	[76]
Chitosan nanoparticles	Films of hydroxypropyl methylcellulose strengthened with nanoparticles of chitosan	Grapes and plums had longer shelf lives	[71]
Carbon nanotubes	Multi-walled carbon nanotubes doped with ZnO were used to modify a polyvinyl alcohol film	Tensile strength of the modified films was 116% higher than polyvinyl alcohol film. Additionally, enhancement was measured in thermal stability, water vapor transmission rate, hydrophobicity, and antibacterial activity	[88]
Nanoclay	Nanoclay reinforced poly(lactic acid) acid	The surface-modified nanoclay had the highest tensile moduli. Nanoclay damages the degradation's initial temperature	[55]

3.4.2 Biomedical Applications

In the field of biological sciences, nanotechnology creates new possibilities. The ability of nanotechnology to manipulate things at the atomic level holds significant potential for changing many aspects of medical care, including diagnostics, disease monitoring, equipment operation, regenerative medicine, vaccine development, and drug delivery. It also paves the door for the development of medications to enhance therapies for a range of illnesses through the use of advanced research tools. By targeting specific cells in the body with nanotechnology, the chances of failure and rejection can be decreased. The findings of nanotechnology and nanomedicines are so extensive and diverse. Impressive advancements in nanomedicine have elevated the medication to a new level with noteworthy medical results. Research on the substantial potential of nanotechnology in healthcare is necessary. Numerous studies are being conducted in the medical field to investigate optimal techniques and approaches, such as cancer treatment, cardiovascular disease therapeutic gene, and nephrology [28]. Traditional treatment has advanced significantly, and nanotechnology and nanoparticle quality have both improved and produced promising outcomes. Nanotechnology expedites the procedure by enabling nearly instantaneous processing and analysis through a small, portable device that collects samples. Biosensors and sample sizes will continue to grow as a result of in vitro diagnostic testing. By employing smaller and more efficient dosages of the diagnostic chemicals, the nanoparticle formulations for iron oxides and specialty polymers will improve their imaging capabilities, enabling the early identification of tumors, genetic abnormalities, and the whole range of disease states. Similar to biotechnology, nanomedicine has raised some concerns, especially with regard to privacy and safety.

Adsorbed polymer matrices and nanoparticles containing one or more therapeutic medicines that can bind or scatter are the first components of nanotechnology-based drug delivery systems. The manufacturing of nanodrugs employing imaging, therapies, and diagnostics has advanced significantly in recent years. The main goals of nano-drug systems are to increase the bioavailability of targeted tissue delivery, prolong the half-life of injectable medications, and administer pharmaceuticals orally. Lower dosages of nanomedicines result in notable improvements in their pharmacological effects, as well as a decreased risk of side effects and health hazards. Other possible applications of nanomedicine include regenerative medicine, including tissue engineering and cell therapy. The structural "scaffolding" that nanotechnology offers is crucial for creating novel functional structures that resemble natural tissues and allow for the introduction and growth of live cells. The biocompatibility of the cells and nanoparticles with the sample source is advantageous for the emerging tissues. Doctors can then control the processes of cell formation, development, and repair in the resulting tissue. Nanotechnology is thought to be one of the most promising developments in drug delivery in the current battle against cancer. Usually not produced by pharmaceutical companies, nanoparticles have the potential to deliver conventional cancer medications to tumors with fewer side effects and enable the targeted destruction of cancer cells in unconventional treatments. More cytotoxic medication can be administered at the intended site using nanoparticle-permitted chemotherapeutic delivery, and off-target toxicity and efficacy are reduced.

Nanobots are the most important development in nanomedicine. Nanobots can be used to replace entire intracellular components and repair damaged cells. Additionally, they can be duplicated to fix a genetic flaw or swap out a DNA molecule to eradicate an illness. In medicine, nanobots can offer lucrative healthcare opportunities like clearing clogged arteries or replacing an organ completely. Despite their ability to deliver precise dose concentrations to targeted areas, nanocarriers' small size forces a paradoxical trade-off between manufacturing and volume. Artificial intelligence-enabled coordinated behavior and communication will enable nanobots to do their missions methodically and effectively. Individual nanobots could be programmed, searched, and moved in different directions. To more effectively monitor the presence, diagnosis, or targeting of diseased tissues, a single nanobot may link with friends and colleagues, each with their own set of rules, and request that they band their coatings, loads, or sizes together.

3.4.3 Energy Storage

Given the limited energy sources and rising global energy demand, energy storage plays a critical role in energy conservation. Different kinds of materials and media can be utilized for storage, depending on the kind of energy that is needed. Several physical and chemical techniques can be used to store heat. Thermal energy is often stored using thermochemical storage and sensible and latent heats of material. It would be feasible to modify the characteristics of materials used for energy storage and other applications by utilizing nanotechnology. In addition to the kind of phase change material, the most obvious influences on the properties of the nano-incorporated phase change materials are the type and concentration of nano-sized material. The capacity of nanotechnology to reduce device sizes, increase efficiency, and improve material qualities has drawn attention in recent years. Recent research has demonstrated that materials with low thermal conductivity can have their heat transmission improved by using nanotechnology. One viable strategy for enhancing heat transfer is to increase the thermal conductivity of materials. It is possible to create stable media with enhanced heat conductivity by dispersing solid objects with nanometer-scale dimensions. In order to enhance heat transmission, it is desirable to incorporate nanostructures with higher thermal conductivity because the thermophysical properties of the materials depend on the characteristics of their constituent parts. Nevertheless, some of these conductive materials are not economically viable. Apart from thermal conductivity, the total specific heat is strongly influenced by the component specific heats. The specific heat of the nanofluid and the nano-included phase transition material is often decreased when conductive nanostructures, such as metals and carbonic materials, are incorporated. Tables 3.7 explores some bio-based nanocomposites for energy storage.

Table 3.7 Bio-based nanocomposites for thermal energy storage

Nanocomposites	Source	Improvement	References
Nano tubular ZnO	Paraffin	7% decrease in latent heat capacity at 10% concentration of nanostructures.	[66]
TiO_2	Palmitic acid	By adding 5% nanoparticles, the phase change material thermal conductivity can increase by up to 80%	[72]
Al_2O_3 and graphite	Myristic acid/high density polyethylene	Using graphite and Al_2O_3 nanostructures at 30 °C increased the phase transition material's thermal conductivity by 121% and 95%, respectively	[83]
Graphene	Beeswax	Thermal conductivity increased from 0.25 to over 2.8 W/mK	[3]
Graphene	Capric acid, $CaCl_26H_2O$, and n-octadecane	Applying nanostructures to all phase-change materials under consideration improves heat transmission	[38]

3.4.4 Construction and Infrastructure

The usage of various nanoscale materials as additives in home construction highlights their uses as moisture and pathogenic microbe repellents, air purifiers, self-cleaning surfaces, thermal insulators, and anti-corrosives. Furthermore, certain nanoparticles can be employed as cement binders, resulting in increased yields and, as a result, reduced expenses in the building industry. Nanomaterials add useful qualities to conventional home building techniques from an environmental, financial, and quality standpoint. For instance, cement-based building strength and endurance can be increased by up to 20% by adding SiO_2 and Al_2O_3. Carbon nanotubes enhance thermal and electrical conductivity while optimizing strength with high moduli of elasticity. The environment can be self-cleaned and the photodegradation of gasses like NO_X, CO_X, and other pollutants produced within a home can be catalyzed by semiconducting nanoparticles like TiO_2 and other photocatalysts that can be added during the concrete preparation stage [46]. By reducing permeability and shrinkage, nanoclay has also been used to enhance the mechanical performance of concrete, reducing the likelihood of cracks forming and causing damage to architectural structures. Furthermore, some of these nanomaterials can be acquired in environmentally friendly ways, such as by being extracted from industrial and agricultural waste or extremely abundant natural resources. This enables the revaluation of these products and, consequently, lessens the environmental impact that the construction industry produces. Although the benefits and drawbacks of using these materials as additives in housing construction mixtures, their application levels, and their effects at the technical-structural level still require discussion and further research, nanotechnology is a promising field with intriguing applications in the housing construction industry.

In recent decades, nanotechnology has grown rapidly, transforming a wide range of industries, including electronics, construction, and medicine. This is predicated on the creation of materials with special chemical and physical characteristics that have the potential to revolutionize the housing construction sector. The most recent developments directly address the problems of climate change and global urbanization by making it possible to build stronger, more resilient, effective, and sustainable structures. With the creation of carbon-based structures in the middle of the 1980s, nanomaterials began to appear in the building industry. However, the discovery of fullerenes in 1985 and carbon nanotubes in 1991 created a strong basis for their use on a broader scale, making their successful integration in construction processes more significant. According to the advancements in research, invention, and development, the practical application of these materials started to take shape in the 2000s. The development of more sophisticated, effective, and selective nanomaterials is being fueled by the significant advancements in nanotechnology research. The advent of cutting-edge materials like silica, titanium dioxide, and nanoclays has increased their potential in the building industry and made it possible to employ these cutting-edge nanomaterials to reorient trends in environmentally friendly building projects [46]. Table 3.8 explores bio-based nanomaterials in construction application.

Table 3.8 Applications of bio-based nanomaterials in construction application

Nanomaterials	Properties	Enhancement	References
TiO_2	Particle size: 10–100 nm Density: 3.8–4.25 g/cm^3 Electrical permeability: 48–115 Refractive index: 2.5–2.6 Surface area: 50–300 m^2/g	Decrease water absorption by 5–10% Increase thickness swelling by 2–5% Improved early compressive strength by 10–15% Self-cleaning of contamination	[82]
Carbon nanotube	Particle size: 3–15 nm Surface area: >1500 m^2/g Conductivity: 5000 S/cm	Increased compressive strength by 83% Increased tensile strength by 98% Improved cracking toughness by 31% Reduced self-shrinkage by over 40%. Enhanced flexural strength by 104%	[57]
Nanosilica	Particle size: 10–500 nm Pore size: 2–50 nm Surface area: >500 m^2/g	Increased compressive strength by 20–25% Reduces drying shrinkage by 25% Decreased porosity. Enhanced bubble stability	[1]
Nanocellulose	Particle size: 2–20 nm Young's modulus: 100–200 GPa)	Developed high elastic modulus, biocompatibility, and biodegradability Increased tensile strength and modulus by 50% Offers potential for cost-effective production via alternative feedstocks	[59]
Nanoclay	Particle size: 1–10 nm Plate or sheet structures Mechanical resistance Uniform polymeric matrices High thermal stability and hydrophilic nature	Increased dynamic and static yield stress by up to 50% Increased static yield stress by 80% More robust and durable concrete	[40]

3.5 Environmental and Sustainability Perspectives

Bio-based nanomaterials offer eco-friendly alternatives to synthetic materials, enhancing sustainability through renewable resources, biodegradability, and reduced environmental impact. Their applications span biomedicine, packaging, and energy storage, promoting circular economy principles. However, challenges like scalability, toxicity, and lifecycle assessment must be addressed to ensure their responsible and sustainable integration into various industries.

3.5.1 Eco-Friendly Synthesis of Nanomaterials

The broad field of green chemistry and green nanotechnology are paving the way for fresh scientific vision and inventiveness. The broad scientific vision of environmental engineering and petroleum engineering is being severely challenged by global warming, climate change, and the depletion of fossil fuel resources. The scientific origins and nuances of heavy metal contamination of drinking water in both industrialized and developing countries worldwide are difficult for technology to address today. Green nanomaterials, green nanotechnology, and green chemistry are therefore necessary. At this pivotal point in history, green engineering, green chemistry, and nanotechnology can address a wide range of scientific problems worldwide. One of the most important topics in the scientific community today is environmental cleanup. In order to truly emancipate science and technology on a worldwide scale, civilization's tremendous scientific grit, perseverance, comprehension of the demands of human progress, and vision for nanotechnology will all play a significant role

The science of sustainability faces numerous scientific challenges and technological setbacks. The cutting-edge research questions of today are in nanoscience and nanotechnology. A new era of vision in nanotechnology and green engineering is being ushered in by scientific articulation and inventiveness. Uncovering the scientific reality of nanoscience and nanotechnology will be greatly aided by environmental protection, chemical engineering, and environmental engineering deep scientific vision. As society and science advance, the provision of essential human needs, such as food, water, energy, electricity, and shelter, becomes increasingly important. In a similar vein, the science of water purification faces enormous challenges as countries all over the world deal with the devastating effects of groundwater and drinking water contamination from heavy metals. As a result, scientists, engineers, and civic society must work together. These days, there is a connection between environmental engineering science and green engineering. Green nanomaterials and their sustainable solutions are therefore desperately needed in today human society. Space technology and nuclear science are developing at a rapid rate, breaking through barriers one after another. Similarly, nanotechnology is currently undergoing a new phase of development.

The foundations of modern human civilization technical drive, scientific liberation, and realization are green nanomaterials and green nanotechnology. The increased commercialization of green chemistry and green nanotechnology is currently needed by human society. As science and engineering advance, the application and commercialization of nanotechnology to human society become increasingly significant. Scientific breakthroughs in sustainable green nanomaterials have a huge and revolutionary vision and goals. As environmental engineering advances toward a new knowledge dimension, there are numerous and varied hurdles in terms of use and commercialization. The success of sustainability concerns in the application of green nanomaterials in many scientific and technical fields, such as environmental protection and groundwater remediation, is the primary focus. The sustainability of the environment and energy are two critical requirements for scientific advancement in modern society. The foundation of modern scientific renewal is the use of nanotechnology in sustainability, including energy, the environment, society, and the economy. Today, the goal of science should be to apply lab-scale research to human society. Under such circumstances, the use of green nanomaterials in human civilization and its sustainable solutions are the true precursors to a time when science will validate them. In order to truly realize nanotechnology, green engineering, green chemistry, and green nanomaterials, Research and development projects are required to apply nanotechnology in the various spectrums of applications [56].

3.5.2 *Life Cycle Analysis of Nano-Bio-based Products*

The effects of nanomaterials on the environment and human health vary throughout their life cycle, from manufacture to use and eventual disposal. Exposure to the environment as a result of nanomaterial leakage is a major source of variance. In turn, this depends on how nanomaterials are transformed from raw materials into a nanoproduct (a commercial product that contains nanoparticles) and how the environment impacts this product during the course of its life (from use to disposal stage) [51]. The exposure also depends on how the nanomaterials act in the environment before affecting humans and biota. Life cycle assessment is one comprehensive, yet rigorous, tool for including the entire life cycle into comparison and evaluation of systems and products. Although various nanomaterials release from various products, there is presently no method available to quantitatively analyze the environmental repercussions resulting from the release of nanomaterials as depicted in Fig. 3.5. This is in spite of the fact that life cycle evaluation of nanoproducts addresses all environmental impacts. By calculating, measuring, or modeling the exposure concentration and comparing it to the impact concentrations seen in laboratory research, one can determine the eco-toxicological risk of a substance in

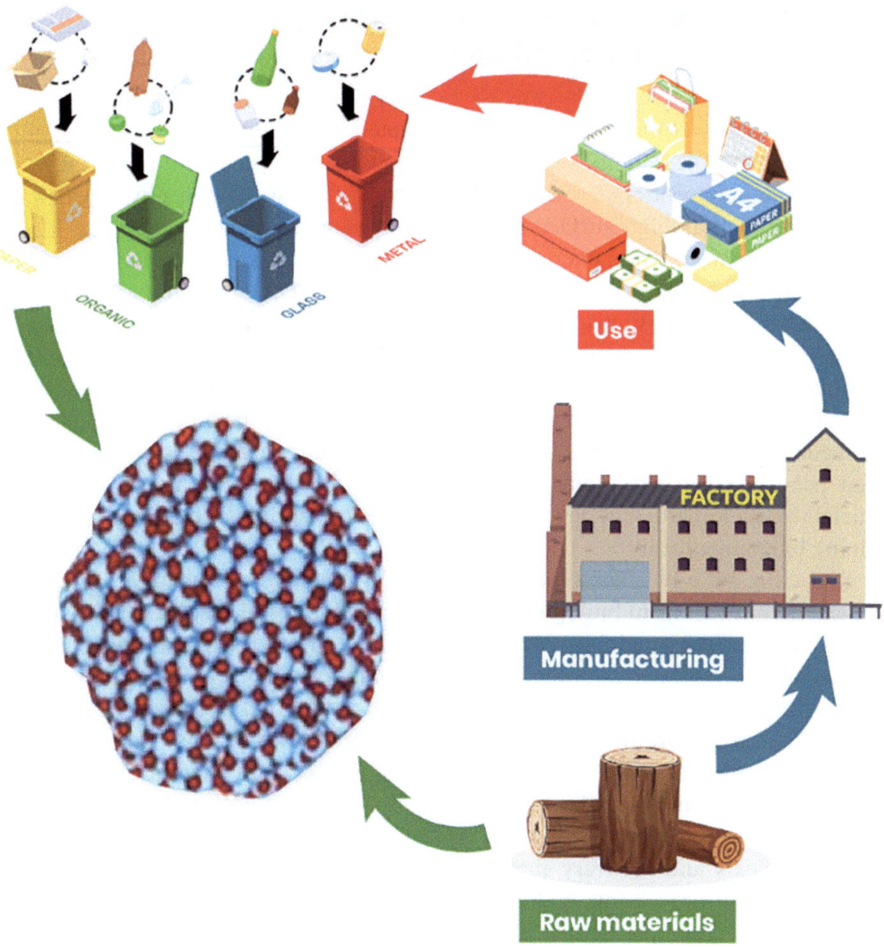

Fig. 3.5 Nanomaterials obtained from the cycle of products from ram materials to the disposal

chemicals risk assessment (if the data are accessible). Conversely, life cycle assessment is a technique for evaluating relative impact. A defined functional unit (a product system's quantifiable performance) serves as the assessment object, allowing for comparisons between product systems but also precluding risk estimation. Additionally, a variety of possible environmental effects are considered. Although life cycle assessment can be applied to nanoproducts in theory, it is not possible to sufficiently evaluate the possible effects of nanomaterials in practice. One major issue is the lack of understanding of the physicochemical properties of nanomaterials in connection to their behavior and effects on the environment, which affects risk assessment.

3.5.3 *Challenges in Disposal and Recycling of Nano-Bio-based Materials*

Disposal and recycling of nano-bio-based materials present significant challenges due to their complex composition, environmental persistence, and potential health risks. Unlike conventional waste, nanomaterials possess unique physicochemical properties, such as high surface reactivity and nanoscale size, which can lead to unforeseen environmental interactions. One primary concern is the safe degradation of these materials after use, as their small size allows them to accumulate in biological systems and ecosystems, potentially causing cytotoxicity or long-term ecological harm. Additionally, the lack of standardized disposal methods for nanomaterials complicates waste management efforts, as existing protocols for hazardous materials may not fully account for nanoparticle-specific risks. Furthermore, their use in biomedical applications, such as drug delivery and imaging, raises concerns about their persistence in the human body and unintended bioaccumulation, making it crucial to develop efficient degradation strategies that mitigate toxicity while maintaining biocompatibility.

Recycling nano-bio-based materials is equally challenging due to the difficulties in separating them from biological and chemical residues in medical and industrial waste streams. The small size and strong binding affinity of nanoparticles to biological tissues, polymers, or other substrates complicate their retrieval and purification for reuse. Current recycling methods, such as thermal treatment or chemical degradation, may alter or destroy the functional properties of nanomaterials, limiting their potential for reuse. Moreover, sustainable recycling approaches must consider the economic feasibility of recovering valuable nanomaterials while minimizing energy consumption and waste generation [35]. The development of green synthesis and recovery methods, including bio-based extraction and eco-friendly solvent processing, could provide viable alternatives to conventional disposal. However, comprehensive life-cycle assessments and regulatory frameworks are essential to ensure the safe and efficient recycling of these materials while mitigating potential environmental and health hazards.

3.6 Safety and Toxicological Aspects

The rapid advancement and integration of nanomaterials into a broad spectrum of industrial, medical, agricultural, and environmental applications have raised critical concerns regarding their safety, toxicological profiles, and long-term environmental fate [50]. Despite their significant functional advantages including high surface area-to-volume ratio, tunable physicochemical properties, and enhanced reactivity the unique attributes that make nanomaterials highly effective also introduce potential risks when interacting with biological and ecological systems. Understanding the toxicological behavior of nanomaterials is essential for ensuring their responsible development and deployment. Unlike conventional chemicals, nanomaterials exhibit size-dependent and shape-dependent interactions with cells and tissues, often penetrating biological membranes and accumulating in organs. Several studies have reported that certain nanoparticles can induce oxidative stress, inflammation, DNA damage, and even apoptosis, depending on their composition, surface chemistry, solubility, and exposure dose. For instance, metal oxide nanoparticles such as TiO_2 and ZnO have demonstrated photocatalytic reactivity that may lead to reactive oxygen species (ROS) generation in living organisms, posing cytotoxic and genotoxic effects.

Moreover, the route of exposure whether inhalation, ingestion, dermal contact, or injection, greatly influences the toxicokinetics and biodistribution of nanoparticles. Inhaled nanoparticles can translocate from the respiratory tract into systemic circulation and reach secondary organs such as the brain, liver, and spleen. This phenomenon underscores the importance of comprehensive risk assessment models that consider particle size, agglomeration behavior, and surface modification, which collectively influence their biological uptake and clearance mechanisms [17]. Environmental toxicity is another crucial area of concern. The release of engineered nanomaterials (ENMs) into aquatic or terrestrial ecosystems during manufacturing, usage, or disposal can lead to bioaccumulation in aquatic organisms and trophic transfer in food chains. For example, silver nanoparticles (AgNPs), widely used for their antimicrobial properties, have shown toxicity in algae, daphnia, and fish species, often disrupting photosynthetic activity and reproductive performance. The ecotoxicological assessment of nanomaterials therefore requires multidisciplinary approaches integrating nanomaterial characterization, environmental chemistry, and ecotoxicology.

In addition to biological effects, occupational exposure to nanomaterials in laboratory and industrial settings necessitates stringent safety protocols. Engineering controls such as fume hoods, glove boxes, and filtration systems, along with personal protective equipment (PPE) and material safety data sheets (MSDS), are critical to safeguard workers handling nanomaterials. Moreover, standardization efforts are underway to develop internationally accepted guidelines for nanomaterial labeling, classification, and disposal, led by regulatory bodies such as the OECD, ISO, and REACH (Registration, Evaluation, Authorisation and Restriction of Chemicals).

Advancing the safe design and application of nanomaterials requires the adoption of a "safety-by-design" approach, where toxicity screening and hazard assessment are embedded in the early stages of material synthesis and functionalization [22]. Computational modeling tools such as Quantitative Structure-Activity Relationship (QSAR) models and machine learning-based toxicity prediction algorithms are emerging as valuable tools for predicting nanotoxicity and guiding safer alternatives. Therefore, while nanotechnology offers transformative potential across disciplines, its sustainable development hinges on a robust understanding of the safety and toxicological aspects of nanomaterials. Continued research into the interaction mechanisms, standardized testing protocols, and long-term exposure scenarios is essential to mitigate risks and promote societal acceptance of nanotechnology-based innovations.

3.6.1 Nano-safety Guidelines

As the use of nanomaterials continues to expand across various industries, ensuring their safe application becomes crucial. Nanomaterials possess unique physicochemical properties that enhance their functionality but may also pose potential health and environmental risks. Regulatory bodies such as the World Health Organization (WHO), the International Organization for Standardization (ISO), and the Organization for Economic Cooperation and Development (OECD) have developed guidelines to monitor and regulate their use [80]. These guidelines focus on assessing exposure levels, potential toxicity, and proper handling procedures. Industries incorporating nanotechnology are encouraged to conduct thorough risk assessments to determine the impact of nanomaterials on human health and ecosystems. Occupational safety measures, including protective equipment and controlled environments, are recommended to minimize direct exposure to nanoparticles. Additionally, manufacturers must comply with labeling and documentation standards to ensure transparency about the composition and potential risks of nanomaterial-containing products.

Environmental safety is another critical aspect of nano-safety guidelines. Due to their small size and high reactivity, nanoparticles can easily disperse into air, water, and soil, potentially leading to bioaccumulation and long-term ecological effects. Regulatory frameworks advocate for proper waste disposal and recycling practices to limit environmental contamination. Researchers are also encouraged to develop sustainable nanomaterials with lower toxicity profiles and enhanced biodegradability. Consumer safety is addressed through standardized testing methods that evaluate the effects of nanomaterials in everyday products such as cosmetics, food packaging, and medical devices. Public awareness campaigns and educational initiatives further reinforce safe handling and usage practices. By adhering to these nano-safety guidelines, industries can maximize the benefits of nanotechnology while mitigating its risks, ensuring a safer and more sustainable future.

3.6.2 Impact on Human Health and Environment

Nanomaterials get into the human body by many exposure routes, including ingestion, inhalation, and dermal penetration. Because of their size, these materials can also disrupt organisms at the cell level, leading to cell death. The adsorption rate of nanomaterials is influenced by their size, as are the smaller particles that endocytosis can carry upward. After entering the bloodstream, each of these nanomaterials exhibits a similar pattern in the inhalation exposure route and, by creating chemically sensitive modes, causes chaos in the organ structures at the cellular and subcellular levels. There are few detailed observations on the effects of nanomaterials on human health. Additionally, there is some information on worker exposure in nanomaterials-using sectors. However, exposure level, exposure pathways, and the size, type, distribution, reactivity, and form of the manufactured nanomaterials all influence human risk factors [6]. The size, shape, properties, and breathing rate of the nanomaterials determine their effects when inhaled through the respiratory system. While nanomaterials smaller than 10 nm can congregate in the thoracic zone, those between 10 and 100 nm in size tend to gather in the alveolar region. Long multiwalled carbon nanotubes may cause the lung's clearing mechanism to deteriorate.

Skin cells can be penetrated by nanomaterials smaller than 100 nm, nuclei can be traversed by those near 40 nm, and the blood-brain barrier can be crossed by those smaller than 35 nm [42]. Additionally, smaller nanoparticles may have improved binding capacity, adsorption rates, and catalytic mobility, which could affect the dwelling time in the body. The spleen, lungs, bone marrow, kidneys, liver, parathyroid, thyroid, brain, skin, eyes, heart, muscles, small intestine, stomach, prostate, tongue, blood, teeth, duodenum, and pancreas are just a few of the organs to which studies have shown that oral exposure to silver nanoparticles can direct their transference. Studies involving albino mice administered dose-dependent silver nanoparticles orally for 21 days showed weight loss and negatively affected intestinal glands and microvilli, resulting in completely decreased intestinal intake.

Because of their distinct physical characteristics, including their structural and chemical components, carbon nanotubes have been the subject of extensive research. They are anticipated to find widespread application in a variety of industries, including electronics and medicine. Furthermore, because carbon nanotubes are insoluble in the lungs, there have been worries about potential safety hazards. The harmful effects of carbon nanotubes include fibrosis, cancer, lung inflammation, and mesothelioma. Carbon nanotube toxicity has been described as dependent on rigidity, route, surface functionalization, dispersion method, size, contaminants, and exposure duration [81]. Longer carbon nanotubes were found to be more harmful than shorter ones in a number of studies.

It was shown that silica nanomaterials of titanium dioxide were harmful to the immune system. When titanium dioxide nanoparticles were given orally to mother mice, the fetus's DNA deletion increased, indicating that it may be inherited by the offspring. Myocardial and kidney damage were revealed as a result of oral exposure of female mice to titanium dioxide nanoparticles ranging in size from 25 to 80 nm at a dose of approximately 5 g/kg [30]. This resulted in a significantly higher level of liver inflammation compared to male mice. According to studies, the harmful effects of gold nanoparticles differ depending on their size, with exposure to smaller nanoparticles producing a more noticeable effect. Although gold nanoparticles are used extensively in many medical fields, because of their size, they have been shown to have an impact on human embryonic stem cells. Smaller nanoparticles are a major factor in promoting toxicity, and scientists generally concur that toxicity will increase with their size. Similar to this, the nanomaterial pattern may be a determining factor in human health characteristics. For instance, gold nanospheres measuring 61.46 nm were more hazardous to fibroblasts than nanostars with a smaller diameter of 33.69 nm.

3.7 Summary

Nanotechnology is recently revolutionizing various fields, offering groundbreaking applications driven by the unique physical and chemical properties of nanostructured materials. These materials are superior to their macro-scale counterparts due to their enhanced functionality, making them highly desirable across industries. Bio-based nanomaterials gained significant attention for their nontoxicity, renewability, biodegradability, and environmental compatibility, further strengthening their role in sustainable innovation. The integration of nanotechnology with green-based practices unlocked invaluable opportunities where bio-nanocomposites are increasingly utilized as cost-effective and highly selective adsorbents, catalysts, membranes, and sensors. Bio-based nanomaterials have found widespread applications in packaging, biomedicine, energy storage, and construction, employing top-down, bottom-up, and green synthesis approaches. However, as nanotechnology advances, it is crucial to assess its potential impact on human health and the ecosystem. Addressing these concerns through rigorous safety assessments and responsible innovation is essential for sustainable development. Despite these challenges, ongoing progress in nanotechnology continues to pave the way for the safe and efficient utilization of bio-based nanomaterials, ensuring a balance between technological advancement and environmental responsibility.

References

1. Abd Elrahman M, Sikora P, Chung S-Y, Stephan D (2021) The performance of ultra-lightweight foamed concrete incorporating nanosilica. Arch Civil Mech Eng 21(2):79. https://doi.org/10.1007/s43452-021-00234-2
2. Alnazza Alhamad A, Zeghoud S, Amor IB, Zaater A, Amor AB, Aouadi A, Hemmami A, Nleonu EC (2023) A short review of nanomaterials: synthesis methods, properties, and applications. Alger J Chem Eng 01(March):1–07. https://doi.org/10.5281/zenodo.7763869
3. Amin M, Putra N, Kosasih EA, Prawiro E, Luanto RA, Mahlia TMI (2017) Thermal properties of beeswax/graphene phase change material as energy storage for building applications. Appl Therm Eng 112:273–280. https://doi.org/10.1016/j.applthermaleng.2016.10.085
4. Annamalai J, Nallamuthu T (2015) Characterization of biosynthesized gold nanoparticles from aqueous extract of Chlorella vulgaris and their anti-pathogenic properties. Appl Nanosci 5(5):603–607. https://doi.org/10.1007/s13204-014-0353-y
5. Ashfaq A, Khursheed N, Fatima S, Anjum Z, Younis K (2022) Application of nanotechnology in food packaging: Pros and cons. Journal of Agriculture & Food Research 7, 100270. https://doi.org/10.1016/j.jafr.2022.100270
6. Asmatulu E, Andalib MN, Subeshan B, Abedin F (2022) Impact of nanomaterials on human health: a review. Environ Chem Lett 20(4):2509–2529. https://doi.org/10.1007/s10311-022-01430-z
7. Bai HJ, Zhang ZM, Guo Y, Yang GE (2009) Biosynthesis of cadmium sulfide nanoparticles by photosynthetic bacteria Rhodopseudomonas palustris. Colloids Surf B Biointerf 70(1):142–146. https://doi.org/10.1016/j.colsurfb.2008.12.025

8. Baig N, Kammakakam I, Falath W (2021) Nanomaterials: a review of synthesis methods, properties, recent progress, and challenges. Mat Adv 2(6):1821–1871. https://doi.org/10.1039/D0MA00807A
9. Beck-Candanedo S, Roman M, Gray DG (2005) Effect of reaction conditions on the properties and behavior of wood cellulose nanocrystal suspensions. Biomacromolecules 6(2):1048–1054. https://doi.org/10.1021/bm049300p
10. Biswas B, Warr LN, Hilder EF, Goswami N, Rahman MM, Churchman JG, Vasilev K, Pan G, Naidu R (2019) Biocompatible functionalisation of nanoclays for improved environmental remediation. Chem Soc Rev 48(14):3740–3770. https://doi.org/10.1039/c8cs01019f
11. Carpenter AW, de Lannoy C-F, Wiesner MR (2015) Cellulose nanomaterials in water treatment technologies. Environ Sci Technol 49(9):5277–5287. https://doi.org/10.1021/es506351r
12. Chai B, Xu M, Yan J, Ren Z (2018) Remarkably enhanced photocatalytic hydrogen evolution over MoS 2 nanosheets loaded on uniform CdS nanospheres. Appl Surf Sci 430:523–530. https://doi.org/10.1016/j.apsusc.2017.07.292
13. Chakrabartty I, Hakeem KR, Mohanta YK, Varma RS (2022) Greener nanomaterials and their diverse applications in the energy sector. Clean Techn Environ Policy 24(10):3237–3252. https://doi.org/10.1007/s10098-022-02368-0
14. Chausali N, Saxena J, Prasad R (2022) Recent trends in nanotechnology applications of bio-based packaging. J Agric Food Res 7:100257. https://doi.org/10.1016/j.jafr.2021.100257
15. Chen Y, Zhang L, Yang Y, Pang B, Xu W, Duan G, Jiang S, Zhang K (2021) Recent progress on nanocellulose aerogels: preparation, modification, composite fabrication, applications. Adv Mater 33(11). https://doi.org/10.1002/adma.202005569
16. Danks AE, Hall SR, Schnepp Z (2016) The evolution of 'sol-gel' chemistry as a technique for materials synthesis. Mat Horiz 3(2):91–112. https://doi.org/10.1039/C5MH00260E
17. De Matteis V (2017) Exposure to inorganic nanoparticles: routes of entry, immune response, biodistribution and in vitro/in vivo toxicity evaluation. Toxics 5(4). https://doi.org/10.3390/toxics5040029
18. Dong Y, Du X, Liang P, Man X (2020) One-pot solvothermal method to fabricate 1D-VS4 nanowires as anode materials for lithium ion batteries. Inorg Chem Commun 115:107883. https://doi.org/10.1016/j.inoche.2020.107883
19. Đorđević L, Arcudi F, Prato M (2019) Preparation, functionalization and characterization of engineered carbon nanodots. Nat Protoc 14(10):2931–2953. https://doi.org/10.1038/s41596-019-0207-x
20. Duque JS, Madrigal BM, Riascos H, Avila YP (2019) Colloidal metal oxide nanoparticles prepared by laser ablation technique and their antibacterial test. Coll Interf 3(1):25. https://doi.org/10.3390/colloids3010025
21. Fardsadegh B, Jafarizadeh-Malmiri H (2019) Aloe vera leaf extract mediated green synthesis of selenium nanoparticles and assessment of their In vitro antimicrobial activity against spoilage fungi and pathogenic bacteria strains. Green Process Synthes 8(1):399–407. https://doi.org/10.1515/gps-2019-0007
22. Furxhi I, Perucca M, Koivisto AJ, Bengalli R, Mantecca P, Nicosia A, Burrueco-Subirà D, Vázquez-Campos S, Lahive E, Blosi M, de Ipiña JL, Oliveira J, Carriere M, Vineis C, Costa A (2024) A roadmap towards safe and sustainable by design nanotechnology: implementation for nano-silver-based antimicrobial textile coatings production by ASINA project. Comput Struct Biotechnol J 25(June):127–142. https://doi.org/10.1016/j.csbj.2024.06.013
23. Gaharwar AK, Cross LM, Peak CW, Gold K, Carrow JK, Brokesh A, Singh KA (2019) 2D nanoclay for biomedical applications: regenerative medicine, therapeutic delivery, and additive manufacturing. Adv Mater 31(23):1–28. https://doi.org/10.1002/adma.201900332
24. Garcia-Garcia D, Lopez-Martinez J, Balart R, Strömberg E, Moriana R (2018) Reinforcing capability of cellulose nanocrystals obtained from pine cones in a biodegradable poly(3-hydroxybutyrate)/poly(ε-caprolactone) (PHB/PCL) thermoplastic blend. Eur Polym J 104:10–18. https://doi.org/10.1016/j.eurpolymj.2018.04.036

25. Geetha Bai R, Muthoosamy K, Manickam S, Hilal-Alnaqbi A (2019) Graphene-based 3D scaffolds in tissue engineering: fabrication, applications, and future scope in liver tissue engineering. Int J Nanomedicine 14:5753–5783. https://doi.org/10.2147/IJN.S192779
26. Grand View Research (2022) Nanomaterials market size, share and growth report, 2030. Grand View Res. https://www.grandviewresearch.com/industry-analysis/nanotechnology-and-nanomaterials-market#
27. Guo F, Aryana S (2016) An experimental investigation of nanoparticle-stabilized CO_2 foam used in enhanced oil recovery. Fuel 186:430–442. https://doi.org/10.1016/j.fuel.2016.08.058
28. Haleem A, Javaid M, Singh RP, Rab S, Suman R (2023) Applications of nanotechnology in medical field: a brief review. Glob Health J 7(2):70–77. https://doi.org/10.1016/j.glohj.2023.02.008
29. Hari N, Francis S, Rajendran Nair AG, Nair AJ (2018) Synthesis, characterization and biological evaluation of chitosan film incorporated with β-Carotene loaded starch nanocrystals. Food Packag Shelf Life 16:69–76. https://doi.org/10.1016/j.fpsl.2018.02.003
30. Hong F, Zhou Y, Zhao X, Sheng L, Wang L (2017) Maternal exposure to nanosized titanium dioxide suppresses embryonic development in mice. Int J Nanomedicine 12:6197–6204. https://doi.org/10.2147/IJN.S143598
31. Hossain N, Mobarak MH, Mimona MA, Islam MA, Hossain A, Zohura FT, Chowdhury MA (2023) Advances and significances of nanoparticles in semiconductor applications – a review. Results Eng 19(June). https://doi.org/10.1016/j.rineng.2023.101347
32. Huston M, DeBella M, DiBella M, Gupta A (2021) Green synthesis of nanomaterials. Nanomaterials 11(8):2130. https://doi.org/10.3390/nano11082130
33. Iravani,R, An C, Adamian Y, Mohammadi M (2022) A review on the use of nanoclay adsorbents in environmental pollution control. Water, Air, & Soil Pollution 233(4), 109. https://doi.org/10.1007/s11270-022-05580-2
34. Ismail RA, Mohsin MH, Ali AK, Hassoon KI, Erten-Ela S (2020) Preparation and characterization of carbon nanotubes by pulsed laser ablation in water for optoelectronic application. Physica E 119:113997. https://doi.org/10.1016/j.physe.2020.113997
35. Jamalipour Soufi G, Iravani S (2020) Eco-friendly and sustainable synthesis of biocompatible nanomaterials for diagnostic imaging: current challenges and future perspectives. Green Chem 22(9):2662–2687. https://doi.org/10.1039/D0GC00734J
36. Jha AK, Prasad K, Prasad K (2009) A green low-cost biosynthesis of Sb2O3 nanoparticles. Biochem Eng J 43(3):303–306. https://doi.org/10.1016/j.bej.2008.10.016
37. Kalia S, Boufi S, Celli A, Kango S (2014) Nanofibrillated cellulose: surface modification and potential applications. Colloid Polym Sci 292(1):5–31. https://doi.org/10.1007/s00396-013-3112-9
38. Kant K, Shukla A, Sharma A, Henry Biwole P (2017) Heat transfer study of phase change materials with graphene nano particle for thermal energy storage. Sol Energy 146:453–463. https://doi.org/10.1016/j.solener.2017.03.013
39. Kargarzadeh H, Mariano M, Gopakumar D, Ahmad I, Thomas S, Dufresne A, Huang J, Lin N (2018) Advances in cellulose nanomaterials. Cellulose 25(4). Springer Netherlands. https://doi.org/10.1007/s10570-018-1723-5
40. Kawashima S, Wang K, Ferron RD, Kim JH, Tregger N, Shah S (2021) A review of the effect of nanoclays on the fresh and hardened properties of cement-based materials. Cem Concr Res 147:106502. https://doi.org/10.1016/j.cemconres.2021.106502
41. Khan I, Saeed K, Khan I (2019) Nanoparticles: properties, applications and toxicities. Arab J Chem 12(7):908–931. https://doi.org/10.1016/j.arabjc.2017.05.011
42. Kulkarni M, Patel K, Patel A, Patel S, Desai J, Patel M, Shah U, Patel A, Solanki N (2024) Nanomaterials as drug delivery agents for overcoming the blood-brain barrier: a comprehensive review. ADMET DMPK 12(1):63–105. https://doi.org/10.5599/admet.2043
43. Liu J, Li R, Yang B (2020) Carbon dots: a new type of carbon-based nanomaterial with wide applications. ACS Cent Sci 6(12):2179–2195. https://doi.org/10.1021/acscentsci.0c01306

44. Lyu H, Gao B, He F, Ding C, Tang J, Crittenden JC (2017) Ball-milled carbon nanomaterials for energy and environmental applications. ACS Sustain Chem Eng 5(11):9568–9585. https://doi.org/10.1021/acssuschemeng.7b02170
45. Machac P, Cichon S, Lapcak L, Fekete L (2020) Graphene prepared by chemical vapour deposition process. Graph Technol 5(1–2):9–17. https://doi.org/10.1007/s41127-019-00029-6
46. Macías-Silva MA, Cedeño-Muñoz JS, Morales-Paredes CA, Tinizaray-Castillo R, Perero-Espinoza GA, Rodríguez-Díaz JM, Jarre-Castro CM (2024) Nanomaterials in construction industry: an overview of their properties and contributions in building house. Case Stud Chem Environ Eng 10:100863. https://doi.org/10.1016/j.cscee.2024.100863
47. Mahfoudhi N, Boufi S (2017) Nanocellulose as a novel nanostructured adsorbent for environmental remediation: a review. Cellulose 24(3):1171–1197. https://doi.org/10.1007/s10570-017-1194-0
48. Martínez G, Merinero M, Pérez-Aranda M, Pérez-Soriano EM, Ortiz T, Villamor E, Begines B, Alcudia A (2021) Erratum: Guillermo M., et al. environmental impact of nanoparticles' application as an emerging technology: a review. Materials 14(7):1–26. https://doi.org/10.3390/ma14071710
49. Matsumoto R, Adachi S, Sadki EHS, Yamamoto S, Tanaka H, Takeya H, Takano Y (2020) Maskless patterning of gallium-irradiated superconducting silicon using focused ion beam. ACS Appl Electron Mat 2(3):677–682. https://doi.org/10.1021/acsaelm.9b00781
50. Meng Y, Feng Y, Bai X, Yu Q, Zhou J, Wang J (2025) Application of nanotechnology in agricultural sustainability: absorption, translocation, and challenges of nanoparticles. Curr Plant Biol 42(December 2024):100492. https://doi.org/10.1016/j.cpb.2025.100492
51. Miseljic M, Olsen SI (2014) Life-cycle assessment of engineered nanomaterials: a literature review of assessment status. J Nanopart Res 16(6):2427. https://doi.org/10.1007/s11051-014-2427-x
52. Missoum K, Belgacem M, Bras J (2013) Nanofibrillated cellulose surface modification: a review. Materials 6(5):1745–1766. https://doi.org/10.3390/ma6051745
53. Momeni S, Nabipour I (2015) A simple green synthesis of palladium nanoparticles with Sargassum alga and their electrocatalytic activities towards hydrogen peroxide. Appl Biochem Biotechnol 176(7):1937–1949. https://doi.org/10.1007/s12010-015-1690-3
54. Nam JH, Jang MJ, Jang HY, Park W, Wang X, Choi SM, Cho B (2020) Room-temperature sputtered electrocatalyst WSe2 nanomaterials for hydrogen evolution reaction. J Energy Chem 47:107–111. https://doi.org/10.1016/j.jechem.2019.11.027
55. Oliver-Ortega H, Tresserras J, Julian F, Alcalà M, Bala A, Espinach FX, Méndez JA (2021) Nanocomposites materials of PLA reinforced with Nanoclays using a masterbatch technology: a study of the mechanical performance and its sustainability. Polymers 13(13):2133. https://doi.org/10.3390/polym13132133
56. Palit S, Hussain CM (2020) Green nanomaterials: a sustainable perspective, pp 23–41. https://doi.org/10.1007/978-981-15-3560-4_2
57. Pan Y (2023) Application of carbon nanotubes in various fields. MATEC Web Conf 386:01008. https://doi.org/10.1051/matecconf/202338601008
58. Parashar M, Shukla VK, Singh R (2020) Metal oxides nanoparticles via sol-gel method: a review on synthesis, characterization and applications. J Mater Sci Mater Electron 31(5):3729–3749. https://doi.org/10.1007/s10854-020-02994-8
59. Peng N, Huang D, Gong C, Wang Y, Zhou J, Chang C (2020) Controlled arrangement of Nanocellulose in polymeric matrix: from reinforcement to functionality. ACS Nano 14(12):16169–16179. https://doi.org/10.1021/acsnano.0c08906
60. Perumal AB, Sellamuthu PS, Nambiar RB, Sadiku ER (2018) Development of polyvinyl alcohol/chitosan bio-nanocomposite films reinforced with cellulose nanocrystals isolated from rice straw. Appl Surf Sci 449:591–602. https://doi.org/10.1016/j.apsusc.2018.01.022
61. Poolakkandy RR, Menamparambath MM (2020) Soft-template-assisted synthesis: a promising approach for the fabrication of transition metal oxides. Nanoscale Adv 2(11):5015–5045. https://doi.org/10.1039/D0NA00599A

62. Rahman G, Najaf Z, Mehmood A, Bilal S, Shah A, Mian S, Ali G (2019) An overview of the recent progress in the synthesis and applications of carbon nanotubes. C 5(1):3. https://doi.org/10.3390/c5010003
63. Rezania S, Darajeh N, Rupani PF, Mojiri A, Kamyab H, Taghavijeloudar M (2025) Recent advances in the adsorption of different pollutants from wastewater using carbon-based and metal-oxide nanoparticles. European Commission. https://civil-protection-humanitarian-aid.ec.europa.eu/news-stories/news/new-eu148-million-eu-humanitarian-aid-package-ukraine-2025-01-13_en
64. Rhim JW, Park HM, Ha CS (2013) Bio-nanocomposites for food packaging applications. Prog Polym Sci 38(10–11):1629–1652. https://doi.org/10.1016/j.progpolymsci.2013.05.008
65. Romero R (2025) Application of pillared clays for water recovery. Catalysts 15(2). https://doi.org/10.3390/catal15020159
66. Şahan N, Paksoy H (2017) Investigating thermal properties of using nano-tubular ZnO powder in paraffin as phase change material composite for thermal energy storage. Compos Part B 126:88–93. https://doi.org/10.1016/j.compositesb.2017.06.006
67. Sajid M (2022) Nanomaterials: types, properties, recent advances, and toxicity concerns. Curr Opin Environ Sci Health 25:100319. https://doi.org/10.1016/j.coesh.2021.100319
68. Saleh TA (2020) Nanomaterials: classification, properties, and environmental toxicities. Environ Technol Innov 20:101067. https://doi.org/10.1016/j.eti.2020.101067
69. Saraswat P, Singh S, Prasad M, Misra R, Rajput VD, Ranjan R (2023) Applications of bio-based nanomaterials in environment and agriculture: a review on recent progresses. Hybr Adv 4(June):100097. https://doi.org/10.1016/j.hybadv.2023.100097
70. Shah KA, Tali BA (2016) Synthesis of carbon nanotubes by catalytic chemical vapour deposition: a review on carbon sources, catalysts and substrates. Mater Sci Semicond Process 41:67–82. https://doi.org/10.1016/j.mssp.2015.08.013
71. Shanmuga Priya D, Suriyaprabha R, Yuvakkumar R, Rajendran V (2014) Chitosan-incorporated different nanocomposite HPMC films for food preservation. J Nanopart Res 16(2):2248. https://doi.org/10.1007/s11051-014-2248-y
72. Sharma RK, Ganesan P, Tyagi VV, Metselaar HSC, Sandaran SC (2016) Thermal properties and heat storage analysis of palmitic acid-TiO 2 composite as nano-enhanced organic phase change material (NEOPCM). Appl Therm Eng 99:1254–1262. https://doi.org/10.1016/j.applthermaleng.2016.01.130
73. Sharma G, Kumar D, Kumar A, Al-Muhtaseb AH, Pathania D, Naushad M, Mola GT (2017) Revolution from monometallic to trimetallic nanoparticle composites, various synthesis methods and their applications: a review. Materials Science and Engineering: C 71:1216–1230. https://doi.org/10.1016/j.msec.2016.11.002
74. Sharma G, Kumar A, Sharma S, Naushad M, Prakash Dwivedi R, ALOthman ZA, Mola GT (2019) Novel development of nanoparticles to bimetallic nanoparticles and their composites: a review. J King Saud Univ Sci 31(2):257–269. https://doi.org/10.1016/j.jksus.2017.06.012
75. Shiri MSZ, Henderson W, Mucalo MR (2019) A review of the lesser-studied microemulsion-based synthesis methodologies used for preparing nanoparticle Systems of the Noble Metals, Os, Re, Ir and Rh. Materials 12(12):1896. https://doi.org/10.3390/ma12121896
76. Silva APM, Oliveira AV, Pontes SMA, Pereira ALS, Souza Filho M d s M, Rosa MF, Azeredo HMC (2019) Mango kernel starch films as affected by starch nanocrystals and cellulose nanocrystals. Carbohydr Polym 211:209–216. https://doi.org/10.1016/j.carbpol.2019.02.013
77. Slavin YN, Asnis J, Häfeli UO, Bach H (2017) Metal nanoparticles: understanding the mechanisms behind antibacterial activity. J Nanobiotechnology 15(1):65. https://doi.org/10.1186/s12951-017-0308-z
78. Somwanshi SB, Somvanshi SB, Kharat PB (2020) Nanocatalyst: a brief review on synthesis to applications. J Phys Conf Ser 1644(1):012046. https://doi.org/10.1088/1742-6596/1644/1/012046
79. Son HH, Seo GH, Jeong U, Shin DY, Kim SJ (2017) Capillary wicking effect of a Cr-sputtered superhydrophilic surface on enhancement of pool boiling critical heat flux. Int J Heat Mass Transf 113:115–128. https://doi.org/10.1016/j.ijheatmasstransfer.2017.05.055

80. Strotmann U, Durand MJ, Thouand G, Eberlein C, Heipieper HJ, Gartiser S, Pagga U (2024) Microbiological toxicity tests using standardized ISO/OECD methods—current state and outlook. Appl Microbiol Biotechnol 108(1). https://doi.org/10.1007/s00253-024-13286-0
81. Sudjit Luanpitpong, Liying Wang, Y. R. (2019) The effects of carbon nanotubes on lung and dermal cellular behaviors. Physiol Behav 176(3):139–148. https://doi.org/10.2217/nnm.14.42.The
82. Taiwo LA, Obianyo II, Amu OO, Omoniyi AO, Onwualu AP, Soboyejo ABO (2022) Effects of titanium dioxide coatings on building composites for sustainable construction applications. Cogent Eng 9(1). https://doi.org/10.1080/23311916.2022.2151168
83. Tang Y, Su D, Huang X, Alva G, Liu L, Fang G (2016) Synthesis and thermal properties of the MA/HDPE composites with nano-additives as form-stable PCM with improved thermal conductivity. Appl Energy 180:116–129. https://doi.org/10.1016/j.apenergy.2016.07.106
84. Tarcan R, Todor-Boer O, Petrovai I, Leordean C, Astilean S, Botiz I (2020) Reduced graphene oxide today. J Mater Chem C 8(4):1198–1224. https://doi.org/10.1039/C9TC04916A
85. Tipa C, Cidade MT, Borges JP, Costa LC, Silva JC, Soares PIP (2022) Clay-based nanocomposite hydrogels for biomedical applications: a review. Nanomaterials 12(19). https://doi.org/10.3390/nano12193308
86. Uddin MN, Hossain MT, Mahmud N, Alam S, Jobaer M, Mahedi SI, Ali A (2024) Research and applications of nanoclays: a review. SPE Polym 5(4):507–535. https://doi.org/10.1002/pls2.10146
87. Wang D (2019) A critical review of cellulose-based nanomaterials for water purification in industrial processes. Cellulose 26(2):687–701. https://doi.org/10.1007/s10570-018-2143-2
88. Wen Y-H, Tsou C-H, de Guzman MR, Huang D, Yu Y-Q, Gao C, Zhang X-M, Du J, Zheng Y-T, Zhu H, Wang Z-H (2022) Antibacterial nanocomposite films of poly(vinyl alcohol) modified with zinc oxide-doped multiwalled carbon nanotubes as food packaging. Polym Bull 79(6):3847–3866. https://doi.org/10.1007/s00289-021-03666-1
89. Xu K, Chen J (2020) High-resolution scanning probe lithography technology: a review. Appl Nanosci 10(4):1013–1022. https://doi.org/10.1007/s13204-019-01229-5
90. Yamaguchi T, Tsuruda Y, Furukawa T, Negishi L, Imura Y, Sakuda S, Yoshimura E, Suzuki M (2016) Synthesis of CdSe quantum dots using Fusarium oxysporum. Materials 9(10):855. https://doi.org/10.3390/ma9100855
91. Yap YH, Azmi AA, Mohd NK, Yong FSJ, Kan S-Y, Thirmizir MZA, Chia PW (2020) Green synthesis of silver nanoparticle using water extract of onion peel and application in the acetylation reaction. Arab J Sci Eng 45(6):4797–4807. https://doi.org/10.1007/s13369-020-04595-3
92. Zahid M, Abd-Elsalam KA (2021) Applications of nanomaterials in water remediation: a note from the editors. In: Aquananotechnology. Elsevier, pp 1–10. https://doi.org/10.1016/B978-0-12-821141-0.00021-5
93. Zhang D, Ye K, Yao Y, Liang F, Qu T, Ma W, Yang B, Dai Y, Watanabe T (2019) Controllable synthesis of carbon nanomaterials by direct current arc discharge from the inner wall of the chamber. Carbon 142:278–284. https://doi.org/10.1016/j.carbon.2018.10.062
94. Zhang J, Xia A, Chen H, Nizami A-S, Huang Y, Zhu X, Zhu X, Liao Q (2022) Biobased carbon dots production via hydrothermal conversion of microalgae Chlorella pyrenoidosa. Sci Total Environ 839:156144. https://doi.org/10.1016/j.scitotenv.2022.156144
95. Zhu Q, Wang X, Miller JD (2019) Advanced nanoclay-based nanocomposite solid polymer electrolyte for lithium iron phosphate batteries. ACS Appl Mater Interfaces 11(9):8954–8960. https://doi.org/10.1021/acsami.8b13735

Chapter 4
Environmental Pollution Control Using Micro and Nano Bio-based Materials

4.1 Overview of Environmental Pollution Challenges and Pollution Control

The accelerated pace of industrialization, urbanization, and intensive agricultural practices has significantly elevated levels of environmental contamination across various ecosystems, including air, water, and soil. Major pollutants include heavy metals such as lead, cadmium, and chromium; synthetic organic compounds such as dyes, pesticides, and pharmaceuticals; as well as emerging pollutants like micro- and nanoplastics and greenhouse gases. These contaminants pose long-lasting and often irreversible risks to biodiversity, public health, and ecological stability due to their persistence, toxicity and bioaccumulative nature. Conventional remediation methods, including physical separation, chemical coagulation, incineration, and advanced oxidation, though effective to an extent, are typically energy-intensive, costly, and can generate harmful secondary pollutants. As global environmental challenges grow more complex, there is a critical need for sustainable, green alternatives that align with circular economy principles and minimize ecological disruption.

Given these drawbacks, the scientific community is progressively turning toward sustainable and eco-efficient alternatives. Among these, bio-based materials synthesized at micro- or nanoscale stand out due to their unique physicochemical advantages: high specific surface area, tunable surface chemistry, inherent biocompatibility and renewable origin. These materials are often produced using green synthesis methods, including microbial fermentation, phyto-extraction, and the valorization of agricultural or industrial waste streams, making them inherently more sustainable than conventional synthetic nanomaterials. Their environmentally friendly synthesis, coupled with their functional diversity, enables the development of multi-functional adsorbents, catalysts, and filtration media for broad-spectrum pollution control. The integration of such materials into environmental technologies

M. I. Ahmad, A. K. H. P. S., *Environmental and Energy Technology*, Green Energy and Technology, https://doi.org/10.1007/978-3-032-10706-0_4

represents a significant step toward greener, circular and low-impact remediation practices that support global sustainability targets and regulatory frameworks.

In water treatment applications, bio-based nanomaterials such as cellulose nanocrystals (CNCs), chitosan nanoparticles, and lignin-derived nanostructures have demonstrated high performance in removing various pollutants. Cellulose nanocrystals, derived from plant biomass, possess high surface area, abundant hydroxyl groups, and strong mechanical properties, making them suitable for adsorbing dyes, heavy metals, and persistent organic pollutants. Chitosan-based nanomaterials, on the other hand, offer additional benefits such as natural antimicrobial activity, biocompatibility, and ability to chelate metal ions due to the presence of amino groups. Studies by Khedulkar et al. [43] and Saraswat et al. [84], show that these nanomaterials can achieve adsorption capacities comparable to or exceeding those of traditional nanomaterials, while also reducing environmental burden. Furthermore, bacterial cellulose and biochar produced from agro-waste offer versatile and scalable solutions, with effective performance under varying water chemistries. These developments position bio-based nanomaterials as transformative agents in sustainable water remediation technologies.

Extending beyond aqueous systems, micro- and nano-bio-based materials are emerging as innovative tools for addressing micro- and nanoplastic pollution a growing concern due to the limitations of conventional removal methods. Traditional filtration and sedimentation methods often fail to capture ultrafine plastic particles, leading to their accumulation in aquatic ecosystems and food chains. Engineered bio-based materials, such as biodegradable polymers, modified biochar composites, and functionalized cellulose membranes, provide selective adsorption, biodegradability, and cost-efficiency. Research by Enyoh et al. [20] and Kumar et al. [45] highlights how these materials can trap, isolate, and degrade microplastics under environmentally relevant conditions. Policy frameworks promoting the use of bio-based packaging and microbead replacements are further accelerating innovation in this domain. However, real-world assessments of biodegradability under different environmental conditions remain a research priority, particularly regarding fragmentation, ecotoxicity, and life-cycle impacts.

Lastly, heavy metal remediation has also greatly benefited from advances in micro/nano bio-based material technologies. Bio-based nanomaterials such as nano zero-valent iron (nZVI), biogenic metal nanoparticles, and engineered chitosan-metal composites have shown exceptional capabilities in adsorbing, immobilizing, or chemically transforming toxic metal ions. For instance, green-synthesized metal oxide nanoparticles like ZnO and TiO_2 produced via plant extract routes exhibit efficient photocatalytic and redox activity for metal ion reduction. Moreover, functional hybrid materials combining organic matrices such as lignin or cellulose with metallic nanoparticles further enhance pollutant binding through synergistic mechanisms. Yu et al. [112] and Meftah et al. [60] demonstrate that these bio-derived alternatives often achieve comparable or superior results to synthetic counterparts with reduced environmental and human toxicity. The low energy demands of their synthesis, coupled with high efficacy and minimal secondary waste, position these materials as future pillars in green heavy metal remediation technologies.

4.2 Role of Micro and Nano Bio-based Materials in Pollution Control

Micro and nano-scale bio-based materials, including cellulose nanofibers (CNFs), chitosan nanoparticles, nano-biochar, and fungal/bacterial biomass, are gaining recognition as highly effective and sustainable solutions for environmental pollution control. Derived from renewable resources such as plants, marine organisms, and microorganisms, these materials possess a range of advantageous physicochemical properties. Their nanoscale dimensions contribute to a high surface area-to-volume ratio, which enhances reactivity, diffusion, and pollutant interactions. Additionally, these materials offer tunable surface chemistry and eco-compatibility, making them suitable for deployment in soil, water, and air remediation applications. The combination of biodegradability, abundance and low toxicity further supports their potential in environmental technologies. These properties enable their integration into advanced systems such as filters, membranes, and reactive composites. As the demand for green and cost-effective pollution control technologies increases, micro- and nano-bio-based materials present a compelling platform to replace or augment conventional adsorbents and catalysts, especially in areas requiring selective and efficient pollutant removal. Figure 4.1 shows the overview of micro- and nano-bio-based materials in environmental pollution control. This figure summarizes their sources, material types, physicochemical properties, mechanisms of action, applications, benefits, challenges and future prospects, highlighting their multifunctionality and sustainability.

Cellulose nanofibers (CNFs), derived from both plant-based cellulose and microbial fermentation, represent one of the most extensively studied classes of bio-based nanomaterials for pollution control. Their fibrous morphology and high aspect ratio, coupled with abundant hydroxyl functional groups, allow for versatile chemical modifications to tailor surface interactions with specific contaminants. Techniques such as TEMPO-mediated oxidation and carboxymethylation enhance CNF functionality, enabling them to bind effectively with heavy metals like Pb(II), Cd(II), and Cu(II), as well as with various organic pollutants including dyes and pharmaceuticals [88]. Modified cellulose nanofibers (CNFs) have demonstrated excellent performance in pollutant adsorption, particularly in water treatment applications [16, 19]. CNFs are also known for their robust mechanical properties, making them ideal candidates for incorporation into composite membranes and filtration systems [7]. These hybrid systems can be deployed for water purification in both centralized and decentralized settings. In addition to pollutant removal, CNFs offer the benefits of biodegradability, renewability, and compatibility with other biopolymers, making them a sustainable option for scalable pollution control technologies.

Chitosan nanoparticles, synthesized via deacetylation of chitin obtained from crustacean shells, have emerged as a powerful class of bio-based materials for environmental remediation. Rich in amino and hydroxyl groups, these nanoparticles exhibit excellent affinity toward a broad spectrum of pollutants, including divalent and trivalent metal ions [71]. Their binding mechanisms, primarily through

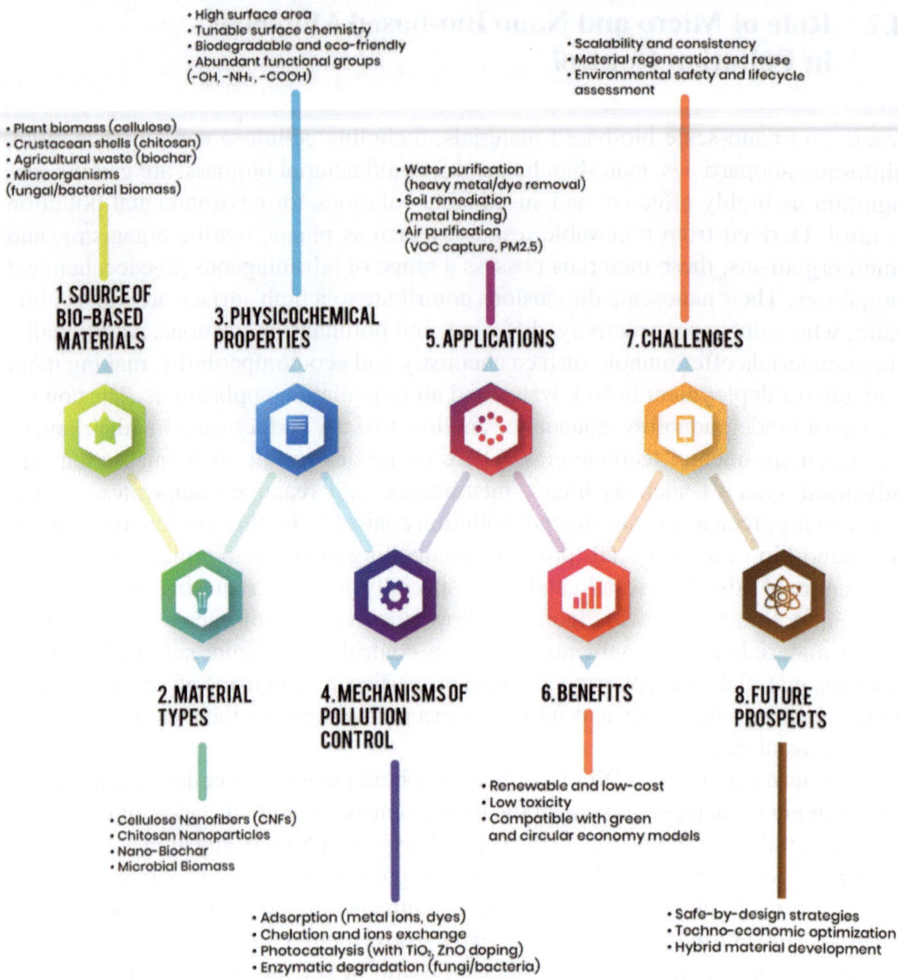

Fig. 4.1 Overview of micro- and nano-bio-based materials in pollution control

chelation and electrostatic interactions, result in high adsorption capacities often exceeding 1 mmol/g, even under competitive conditions. Chitosan nanoparticles are also effective against organic dyes like methylene blue and Congo red, offering multifunctional performance in wastewater treatment [43]. Their biocompatibility, non-toxic nature, and biodegradability make them particularly attractive for use in eco-sensitive areas. Furthermore, their low cost and scalability enhance their industrial applicability. Current research is focused on improving their stability and regeneration capabilities through cross-linking and hybridization with other nanomaterials. These advancements aim to extend their operational lifespan and broaden their utility in complex wastewater matrices, thereby supporting sustainable and efficient pollution management strategies.

Nano-biochar, produced via pyrolysis of agricultural and forestry biomass under oxygen-limited conditions, is gaining attention for its multifunctional role in pollution remediation. Its inherently porous structure, high surface area, and surface functional groups make it a promising material for adsorbing a range of contaminants. When engineered at the nanoscale and doped with photocatalysts like TiO_2 or ZnO, nano-biochar demonstrates dual functionality which combining physical adsorption with photocatalytic degradation. These composite systems can operate under visible light to generate reactive oxygen species (ROS) that degrade persistent organic pollutants (POPs), including pesticides and pharmaceuticals [87, 91]. Despite its potential, concerns about the stability of embedded nanoparticles, leaching risks, and long-term environmental impact must be addressed. Current research is exploring methods to immobilize nanoparticles more securely within the biochar matrix and reduce ecotoxicological risks. With proper risk assessment and lifecycle evaluation, nano-biochar can offer an innovative, low-cost, and scalable option for addressing diverse pollution challenges in both urban and rural contexts.

Fungal and bacterial biomass offer biologically driven pathways for pollution control through enzymatic degradation and metal bioaccumulation. White-rot fungi, for example, produce oxidative enzymes such as laccases and peroxidases capable of degrading complex organic pollutants like polycyclic aromatic hydrocarbons (PAHs), synthetic dyes, and endocrine-disrupting chemicals [86]. Bacterial species contribute by transforming nitrates, sulfates, and heavy metals through metabolic processes. These organisms can be immobilized on nano-structured carriers to enhance their stability, reusability, and pollutant interaction. Immobilization also allows simultaneous adsorption and biodegradation, increasing the efficiency of treatment systems. While promising, biological degradation is often limited by slow kinetics and sensitivity to environmental variables such as pH, temperature, and pollutant concentration. Advances in genetic engineering and process optimization are being investigated to improve enzymatic activity and resilience. Despite these limitations, microbial biomass systems represent a renewable and low-energy option, particularly for treating low-concentration and complex pollutant mixtures in wastewater and industrial effluents [25].

Despite their advantages, several challenges must be addressed before micro and nano-bio-based materials can be widely adopted in pollution control. Variability in raw material sources, energy-intensive processing techniques and difficulties in achieving large-scale uniformity hinder their commercial scalability [29]. Additionally, concerns regarding their environmental fate, toxicity, and regeneration after pollutant saturation raise questions about long-term sustainability. Lifecycle assessment (LCA) and techno-economic analysis are critical to evaluate the feasibility of integrating these materials into existing environmental management systems. Safe-by-design approaches, which consider environmental and health risks during the material development phase, are being encouraged [66]. Regulatory frameworks that support innovation while ensuring safety are also essential to guide industrial adoption. Future research should focus on improving synthesis efficiency, exploring circular economy models for material reuse, and establishing standardized evaluation metrics. Through these efforts, micro and nano bio-based materials can transition from experimental innovations to practical

Table 4.1 Nano bio-based materials used in environmental pollution control including their sources, mechanisms, advantages and limitations

Material	Source	Primary mechanisms	Strengths	Limitations
Cellulose nanofibers (CNF)	Plant/bacterial cellulose	Adsorption, functional-group chelation	High surface area; modifiable; biodegradable	Energy-intensive prep; source variability
Chitosan nanoparticles	Deacetylated crustacean chitin	Metal chelation, dye adsorption	Biocompatible; selective and regenerable	Shellfish allergen; effectiveness depends on pH
Nano-modified biochar	Pyrolysed biomass + nano-additives	Adsorption, photocatalysis, redox reactions	Low cost; multi-functionality; carbon footprint	Potential nanoparticle leaching/toxicity
Fungal/ bacterial biomass	Myco−/ microorganisms	Enzymatic degradation, biosorption	Low-cost, enzyme-based degradation	Slow degradation rate; environmental sensitivity

solutions in the global fight against environmental pollution [87, 91]. Table 4.1 shows the summary of the nano bio-based materials used in environmental pollution control including their sources, mechanisms, advantages and limitations.

4.3 Mechanisms of Pollution Control

Pollution control is a cornerstone of environmental protection, and recent advances in micro- and nano-scale bio-based materials have introduced novel mechanisms for more efficient and sustainable remediation. These materials derived from renewable biological sources such as cellulose, chitosan, lignin, and proteins offer tunable surface properties, high surface area-to-volume ratios, and functional versatility, enabling enhanced interaction with a wide range of pollutants. The mechanisms of pollution control using such materials span physical adsorption, ion exchange, electrostatic attraction, catalytic degradation, and bioremediation each contributing to the removal or neutralization of contaminants in air, water, and soil. At the nanoscale, the unique physicochemical characteristics of bio-based structures amplify reaction kinetics and pollutant binding affinity, allowing for improved selectivity and efficiency even at trace concentrations.

4.3.1 Adsorption Mechanism

Adsorption is one of the most extensively studied and applied mechanisms for pollution mitigation using bio-based micro- and nano-materials. These materials, including nanocellulose, biochar, chitosan nanoparticles, and fungal biosorbents, exhibit a high surface area and contain various functional groups that interact with pollutants via physical and chemical adsorption processes [99]. Key mechanisms such as ion exchange, electrostatic attraction, hydrogen bonding, surface complexation, and π-π interactions allow effective removal of a wide array of contaminants from water and air matrices [14, 108].

Nanocellulose, particularly in oxidized or functionalized forms, has demonstrated high adsorption capacities for heavy metal ions such as Pb(II), Cd(II), and Cr(VI), as well as cationic dyes. The abundance of hydroxyl and carboxyl groups on nanocellulose surfaces facilitates strong coordination and chelation of metal ions. In one study, TEMPO-oxidized nanocellulose removed more than 95% of Pb(II) from aqueous solutions at pH 5–6, showcasing its selective affinity and efficiency under mildly acidic conditions [5]. Biochar and nano-biochar, produced from biomass pyrolysis, offer another efficient and low-cost adsorbent platform. The porous nature of biochar allows for significant surface area exposure, while surface functionalities such as carbonyl, hydroxyl, and carboxylic groups enable multiple interaction pathways with pollutants. Additionally, modification of biochar with magnetic nanoparticles or metal oxides (e.g., Fe_3O_4, TiO_2) enhances its adsorption performance and facilitates recovery from aqueous media [73, 91].

Chitosan nanoparticles are widely known for their excellent metal chelation ability due to the presence of abundant amino and hydroxyl functional groups on their surface. These functionalities facilitate strong binding with various metal ions, particularly Cu(II), Zn(II), and Pb(II), through mechanisms involving chelation, ion exchange, and electrostatic attraction. Numerous studies have demonstrated their high adsorption capacities under a range of ionic strengths and pH conditions, making them suitable for diverse wastewater treatment applications. In addition to heavy metals, chitosan-based nanocomposites have shown remarkable effectiveness in the removal of anionic dyes such as Congo red and methylene blue, as well as pharmaceutical contaminants like diclofenac and ibuprofen [87]. These interactions are primarily driven by electrostatic interactions and hydrogen bonding, enabling selective and efficient adsorption. Moreover, chitosan nanoparticles are biodegradable, non-toxic, and can be regenerated using mild desorption agents, enhancing their practicality for sustainable pollution remediation systems.

Fungal and bacterial biomass also play a vital role in bio-based adsorption through biosorption mechanisms. The cell walls of fungi and bacteria contain polysaccharides, proteins, and lipids with functional groups such as carboxyl, amino, and phosphate groups, which actively bind to pollutants via ion exchange, complexation, and electrostatic interactions. White-rot fungi, for instance, produce extracellular enzymes like laccase and manganese peroxidase that assist in degrading complex organic molecules such as dyes and PAHs. Additionally, immobilizing microbial biomass on nanostructured supports improves both biosorption efficiency and stability in continuous systems. This approach has shown great potential for industrial effluent treatment, though slow kinetics and sensitivity to environmental fluctuations remain limiting factors [80]. Table 4.2 present the summary of the bio-based materials used for adsorption-based pollution control, including adsorption mechanisms and target pollutants.

Table 4.2 Bio-based materials used for adsorption-based pollution control, including adsorption mechanisms and target pollutants

Material	Functional groups involved	Target pollutants	Adsorption mechanisms	Notable features
Nanocellulose	-OH, -COOH	Heavy metals, dyes	Hydrogen bonding, ion exchange, chelation	High selectivity; biodegradable
Biochar/ nano-biochar	-COOH, -C=O, -OH	Metals, organics, pharmaceuticals	Surface complexation, π-π stacking, redox	Low-cost; tunable surface; porous structure
Chitosan nanoparticles	$-NH_2$, -OH	Metals, dyes, drugs	Electrostatic attraction, chelation, H-bonding	Regenerable; biocompatible
Fungal/ bacterial biomass	Polysaccharides, proteins	Heavy metals, synthetic dyes	Biosorption, ion exchange, coordination	Enzymatic synergy; renewable; eco-friendly

4.3.2 Catalytic Degradation of Pollutants

Catalytic degradation represents a transformative and environmentally sustainable strategy in pollution mitigation, particularly for the breakdown of persistent organic pollutants (POPs), synthetic dyes, pharmaceutical residues, and hydrocarbons. Micro and nano-scale bio-based catalysts have shown exceptional potential due to their high surface reactivity, favorable redox potential, and compatibility with environmentally benign processes. These catalysts operate primarily through the in-situ generation of reactive oxygen species (ROS), including hydroxyl radicals (OH), superoxide anions (O_2-), and singlet oxygen ($^1O_2$). These ROS possess strong oxidative power and are capable of decomposing complex and resistant pollutants into simpler, non-toxic end-products such as carbon dioxide and water. The application of photocatalytic materials such as TiO_2, ZnO, and Fe_3O_4 supported on biodegradable matrices not only amplifies their catalytic efficacy but also minimizes ecological risks by preventing nanoparticle release. This mechanism supports green chemistry initiatives and is pivotal to the development of next-generation pollution control technologies.

Metal oxide nanoparticles such as titanium dioxide (TiO_2), zinc oxide (ZnO), and magnetite (Fe_3O_4), when supported on environmentally friendly matrices like biochar, chitosan, or cellulose, have gained significant attention for their application in pollution remediation. These materials are especially effective in water, soil, and air due to their light-responsive catalytic activity. When irradiated by UV or visible light, these catalysts generate electron-hole pairs that drive the formation of reactive oxygen species (ROS), such as hydroxyl radicals (•OH) and superoxide anions (O_2-), which facilitate the oxidative breakdown of persistent pollutants [54]. The use of bio-based supports enhances the surface area, dispersibility, and structural stability of the catalysts while simultaneously reducing the risk of nanoparticle leaching. Moreover, functionalized supports like chitosan and cellulose offer additional adsorption capabilities and biocompatibility. These composites are increasingly being developed for real-world use in wastewater treatment plants, soil decontamination systems, and indoor air purification technologies [100].

In water pollution applications, ZnO and TiO_2 nanoparticles immobilized on biochar or biopolymer matrices have demonstrated outstanding efficiency in the removal of various waterborne contaminants. These composites harness the dual advantage of adsorption and photocatalysis, facilitating the degradation of dyes, pharmaceuticals, and antibiotics from complex wastewater matrices. For instance, ZnO/biochar composites derived from bamboo and reed straw were shown to degrade over 90% of methylene blue and tetracycline under simulated sunlight, benefiting from the synergistic interaction between ZnO's photocatalytic action and the biochar's porous adsorptive surface [70]. Similarly, TiO_2 embedded in bamboo pulp aerogels has demonstrated significant visible-light-driven degradation of organic dyes, confirming the role of the carbon framework in enhancing light absorption and ROS generation [50].

These materials present a promising green approach to advanced wastewater treatment technologies due to their high activity, reusability, and compatibility with solar-driven processes. In the context of soil remediation, bio-supported TiO_2 and Fe_3O_4 nanoparticles have been effectively applied to degrade agrochemical residues, including chlorpyrifos, atrazine, and various herbicides. These nanoparticles, when embedded into cellulose or carbon-based supports, initiate redox reactions under light irradiation or in reactive soil microenvironments. This activity results in the oxidative breakdown of persistent organic compounds that are otherwise resistant to natural degradation. Furthermore, these catalysts can improve soil porosity and influence microbial diversity by reducing the bioavailability of toxic compounds. For instance, grapefruit peel-derived TiO_2-biochar composites achieved over 99% removal of rhodamine B from contaminated soil and demonstrated stability across multiple remediation cycles [107]. Their porous structures also promote oxygen diffusion and microbial colonization, leading to enhanced biodegradation synergy. These characteristics make TiO_2 and Fe_3O_4 biohybrids particularly valuable for sustainable soil restoration practices in agriculture and industrial land reclamation.). Their porous structures also aid in enhancing soil aeration and microbial activity, contributing to holistic soil health.

Air quality improvement technologies also utilize these catalysts due to their effective photocatalytic properties under ambient lighting conditions. TiO_2 and ZnO nanoparticles, when incorporated into cellulose-based filters, membranes, or coatings, are capable of degrading a wide range of volatile organic compounds (VOCs), including benzene, formaldehyde, and toluene. These pollutants are commonly present in indoor environments and contribute significantly to poor air quality and health issues. For instance, TiO_2 chitosan composites have achieved over 90% degradation of diclofenac under visible light within 90 min [93]. The chitosan not only provides structural support but also enhances the dispersion of TiO_2 particles, increasing light absorption and active site exposure. These photocatalytic systems have demonstrated scalability for incorporation into building materials, HVAC systems, and surface coatings, offering a passive, energy-efficient means to purify indoor air. Their low toxicity and renewable base materials further support sustainable implementation [33]. These materials are also being explored for their use in building facades and HVAC systems to passively improve indoor air quality.

The incorporation of TiO_2, ZnO, and Fe_3O_4 nanoparticles onto biodegradable supports such as biochar and cellulose offers a versatile and sustainable approach to pollution control across water, soil, and air environments. These hybrid systems are highly adaptable, capable of functioning efficiently under both UV and visible light conditions, and can be tailored to specific pollutant types and environmental matrices. Their multifunctionality enables simultaneous adsorption and catalytic degradation, improving overall treatment efficiency. Additionally, the use of biodegradable supports enhances material stability and reduces environmental impact by limiting secondary pollution. However, the long-term application of these systems in real-world scenarios demands further research and development. The challenges include ensuring consistent catalyst regeneration, maintaining stability over repeated cycles, and minimizing the ecological risks of nanoparticle leaching [6]. Therefore, future advancements should focus on lifecycle assessments, material recyclability, and safety optimization to support their broader implementation in large-scale remediation projects.

Enzyme-based catalytic degradation utilizes oxidoreductases such as laccases, peroxidases, and dehydrogenases. Enzymes secreted naturally by fungi and bacteria for the transformation of organic pollutants into less harmful substances. These enzymes are especially effective in degrading aromatic, phenolic, and pharmaceutical compounds under mild environmental conditions, making them suitable for sustainable applications. In water treatment, white-rot fungi-derived laccase immobilized on magnetic nanocellulose has demonstrated over 85% degradation of bisphenol A within 2 h, highlighting its potential for endocrine-disruptor removal in wastewater systems [22]. In soil remediation, these enzymes can break down persistent agrochemicals like phenols, chlorinated compounds, and lignin-derived pollutants, improving soil health while minimizing chemical use. For air purification, immobilized laccases integrated into biofilters or membrane systems have been explored for degrading volatile organic compounds (VOCs) such as toluene and formaldehyde. Immobilization on nano-cellulose or silica supports enhances enzyme durability, activity retention and operational reusability across diverse environmental matrices.

Nano zero-valent iron (nZVI) and bio-based redox-active nanocomposites are increasingly recognized for their powerful reductive catalytic capabilities in pollution remediation for air, water, and soil matrices. These materials drive contaminant transformation through electron transfer mechanisms, breaking down halogenated organics like trichloroethylene (TCE), chlorinated phenols, and nitrobenzene into less harmful compounds. For instance, tea polyphenol-modified sludge biochar loaded with nZVI (TP-nZVI/BC) efficiently removed hexavalent chromium

(Cr(VI)) from aqueous solutions in fixed-bed tests, achieving capacities up to ~17 mg/g and demonstrating versatility under both aerobic and anaerobic conditions [113]. Similarly, corn-straw biochar composites have achieved ~46–64 mg/g for Cr(VI), highlighting their potential in water purification [103]. Beyond water systems, biochar-supported nZVI has been successfully applied in soil remediation, enhancing electron transfer and immobilizing contaminants [115]. In the atmosphere, permeable reactive barriers (PRBs) constructed from ZVI intercept and reductively degrade volatile chlorinated compounds, with pilot sites like Moffett Field confirming sustained performance [31]. Overall, nZVI-based composites deliver low-cost, green, and site-adaptable solutions across multiple environmental compartments.

Despite their high efficiency, catalytic systems necessitate thorough evaluation concerning their long-term stability, environmental resilience, and reusability under operational conditions. A critical concern involves the potential leaching of active components such as metal nanoparticles or redox agents, which may introduce secondary contamination. Ensuring immobilization of these components within a stable matrix is vital for sustainable application. Additionally, the integration of catalytic systems with adsorption-based technologies has emerged as a promising approach for synergistic pollutant removal. These dual-function materials first adsorb target contaminants onto the surface and subsequently catalyze their degradation, thereby enhancing overall treatment efficacy and extending material life span. For example, composite systems combining nZVI or TiO2 with activated carbon or biochar have shown improved removal efficiencies for both organic and inorganic pollutants. Future development should emphasize safe-by-design approaches, supported by lifecycle assessments and regeneration studies, to facilitate reliable deployment in real-world environmental remediation scenarios. Table 4.3 represent the micro and nano bio-based materials for catalytic degradation of pollutants.

Table 4.3 Micro and nano bio-based materials for catalytic degradation of pollutants

Catalyst type	Support material	Target pollutants	Mechanism	Advantages
ZnO, TiO_2 NPs	Biochar, chitosan	Dyes, pharmaceuticals	Photocatalysis (UV/visible)	Renewable, reusable, sunlight activation
Laccase, peroxidase	Nanocellulose, silica	Aromatics, phenolics	Enzymatic oxidation	Biodegradable, selective, mild conditions
Zero-valent iron (nZVI)	Biochar, cellulose	Chlorinated solvents, Cr(VI)	Reductive degradation	Strong redox; effective in anaerobic zones
MnO_2, Fe_3O_4	Bacterial cellulose	Pesticides, antibiotics	Redox, Fenton-like reactions	High reactivity, magnetically recoverable

Certain bio-based nanomaterials serve not only as adsorbents but also as active catalysts or catalyst supports in the degradation of recalcitrant pollutants. These materials, including enzyme-immobilized nanofibers and metallic nanoparticle-functionalized biopolymers, enable the transformation of persistent organic contaminants such as phenols, dyes, and pesticides under relatively mild environmental conditions. The catalytic mechanisms involved often encompass photocatalysis, Fenton-like oxidation, or enzymatic redox reactions that contribute to the mineralization of complex compounds into less harmful or biodegradable forms. For example, laccase-immobilized cellulose nanofibers have demonstrated high degradation efficiency for bisphenol A and azo dyes in water, while ZnO-decorated chitosan composites effectively break down pesticides through visible light-induced photocatalysis. These systems offer environmental compatibility, reusability, and adaptability across various pollution contexts. Continued innovation in material synthesis and catalytic pathway optimization will further enhance the applicability of these bio-based catalysts in integrated environmental treatment systems.

4.3.3 *Microbial Degradation of Organic Wastes*

Microbial degradation or biodegradation, is a biologically driven process wherein microorganisms decompose organic waste into simpler inorganic compounds such as carbon dioxide, methane, and water. This process plays a pivotal role in the global carbon cycle, sustainable waste management, and ecosystem functioning. Bacteria and fungi are the principal agents of microbial degradation, secreting extracellular enzymes that catalyze the breakdown of complex polymers like cellulose, lignin and proteins into assimilable units. This degradation can proceed aerobically which yielding CO_2 and water or anaerobically producing methane and other reduced compounds, depending on oxygen availability [56]. Key applications of microbial degradation include composting, wastewater treatment and bioremediation. In composting, microbial communities convert organic waste into stable humus-like material rich in nutrients. In wastewater treatment facilities, activated sludge systems rely on microbial consortia to mineralize organic pollutants. In bioremediation, native or engineered microbes degrade hazardous contaminants in soil and groundwater enhancing site recovery [75]. Furthermore, microbial degradation contributes significantly to soil health by releasing essential nutrients such as nitrogen, phosphorus and sulfur, promoting plant growth and soil fertility. The eco-friendliness of microbial degradation based on its natural, low-energy inputs makes it a promising alternative to chemical-intensive waste treatment technologies. Ongoing research into microbial metabolic pathways, bioaugmentation and integration with bio-nano systems is expanding its applicability for circular bioeconomy and sustainable environmental management.

The composting process progresses through four distinct biological and physicochemical stages: mesophilic, thermophilic, cooling, and maturation. In the initial mesophilic phase, moderate-temperature microbes (~10–45 °C), such as bacteria, fungi, and actinomycetes, rapidly decompose labile substrates like simple sugars, amino acids, and lipids. Their metabolism generates significant heat, raising the internal pile temperature to around 40 °C within a few days. Concurrently, early fermentation produces organic acids, causing a transient pH drop; however, this acidity is short-lived, as acids are oxidized or volatilized, and ammonia released from protein breakdown buffers the system toward neutrality or slight alkalinity [55]. This mesophilic activity lays the groundwork for the thermophilic stage by ensuring sufficient thermal buildup and microbial priming.

As temperatures surpass approximately 55 °C, the thermophilic phase becomes dominant. Thermophilic bacteria which often *Bacillus*, *Thermus*, and various Actinobacteria that alongside fungi adapted to high heat, proliferate and efficiently degrade complex biomolecules such as cellulose, hemicellulose, proteins and lipids. Metabolic processes release ammonia, raising the pH to alkaline levels. Sustained temperatures between 55–65 °C over several days are critical for microbiological safety, effectively inactivating human and plant pathogens such as *Escherichia coli* and *Salmonella*, and destroying weed seeds [69, 104] demand peaks during this phase, fueled by intense microbial respiration and enzymatic degradation.

Following substrate depletion, the composting process transitions into the cooling phase, also referred to as the second mesophilic phase. During this stage, internal pile temperatures decline gradually to around 30–45 °C as microbial respiration decreases due to the reduced availability of easily degradable organic matter. However, biological activity persists as mesophilic bacteria and fungi recolonize the material, often in association with macrofungi capable of breaking down more recalcitrant compounds such as lignin, cellulose, and polyphenolics. These organisms play an essential role in the continued transformation of intermediate degradation products into more stable organic matter. The compost's pH, which had previously increased due to ammonia accumulation during the thermophilic phase, begins to shift toward neutrality as ammonia volatilizes and nitrification processes begin. This cooling phase can span several weeks and frequently overlaps with the early stages of compost maturation. Though visually the compost may begin to resemble the final product, physicochemical parameters still indicate that biological stabilization is incomplete [21].

The maturation or curing phase represents the final and most critical stage in composting, where stability and agronomic safety are fully achieved. At this point, the compost temperature equilibrates with ambient conditions, signaling that microbial metabolism has diminished to background levels. However, microbial activity does not cease entirely. Specialized microbial consortia particularly fungi and actinomycetes which continue to degrade the remaining resistant organic molecules and polymeric compounds, facilitating humification. The result is a steady increase in the content of stable humic and fulvic acids, which contribute to soil fertility and structure. The pH of the compost typically stabilizes between 6 and 8, buffered by humic substances and the gradual loss of residual ammonia. The final compost product is dark, crumbly, and emits an earthy odor, signifying maturity. To verify compost quality and readiness for application, indices such as the carbon-to-nitrogen

(C/N) ratio, seed germination index and oxygen uptake rate are assessed to ensure the absence of phytotoxicity and the presence of microbiological stability [21].

In the maturation (curing) phase, compost aligns with ambient temperature, indicating that microbial metabolism has decelerated to baseline levels. Microbial communities especially fungi and actinomycetes which continue to foster the transformation of remaining organic fragments into stable humic substances. The pH balances out between 6 and 8, buffered by humic compounds and residual ammonia content [21]. The final product is a dark, crumbly, humus-rich material with an earthy aroma. Quality and stability are validated using metrics such as the C/N ratio, germination index and respiration activity ensuring phytotoxic compounds have dissipated, and the compost is safe and agronomically effective. Furthermore, microbial degradation in wastewater treatment encompasses several interconnected biological processes, beginning with bioremediation in the purposeful use of indigenous or introduced microorganisms to metabolize and transform pollutants into less harmful compounds. This ecosystem-driven mechanism involves bacteria, fungi, and algae equipped with metabolic pathways capable of degrading a broad spectrum of organic contaminants, including hydrocarbons, phenolics, and pharmaceuticals. The efficiency of bioremediation is contingent upon optimal environmental conditions such as temperature, pH, and nutrient availability as well as sufficient oxygen or electron acceptors to support aerobic or anaerobic metabolisms. Enhancing this process through bioaugmentation where specific strains or microbial consortia are introduced which allows targeting of recalcitrant pollutants that native populations degrade inefficiently. Recent reviews suggest bioaugmentation, especially when combined with biostimulation, offers significant improvements in treating industrial effluents by enhancing microbial diversity and catabolic potential [11, 67, 110].

Biostimulation, a complementary strategy, provides necessary nutrients or electron acceptors and donors (e.g., nitrogen, phosphorus, molasses, or oxygen) to enrich and activate native microbial populations. This is particularly critical for degradation of hydrocarbon and halogenated contaminants in situ, where environmental conditions limit microbial performance. For instance, adding phosphate and nitrate can accelerate herbicide biodegradation, while providing electron donors enhances reductive dechlorination in subsurface environments [41]. Biostimulation's efficacy hinges on site-specific parameters such as subsurface hydrogeology and microbial community structure; failures often stem from uneven additive distribution or poor survival of stimulated organisms. Nonetheless, when designed appropriately, biostimulation is a cost-effective and scalable method for augmenting the native biodegradation capacity of wastewater systems. Beyond metabolic breakdown, microbial processes also include pollutant removal via biosorption and bioaccumulation. Biosorption leverages microbial cell surfaces rich in functional groups like carboxyl and amino moieties which to passively bind heavy metals and colloidal toxins, facilitating removal from aqueous media. In parallel, bioaccumulation allows living cells to uptake and sequester pollutants intracellularly, offering another route to detoxify waste streams. Specialized enzyme-mediated pathways further contribute to degradation. For example, extracellular enzymes such as laccases, peroxidases, proteases, ureases and lipases catalyze the breakdown of phenolic compounds, proteins, urea and lipids. These enzymatic systems enhance the biodegradation of pollutants that may not be directly metabolized by intact microbial cells [57, 77].

Advancements in bio-electrochemical systems particularly microbial fuel cells (MFCs) have merged pollutant degradation with renewable energy generation. MFCs employ electroactive bacteria to oxidize organic pollutants at the anode, transferring electrons to a cathode and producing electrical current. Recent reviews affirm that while MFCs demonstrate promising COD removal and electricity recovery, their performance in real wastewater is variable due to electrode material limitations, microbial community stability and system scaling challenges [74]. Innovations in electrode design and microbial consortia selection, including strategic use of organisms like *Pseudomonas aeruginosa*, have enhanced both power output and treatment efficiency in pilot applications [111]. Although optimization continues, MFCs hold significant promise for sustainable wastewater treatment by converting biological degradation into usable electricity. These microbial strategies are employed across multiple wastewater treatment configurations. In activated sludge systems, microbial flocs comprising bacteria and protozoa aerobically degrade organic load through respiration. Biofilters support attached microbial communities that oxidize pollutants as effluent percolates through substrates such as sand or ceramics. Membrane bioreactors combine advanced microbial treatment with membrane filtration to achieve high effluent quality and biomass retention. Anaerobic digestion, another key process, relies on anaerobic consortia to decompose organic matter into biogas which mainly methane and carbon dioxide under oxygen-limited conditions. The efficacy of these treatments depends on microbial structure and function, pollutant type and concentration and environmental parameters (temperature, pH, nutrient balance). Integrating bioaugmentation, biostimulation, biosorption, and enzymatic degradation within these systems can significantly improve degradation efficiency, pathogen reduction, and operational resilience [38, 109]. A comprehensive comparison of the advantages and limitations of these microbial mechanisms is summarized in Table 4.4, which highlights their respective roles, operational strengths, and potential constraints in wastewater treatment applications.

Each mechanism presents unique advantages and limitations. Bioremediation and biodegradation are effective and naturally occurring but may be slow for persistent pollutants. Bioaugmentation and biostimulation improve system efficiency yet require precise control and may face ecological barriers. Biosorption and bioaccumulation are particularly useful for metal removal but do not always lead to complete pollutant mineralization. Enzyme-based treatments offer specificity but are often limited by cost and enzyme stability. MFCs provide dual benefits but face operational and scalability constraints. Microbial degradation processes are indispensable to modern environmental management strategies. Their integration into existing treatment infrastructures not only enhances pollutant removal and system resilience but also aligns with the principles of the circular bioeconomy. Future advancements should focus on optimizing microbial consortia, integrating

Table 4.4 Advantages and disadvantages of microbial degradation mechanisms in wastewater treatment

Mechanism	Advantages	Disadvantages
Bioremediation	Cost-effective and eco-friendly Capable of degrading diverse organics Applicable in situ and ex situ	Slow for recalcitrant pollutants Strongly affected by pH, temp, O_2 Native microbes may be insufficient
Bioaugmentation	Enables specific pollutant degradation Enhances microbial diversity Improves treatment of toxic compounds	Introduced microbes may fail to compete Ecological imbalance risk Requires monitoring and optimization
Biostimulation	Activates existing microbes Enhances biodegradation rates Low-cost strategy	Overdosing causes nutrient imbalance Efficiency is site-dependent Difficult to uniformly apply in large systems
Biosorption	Effective for heavy metal removal Can use inactive (dead) biomass Regenerable and reversible	No degradation occurs, only separation Requires periodic regeneration Less effective at low concentrations
Bioaccumulation	Absorbs pollutants inside microbial cells Useful for trace metal uptake Can allow biomass recovery	Depends on living, viable cells Risk of pollutant release upon cell death Not suitable for bulk pollutant loads
Biodegradation	Converts organics into CO_2, H_2O Core process in aerobic/anaerobic systems Versatile for many pollutants	Not all pollutants fully degraded May produce toxic intermediates Needs balanced microbial ecosystems
Enzyme-mediated	Targets specific pollutants Works under mild conditions Enzymes can be immobilized	High production cost Enzymes may denature over time Limited to certain substrate types
Microbial fuel cells	Degrades organics and generates electricity Reduces COD efficiently Suitable in various redox conditions	Low power output Electrode fouling is common Scaling up is complex and costly

microbial degradation with nano- and electrochemical technologies, and tailoring these systems for site-specific environmental conditions. Through interdisciplinary innovation and ecological stewardship, microbial degradation will continue to play a pivotal role in advancing sustainable, low-impact solutions for global waste and pollution challenges.

4.3.4 Water Purification and Filtration Systems

Water purification systems rely on multiple mechanisms working synergistically to remove pollutants including physical barriers, chemical reactions, and biological processes. Filtration a foundational physical process, utilizes materials such as sand, gravel or advanced membranes to trap suspended solids, microorganisms, and particulate matter. Mechanical filtration employs meshes or porous structures with controlled pore sizes to physically exclude contaminants above a specific threshold. Sedimentation complements this process by allowing heavier particles to settle under gravity in clarifiers or settling tanks, thus reducing turbidity and enhancing downstream filtration efficiency [4, 83]. These pre-treatment steps are critical in both point-of-use systems and large-scale treatment plants to prevent clogging and fouling of finer membranes, ensuring sustained performance. By combining sedimentation with filtration, treatment systems can achieve greater contaminant removal with lower maintenance demands.

A more advanced form of physical removal is reverse osmosis (RO), which uses pressure to overcome osmotic resistance and force water through semi-permeable membranes. The process effectively removes dissolved salts, heavy metals, organic pollutants, and microbial contaminants down to the molecular level. Recent literature highlights RO's high rejection rates removing up to 99% of dissolved solids and its adaptability across diverse applications, from residential drinking water systems to large-scale desalination plants. Innovations such as spiral-wound and hollow-fiber membrane modules have improved efficiency and longevity [96, 116]. However, RO systems face notable challenges, including high energy consumption and fouling issues such as membrane scaling by sparingly soluble salts. These limitations are addressed through pre-treatment measures like antiscalant dosing and periodic backwashing [23, 105].

Moving down the membrane hierarchy, nanofiltration (NF) operates with pore sizes in the 1–10 nm range, providing selective removal of divalent ions, organic micro-pollutants, and larger molecules. NF stands out for its ability to soften water removing calcium and magnesium while retaining small monovalent salts and enabling high permeate flux at moderate pressure levels [96, 105]. Its intermediate selectivity and efficiency use less energy than RO and require simpler pretreatment. NF is increasingly used for heavy metal removal in industrial wastewater streams, with modified polymeric membranes achieving over 99% removal of chromium (III). Nonetheless, NF membranes are susceptible to fouling and require periodic cleaning, and their application is limited by c Sedimentation and distillation are other physical separation methods with distinct operational contexts. Sedimentation leverages low-energy gravitational settling in clarifiers and primary treatment tanks, effectively removing coarse solids and biomass before biological treatment stages. Distillation uses heat to evaporate water and condense it back into liquid form, leaving behind non-volatile contaminants. While distillation achieves high purity, it is energy-intensive and more often reserved for specialized applications like laboratory-grade water production rather than large-scale municipal treatment. Reverse osmosis and RO-NF systems are preferred for routine high-efficiency desalination and purification due to superior energy profiles and scalability.

Overall, physical mechanisms involve mechanical filtration, sedimentation, distillation, nanofiltration and reverse osmosis form the backbone of modern water purification. From removing suspended solids and microorganisms to eliminating dissolved contaminants and salts, each process offers unique advantages and limitations. Their integration through pretreatment and staged deployment which allows systems to optimize performance, maintain operational longevity, and ensure consistent water quality. Together, they provide robust, scalable solutions for delivering clean, safe water across domestic, industrial, and environmental applications. Ost and membrane durability in long-term operation. Meanwhile in chemical processes are integral to modern water purification systems which playing a vital role in removing a broad range of contaminants that are not effectively addressed by physical filtration alone. These processes operate through various mechanisms such as particle destabilization, molecular binding, oxidation, and ion displacement, each tailored to target specific classes of pollutants including dissolved metals, organic compounds, pathogens, and excess nutrients. Techniques such as coagulation-flocculation, chemical precipitation, adsorption, ion exchange, and disinfection have been widely implemented in both municipal and industrial water treatment applications due to their high efficacy and adaptability. Their effectiveness depends on factors such as water chemistry, contaminant concentration, pH, and operational design. While some methods like chlorination are established standards and others such as advanced adsorption using nanomaterials which continue to evolve with emerging contaminants and sustainability goals. This section discusses the core chemical treatment mechanisms, highlighting their principles, applications, advantages and limitations within integrated water purification frameworks.

Flocculation and coagulation are fundamental chemical processes in water treatment, often applied as pre-treatment steps to destabilize and aggregate colloidal particles, thereby facilitating their removal. Coagulation involves adding positively charged coagulants such as aluminium or iron salts to neutralize the electrical charges of suspended colloids, reducing repulsive forces and promoting aggregation [62, 94]. Subsequent gentle flocculation encourages these destabilized particles to collide and form flocs, which are then removed via settling or filtration. These processes are not only efficient in reducing turbidity and organic load but also critical in mitigating membrane fouling in downstream ultrafiltration systems. Research shows that optimizing coagulant dosage, pH, and mixing conditions can significantly improve removal efficiency across different water types including challenging industrial and textile effluents [12]. While cost-effective and scalable, the technique produces sludge requiring disposal and demands careful chemical dosing to avoid residual aluminium or iron in treated water.

Chemical precipitation removes dissolved ions by converting them into insoluble solids. For example, the addition of lime or soda ash precipitates hardness-causing calcium and magnesium ions as carbonates, while sulfide precipitation targets heavy metals like lead, cadmium and copper. This process is widely used in industrial wastewater treatment to address metal-laden effluent before discharge or reuse.

Precipitation is advantageous due to its high contaminant removal efficiency and straightforward operation. However, its effectiveness is heavily pH-dependent and often results in large volumes of sludge requiring dewatering and disposal [42, 53]. Moreover, managing the disposal or potential reuse of the precipitated solids is a critical component of process sustainability, especially in contexts where metal recovery is valuable.

Adsorption is a versatile chemical purification technique where pollutants adhere to adsorbent surfaces via physisorption (weak van der Waals forces) or chemisorption (strong chemical bonds) [42]. Activated carbon is the most widely employed adsorbent due to its high surface area and effectiveness in binding organic compounds, taste and odor-causing substances, and certain inorganic pollutants. Chemisorption such as when functionalized carbon or metal oxides chemically bind heavy metals which offers specificity and high capacity but may be irreversible. Conversely, physisorption is mostly reversible and suitable for removing volatile organics. The broad utility of adsorption is evident in its adaptability from household filters to large-scale industrial systems. However, adsorbent regeneration or replacement adds to operational costs and logistics. Emerging bio-based adsorbents like chitosan and nanocellulose shows promise in cost-effective, sustainable adsorption of dyes, metals, and pharmaceuticals, albeit still limited by sorbent stability and production scalability.

Ion exchange resins facilitate purification by swapping undesirable ions in water with benign ones anchored to a resin's functional groups. Cationic resins commonly address hardness (Ca^{2+}/Mg^{2+}) and heavy metals, while anionic resins remove nitrates, arsenic and fluorides [82]. The high selectivity, reusability, and consistent removal performance make ion exchange ideal for demineralization and trace contaminant control. However, high resin and regeneration chemical costs, dependency on brine for regeneration, and reduced lifespan due to fouling or resin degradation are notable drawbacks [42, 72]. In industrial contexts, ion exchange is effective for targeted ion removal and resource recovery, yet its viability hinges on careful system design, efficient regeneration processes, and disposal of regeneration waste streams.

Disinfection is the critical final step in water purification, aimed at inactivating pathogens to ensure safe consumption. Chlorination remains the most widely used method worldwide, utilizing chlorine oxidizing power to damage microbial cell walls and disrupt metabolic processes. One significant advantage of chlorination is the residual chlorine it leaves behind, which maintains disinfection efficacy throughout the distribution system. However, its use can produce harmful by-products, particularly trihalomethanes (THMs) and haloacetic acids (HAAs), that have been epidemiologically linked to increased risks of bladder and colorectal cancer [30]. Studies consistently show correlations between long-term exposure to THMs even below regulatory limits of 80 ppb and elevated cancer risks [30, 47]. Ultraviolet

(UV) irradiation offers a chemical-free alternative that deactivates a broad spectrum of pathogens, including bacteria, viruses, and protozoan cysts such as *Giardia* and *Cryptosporidium*, by damaging their nucleic acids. While UV treatment is highly effective even outperforming chlorination for protozoa. Its efficacy is reduced by water turbidity, which scatters and absorbs UV light and it does not provide residual disinfection [27]. Furthermore, ozonation is another potent oxidation-based method that efficiently neutralizes organic micropollutants and pathogens without forming chlorinated by-products. However, ozonation can convert naturally occurring bromide in water into carcinogenic bromate (BrO_3^-), necessitating strict monitoring and control measures to keep bromate levels within health-safe limits [37, 106]. Because each disinfection method has unique benefits and limitations, hybrid systems such as UV treatment followed by chlorination that are often employed. These combinations leverage the strengths of each approach, ensuring immediate pathogen inactivation, residual protection, and minimized by-product formation. Effective disinfection thus relies on careful optimization of operational parameters like dosage, contact time, water quality, and residual management to balance microbial safety with chemical risk control.

Biological processes play a fundamental role in water purification, offering sustainable, low-energy alternatives to conventional chemical and physical treatment methods. These processes harness the metabolic capabilities of microorganisms and plants to remove, transform, or immobilize contaminants in both natural and engineered environments. Through mechanisms such as biodegradation, biosorption, nitrification-denitrification, and phytoremediation, biological systems can effectively target a broad spectrum of pollutants, including organic matter, nutrients, heavy metals, and pathogens. Unlike physical filtration or chemical disinfection, biological treatment often results in the complete mineralization of pollutants, making it highly desirable for long-term environmental protection and water reuse. Biological approaches are increasingly applied in decentralized treatment settings, constructed wetlands, slow sand filtration units, and biofiltration systems due to their operational simplicity and ecological compatibility. Recent advancements in microbial ecology, biofilm engineering, and plant-microbe interactions have further enhanced the performance and reliability of biological treatment technologies. This section explores the core biological processes used in water purification, highlighting their mechanisms, applications, benefits, and limitations within modern water management systems.

Slow sand filtration (SSF) is a biological process that combines physical straining with microbial biodegradation within a biologically active layer known as the *Schmutzdecke* at the top of the sand bed. This biofilm, formed by bacteria, fungi, protozoa, and algae, traps suspended solids while concurrently metabolizing organic matter and inactivating pathogens such as bacteria and protozoan cysts [26, 98]. SSF typically achieves log reductions of 90–99% in microbial counts, making it highly effective for rural and decentralized water treatment systems [3, 58]. Its major benefits which low energy requirements, minimal chemical use and ease of maintenance make SSF well-suited for resource-limited settings. However, challenges such as channeling, seasonal temperature effects on microbial activity, and limited removal of dissolved contaminants emphasize the need for periodic maintenance and integration with additional treatment stages.

Biologically active carbon (BAC) filters merge adsorption and biodegradation by supporting microbial communities on the surface of activated carbon. Compared to conventional activated carbon, BAC systems leverage the metabolic activity of bacteria (*Nitrosomonas*, *Nitrospira*), fungi, and protozoa to degrade adsorbed organic matter, thereby regenerating adsorption capacity and extending filter lifespan [17, 39]. Full scale BAC implementations have demonstrated robust DOC removal in drinking water systems, reducing disinfection byproduct precursors while maintaining microbial stability [48]. Despite operational considerations like biofilm control, backwashing, and maintaining dissolved oxygen levels, BAC technology offers a sustainable alternative that reduces chemical use and enhances robustness in advanced treatment configurations [39, 90].

Meanwhile, phytobial remediation integrates plant systems with microbial symbionts to remove nutrients, heavy metals, and organic pollutants from aqueous environments. This approach particularly within constructed wetlands leverages the rhizosphere as a biotic removal zone, where root exudates stimulate microbial populations that facilitate pollutant transformation, uptake, or stabilization [78, 97]. For instance, arsenic removal efficacy is enhanced by microbe-assisted plant uptake, enabling simultaneous phytoremediation and microbial degradation, with potential for bioenergy recovery. While phytobial systems provide low-energy and ecologically sustainable treatment, their effectiveness is influenced by plant microbe compatibility, contaminant bioavailability, and site-specific environmental conditions, necessitating tailored designs and long-term monitoring.

Phytoremediation employs plants especially hyperaccumulators like willows and poplars to uptake and sequester dissolved pollutants, including metals and nutrients, from contaminated water through root systems. Techniques such as rhizofiltration utilize deep-rooted vegetation to pump and treat groundwater, significantly reducing nitrate, phosphate and metal concentrations. While phytoremediation is cost-effective, passive, and visually appealing, its performance is constrained by plant growth rates, seasonal variability, and limited removal of non-accumulable substances. To enhance treatment efficacy, phytoremediation is often integrated with other biological or chemical barriers to ensure comprehensive pollutant removal.

Traditional microbial bioremediation relies solely on microorganisms to degrade organic pollutants through aerobic or anaerobic metabolic pathways. Diverse bacteria, fungi, and algae facilitate the breakdown of hydrocarbons, phenols, pesticides, and other persistent contaminants. Although microbial degradation is eco-friendly and scalable, this process may prove slow and inefficient when microbial communities lack specific degradative pathways. Enhancements like bioaugmentation (adding specialized strains), bio stimulation (nutrient amendments), or genetic cooling strategies are employed to boost efficacy [59, 98]. Fungal-based systems, for example, have shown promise in degrading herbicides in reactor conditions via ligninolytic enzymes [32].

In response to increasing water pollution complexity and the emergence of contaminants of concern such as microplastics, pharmaceuticals, and heavy metals, advanced water treatment technologies have been developed to supplement conventional methods. These technologies leverage cutting-edge materials science and electrochemical principles to enhance pollutant removal efficiency, reduce treatment footprints and meet stringent water quality standards. Among these, membrane filtration systems, electrokinetic remediation, and aerogel-based adsorption represent promising innovations due to their tunable properties, modularity and compatibility with hybrid systems. Understanding the underlying mechanisms and material compositions of these technologies is essential for optimizing their performance, integration and scalability in both centralized and decentralized treatment infrastructures.

Structured comparison of three advanced water purification technologies membrane filtration, electrokinetic processes and aerogels each representing a distinct approach to pollutant removal at the micro- and nanoscale level. Membrane technology, widely implemented in modern treatment systems, utilizes physical exclusion via selective pore sizes and pressure differentials to separate contaminants. Its effectiveness across microfiltration (MF), ultrafiltration (UF), nanofiltration (NF), and reverse osmosis (RO) makes it suitable for applications ranging from microbial removal to desalination. However, issues such as membrane fouling and high operational costs, particularly for RO which can limit long-term sustainability without proper maintenance strategies.

Electrokinetic processes, on the other hand, leverage electric fields to mobilize charged pollutants, making them particularly effective for treating soils or wastewaters with high heavy metal concentrations or low permeability. While this technique offers targeted pollutant extraction, energy demands and pH imbalance must be carefully managed. Meanwhile, aerogels emerging as highly porous, lightweight materials offer exceptional adsorption capabilities owing to their expansive surface area and tunable chemistry. These materials are increasingly used for the removal of dyes, metals, and oils, but their fragile structure and limited regenerability remain barriers to full-scale implementation. Overall, the comparative insights in Table 4.5 highlights the advanced water purification technologies and their advantages and disadvantages.

Table 4.5 Advanced water purification technologies

Technology	Definition	Mechanism	Materials Used	Advantages	Disadvantages
Membrane technology	Filtration method using semi-permeable membranes to separate contaminants from water	Physical separation via pore size and pressure; includes microfiltration (MF), ultrafiltration (UF), nanofiltration (NF), and reverse osmosis (RO)	Polymeric membranes (PVDF, PES), ceramic membranes, nanocomposites	High removal efficiency for a wide range of pollutants, modular design, suitable for reuse systems	High energy consumption (RO), membrane fouling, expensive maintenance
Electrokinetic process	Remediation process that uses electric fields to move and separate charged pollutants in water or soil	Electromigration of ions and electroosmosis under applied voltage; enhanced by surfactants or chelators	Electrodes (graphite, titanium), power supply, ionic enhancers	Effective for heavy metals and charged pollutants, suitable for low-permeability environments	Energy-intensive, electrode corrosion, byproduct formation (pH changes)
Aerogels	Ultralight porous materials with high surface area for pollutant adsorption	Adsorption through van der Waals forces, ion exchange, or chemisorption depending on functionalization	Graphene, silica, cellulose, chitosan, carbon aerogels	High adsorption capacity, low density, can be tailored for specific pollutants	Fragility in water, high production cost, limited regeneration cycles

4.4 Applications in Pollution Control

The intensification of environmental pollution across air, water, and soil systems has driven the urgent need for innovative, sustainable approaches to pollution control. Traditional remediation methods, while effective in some contexts, often pose limitations in terms of environmental impact, cost, and long-term viability. In response, the development of micro and nano bio-based materials has emerged as a promising frontier, offering eco-friendly and efficient alternatives rooted in renewable resources. These materials exhibit exceptional physicochemical properties such as high surface reactivity, biodegradability, and functional versatility that make them suitable for a wide range of pollution control applications. This section explores how the integration of these advanced materials into environmental technologies contributes to the mitigation of diverse pollutants, supporting the transition toward more sustainable and resilient ecosystems.

4.4.1 Air Pollution Control

The increasing burden of atmospheric pollution from industrialization, urbanization, and vehicular emissions has prompted the exploration of sustainable, low-impact technologies for air purification. Bio-based materials, derived from renewable biological sources, have emerged as promising candidates for mitigating air pollution through eco-friendly mechanisms that complement or replace traditional chemical and mechanical systems. These materials operate within a diverse range of processes including biofiltration, bioremediation, and nanotechnology-enhanced adsorption by utilizing the natural metabolic or structural properties of microorganisms, plants, and biopolymers to capture, transform, or degrade airborne contaminants. Unlike conventional filters or scrubbers that may generate hazardous waste or require intensive energy input, bio-based technologies offer advantages such as biodegradability, low carbon footprint, and adaptability to specific pollutants such as volatile organic compounds (VOCs), nitrogen oxides (NOx), sulfur oxides (SOx), and fine particulate matter (PM2.5). Furthermore, innovations in microbial ecology and material science have expanded the efficiency, specificity, and durability of these systems in both indoor and ambient environments. This section explores the current landscape and mechanisms of bio-based air pollution control technologies, evaluating their applications, performance and limitations in contemporary environmental engineering contexts.

Biofiltration is a sustainable and cost-effective method for removing gaseous pollutants such as volatile organic compounds (VOCs), odorous gases, and low-level particulates from air streams. In this approach, contaminated air is passed through a packed bed often composed of organic media like compost, wood chips or peat on which diverse microbial communities reside. These microbes, including bacteria and fungi, form a biofilm that metabolizes pollutants into simpler, harmless substances like carbon dioxide and water. This process is grounded in engineering principles of mass transfer and biodegradation: contaminants diffuse into the biofilm, where extracellular enzymes break chemical bonds, effectively removing the pollutants. Performance data indicate that biofilters can reduce VOC concentrations by 70–95%, depending on media composition, temperature, humidity, and residence time [34, 89]. While biofiltration offers low operational costs and minimal chemical inputs, its efficacy depends on maintaining optimal microbial activity. Challenges include seasonal temperature fluctuations, periodic clogging, and potential bioaerosol emissions if airflow disrupts the biofilm.

Bioremediation leverages targeted microorganisms to degrade or neutralize specific airborne contaminants, including smog-causing pollutants such as nitrogen oxides (NOx), sulfur oxides (SOx), and atmospheric VOCs. Using both bacteria and fungi either superior native strains or genetically enhanced cultures. This technique can transform toxic molecules into benign ones through enzymatic processes. Current research explores consortia-based approaches, combining microbes with plant leaf surfaces (*phylloremediation*) to amplify airborne pollutant uptake [36]. Such systems are particularly useful in urban areas plagued by smog, where passive biological treatment can supplement active air handling systems. Moreover, genetic engineering efforts, including CRISPR-mediated gene editing, are enhancing microbial metabolic pathways to improve degradation rates of NOx, SOx, and heavy hydrocarbons. However, practical challenges include environmental variability, containment of modified organisms, and ensuring long-term stability of microbial communities. Nonetheless, bioremediation represents a green, ecosystem-based strategy to reduce airborne contaminants at the source. Table 4.6 shows the classification and evaluation of bioremediation mechanisms for air pollution control, detailing their operational principles, environmental benefits and technical constraints.

Bio-based nanomaterials are increasingly recognized as a hybrid innovation that synergizes nanotechnology with renewable, sustainable feedstocks to address air pollution challenges. These materials commonly derived from sources such as cellulose nanofibers, chitosan, lignin, and microbial polysaccharides are characterized by their high specific surface area, tunable surface chemistry, and environmental compatibility. Their multifunctional properties enable several air purification mechanisms, including the adsorption of particulate matter, photocatalytic degradation of volatile organic compounds (VOCs), and antimicrobial activity against airborne pathogens [8, 46]. For instance, electrospun membranes embedded with plant-derived cellulose nanofibers efficiently capture fine particulate matter ($PM_{2.5}$), while graphene oxide cellulose aerogels have demonstrated photocatalytic oxidation of VOCs under light exposure.

In parallel, bio-based nanomaterials such as chitosan, cellulose nanofibers, and biochar-derived carbon exhibit strong potential for adsorption-based pollutant removal due to their porous structure and abundance of functional groups. These nanomaterials effectively capture airborne contaminants including heavy metal vapors such as mercury, acidic gases (SO_2 and NO_x), and VOCs through mechanisms like electrostatic interactions, hydrogen bonding, and chelation. Chitosan nanoparticles, for example,

Table 4.6 Classification and evaluation of bioremediation mechanisms for air pollution control and their operational principles, environmental benefits and technical constraints

Type of bioremediation	Mechanism and description	Advantages	Disadvantages
Biofiltration	Polluted air is passed through a packed bed of organic material (e.g., compost, wood chips) where microbes degrade VOCs and odorous gases into CO_2 and H_2O	Effective for VOCs and odorous compounds- low operating costs- renewable biofilm regeneration	Sensitive to moisture and temperature changes- risk of clogging and pressure drop
Bioventing	Controlled aeration enhances indigenous microbial activity in the contaminated zone, allowing in-situ degradation of airborne or vapor-phase pollutants	In-situ process with minimal disturbance- cost-effective and simple to implement	Limited to aerobic conditions- may be slow for highly recalcitrant compounds
Biofiltration with fungi	Fungal species such as *Phanerochaete chrysosporium* degrade complex VOCs, including aromatic and chlorinated hydrocarbons, through enzymatic oxidation	Effective for hydrophobic and complex organics. Broader substrate range than bacteria	Slower growth rates- requires controlled humidity and pH
Phylloremediation	Leaf surfaces host microbial biofilms that absorb and degrade pollutants such as SO_2 and NOx from ambient air (used in green wall systems)	Passive treatment in urban spaces Enhances urban greenery and aesthetics	Variable efficiency depending on plant species and climate Limited control over microbial communities
Genetically engineered microbes	Engineered bacteria or fungi are designed to express specific enzymes for degradation of targeted pollutants (e.g., aromatic VOCs, NOx)	High degradation efficiency Target-specific transformation pathways	Biosafety and containment concerns. Regulatory hurdles for environmental release

have achieved over 90% removal efficiency for mercury and VOCs under optimized conditions [43, 81] Similarly, modified bio-waste materials like fly ash have proven effective in binding acidic gases owing to their intrinsic porosity and alkaline nature [24]. The advantages of these materials include biodegradability, low production cost from agro waste, and high specificity toward targeted pollutants. Nevertheless, their practical deployment is constrained by challenges such as limited regeneration capacity, susceptibility to structural degradation, and variable performance under high humidity or mixed-pollutant environments. Current research is focused on enhancing durability and selectivity through functionalization techniques and hybridization with metal oxides to expand their scalability and real-world applicability.

Certain semiconducting bio-based nanomaterials particularly TiO_2 supported on biomass-derived carbon function as photocatalysts to degrade airborne organic and inorganic pollutants under UV or visible light. Photocatalytic oxidation (PCO) initiated by light-generated reactive species such as hydroxyl radicals can mineralize VOCs, NO_x, and even inactivate airborne pathogens [52, 85]. Reviews emphasize that TiO_2-based nanocomposites grown on green carbon substrates achieve high reaction rates and improved charge separation, boosting degradation efficiency in indoor air purification systems. These systems are advantageous due to their self-cleaning properties, low chemical waste generation, and potential for passive operation under ambient light. However, practical deployment faces challenges such as catalyst deactivation due to by-product accumulation, diminished performance under low light or high humidity, and limited visible-light absorption. Addressing these limitations involves modifying the catalyst's morphology, doping with metals, or combining with adsorption layers to create multifunctional air treatment units.

The bio-based nanomaterials can be integrated into electrospun or composite membranes to enhance air filtration efficiency, specifically targeting fine particulate matter ($PM_{2.5}$), VOCs, and pathogens. Cellulose or chitosan nanofibers produce filter media with adjustable porosity (<300 nm), high filtering efficiency (>90%), and low pressure drop when manufactured via sustainable electrospinning processes [10, 79]. Nanofiber membranes are lightweight and biodegradable, offering environmental and performance benefits. They reduce filter fouling by enabling moisture vapor exchange and resisting biofilm formation. Nevertheless, their main limitations include fragility under mechanical stress, scalable production challenges, and inconsistent removal of gaseous pollutants without embedded functionalities. To overcome these issues, composite membranes incorporating nanomaterials like graphene oxide or metal oxides are being developed to add antimicrobial and photocatalytic capabilities, creating hybrid filters which combine physical and chemical removal mechanisms.

Recent advancements in environmental nanotechnology have enabled the development of bio-based nanomaterials with multifaceted roles in air pollution control. Among these innovations, nanomaterials derived from renewable sources such as cellulose, carbon nanotubes, and graphene are being engineered into highly sensitive gas sensors for detecting trace levels of airborne pollutants. These sensors typically operate based on chemiresistive mechanisms, where the adsorption of gas molecules induces measurable changes in the electrical resistance of the sensing matrix. As a result, pollutants like volatile organic compounds (VOCs) including benzene and toluene, as well as nitrogen dioxide (NO_2), sulfur dioxide (SO_2), ammonia (NH_3), and carbon monoxide (CO), can be detected at parts-per-billion (ppb) concentrations. The integration of bio-derived nanofibers with metal oxides like ZnO or TiO_2 enhances selectivity and detection sensitivity, creating hybrid sensor systems suitable for low-power, wearable, or distributed monitoring applications. Despite their promise, these sensors face challenges such as reduced performance under fluctuating humidity and temperature conditions, baseline signal drift, and environmental interferences. Nonetheless, ongoing research into nanostructuring techniques such as patterning graphene nanoribbons and surface functionalization with bio-receptors is addressing these limitations and advancing their applicability for real-time air quality monitoring.

Table 4.7 Bio-based materials in air pollution control

Bio-based material/ approach	Application in air pollution control	Advantages	Challenges
Biochars	Used to remove gaseous pollutants from industrial flue gases (e.g., from biomass and coal combustion)	Renewable, derived from biomass waste Good adsorption properties	May require activation for high adsorption Variable quality depending on feedstock
Cellulose-based materials	Used in air filters to capture particulate matter and other airborne pollutants	Renewable and biodegradable Effective in particulate filtration	Can degrade under humidity Limited mechanical strength
Starch-based nanomaterials	Used to create biodegradable plastics, reducing pollution from conventional plastics	Biodegradable Reduces plastic pollution Made from renewable starch sources	Limited thermal and mechanical stability Less suitable for harsh environments

To complement these emerging sensor technologies, several types of bio-based materials have been utilized in conventional air pollution mitigation systems. These materials include biochars, cellulose-based fibers, and starch-derived nanomaterials, all of which are derived from sustainable sources and exhibit advantageous environmental profiles. As summarized in Table 4.7, these materials offer unique properties such as high adsorption capacities, biodegradability, and compatibility with circular economy principles. Biochars, for example, are effective in capturing industrial gaseous pollutants, while cellulose-based materials have shown efficacy in removing fine particulate matter in air filters. Starch-based nanomaterials, meanwhile, are being incorporated into biodegradable packaging materials to mitigate plastic-related pollution. Despite their benefits, practical implementation at scale presents several obstacles, including reduced material durability under environmental stress, performance variability depending on feedstock quality, and challenges in industrial scalability. Nevertheless, with continued innovation in material synthesis and processing, bio-based solutions are becoming increasingly viable for integrated and sustainable air pollution control strategies.

4.4.2 Water Pollution Control

Water pollution remains a critical global issue, threatening public health, aquatic ecosystems, and sustainable development. Rapid industrialization, urban runoff, agricultural activities, and the discharge of untreated wastewater have introduced a wide range of contaminants into water bodies including heavy metals, dyes, pharmaceuticals, microplastics, and pathogens. Conventional water treatment technologies while effective often rely on synthetic chemicals and high-energy processes, which may be costly, environmentally taxing, and unsustainable in the long term. In response to these challenges, bio-based materials derived from renewable biological sources have emerged as promising alternatives for water purification and pollution control. These materials, including cellulose, chitosan, bacterial cellulose, lignin, and biochar, offer unique advantages such as biodegradability, tunable surface chemistry and low environmental impact. Their functionality spans across key water treatment mechanisms, including adsorption, filtration, flocculation, and microbial remediation. As a result, the integration of bio-based materials into water treatment technologies aligns with circular economy principles, offering a sustainable and scalable pathway for mitigating water contamination while minimizing ecological footprints. The following sections delve into specific types of bio-based materials and their applications in water pollution control, highlighting their mechanisms, advantages and current limitations supported by recent scientific studies.

Bio-based filtration materials including cellulose, chitosan and bacterial cellulose are gaining traction as sustainable and effective agents for water purification. Cellulose the most abundant natural polymer is renowned for its versatility in various water treatment forms such as membranes, aerogels, films, and filters. Functional groups like hydroxyl and carboxyl on cellulose nanofibers can be chemically tailored or hybridized with nanoparticles (e.g., ZnO, graphene oxide) to target pollutants ranging from heavy metals to dyes and microbial contaminants. Studies have demonstrated that modified cellulose nanofibrils can remove over 98% of heavy metals due to increased surface area and adsorption sites [1]. Moreover, recent silk-cellulose composite filters have shown exceptional efficacy in eliminating persistent "forever chemicals" and heavy metals, while resisting biofouling.

Chitosan derived from crustacean shells, is a powerful, biocompatible adsorbent for water contaminants. Its amino and hydroxyl functionalities enable chelation and ionic interactions with metal ions, dyes, and organic compounds. Reviews report that chitosan-based materials (nanoparticles, composites) achieve high removal efficiencies even exceeding 90% for heavy metals like Cu^{2+} and Pb^{2+} and can be regenerated using magnetic separation techniques. Chitosan is sustainable and cost-effective, but challenges include structural stability in acidic environments and limited mechanical strength. Researchers address these by crosslinking chitosan with polymers or nanoparticles to enhance durability and selectivity [2, 9].

Bacterial cellulose (BC) offers unique advantages over plant-derived cellulose. It exhibits higher purity, crystallinity, tensile strength, and porosity. BC-based membranes have demonstrated notable pollutant removal capacities for heavy metals and dyes. Often surpassing conventional materials and maintain performance across multiple usage cycles. BC aerogels, formed via freeze-drying or chemical modification show great promise for oil spill remediation and high-capacity adsorption, though their scalability and production costs remain under development [9, 64]. Collectively, these bio-based filtration systems harness renewable, non-toxic feedstocks and offer high-performance, customizable solutions for water treatment, aligning with sustainability goals and circular economy principles.

Biochar, a porous carbon-rich material derived from biomass pyrolysis, has emerged as a versatile and sustainable solution for water pollution management. Its unique structure which featuring high surface area and tunable functional groups which enables it to adsorb a wide range of contaminants. Additionally, biochar serves multiple roles in treatment systems by enhancing nutrient removal, improving sludge handling, and reducing odors. The following paragraphs explore biochar's applications in adsorption, process enhancement, and sludge management, highlighting mechanisms, benefits, limitations and current research insights. Biochar excels in the adsorption of diverse water pollutants including heavy metals, dyes, and organic compounds due to its large surface area, porous texture, and reactive surface sites. Through mechanisms such as ion exchange, complexation, precipitation, and electrostatic attraction, biochar can capture contaminants like lead, cadmium, azo dyes, and phenolic compounds. A recent review affirmed that both raw and modified biochars can achieve high-efficiency removal of heavy metals, and that modifications such as oxidation or functionalization to enhance adsorption capacity and selectivity. While biochar's adsorption process is effective and low-cost, factors like pyrolysis temperature, feedstock type, solution pH, and competing ions significantly affect performance [13, 102]. Thus, optimizing biochar production methods is crucial for tailoring adsorbents to specific water treatment needs.

Beyond adsorption, biochar can enhance existing treatment technologies by integrating with biological and chemical systems. For instance, adding biochar to biological filters or constructed wetlands can improve nitrogen and phosphorus removal via microbial biofilm support and nutrient sorption [13]. Biochar amended media have shown elevated denitrification rates and phosphorus immobilization due to increased microbial colonization and nutrient adsorption sites [15, 18]. Hybrid systems which biochar enriched with microbial communities that combine physical, chemical, and biological mechanisms to significantly improve effluent quality. Challenges include ensuring consistent biochar quality and avoiding nutrient leaching, which necessitates careful system design and monitoring during implementation.

Biochar also offers practical benefits in sludge handling and odor mitigation, making it a valuable addition to sustainable water treatment systems. When incorporated into wastewater sludge, biochar facilitates flocculation and enhances dewatering efficiency due to its high surface area, porosity, and favorable physicochemical properties. This helps reduce sludge volume, water content, and the associated costs of sludge management. Additionally, biochar demonstrates a strong capacity to adsorb odor-causing volatile organic compounds (VOCs) such as hydrogen sulfide and amines, thereby minimizing unpleasant smells during sludge treatment and storage phases. These attributes not only improve the operational performance of treatment facilities but also contribute to reduced greenhouse gas emissions, particularly methane, from anaerobic sludge decomposition. However, some challenges remain. The effectiveness of biochar depends greatly on its feedstock and pyrolysis conditions, which affect its surface chemistry and adsorption characteristics. Moreover, biochar-laden sludge may accumulate concentrated pollutants, raising concerns over safe disposal or potential leaching. Table 4.8 below outlines the advanced applications of bio-based materials in water treatment, including the use

Table 4.8 Advanced applications of bio-based materials in water treatment

Approach	Description	Advantages	Disadvantages
Eco-friendly coagulants	Cellulose-based or plant-derived coagulants replace traditional aluminum/iron salts to aid in particle aggregation and sedimentation	Biodegradable, non-toxic, renewable, lower sludge production	May have lower coagulation efficiency compared to chemical coagulants Performance may vary
Enhanced adsorption	Functionalized cellulose (e.g., aminated, sulfonated) improves heavy metal ion adsorption from wastewater	High affinity for target pollutants, surface tunability for specificity	Requires chemical modification May increase cost and complexity
IoT integration	Bio-based filters embedded with IoT sensors for real-time contaminant monitoring and performance tracking	Enables smart control, predictive maintenance, improved efficiency	Requires investment in sensors and connectivity Data accuracy may vary with environment
Biochar in sludge management	Biochar added to sludge improves dewatering and flocculation Also adsorbs odor-causing compounds like hydrogen sulfide	Reduces sludge volume, mitigates odor, enhances composting potential	Performance varies with feedstock and pyrolysis Disposal of spent biochar requires attention
Circular economy via bio-waste	Conversion of agricultural or food waste into biochar or bio-materials for filtration, adsorption and coagulant applications	Low-cost, reduces landfill use, promotes sustainability.	Feedstock variability affects performance Requires appropriate processing infrastructure

of cellulose-based eco-coagulants, IoT integration for real-time monitoring, and waste-to-resource conversion approaches like biochar. Each method is evaluated based on its operational description, advantages, and potential drawbacks to offer a comprehensive perspective on innovative and sustainable treatment technologies.

4.4.3 Soil Remediation

The remediation of contaminated soils has increasingly turned to bio-based materials such as biochar, carbon nanotubes, graphene, and modified geotextiles to provide sustainable, eco-friendly solutions. These materials serve multiple functions, including adsorption of pollutants, stabilization of soil structure, facilitation of phytoremediation and microbial activity, and enhancement of composting processes. Their versatility and renewable origins make them attractive alternatives to traditional chemical treatments. The following sections detail how these materials contribute to soil clean-up efforts, supported by recent scientific literature.

Biochar a carbon-rich byproduct of biomass pyrolysis has been widely adopted for its high adsorption capacity for heavy metals, pesticides, and organic pollutants.

Its porous structure, large surface area, and functional groups (e.g., carboxyl, hydroxyl) facilitate pollutant binding via ion exchange, complexation, and physical adsorption [61]. Studies report that optimized biochar additions can remove up to 80% of heavy metals from contaminated soils [63]. Furthermore, engineered biochars modified by oxidation, doping or activation to enhance selectivity and adsorption performance, particularly for heavy metals like lead and cadmium [92]. While effective and cost-efficient, these benefits depend on feedstock and pyrolysis conditions, and the risk of re-desorption under changing environmental conditions persists. Biochar and related materials can significantly enhance phytoremediation a green remediation strategy by immobilizing soil contaminants and improving plant viability. In one study, biochar-amended cadmium-contaminated soils saw increased biomass in grasses (*Paspalum vaginatum* and *Pennisetum alopecuroides*), reflecting reduced metal toxicity and improved metal uptake [51]. Such enhancements improve both phytostability and phytoextraction by hyperaccumulator plants. Additionally, biochar can serve as a slow-release matrix for nutrient amendments or microbial inocula, fostering rhizosphere health and remediation efficacy.

Biochar also functions as a scaffold for microbial communities to degrade organic contaminants. Its porous structure provides shelter for microbes, improving stability and activity of biodegrading strains. Biochar-microbe combinations have accelerated the degradation of chlorinated hydrocarbons and other persistent pollutants in soil systems [65]. Such synergistic action combines adsorption, microbial degradation, and nutrient cycling. However, compatibility between biochar properties and microbial communities is essential; poorly matched combinations may inhibit microbial activity or compromise treatment outcomes. Composting contaminated soil using bio-based materials such as wood chips, manure, and plant residues provides a low-cost method for degrading organic pollutants while improving soil quality. The composting process promotes aerobic microbial activity, transforming organic toxins into CO_2, water, and biomass [114]. Biochar added to compost enhances nutrient retention, odor control, and pollutant adsorption, resulting in stable organic amendments suitable for soil restoration. Nevertheless, the process requires careful monitoring of temperature, moisture, and contaminant concentrations to ensure complete pathogen and toxin removal.

In addition to biochar, carbon-based nanomaterials such as carbon nanotubes (CNTs), graphene oxide (GO), and nanofibers have shown exceptional promise in soil remediation. Their nanoscale size and high surface area enable rapid adsorption of heavy metals and organic pollutants through mechanisms such as π-π interactions, electrostatic attraction, and redox reactions. For instance, GO has been effective in immobilizing arsenic, chromium, and organic dyes in soil matrices while maintaining low toxicity to soil microbiota [76]. Functionalized CNTs have demonstrated high affinity for lead and phenolic compounds, enhancing the removal efficiency even at low concentrations [35]. However, environmental persistence and potential ecotoxicity of these materials remain key concerns, prompting the development of biodegradable carbon nanomaterials designed to combine high reactivity with reduced environmental risk [49]. These nanomaterials also enhance electrokinetic remediation techniques, which use low-voltage electric fields to mobilize contaminants, thereby accelerating the clean-up of heterogeneous and low-permeability soils [101].

Another emerging approach involves the use of modified geotextiles made from natural fibers such as jute, flax, or coir, functionalized with nanoparticles or polymer coatings to increase reactivity and durability. These geotextiles act as passive barriers or reactive liners, intercepting pollutants and reducing their mobility while enhancing vegetation support [40]. For example, geotextiles infused with zeolite or biochar have demonstrated capacity to adsorb ammonium, nitrate, phosphate, and heavy metals from leachate and stormwater runoff, limiting groundwater contamination [68]. Additionally, bio-enhanced geotextiles serve as substrates for beneficial microbial consortia, facilitating in situ biodegradation of petroleum hydrocarbons and agrochemicals [44]. Their use in phytoremediation settings not only supports plant anchorage but also promotes synergistic contaminant uptake, erosion control, and long-term soil rehabilitation. These systems provide cost-effective and biodegradable alternatives to synthetic geomembranes, aligning well with sustainable development goals and circular bioeconomy principles.

4.4.4 Plastic and Microplastic Pollution Mitigation

The escalating problem of plastic pollution, particularly microplastics has prompted urgent exploration of eco-friendly alternatives. Bio-based materials and technologies offer promising means to both replace traditional plastics and remediate existing plastic contaminants in the environment. Derived from renewable sources like plants, algae, and microbial biomass, these materials not only reduce reliance on fossil fuels but also promote biodegradability and circularity. Additionally, innovative bio-based removal strategies including phytoremediation and microbial degradation hold potential for extracting microplastics from terrestrial and aquatic ecosystems. The following sections explore these solutions in depth, emphasizing recent advancements and challenges in both material substitution and pollutant remediation.

Bio-based polymers such as polylactic acid (PLA) and polyhydroxyalkanoates (PHA) provide compostable alternatives to conventional plastics, particularly in packaging applications. PLA synthesized from renewable resources like corn starch is increasingly used for disposable cups, food containers, and agricultural mulch films due to its biodegradability under industrial composting conditions. Similarly, PHAs produced by microbial fermentation degrade into CO_2 and water, offering sustainable life cycles for disposable items. Biodegradable microbeads crafted from these polymers are now replacing microplastics in personal care products, substantially lowering their environmental footprint. Furthermore, biodegradable fishing gear made from such polymers minimizes ocean entanglement risks and ingestion hazards. Despite these benefits, commercial adoption faces challenges, including higher material costs, variable mechanical and thermal properties, and the need for industrial composting infrastructure.

Emerging methods such as phytoremediation and microbial degradation are being evaluated for their capacity to remove microplastics from the environment. Aquatic plants like *Eichhornia crassipes* (water hyacinth) have demonstrated significant uptake of polystyrene microplastics (55–70% removal within 48 h), indicating their potential as passive biofilters [111]. However, while these plants effectively intercept microplastics, evidence of their biodegradation within plant tissues remains limited [95]. On the microbial front, recent discoveries include bacteria from Antarctic soils capable of degrading polypropylene microplastics, achieving up to 23% weight loss in 20 days [28]. These findings suggest novel biocatalytic pathways for long-term remediation. Nevertheless, practical scalability, pollutant recovery, and ecological risks require comprehensive investigation before field implementation.

A holistic approach to reducing plastic pollution also involves improving recycling infrastructure, raising public awareness, and enacting supportive policies. Refinement of recycling technologies to include bio-based plastics such as PLA and PHA. It can help establish circular material flows and minimize landfill contributions. Regulatory measures aimed at banning single-use plastics, alongside educational campaigns, can drive consumer behavior toward sustainable alternatives. Furthermore, integrating bio-based materials into waste management through composting, industrial-scale biodegradable plastic processing, and microplastic capture within recycling plants which helps prevent environmental release. However, these strategies depend on harmonized legal frameworks, robust composting systems, and transparent labeling to avoid unintended persistence of bio-based residues in nature.

Bio-based solutions represent a multifaceted strategy against plastic and microplastic pollution, combining sustainable material substitution with innovative environmental remediation methods. While materials like PLA and PHA offer promising replacements for conventional plastics, complete mitigation also requires effective systems for waste collection, composting, and recycling. Meanwhile, phytoremediation and microbial degradation are emerging as novel, complementary tools for removing existing microplastic contamination. A successful transition to a circular, low-impact plastic economy will depend on interdisciplinary research, scalable biotechnologies, policy support, and public engagement. Together, these integrated strategies can substantially reduce plastic pollution and restore environmental integrity over time.

4.5 Summary

The utilization of micro- and nano-scale bio-based materials in environmental pollution control presents a transformative approach rooted in sustainability, renewability, and ecological compatibility. These materials derived from cellulose, chitosan, lignin, bacterial biopolymers, and agro-waste demonstrate exceptional capabilities in removing a broad spectrum of pollutants such as heavy metals, dyes, pharmaceutical residues, volatile organic compounds (VOCs), and microplastics. Their high surface area, tailored surface functionality, and innate biodegradability make them especially suited for water filtration, biosorption, catalysis, and pollutant degradation processes. In wastewater treatment, cellulose nanocrystals, chitosan nanoparticles, and bacterial cellulose exhibit remarkable adsorption and antimicrobial activities, often outperforming synthetic alternatives with significantly lower environmental footprints. Likewise, in air purification, biochar, lignin-based filters, and photocatalytic nanomaterials derived from bio-resources have been effectively employed for VOC degradation, particulate matter capture, and gaseous contaminant removal. These mechanisms collectively address the limitations of conventional methods by reducing energy input, toxic by-products, and secondary pollution, while supporting circular bioeconomy principles. In parallel, bio-based materials have shown considerable promise in soil remediation and plastic waste management. Biochar and modified biomass derivatives facilitate the immobilization and degradation of toxic compounds in soil while simultaneously enhancing soil structure and fertility. Additionally, bio-based alternatives to plastics such as polylactic acid (PLA), polyhydroxyalkanoates (PHA) and algae-based bioplastics are contributing to reducing microplastic contamination and fossil-based plastic dependency. The convergence of bio-based materials with modern innovations such as microbial fuel cells, membrane bioreactors, enzymatic catalysis, and IoT-based monitoring has enabled the creation of multifunctional treatment systems that are adaptive, scalable, and efficient. While current limitations include challenges in mass production, long-term material stability, and variability in environmental performance, these are steadily being addressed through advances in green synthesis, functionalization techniques, and regulatory support. Overall, the integration of micro and nano bio-based materials into pollution mitigation frameworks is paving the way for cleaner, safer, and more sustainable ecosystems across global industrial and municipal infrastructures.

References

1. Abdelhamid HN, Mathew AP (2021) Cellulose-based materials for water remediation: adsorption, catalysis, and antifouling. Front Chem Eng 3(December 2021):1–23. https://doi.org/10.3389/fceng.2021.790314
2. Abdel-Raouf MES, Farag RK, Farag AA, Keshawy M, Abdel-Aziz A, Hasan A (2023) Chitosan-based architectures as an effective approach for the removal of some toxic species from aqueous media. ACS Omega 8(11):10086–10099. https://doi.org/10.1021/acsomega.2c07264
3. Abdiyev K, Azat S, Kuldeyev E, Ybyraiymkul D, Kabdrakhmanova S, Berndtsson R, Khalkhabai B, Kabdrakhmanova A, Sultakhan S (2023) Review of slow sand filtration for raw water treatment with potential application in less-developed countries. Water (Switzerland) 15(11). https://doi.org/10.3390/w15112007
4. Ahuchaogu AA, Chukwu OJ, Obike AI, Igara CE (2018) Reverse osmosis technology, its applications and nano-enabled membrane. Int J Adv Res Chem Sci 5(2):20–26. https://doi.org/10.20431/2349-0403.0502005
5. Al-Gethami W, Qamar MA, Shariq M, Alaghaz ANMA, Farhan A, Areshi AA, Alnasir MH (2024) Emerging environmentally friendly bio-based nanocomposites for the efficient removal of dyes and micropollutants from wastewater by adsorption: a comprehensive review. RSC Adv 14(4):2804–2834. https://doi.org/10.1039/d3ra06501d

6. Alsawy T, Rashad E, El-Qelish M, Mohammed RH (2022) A comprehensive review on the chemical regeneration of biochar adsorbent for sustainable wastewater treatment. npj Clean Water 5:Article 29. https://doi.org/10.1038/s41545-022-00172-3
7. Antony Jose S, Cowan N, Davidson M, Godina G, Smith I, Xin J, Menezes PL (2025) A comprehensive review on cellulose nanofibers, nanomaterials, and composites: manufacturing, properties, and applications. Nano 15(5):1–33. https://doi.org/10.3390/nano15050356
8. Asghar N, Hussain A, Nguyen DA, Ali S, Hussain I, Junejo A, Ali A (2024) Advancement in nanomaterials for environmental pollutants remediation: a systematic review on bibliometrics analysis, material types, synthesis pathways, and related mechanisms. J Nanobiotechnol 22(1). https://doi.org/10.1186/s12951-023-02151-3
9. Ashraf A, Dutta J, Farooq A, Rafatullah M, Pal K, Kyzas GZ (2024) Chitosan-based materials for heavy metal adsorption: recent advancements, challenges and limitations. J Mol Struct 1309:138225. https://doi.org/10.1016/j.molstruc.2024.138225
10. Ayappan C, Xing R, Zhang X, Luo D, Liu S, Fujishima A (2024) TiO2-based photocatalysts for emerging gaseous pollutants removal: from photocatalysts to reactors design. Coord Chem Rev 515(May):215960. https://doi.org/10.1016/j.ccr.2024.215960
11. Ayilara MS, Babalola OO (2023) Bioremediation of environmental wastes: the role of microorganisms. Front Agron 5(May):1–15. https://doi.org/10.3389/fagro.2023.1183691
12. Badawi AK, Salama RS, Mostafa MMM (2023) Natural-based coagulants/flocculants as sustainable market-valued products for industrial wastewater treatment: a review of recent developments. RSC Adv 13(28):19335–19355. https://doi.org/10.1039/d3ra01999c
13. Bayar J, Ali N, Dong Y, Ahmad U, Anjum MM, Khan GR, Zaib M, Jalal A, Ali R, Ali L (2024) Biochar-based adsorption for heavy metal removal in water: a sustainable and cost-effective approach. Environ Geochem Health 46(11):1–25. https://doi.org/10.1007/s10653-024-02214-w
14. Bhattarai G, Sim HJ, So HS, Lee JC, Kook SH (2023) Exposure of newborns to atmospherically relevant artificial particulate matter induces hematopoietic stem cell senescence. J Hazard Mater 452(March):131293. https://doi.org/10.1016/j.jhazmat.2023.131293
15. Bocşa M, Pintea S, Lung I, Opriş O, Stegarescu A, Humayun M, Bououdina M, Soran ML, Bellucci S (2023) Biochar-based adsorbents for pesticides, drugs, phosphorus, and heavy metal removal from polluted water. Separations 10(10):1–27. https://doi.org/10.3390/separations10100533
16. Chinthalapudi N, Kommaraju VVD, Kannan MK, Nalluri CB, Varanasi S (2021) Composites of cellulose nanofibers and silver nanoparticles for malachite green dye removal from water. Carbohydr Polym Technol Appl 2:100098. https://doi.org/10.1016/j.carpta.2021.100098
17. Dong X, Bäcker LE, Rahmatullah M, Schunk D, Lens G, Meckenstock RU (2019) Quantification of microbial degradation activities in biological activated carbon filters by reverse stable isotope labelling. AMB Express 9(1). https://doi.org/10.1186/s13568-019-0827-0
18. Duwiejuah AB, Abubakari AH, Quainoo AK, Amadu Y (2020) Review of biochar properties and remediation of metal pollution of water and soil. J Health Pollut 10(27). https://doi.org/10.5696/2156-9614-10.27.200902
19. Elizabeth A, Prabu HJ, Sahayaraj AF, Johnson I (2025) A review on cellulose-based nanocomposites for the removal of contaminants for the treatment of wastewater. Polym Bull (Issue 0123456789). Springer Berlin Heidelberg. https://doi.org/10.1007/s00289-025-05826-z
20. Enyoh CE, Devi A, Maduka TO, Tyagi L, Rana S, Akuwudike IS, Wang Q (2025) A review of materials for the removal of micro- and nanoplastics from different environments. Micro 5(2):1–26
21. Finore I, Feola A, Russo L, Cattaneo A, Di Donato P, Nicolaus B, Poli A, Romano I (2023) Thermophilic bacteria and their thermozymes in composting processes: a review. Chem Biol Technol Agric 10(1):1–22. https://doi.org/10.1186/s40538-023-00381-z
22. Gałązka A, Jankiewicz U (2022) Endocrine disrupting compounds (Nonylphenol and bisphenol A)-sources, harmfulness and laccase-assisted degradation in the aquatic environment. Microorganisms 10(11). https://doi.org/10.3390/microorganisms10112236

23. Garud RM, Kore SV, Kore VS, Kulkarni GS (2011) A short review on process and applications of reverse osmosis. Univers J Environ Res Technol 1(3):233–238
24. Ge JC, Yoon SK, Choi NJ (2018) Application of Fly ash as an adsorbent for removal of air and water pollutants. Appl Sci (Switzerland) 8(7). https://doi.org/10.3390/app8071116
25. Gonçalves EV, De Jesús E, Roqueme B, Martins MD (2024) Explorando o potencial dos fungos para a melhoria do meio ambiente. 56(1):1–8
26. Guchi E (2015) Review on slow sand filtration in removing microbial contamination and particles from drinking water. Am J Food Nutr 3(2):47–55. https://doi.org/10.12691/ajfn-3-2-3
27. Guerrero-granados KF, Bellandi G, Mante J, Rehman U, Boergers A (2025) Variability by the assessment of bromate formation in wastewater treatment. Chem Eng J 519(June):164904. https://doi.org/10.1016/j.cej.2025.164904
28. Habib S, Iruthayam A, Shukor MYA, Alias SA, Smykla J, Yasid NA (2020) Biodeterioration of untreated polypropylene microplastic particles by antarctic bacteria. Polymers 12(11):2616. https://doi.org/10.3390/polym12112616
29. Hachhach M, Bayou S, El Kasmi A, Saidi MZ, Akram H, Hanafi M, Achak O, El Moujahid C, Chafik T (2025) Towards sustainable scaling-up of nanomaterials fabrication: current situation, challenges, and future perspectives. Eng 6(7):149. https://doi.org/10.3390/ENG6070149
30. Helte E, Söderlund F, Säve-Söderbergh M, Larsson SC, Åkesson A (2025) Exposure to drinking water trihalomethanes and risk of cancer: a systematic review of the epidemiologic evidence and dose-response meta-analysis. Environ Health Perspect 133(1):16001. https://doi.org/10.1289/EHP14505
31. Henderson AD, Demond AH (2007) Long-term performance of zero-valent iron permeable reactive barriers: a critical review. Environ Eng Sci 24(4):401–423. https://doi.org/10.1089/ees.2006.0071
32. Hu K, Torán J, López-García E, Barbieri MV, Postigo C, de Alda ML, Caminal G, Sarrà M, Blánquez P (2020) Fungal bioremediation of diuron-contaminated waters: evaluation of its degradation and the effect of amendable factors on its removal in a trickle-bed reactor under non-sterile conditions. Sci Total Environ 743. https://doi.org/10.1016/j.scitotenv.2020.140628
33. Huang Y, Lu K, Guo J, Wang Y, Zhao C, Zhai Y (2020) Study on ventilation performance of lateral exhaust hood under the influence of two high-temperature buoyant jets. Build Environ 177(April). https://doi.org/10.1016/j.buildenv.2020.106849
34. Hudock DM (2007) Biofiltration as a viable alternative for air pollution control at department of defense surface coating facilities. https://scholar.afit.edu/etd/3032/
35. Hussain B, Umer MJ, Li J, Ma Y, Abbas Y, Ashraf MN, Tahir N, Ullah A, Gogoi N, Farooq M (2021) Strategies for reducing cadmium accumulation in rice grains. J Clean Prod 286:Article 125557. https://doi.org/10.1016/j.jclepro.2020.125557
36. Isha, Ali S, Khalid A, Naseer IA, Raza H, Chang YC (2024) Bioremediation of smog: current trends and future perspectives. PRO 12(10). https://doi.org/10.3390/pr12102266
37. Jahan BN, Li L, Pagilla KR (2021) Fate and reduction of bromate formed in advanced water treatment ozonation systems: a critical review. Chemosphere 266:128964. https://doi.org/10.1016/j.chemosphere.2020.128964
38. Jalili P, Ala A, Nazari P, Jalili B, Ganji DD (2024) A comprehensive review of microbial fuel cells considering materials, methods, structures, and microorganisms. Heliyon 10(3):e25439. https://doi.org/10.1016/j.heliyon.2024.e25439
39. Jin P, Jin X, Wang X, Feng Y, Wang XC (2013) Biological activated carbon treatment process for advanced water and wastewater treatment. In: Biomass now – cultivation and utilization. https://doi.org/10.5772/52021
40. Kadam P, Dwivedi P, Cook S, Graham B, Perry A, Megalos M, Height T (2024) Qualifying gaps between scientific prescriptions and social perceptions about the role of forest management practices for increasing carbon stocks on family forestlands in the US South. Trees, Forests and People 16:Article 100545. https://doi.org/10.1016/j.tfp.2024.100545
41. Kanissery RG, Sims GK (2011) Biostimulation for the enhanced degradation of herbicides in soil. Appl Environ Soil Sci 2011:1–10. https://doi.org/10.1155/2011/843450

42. Kato S, Kansha Y (2024) Comprehensive review of industrial wastewater treatment techniques. Environ Sci Pollut Res 31(39). https://doi.org/10.1007/s11356-024-34584-0. Springer Berlin Heidelberg
43. Khedulkar AP, Bobade RG, Doong R, Pandit B, Nguyen MK, Ambare RC, Hoang TD, Kumar KJ (2025) Bio-based nanomaterials as effective, friendly solutions and their applications for protecting water, soil, and air. Mat Today Chem 46(April):102688. https://doi.org/10.1016/j.mtchem.2025.102688
44. Kim M-S, Lee S-H, Kim J-G (2021) Evaluation of factors affecting arsenic uptake by Brassica juncea in alkali soil after biochar application using partial least squares path modelling (PLS-PM). Chemosphere 275:130095. https://doi.org/10.1016/j.chemosphere.2021.130095
45. Kumar S, Dubey N, Kumar V, Choi I, Jeon J, Kim M (2024) Combating micro/nano plastic pollution with bioplastic: sustainable food packaging, challenges, and future perspectives. Environ Pollut 363. https://doi.org/10.1016/j.envpol.2024.125077
46. Kuppan N, Padman M, Mahadeva M, Srinivasan S, Devarajan R (2024) A comprehensive review of sustainable bioremediation techniques: eco friendly solutions for waste and pollution management. Waste Manag Bull 2(3):154–171. https://doi.org/10.1016/j.wmb.2024.07.005
47. Kvaternik M, Protić M (2022) Disinfection of drinking water and trihalomethanes. Qual Life (Banja Luka) APEIRON 22(1). https://doi.org/10.7251/qol2201022k
48. LaPara TM, Wilkinson KH, Strait JM, Hozalski RM, Sadowksy MJ, Hamilton MJ (2015) The bacterial communities of full-scale biologically active, granular activated carbon filters are stable and diverse and potentially contain novel ammonia-oxidizing microorganisms. Appl Environ Microbiol 81(19):6864–6872. https://doi.org/10.1128/AEM.01692-15
49. Li C, Zhou K, Qin W, Tian C, Qi M, Yan X, Han W (2019) A review on heavy metals contamination in soil: effects, sources, and remediation techniques. Soil Sediment Contam Int J 28(4):380–394. https://doi.org/10.1080/15320383.2019.1592108
50. Li F, Liu G, Liu F, Wu J, Yang S (2023) Synergetic effect of CQD and oxygen vacancy to TiO2 photocatalyst for boosting visible photocatalytic NO removal. J Hazard Mater 452(March):131237. https://doi.org/10.1016/j.jhazmat.2023.131237
51. Liang J, Chang J, Xie J, Yang L, Sheteiwy MS, Moustafa A-RA, Zaghloul MS, Ren H (2023) Microorganisms and biochar improve the remediation efficiency of Paspalum vaginatum and Pennisetum alopecuroides on cadmium-contaminated soil. Toxics 11(7):582. https://doi.org/10.3390/toxics11070582
52. Lin Y, Cao Y, Yao Q, Chai OJH, Xie J (2020) Engineering Noble metal nanomaterials for pollutant decomposition. Ind Eng Chem Res 59(47):20561–20581. https://doi.org/10.1021/acs.iecr.0c04258
53. Liu Y, Wang H, Cui Y, Chen N (2023) Removal of copper ions from wastewater: a review. Int J Environ Res Public Health 20(5). https://doi.org/10.3390/ijerph20053885
54. Liu Y, Dai X, Li J, Cheng S, Zhang J, Ma Y (2024) Recent progress in TiO2-biochar-based photocatalysts for water contaminants treatment: strategies to improve photocatalytic performance. RSC Adv 14(1):478–491. https://doi.org/10.1039/d3ra06910a
55. Ma Q, Li Y, Xue J, Cheng D, Li Z (2022) Effects of turning frequency on ammonia emission during the composting of chicken manure and soybean straw. Molecules 27(2):1–21. https://doi.org/10.3390/molecules27020472
56. Madsen AM, Frederiksen MW, Jacobsen MH, Tendal K (2020) Towards a risk evaluation of workers' exposure to handborne and airborne microbial species as exemplified with waste collection workers. Environ Res 183(November 2019):109177. https://doi.org/10.1016/j.envres.2020.109177
57. Mahurede TP, Chihobo CH, Utete B, Taru P (2023) A review of microbial fuel cell prototypes, their efficacy in wastewater treatment and the contextual situation for Zimbabwe. Fuel Commun 17(April):100094. https://doi.org/10.1016/j.jfueco.2023.100094
58. Maiyo JK, Dasika S, Jafvert CT (2023) Slow sand filters for the 21st century: a review. Int J Environ Res Public Health 20(2):1–25. https://doi.org/10.3390/ijerph20021019

59. Mandal RR, Bashir Z, Raj D (2025) Microbe-assisted phytoremediation for sustainable management of heavy metal in wastewater – a green approach to escalate the remediation of heavy metals. J Environ Manage 375(August 2024):124199. https://doi.org/10.1016/j.jenvman.2025.124199
60. Meftah S, Meftah K, Drissi M, Radah I, Malous K, Amahrous A, Chahid A, Tamri T, Rayyad A, Darkaoui B, Hanine S, El-Hassan O, Bouyazza L (2025) Heavy metal polluted water: effects and sustainable treatment solutions using bio-adsorbents aligned with the SDGs. Discov Sustain 6(1). https://doi.org/10.1007/s43621-025-00895-6
61. Meng F, Wang Y, Wei Y (2025) Advancements in biochar for soil remediation of heavy metals and/or organic pollutants. Materials 18(7). https://doi.org/10.3390/ma18071524
62. Mustereţ CP, Morosanu I, Ciobanu R, Plavan O, Gherghel A, Al-Refai M, Roman I, Teodosiu C (2021) Assessment of coagulation-flocculation process efficiency for the natural organic matter removal in drinking water treatment. Water (Switzerland) 13(21). https://doi.org/10.3390/w13213073
63. Nguyen TB, Sherpa K, Bui XT, Nguyen VT, Vo TDH, Ho HTT, Chen CW, Dong CD (2023) Biochar for soil remediation: a comprehensive review of current research on pollutant removal. Environ Pollut 337(June):122571. https://doi.org/10.1016/j.envpol.2023.122571
64. Nguyen, N. T. T., Nguyen, L. M., Nguyen, T. T. T., Nguyen, D. T. C., & Tran, T. Van. (2024). Synthesis strategies, regeneration, cost analysis, challenges and future prospects of bacterial cellulose-based aerogels for water treatment: a review. Chemosphere, 362(June 2023), 142654. https://doi.org/10.1016/j.chemosphere.2024.142654
65. Niu S, Li C, Gao S, Tian J, Zhang C, Li L, Huang Y, Lyu H (2024) Biochar, microbes, and biochar-microbe synergistic treatment of chlorinated hydrocarbons in groundwater: a review. Front Microbiol 15(July). https://doi.org/10.3389/fmicb.2024.1443682
66. Nizam NUM, Hanafiah MM, Woon KS (2021) A content review of life cycle assessment of nanomaterials: current practices, challenges, and future prospects. Nanomaterials 11(12):3324. https://doi.org/10.3390/NANO11123324
67. Nzila A, Razzak SA, Zhu J (2016) Bioaugmentation: an emerging strategy of industrial wastewater treatment for reuse and discharge. Int J Environ Res Public Health 13(9). https://doi.org/10.3390/ijerph13090846
68. Oluwasola HO, Asegbeloyin JN, Ochonogor AE, Ani JU, Ibeji CU, Oyeka EE (2019) Cadmium and lead adsorption capacities of Nigerian Ultisol soil of tropics. Orient J Chem 35(3):1004–1012. https://doi.org/10.13005/ojc/350312
69. Oshins C, Michel F, Louis P, Richard TL, Rynk R (2021) The composting process. In: The composting handbook: a how-to and why manual for farm, municipal, institutional and commercial composters. Elsevier Inc. https://doi.org/10.1016/B978-0-323-85602-7.00008-X
70. Pei X, Zhao M, Yan M, Luo S, Xu W, Liu S, Xing W, Shao R, Xu Z (2023) Sustainable green reinforcement engineering: high-performance GTRM reinforced composites with interfacial gradient structure. J Clean Prod 415(May):137813. https://doi.org/10.1016/j.jclepro.2023.137813
71. Płotka-Wasylka J, Chabowska A, Pantanit S, Bunkoed O, Fares MY, Sajid M, Lambropoulou D, Kurowska-Susdorf A, Jatkowska N (2024) Natural/bio-based sorbents as greener extractive materials for endocrine disrupting compounds in samples of different matrix composition. TrAC Trends Anal Chem 176(April). https://doi.org/10.1016/j.trac.2024.117773
72. Qasem NAA, Mohammed RH, Lawal DU (2021) Removal of heavy metal ions from wastewater: a comprehensive and critical review. NPJ Clean Water 4(1). https://doi.org/10.1038/s41545-021-00127-0
73. Ravindran A, Panneerselvan L, Palanisami T, Sathish CI (2025) Eco-friendly biopolymers and composites: a sustainable development of adsorbents for the removal of pollutants from wastewater. NPJ Mater Sustain. https://doi.org/10.1038/s44296-025-00057-9

74. Reda S, Safwat SM, Elawwad A, Abdel-Halim H (2025) Dual-chamber microbial fuel cells for enhanced bioelectricity generation and chemical oxygen demand removal from sludges: a comparative study of different real sludge generated from a wastewater treatment plant. Biomass Bioenergy 201(June). https://doi.org/10.1016/j.biombioe.2025.108063
75. Ren L, Peng C, Hu X, Han Y, Huang H (2020) Microbial production of vitamin K2: current status and future prospects. Biotechnol Adv 39(30):107453. https://doi.org/10.1016/j.biotechadv.2019.107453
76. Rizwan M, Murtaza G, Zulfiqar F, Moosa A, Iqbal R, Ahmed Z, Irshad S, Khan I, Li T, Chen J, Zhang M, Siddique KHM, Leng L, Li H (2023) Sustainable manufacture and application of biochar to improve soil properties and remediate soil contaminated with organic impurities: a systematic review. Front Environ Sci 11:1277240. https://doi.org/10.3389/fenvs.2023.1277240
77. Romantschuk M, Lahti-Leikas K, Kontro M, Galitskaya P, Talvenmäki H, Simpanen S, Allen JA, Sinkkonen A (2023) Bioremediation of contaminated soil and groundwater by in situ biostimulation. Front Microbiol 14(November):1–17. https://doi.org/10.3389/fmicb.2023.1258148
78. Roy M, Giri AK, Dutta S, Mukherjee P (2015) Integrated phytobial remediation for sustainable management of arsenic in soil and water. Environ Int 75:180–198. https://doi.org/10.1016/j.envint.2014.11.010
79. Russo F, Castro-Muñoz R, Santoro S, Galiano F, Figoli A (2022) A review on electrospun membranes for potential air filtration application. J Environ Chem Eng 10(5). https://doi.org/10.1016/j.jece.2022.108452
80. Saka D, Adu-Gyamfi J, Skrzypek G, Antwi EO, Heng L, Torres- Martínez JA (2023) Disentangling nitrate pollution sources and apportionment in a tropical agricultural ecosystem using a multi-stable isotope model. Environ Pollut 328(April):121589. https://doi.org/10.1016/j.envpol.2023.121589
81. Saleem H, Zaidi SJ, Ismail AF, Goh PS (2022) Advances of nanomaterials for air pollution remediation and their impacts on the environment. Chemosphere 287(July 2021):132083. https://doi.org/10.1016/j.chemosphere.2021.132083
82. Saleh TA, Mustaqeem M, Khaled M (2022) Water treatment technologies in removing heavy metal ions from wastewater: a review. Environ Nanotechnol Monit Manag 17(November 2020):100617. https://doi.org/10.1016/j.enmm.2021.100617
83. Samavati Z, Samavati A, Goh PS, Fauzi Ismail A, Sohaimi Abdullah M (2023) A comprehensive review of recent advances in nanofiltration membranes for heavy metal removal from wastewater. Chem Eng Res Des 189:530–571. https://doi.org/10.1016/j.cherd.2022.11.042
84. Saraswat P, Singh S, Prasad M, Misra R, Rajput VD, Ranjan R (2023) Applications of bio-based nanomaterials in environment and agriculture: a review on recent progresses. Hybr Adv 4(June):100097. https://doi.org/10.1016/j.hybadv.2023.100097
85. Scaffaro R, Citarrella MC (2023) Nanofibrous polymeric membranes for air filtration application: a review of progress after the COVID-19 pandemic. Macromol Mater Eng 308(9):1–18. https://doi.org/10.1002/mame.202300072
86. Serbent MP, Magario I, Saux C (2024) Immobilizing white-rot fungi laccase: toward bio-derived supports as a circular economy approach in organochlorine removal. Biotechnol Bioeng 121(2):434–455. https://doi.org/10.1002/BIT.28591
87. Shamloo E, Sadeghi E, Soltani M, Mahmoudifar K, Taghizadeh M, Arab M, Abdi-Moghadam Z, Rezagholizade-Shirvan A, Shokri S (2025) Encapsulation of natural/bio-based materials as nano bio-adsorbents for removal of toxic pollutants from food products: a review. Carbohydr Polym Technol Appl 10(May):100870. https://doi.org/10.1016/j.carpta.2025.100870
88. Sharma K, Choudhary P, Majeed A, Guleria S, Kumar M, Rana AK, Rajauria G (2025) Cellulose based membranes, hydrogels and aerogels for water treatment application. Ind Crop Prod 225(October 2024):120474. https://doi.org/10.1016/j.indcrop.2025.120474
89. Sheoran K, Siwal SS, Kapoor D, Singh N, Saini AK, Alsanie WF, Thakur VK (2022) Air pollutants removal using biofiltration technique: a challenge at the Frontiers of sustainable environment. ACS Eng Au 2(5):378–396. https://doi.org/10.1021/acsengineeringau.2c00020

90. Shi S, Wang F, Hu Y, Zhou J, Zhang H, He C (2024) Effects of running time on biological activated carbon filters: water purification performance and microbial community evolution. Environ Sci Pollut Res 31(14):21509–21523. https://doi.org/10.1007/s11356-024-32421-y
91. Si R, Pu J, Luo H, Wu C, Duan G (2022) Nanocellulose-based adsorbents for heavy metal ion. Polymers 14(24):1–23. https://doi.org/10.3390/polym14245479
92. Subramanian P, Pandian K, Pakkiyam S, Dhanuskodi K v, Annamalai S, Padanillay Chidambaram P, Mustaffa MRAF (2024) Biochar for heavy metal cleanup in soil and water: a review. Biomass Convers Biorefin:11421–11441. https://doi.org/10.1007/s13399-024-05989-1
93. Sun Q, Hu X, He Q, Sun M, Dai L, Khan S, Zheng H (2022) Polydopamine-chitosan modified TiO2 nanoparticles for temperature-response removal of diclofenac sodium under visible light irradiation. J Taiwan Inst Chem Eng 131:104151. https://doi.org/10.1016/j.jtice.2021.11.018
94. Tahraoui H, Toumi S, Boudoukhani M, Touzout N, Sid ANEH, Amrane A, Belhadj AE, Hadjadj M, Laichi Y, Aboumustapha M, Kebir M, Bouguettoucha A, Chebli D, Assadi AA, Zhang J (2024) Evaluating the effectiveness of coagulation-flocculation treatment using aluminum sulfate on a polluted surface water source: a year-long study. Water (Switzerland) 16(3). https://doi.org/10.3390/w16030400
95. Tang KHD (2023) Phytoremediation of microplastics: a perspective on its practicality. Indus Domest Waste Manag 3(2):90–102. https://doi.org/10.53623/idwm.v3i2.291
96. Tayeh YA (2024) A comprehensive review of reverse osmosis desalination: technology, water sources, membrane processes, fouling, and cleaning. Desalin Water Treat 320(October). https://doi.org/10.1016/j.dwt.2024.100882
97. Thakur TK, Barya MP, Dutta J, Mukherjee P, Thakur A, Swamy SL, Anderson JT (2023) Integrated Phytobial remediation of dissolved pollutants from domestic wastewater through constructed wetlands: an interactive macrophyte-microbe-based green and low-cost decontamination technology with prospective resource recovery. Water (Switzerland) 15(22). https://doi.org/10.3390/w15223877
98. Truu J, Truu M, Espenberg M, Nõlvak H, Juhanson J (2015) Send orders for reprints to reprints@benthamscience.ae phytoremediation and plant-assisted bioremediation in soil and treatment wetlands: a review. Open Biotechnol J 9:85–92
99. Ullah MH, Rahman MJ (2024) Adsorptive removal of toxic heavy metals from wastewater using water hyacinth and its biochar: a review. Heliyon 10(17):e36869. https://doi.org/10.1016/J.HELIYON.2024.E36869/ASSET/AB34E97A-D44A-459E-88FE-1C371B4D79E3/MAIN.ASSETS/GR8.JPG
100. Van Hung N, Nguyet BTM, Nghi NH, Luon NM, Bich NN, Son LVT, Kien NT, Nhiem DN, Tuoi NT, Khieu DQ (2024) Visible light-driven photocatalytic degradation of doxycycline using ZnO/g-C3N4/biochar composite in aqueous solution. Environ Eng Res 29(4):0–3. https://doi.org/10.4491/eer.2023.596
101. Wang L, Rinklebe J, Tack F, Hou D (2021) A review of green remediation strategies for heavy metal contaminated soil. Soil Use Manag 37(4):936–963. https://doi.org/10.1111/sum.12717
102. Wang Y, Chen L, Zhu Y, Fang W, Tan Y, He Z, Liao H (2024) Research status, trends, and mechanisms of biochar adsorption for wastewater treatment: a scientometric review. Environ Sci Eur 36(1). https://doi.org/10.1186/s12302-024-00859-z
103. Wei Y, Chu R, Zhang Q, Usman M, Haider FU, Cai L (2022) Nano zero-valent iron loaded corn-straw biochar for efficient removal of hexavalent chromium: remediation performance and interfacial chemical behaviour. RSC Adv 12(41):26953–26965. https://doi.org/10.1039/d2ra04650d
104. Werner KA, Poehlein A, Schneider D, El-Said K, Wöhrmann M, Linkert I, Hübner T, Brüggemann N, Prost K, Daniel R, Grohmann E (2022) Thermophilic composting of human feces: development of bacterial community composition and antimicrobial resistance gene pool. Front Microbiol 13. https://doi.org/10.3389/fmicb.2022.824834

105. Wu J, Cao M, Tong D, Finkelstein Z, Hoek EMV (2021) A critical review of point-of-use drinking water treatment in the United States. NPJ Clean Water 4(1):1–25. https://doi.org/10.1038/s41545-021-00128-z
106. Xie SG, Shi DW, Wen DH, Wang R, Xi DL (2007) Control of bromate and THM precursors using ozonation combined system. Biomed Environ Sci 20(3):217–225
107. Xie Q, Guan Q, Li L, Pan X, Ho CL, Liu X, Hou S, Chen D (2021) Exposure of children and mothers to organophosphate esters: prediction by house dust and silicone wristbands. Environ Pollut 282:117011. https://doi.org/10.1016/j.envpol.2021.117011
108. Xie Y, Ye H, Wen Z, Dang Z, Lu G (2022) Sulfide-induced repartition of chromium associated with schwertmannite in acid mine drainage: impacts and mechanisms. Sci Total Environ 848:157863. https://doi.org/10.1016/j.scitotenv.2022.157863
109. Yaqoob AA, Khatoon A, Setapar SHM, Umar K, Parveen T, Ibrahim MNM, Ahmad A, Rafatullah M (2020) Outlook on the role of microbial fuel cells in remediation of environmental pollutants with electricity generation. Catalysts 10(8). https://doi.org/10.3390/catal10080819
110. Yin X, Wang F, Ge M, Zhang F, Liang G (2025a) Pseudomonas aeruginosa promoted microbial fuel cells for cytidine acid production wastewater treatment. Sci Rep 15(1):1–14. https://doi.org/10.1038/s41598-025-90361-0
111. Yin J, Zhu T, Li X, Wang F, Xu G (2025b) Phytoremediation of microplastics by water hyacinth. Environ Sci Ecotechnol 24:100540. https://doi.org/10.1016/j.ese.2025.100540
112. Yu G, Wang X, Liu J, Jiang P, You S, Ding N, Guo Q, Lin F (2021) Applications of nanomaterials for heavy metal removal from water and soil: a review. Sustainability (Switzerland) 13(2):1–14. https://doi.org/10.3390/su13020713
113. Zhao H, Ma F, Ren X, Zhao B, Jiang Y, Zhang J (2025) Green synthesis of nZVI-modified sludge biochar for Cr(VI) removal in water: fixed-bed experiments and artificial neural network model prediction. Water (Switzerland) 17(3):1–17. https://doi.org/10.3390/w17030341
114. Zheng X, Xu W, Dong J, Yang T, Shangguan Z, Qu J, Li X, Tan X (2022) The effects of biochar and its applications in the microbial remediation of contaminated soil: a review. J Hazard Mater 438(May):129557. https://doi.org/10.1016/j.jhazmat.2022.129557
115. Zhou Y, Li X (2022) Green synthesis of modified polyethylene packing supported tea polyphenols-NZVI for nitrate removal from wastewater: characterization and mechanisms. Sci Total Environ 806:150596. https://doi.org/10.1016/j.scitotenv.2021.150596
116. Zubair MM, Saleem H, Zaidi SJ (2023) Recent progress in reverse osmosis modeling: an overview. Desalination 564(May 2023):116705. https://doi.org/10.1016/j.desal.2023.116705

Chapter 5
Applications of Micro to Nano Bio-based Materials in Renewable Energy Systems

5.1 Overview of Renewable Energy Systems

Renewable energy systems harness naturally replenished resources to meet global energy demands sustainably. Among them, solar energy uses photonic radiation via photovoltaic cells or concentrated solar power for electricity and solar thermal applications. Wind energy captures kinetic energy of moving air and transforms it into electricity through turbine systems. Hydropower leverages the gravitational potential of flowing water most commonly via dams and reservoirs to generate electrical energy, providing both base-load and peak-load supply. Geothermal energy exploits heat stored beneath the Earth's surface through steam or hot fluids to produce electricity or direct-use thermal applications. Lastly, bioenergy derives usable energy from organic matter including agricultural residues, forestry waste, micro–/macro-algae, and waste biomass. Diverse conversion technologies combustion, anaerobic digestion/fermentation, and thermochemical routes enable biomass conversion into heat, power, or biofuels. Each source presents unique operational, infrastructural, or environmental considerations such as intermittency, regional suitability, or feedstock logistics in making portfolio diversification crucial for a sustainable energy transition.

Bioenergy plays a pivotal role in displacing fossil fuels, offering carbon-neutral or low-carbon alternatives through the growth and combustion of biomass. This energy stream supports waste valorization from agricultural residues, forestry by-products, energy crops, and municipal or industrial organic waste into valuable fuels and electricity. Thermochemical pathways such as pyrolysis, gasification and hydrothermal liquefaction (HTL) are fundamental, converting biomass into syngas, bio-oil, biochar or HTL-derived crude with high energy density. Pyrolysis, performed under inert conditions, yields bio-oil, biochar and non-condensable gases, with product ratios controlled by heating rate and temperature [8]. Gasification partially oxidizes biomass to produce syngas and heat. HTL uses pressurized hot water

M. I. Ahmad, A. K. H. P. S., *Environmental and Energy Technology*, Green Energy and Technology, https://doi.org/10.1007/978-3-032-10706-0_5

to process wet biomass into H-rich bio-oil with minimal drying needs, achieving energy recovery comparable to traditional fossil oil [63]. These approaches, grounded in lignocellulosic biochemistry, not only enhance resource circularity but also leverage existing infrastructure for biofuels significantly contributing to decarbonizing transportation, power and heating sectors.

Nanomaterials especially those derived from biomass exhibit exceptional surface area, tunable porosity, and catalytic effectiveness. They are instrumental in improving efficiency across energy systems. In energy storage, bio-based nanocellulose and carbon based nanomaterials are increasingly utilized in electrode materials for batteries and supercapacitors, greatly enhancing capacity, cycle stability and charge-discharge rates [57]. For biofuel production, carbon nanomaterials serve as high-surface-area catalysts or supports in thermochemical reactors, accelerating reaction rates and improving yields while showing enhanced resistance to deactivation [28, 58]. Additionally, they facilitate efficient biomass conversion and separation processes by enabling better heat distribution, catalysis, and product extraction. These advancements are in line with green chemistry ideals, offering biodegradable, renewable alternatives that contribute to lowering greenhouse gas emissions and potentially reducing life-cycle impacts.

Renewable energy systems are associated with lower life-cycle greenhouse gas emissions than fossil fuel systems; however, environmental trade-offs remain. Land-use change, water utilization, feedstock cultivation and infrastructure deployment can impact ecosystems if poorly managed. Life-cycle assessments (LCAs) are critical to quantitatively evaluate total impacts from raw material extraction and processing to end-of-life disposal or recycling. Moreover, nanomaterials introduce concerns regarding resource sourcing, potential ecotoxicity, production energy intensity, and end-of-life disposal. Understanding these dynamics is essential to design sustainable systems. Bio-based nanomaterials like cellulose or lignin-derived nanostructures offer enhanced biodegradability and circularity, mitigating environmental burdens when designed for recyclability or compostability. Through careful LCA-informed design, renewable energy technologies can achieve not only operational sustainability but also holistic environmental stewardship.

Bio-based materials which including cellulose, lignin, chitin and PHA offer sustainable alternatives to petroleum-derived components in energy applications. For battery and supercapacitor electrodes, nanocellulose aerogels, carbon nanofibers, and lignin-derived carbon have shown promise due to mechanical flexibility, porosity, and electrical conductivity. In catalysis, bio-sourced carbon supports and nanostructured lignin-based materials act as robust scaffolds for metal catalysts, contributing to efficient biomass conversion and hydrogen production. Emerging biohybrid photovoltaic systems integrating biological photosystems with semiconductors are pushing or approaching the limits of conventional solar-cell efficiencies. The utilization of such materials fosters circular economy principles: derived from renewable feedstocks, designed for recyclability or benign degradation, supplying high-performance functionality without compromising sustainability.

5.2 Role of Bio-based Materials in Energy Sustainability

Bio-based materials derived from agricultural residues and micro–/macro-algae serve as sustainable feedstocks for biofuel production, offering a renewable path to replace fossil fuels and mitigate carbon emissions. Agricultural residues such as straw and husk. Meanwhile, algal biomass contains high lignocellulosic or lipid content, which can be converted into biofuels via thermochemical methods such as pyrolysis, gasification, and hydrothermal liquefaction. Pyrolysis yields bio-oil and biochar under inert conditions while gasification produces syngas that can feed into Fischer-Tropsch synthesis. Hydrothermal liquefaction, in particular is effective for wet biomass, bypassing the need for energy-intensive drying. Bio-based nanomaterials such as lignin-derived carbon or biochar can function as catalysts or catalyst supports, improving reaction kinetics and bio-oil quality. This enables improved yields and energy recovery, enhances feedstock utilization efficiency and supports sustainable energy systems. Integrating wastes into energy production fosters circularity, reduces environmental impact, and builds resilience against fossil-fuel volatility.

Notably biopolymer hydrogel electrolytes, have emerged as eco-conscious alternatives in batteries and supercapacitors. Derived from cellulose, chitosan, or alginate, these hydrogels offer biodegradability, biocompatibility, low toxicity and relatively low cost. They can serve as flexible solid-state electrolytes or even electrodes in electrochemical storage devices. Recent progress includes conductive biopolymer-based hydrogel electrolytes enabling flexible, high-performance supercapacitors suitable for wearable electronics offering durability, stretchability and self-healing properties [16, 56]. Cellulose-based hydrogels have also been successfully carbonized into porous electrodes, demonstrating high mechanical strength and excellent electrochemical performance [65]. These developments advance sustainable energy storage by replacing conventional, often toxic, electrolyte materials with bio-based, recyclable alternatives, enhancing environmental friendliness and aligning with circular economy principles.

Nanomaterials such as cellulose or aerogel nanoparticles can dramatically improve energy efficiency through advanced insulation and enhanced lighting systems. Nanocellulose aerogels, with their exceptionally low thermal conductivity and high porosity, serve as lightweight, high-performance thermal insulators, outperforming traditional foam materials. These aerogels reduce heat transfer in buildings, decreasing heating and cooling demands. Additionally, bio-inspired photoluminescent nanomaterials can be integrated into lighting systems, boosting luminous efficiency and enabling energy savings. Though research is evolving, early studies on bio-derived cellulose nanocrystals for luminescent coatings suggest potential for enhanced light distribution and reduced electrical consumption. These materials provide multifunctional properties in thermal insulation, light management helping decarbonize built environments while leveraging renewable feedstocks.

Biomass-derived activated carbon electrodes enable eco-efficient capacitive deionization (CDI) systems, offering sustainable solutions for desalination and water purification. In CDI, water flows between charged carbon electrodes that adsorb ions, removing salts with minimal energy compared to reverse osmosis. Biomass sources like agricultural and forestry residues can be transformed into porous activated carbons with high surface areas, tailored pore structures, and excellent conductivity for CDI electrodes [7, 16, 17]. Such bio-carbon electrodes deliver strong electrosorption performance while relying on renewable or waste biomass, supporting low-cost production and reduced environmental impact. This biodegradability and sustainability make bio-based CDI particularly attractive for decentralized water treatment, benefiting remote or resource-limited regions.

Integrating bio-based materials into energy systems fosters circular economy principles by using waste biomass as feedstock and facilitating reuse and recycling. For instance, agricultural residues or algal by-products can be converted into biochar for electrodes, hydrogels for energy storage, or insulation materials. After their functional lifespan, many bio-based components are biodegradable or compostable, closing material loops. Hydrogel materials derived from cellulose or proteins can be safely returned to soil systems or biorefineries. This closed-loop approach conserves resources, reduces waste, and minimizes carbon footprints. By designing renewable energy systems that incorporate bio-based materials from inception through end-of-life, a truly circular energy economy becomes achievable where energy systems are regenerative, resource-efficient and ecologically balanced.

Bio-inspired micro- and nanodevices such as biodegradable sensors, nanocellulose-based electrodes and self-assembling peptide supercapacitors are advancing energy production and environmental monitoring at micro scales. Examples include paper-like batteries with cellulose electrodes and ionic liquid electrolytes that are flexible, biocompatible, and eco-friendly. Biodegradable peptide-based supercapacitors exploit self-organizing peptides for high capacitance with low toxicity. Microfluidic devices incorporating nanocellulose membranes detect and manage environmental pollutants or energy-relevant chemicals. These devices emphasize synergy between bio-based materials and microfabrication, enabling low-waste, sensitive, and distributed technologies ideal for Internet of Things (IoT) and environmental sensing. By harnessing bio-derived nanomaterials in microdevices, miniaturized energy solutions with sustainability, driving both technological and ecological innovation. Table 5.1 shows the roles of Bio-Based Materials in Energy Sustainability, highlighting their applications in biofuels, energy storage, energy efficiency, water desalination, circular economy, and emerging micro/nanodevices. These materials including lignocellulosic biomass, cellulose hydrogels, biochar, and nanocellulose demonstrate multifunctionality, biodegradability, and compatibility with advanced technologies such as capacitive deionization and IoT-enabled environmental sensing.

Lignocellulosic biomass serves as a vital feedstock for biofuel production, helping to reduce dependence on fossil fuels through thermochemical and biochemical processes [29, 47]. In energy storage, cellulose hydrogels offer eco-friendly alternatives for use in batteries and supercapacitors due to their flexibility and safety [65]. For enhancing energy efficiency, lightweight cellulose aerogels provide excellent thermal insulation in buildings and devices [12]. In water desalination, activated

Table 5.1 Roles of bio-based materials in energy sustainability

Role of bio-based materials	Material example	Description	References
Biofuels	Lignocellulosic biomass	Used as feedstock for biofuel production through thermochemical and biochemical methods	[29]
Energy storage	Cellulose hydrogels	Eco-friendly hydrogel electrolytes in flexible supercapacitors and batteries	[65]
Energy efficiency	Cellulose aerogels	Lightweight, low thermal conductivity insulators for buildings and devices	[12]
Water desalination	Activated carbon (rice husk)	Porous carbon electrodes derived from rice husk biomass for capacitive deionization (CDI)	[61]
Circular economy	Biochar	Carbon-rich residue used for electrodes, soil amendment, and resource recycling	[21]
Micro/ Nanodevices	Nanocellulose membranes	Biodegradable nanocellulose used in filtration, actuators, and energy-harvesting devices	[38]

carbon derived from rice husk functions as a porous electrode in capacitive deionization systems, enabling sustainable and low-energy purification [61]. Supporting the circular economy, biochar produced from biomass can be reused for electrodes and soil amendment, promoting resource recovery [21]. Furthermore, nanocellulose membranes are applied in micro/nanodevices for energy harvesting, filtration, and actuation, reflecting their adaptability and biodegradability [38]. Together, these materials illustrate the transformative potential of bio-based innovations in achieving a cleaner and more sustainable energy future.

Bio-based materials derived from agricultural residues and micro−/macro-algae present significant opportunities for sustainable energy systems, offering renewable alternatives to fossil fuels and contributing to carbon emission mitigation. Agricultural residues and algal biomass serve as sustainable feedstocks, transformed via thermochemical processes like pyrolysis, gasification and hydrothermal liquefaction into valuable biofuels such as bio-oil, syngas, and biochar. These conversion pathways enhance energy recovery, optimize feedstock utilization and foster environmental sustainability by integrating waste streams into energy production.

In energy storage applications, bio-derived hydrogels from cellulose, chitosan, and alginate have emerged as promising, eco-conscious alternatives to conventional materials. These hydrogels exhibit biodegradability, biocompatibility, flexibility, and low toxicity, positioning them ideally for use in flexible, high-performance supercapacitors and batteries. Innovations such as carbonized cellulose-based porous electrodes further illustrate the potential for advanced, sustainable electrochemical storage solutions aligned with circular economy principles. Additionally, bio-based nanomaterials such as cellulose aerogels significantly contribute to energy efficiency through enhanced insulation and advanced lighting systems. Their exceptional thermal insulation properties effectively reduce energy consumption in buildings, while bio-inspired photoluminescent nanomaterials facilitate energy-efficient lighting.

These developments represent meaningful progress in decarbonizing-built environments through multifunctional, renewable materials. Activated carbon derived from biomass residues exemplifies sustainable water desalination technologies, capacitive deionization (CDI). These bio-carbon electrodes offer improved electro sorption performance, energy efficiency and environmental compatibility, especially beneficial in resource-limited or decentralized water purification applications.

Furthermore, integrating bio-based materials into energy systems reinforces circular economy strategies, enabling resource recovery, waste reduction, and minimized carbon footprints. Materials like biochar and biodegradable hydrogels close the material lifecycle loop by serving multiple functions from energy storage electrodes to soil amendments demonstrating the feasibility of regenerative, ecologically balanced energy systems. Finally, bio-inspired micro- and nanodevices leverage biodegradable materials such as nanocellulose membranes and self-assembling peptides to advance environmental monitoring and micro-scale energy solutions. These innovations underscore the synergy between bio-derived nanomaterials and advanced microfabrication techniques, facilitating sustainable, low-impact and highly efficient technological solutions essential for future applications, including the Internet of Things (IoT). Collectively, these bio-based materials demonstrate multifunctionality, sustainability, and transformative potential, significantly advancing the development of clean, resilient and sustainable energy systems.

5.3 Types of Micro to Nano Bio-based Materials in Energy Applications

The transition to sustainable energy systems has intensified the search for alternative materials that are efficient, renewable, and environmentally benign. Among these, micro- to nano-scale bio-based materials have garnered significant attention due to their unique physicochemical properties, high surface area-to-volume ratios, and biocompatibility. Derived from natural sources such as cellulose, lignin, chitosan, starch, and microbial biomass, these materials offer promising functionalities in various energy-related applications including bioenergy conversion, energy storage, and catalysis. Their tunable morphology and chemical versatility allow them to act as bio-electrodes, membranes, scaffolds, or catalysts in advanced energy systems such as fuel cells, supercapacitors, and solar energy devices. As the field of green energy technology evolves, understanding the types and roles of bio-derived micro- and nanomaterials becomes essential for developing high-performance, low-carbon energy solutions.

5.3.1 Nanocellulose for Energy Storage

Nanocellulose derived from natural sources such as wood pulp, agricultural residues, or microbial cultures, has garnered significant attention as a functional and sustainable material for energy storage systems. Classified into cellulose nanocrystals (CNCs), cellulose nanofibrils (CNFs), and bacterial nanocellulose (BNC), these materials exhibit outstanding properties such as high surface area, excellent mechanical strength, biodegradability, and compatibility with conductive composites. Their eco-friendly nature aligns with the global transition toward green energy technologies, making them attractive for applications in supercapacitors, lithium-ion batteries (LIBs), and sodium-ion batteries (SIBs) [32].

Recent studies highlight the utility of bacterial nanocellulose as a precursor for carbon materials used in electrochemical energy storage. For instance, Villarreal-Rueda et al. [69] reported the synthesis of activated carbon from BNC cultured in fique juice, activated via KOH and H_3PO_4, and pyrolyzed at 700–800 °C. The resulting material achieved a high specific surface area of up to 893 m^2/g and specific capacitance between 97.5 and 220 F/g in acidic electrolytes, confirming its potential for use in supercapacitor electrodes [69]. In another study, [55] developed a nanocellulose/nitrogen-fluorine co-doped graphene composite hydrogel, which exhibited a remarkable specific capacitance of 286.6 F/g and outstanding volumetric performance of 421.3 F/cm^3. The device showed excellent cycling stability beyond 10,000 cycles, positioning the nanocellulose-based hydrogel as a competitive electrode material for high-performance supercapacitors [84].

Further, a study published in Penchev et al. [49] explored the performance of BNC/PBI® composite membranes as separators in supercapacitor cells. These membranes demonstrated excellent electrochemical stability and mechanical integrity, confirming BNC's applicability not only as an electrode material but also as a sustainable separator [49]. A comprehensive review by Choudhary et al. [14] details advancements in nanocellulose-based supercapacitors, covering the integration of CNFs and CNCs with conductive fillers like graphene and carbon nanotubes to enhance ionic and electronic conductivity. The review emphasizes structural tunability, chemical modification strategies, and the potential of heteroatom doping to overcome intrinsic conductivity limitations.

Lastly, an article by Xu et al. [75] discusses flexible nanocellulose-based composite electrodes, which leverage the synergy between nanocellulose and carbonaceous materials to fabricate bendable, high-capacitance energy storage devices. In summary, nanocellulose presents a biodegradable, mechanically stable, and chemically versatile platform for developing sustainable energy storage devices. Despite some challenges such as low intrinsic conductivity and moisture sensitivity, advances in doping, functionalization, and composite integration continue to enhance its performance and commercial viability in LIBs, SIBs and SCs.

The nanocellulose nanomaterials available in three primary forms cellulose nanocrystals (CNCs), cellulose nanofibrils (CNFs) and bacterial nanocellulose (BNC) these materials can serve various functions such as electrodes, separators, and structural supports in supercapacitors and batteries. Their compatibility with carbon-based nanomaterials and conductive polymers enables the fabrication of hybrid composites with improved electrochemical performance. Recent developments have demonstrated significant promise. For example, bacterial nanocellulose-derived activated carbon, processed through chemical activation and pyrolysis, achieved surface areas up to 893 m^2/g and specific capacitance ranging from 97.5 to 220 F/g [69]. Similarly, flexible nanocellulose-based electrodes, reinforced with carbon nanotubes and conductive polymers, have been employed in bendable supercapacitor devices with high retention and flexibility [69]. To illustrate the current advancements and functional versatility of nanocellulose in energy applications, Table 5.2 summarizes recent studies involving different nanocellulose-derived materials. It highlights their processing methods, electrochemical properties, device relevance, and reliable sources published between 2020 and 2025.

The unique advantages of nanocellulose stem from its renewable nature, high aspect ratio, large specific surface area, tunable surface chemistry, and mechanical robustness. These attributes make nanocellulose a versatile candidate for use as a scaffold, binder, separator, or even active electrode component. For instance, BNC-derived activated carbon, when processed via KOH or H_3PO_4 activation followed by pyrolysis, demonstrated high surface area (up to 893 m^2/g) and specific capacitance values reaching 220 F/g, as reported by Villarreal-Rueda et al. [69].

Additionally, nanocellulose-based hydrogels have been successfully integrated with conductive fillers to form composite electrodes. For example, TOCN/CNT/polypyrrole hydrogels developed via in-situ polymerization have shown excellent flexibility, electrochemical stability, and high specific capacitance [81]. These hydrogels are particularly promising for next-generation flexible and wearable supercapacitors due to their high mechanical resilience and solid-state compatibility. Despite its promise, nanocellulose also presents notable technical challenges. One primary limitation is its inherently low electrical conductivity, which necessitates its integration with conductive additives such as carbon nanotubes, graphene, or conductive polymers to realize practical electrochemical applications. Moreover, nanocellulose's hydrophilic nature makes it highly moisture-sensitive, potentially

Table 5.2 Nanocellulose-based materials for energy storage applications

Material type	Precursor/processing	Performance	Application	References
Bacterial nanocellulose (BNC)-derived carbon	Cultured in fique juice, activated using KOH and H_3PO_4, pyrolyzed at 700–800 °C	Specific surface area: 893 m^2/g; capacitance: 97.5–220 F/g	Supercapacitor electrode	[69]
Nanocellulose/N,F-co-doped graphene hydrogel	Nanocellulose and graphene oxide hydrothermally treated with NH_4F	Capacitance: 286.6 F/g; volumetric capacitance: 421.3 F/cm^3; stability >10,000 cycles	High-performance supercapacitor	[65]
Nanocellulose-based flexible electrodes	Embedded with polypyrrole and carbon nanotubes	Capacitance: ~230 F/g; excellent bending stability	Flexible energy storage device	[13]
CNF/graphene composite	Vacuum filtration of CNF and reduced graphene oxide blend	Capacitance: 280–320 F/g; energy density: ~18 Wh/kg	Flexible supercapacitor	[14]
CNC/carbon nanotube composite	Freeze-dried self-assembled CNC and CNT mixture	Capacitance: ~200 F/g; stable over 5000 cycles	Lightweight supercapacitor electrode	[68]

Table 5.3 Advantages and disadvantages of nanocellulose in energy storage applications

Advantages	Disadvantages
Renewable and biodegradable	Low intrinsic electrical conductivity
High specific surface area and tunable porosity	Moisture-sensitive and hygroscopic behavior
Excellent mechanical strength and flexibility	Limited thermal and dimensional stability under humid conditions
Good compatibility with conductive additives (e.g., CNTs, graphene)	Requires composite formulation and advanced processing techniques
Scalable from plant or bacterial sources	Surface modification may require energy-intensive or chemical-intensive methods

affecting the dimensional stability and ionic transport properties of the device during operation [14].

Scalability and processing complexity also remain concerns, particularly in controlling nanocellulose's surface chemistry and pore structure at industrial levels. Furthermore, the energy and chemical demand for some surface modification and drying techniques may counterbalance its otherwise green profile if not optimized properly. Nevertheless, the ongoing development of surface functionalization, heteroatom doping, and 3D printable nanocellulose-based inks is gradually overcoming these limitations, enhancing both performance and processability. When judiciously engineered, nanocellulose offers a compelling pathway to bridge sustainability and high-performance energy storage. Table 5.3 presents a comparative overview of the major advantages and disadvantages of nanocellulose as a material for energy storage.

Numerous studies have demonstrated the feasibility of nanocellulose-based materials particularly when hybridized with conductive fillers or doped with heteroatoms to deliver significant improvements in specific capacitance, energy density, and device flexibility. Composite systems such as nanocellulose-polypyrrole hydrogels and activated BNC-derived carbons have shown notable electrochemical performance while maintaining mechanical integrity and environmental friendliness. Nevertheless, challenges such as low intrinsic conductivity, moisture sensitivity and scale-up complexity must be systematically addressed. Advancements in surface functionalization, hybrid composite design, and low-energy processing routes are crucial to unlocking the full commercial potential of nanocellulose in energy storage technologies. In summary, nanocellulose serves as a compelling bridge between high-performance electrochemical functionality and sustainable material development positioning it as a vital component in the pursuit of greener, more resilient energy systems.

5.3.2 *Biochar and Carbon-Based Nanomaterials*

Biochar produced via the pyrolysis of biomass in oxygen-limited environments is a high-carbon, porous material increasingly explored as an advanced bio-based electrode in energy storage systems. When engineered at the micro- to nano-scale, it exhibits high surface area, tunable porosity and diverse surface functionalities, making it a promising candidate for supercapacitors, lithium-ion batteries (LIBs) and sodium-ion batteries (SIBs). Similarly, carbon-based nanomaterials such as activated carbon, carbon nanotubes and graphene often derived from biomass precursors demonstrate superior electrical conductivity and structural tunability, enabling high power and energy densities in electrochemical devices [43]. A recent Prabakar et al. [52] highlights that operationalizing biochar for energy storage involves carefully selecting feedstock and pyrolysis conditions, followed by activation, doping, or hybridization to enhance conductivity and electrochemical performance. Optimization techniques such as heteroatom doping (e.g., nitrogen, phosphorus), hierarchical porosity control and carbonization under varying temperatures have been shown to significantly improve specific capacitance, energy density and cycling stability. Another study in Yan et al. [78] corroborates these strategies and emphasizes the sustainability of biochar-based energy storage systems. To structure this overview effectively, the materials are organized into three main types:

(i) Stimuli-activated biochar

Derived from agricultural waste or biomass, activated via KOH or CO_2 treatments to produce high surface area carbons suitable for electrochemical double-layer capacitors (EDLCs).

(ii) Doped biochar nanocomposites
(iii) Modified with heteroatoms and/or combined with conductive polymers or metal oxides, these materials provide pseudocapacitance and improved conductivity.
(iv) Biomass-derived carbon nanomaterials

Including activated carbons, graphene, and CNTs produced from agricultural residues or biochar-derived feedstocks, these materials offer hierarchical porosity and excellent thermal and mechanical stability (e.g., carbon aerogels and carbon nanofibers).

The performance of biochar and carbon-based nanomaterials in energy storage systems is intricately tied to the fabrication techniques used during their synthesis and post-processing. The ability to tailor surface area, pore distribution, conductivity, and functional group composition relies heavily on precise control over material processing parameters. Consequently, the optimization of synthesis routes including pyrolysis, activation, doping, and composite formation is critical to enhancing the electrochemical properties and overall functionality of these materials in supercapacitors, lithium-ion batteries (LIBs), sodium-ion batteries (SIBs), and hybrid devices. Pyrolysis, the thermal decomposition of biomass in an inert or oxygen-limited environment, is the foundational step in converting organic precursors into biochar. The temperature, heating rate, residence time, and atmospheric conditions significantly influence the physicochemical characteristics of the resulting carbon material. Low-temperature pyrolysis (300–500 °C) generally yields materials with high oxygen content and poor conductivity, while high-temperature pyrolysis (600–900 °C) facilitates the development of a graphitized carbon structure, enhancing electrical conductivity and thermal stability. A slow heating rate and prolonged dwell time are preferred for developing stable pore architectures and reducing structural collapse during devolatilization.

Meanwhile, to improve the surface area and pore volume of biochar, activation is often employed post-pyrolysis. There are two principal strategies: physical activation and chemical activation. Physical activation involves the use of gases such as CO_2 or steam at elevated temperatures (700–900 °C), promoting the development of mesopores through gasification reactions. In contrast, chemical activation, typically using agents like KOH, H_3PO_4, $ZnCl_2$, or NaOH, enhances microporosity and introduces surface functional groups that can contribute to wettability and electrochemical activity. KOH activation, in particular, is widely used due to its ability to generate interconnected porous networks and high specific surface areas (>2000 m^2/g), ideal for electric double-layer capacitors (EDLCs) [83].

Surface modification through heteroatom doping is a widely adopted technique to enhance the redox activity, conductivity, and charge storage capacity of biochar. Doping with nitrogen (N), phosphorus (P), sulfur (S) or boron (B) introduces defects and polar sites in the carbon lattice, facilitating Faradaic reactions and improving electron/ion transport kinetics. Nitrogen doping, for example, incorporates pyridinic, pyrrolic, or graphitic nitrogen functionalities, which increase surface polarity and electrical conductivity. Co-doping strategies such as N and P, also shown synergistic effects, offering enhanced electrochemical performance compared to single-element doping due to multiple redox-active sites [86]. To overcome the limitations of low intrinsic conductivity and structural fragility, biochar is frequently integrated into composite systems with conductive polymers (e.g., polypyrrole, polyaniline) or transition metal oxides (e.g., MnO_2, Fe_3O_4, Co_3O_4). These composites exhibit hybrid charge storage mechanisms, combining capacitive behavior of the carbon matrix with pseudocapacitance from the redox-active components. The uniform dispersion of active components onto the porous biochar framework provides a large electrochemical interface and short diffusion pathways, thereby improving both energy and power density. In situ polymerization, hydrothermal synthesis and sol-gel methods are commonly used to fabricate such composites.

To achieve hierarchical pore structures or specific morphologies such as hollow spheres, nanotubes, aerogels, template-assisted fabrication techniques are employed. Soft templates (e.g., surfactants, block copolymers) and hard templates (e.g., silica spheres, calcium carbonate) guide the assembly of biomass-derived carbon precursors into ordered structures. After carbonization, the templates are removed via acid etching or calcination. These methods enable precise control over pore geometry and structural orientation, which are critical for optimizing ion diffusion and mechanical strength in flexible or structural energy devices. Furthermore, emerging techniques such as 3D printing of biochar-based inks are being explored to develop customized architectures for micro-supercapacitors and wearable electronics, offering both scalability and design flexibility [70, 71]. The functional performance of biochar and carbon-based nanomaterials in energy applications is highly dependent on the synthesis and post-treatment strategies employed. Each technique from pyrolysis to doping and composite formation plays a pivotal role in tuning the material's physicochemical properties to meet the demands of modern energy storage technologies. As research advances, integrating green, scalable, and low-energy fabrication routes will be essential for aligning material performance with environmental sustainability and industrial applicability.

To critically evaluate the viability of biochar and biomass-derived carbon nanomaterials in energy storage systems, it is essential to compare their electrochemical performance against conventional carbonaceous materials such as commercial activated carbon, graphene, and carbon nanotubes (CNTs). Key performance indicators typically include specific capacitance, energy density, power density, rate capability, and cycle stability. Biochar materials activated via chemical methods (e.g., KOH, H_3PO_4) routinely achieve specific surface areas exceeding 1500–2000 m^2/g and specific capacitance values ranging from 200 to 350 F/g, depending on pore structure and surface functionality. In contrast, commercial activated carbons typically demonstrate capacitance values between 100 and 200 F/g, owing to their less tailored porosity and limited surface redox activity.

Graphene-based materials and CNTs often outperform both biochar and activated carbon in terms of electrical conductivity and specific capacitance, with values reaching 500–700 F/g and superior power density due to their inherently delocalized π-electron systems and high aspect ratio. However, their production involves high-cost precursors, complex synthesis routes (e.g., chemical vapor deposition or Hummers' method), and energy-intensive conditions that raise environmental and economic concerns [70, 71]. Moreover, the scalability of graphene and CNTs remains a bottleneck, limiting their practical integration into commercial energy devices.

In contrast, biochar exhibits moderate electrochemical performance but offers distinct advantages in terms of sustainability, scalability, and cost-effectiveness. Its raw materials are abundantly available as agricultural or municipal waste, and its synthesis can be conducted using low-cost pyrolysis techniques with minimal environmental impact. In terms of energy density, doped biochar nanocomposites (e.g., N-, P-, or S-doped) have demonstrated values ranging from 10 to 30 Wh/kg, while CNT and graphene composites may reach up to 40–60 Wh/kg, depending on the electrolyte and cell configuration. For cycle stability, biochar-based electrodes often maintain 80–95% capacitance retention over 5000–10,000 cycles, comparable to or better than some commercial carbons when properly engineered.

When assessed through a sustainability-performance lens, biochar materials outperform conventional carbons in terms of environmental footprint, resource circularity, and cost per functional unit, making them highly attractive for decentralized, low-cost energy storage systems in off-grid or rural applications. As such, although graphene and CNTs continue to set the upper bounds of performance, biochar represents a practical and eco-efficient compromise, especially in applications where material cost, availability, and carbon neutrality are of equal or greater importance than absolute capacitance values.

The electrochemical performance of biochar and carbon-based nanomaterials in energy storage systems is largely governed by their structural morphology, surface chemistry, and charge transport mechanisms. Stimuli-activated biochar, typically derived from biomass sources such as coconut shells, rice husks, or durian rinds, is processed through chemical (e.g., KOH, H_3PO_4) or physical (e.g., CO_2, steam) activation. This activation results in a high degree of microporosity and surface area, which promotes electric double-layer capacitance (EDLC). In this mechanism, energy is stored through the non-Faradaic accumulation of charge at the electrode-electrolyte interface, allowing for rapid ion adsorption and desorption during charge/discharge cycles. The narrow pore size distribution and large surface area enhance power density and enable fast ion transport, making these materials highly suitable for electrochemical double-layer capacitors (EDLCs) and high-power supercapacitors [1].

In contrast, doped biochar nanocomposites operate through Faradaic mechanisms and offer pseudocapacitive behavior due to their redox-active surface functionalities. Doping with heteroatoms such as nitrogen, phosphorus, or sulfur introduces electronic defects and polar sites that facilitate surface redox reactions, thereby increasing charge storage capacity. Nitrogen-doped biochar for example, introduces pyridinic and graphitic nitrogen moieties that enhance electrical conductivity and improve wettability of the carbon surface. When integrated with conductive polymers such as polypyrrole or metal oxides like MnO_2, these composites exhibit hybrid charge storage behavior combining capacitive and Faradaic contributions to significantly boost energy density while retaining fast response times [86]. Such composites are ideal candidates for hybrid supercapacitors and anodes in sodium- or lithium-ion batteries.

Biomass-derived carbon nanomaterials, including carbon nanotubes (CNTs), graphene, carbon nanofibers, and carbon aerogels, further enhance energy storage through hierarchical charge transport pathways. These materials are typically produced from biochar feedstocks or directly from biomass via pyrolysis or hydrothermal carbonization followed by carbonization. Their structure features interconnected micropores, mesopores, and macropores that collectively support rapid ion diffusion and electrolyte accessibility. At the same time, their highly conjugated carbon networks provide excellent electron mobility, enabling low internal resistance and high conductivity. These materials exhibit mixed storage behavior, involving both EDLC and Faradaic mechanisms, and are particularly effective in flexible and high-capacity energy devices, such as wearable supercapacitors, solid-state batteries, and lightweight structural energy storage systems [70, 71].

These three categories of carbon-based materials demonstrate distinct but complementary mechanisms of action. Stimuli-activated biochar functions primarily via physical ion accumulation, doped biochar composites rely on surface redox activity, and biomass-derived nanomaterials offer multidimensional charge transport and hybrid storage behavior. Together, they provide a highly adaptable platform for engineering sustainable, high-performance energy storage systems tailored to specific application demands. Micro- and nanostructured biochar and carbon-based nanomaterials derived from biomass are establishing a sustainable and eco-conscious foundation for advanced energy storage technologies. Their success lies in striking a balance between intricate structural control and scalable, green manufacturing processes.

Biochar and biomass-derived carbon nanomaterials present a unique combination of sustainability, structural tunability, and electrochemical performance, positioning them as valuable components in energy storage technologies such as supercapacitors, lithium-ion batteries (LIBs), sodium-ion batteries (SIBs), and flexible energy devices. One of the principal advantages of these materials is their renewable and low-cost nature as they are derived from agricultural and forestry residues, including coconut shells, pine cones, durian rind, sawdust, and algae biomass [74]. In addition, chemical or physical activation of biochar (e.g., KOH or CO_2 treatments) yields a high specific surface area and well-developed micro- to mesoporous networks, which significantly enhance ion transport and enable rapid charge discharge cycles in electric double-layer capacitors (EDLCs) [1].

When further functionalized through heteroatom doping typically with nitrogen, phosphorus or sulfur. Biochar can also exhibit pseudocapacitive behavior where reversible redox reactions on the carbon surface contribute to higher energy density. This makes doped biochar nanocomposites suitable for hybrid supercapacitors and as active anode materials in LIBs and SIBs [73]. Moreover, advanced bio-derived carbon nanomaterials, such as graphene-like sheets, carbon nanotubes (CNTs), and carbon nanofibers, exhibit exceptional electrical conductivity, mechanical integrity and hierarchical porosity, enabling superior performance in flexible and solid-state energy storage systems [20].

Despite these strengths, several limitations hinder the full-scale deployment of biochar-based materials in commercial devices. First, intrinsic electrical conductivity is generally low in undoped or unmodified biochar, requiring the addition of conductive additives to achieve competitive performance. Second, the moisture sensitivity and hydrophilicity of biochar due to abundant oxygen-containing functional groups can negatively impact electrode stability and electrolyte compatibility over extended cycling. Third, achieving consistent porosity and surface functionality from batch to batch is challenging due to the heterogeneity of biomass feedstocks. Additionally, post-treatment processes such as doping, carbonization, and activation can be energy-intensive, which may reduce the overall environmental benefit if not optimized.

Despite the promising potential of biochar and biomass-derived carbon nanomaterials in energy storage applications, several critical challenges hinder their scalability and long-term deployment. One of the foremost issues is the lack of standardization and scalability in processing techniques. The production of biochar at a large scale is often inconsistent due to the absence of optimized and reproducible activation and doping protocols, which affects the uniformity and reliability of the resulting materials. This inconsistency is compounded by batch-to-batch variability of biomass feedstocks, where natural differences in lignocellulosic content, ash composition, and moisture levels can result in fluctuating electrochemical properties, including pore structure, conductivity, and surface chemistry. Another major limitation lies in the instability at the electrode-electrolyte interface, particularly in hybrid composite systems. Materials such as doped biochar integrated with conductive polymers or metal oxides often suffer from poor adhesion or delamination under cycling conditions, which can reduce capacitance retention and overall device durability. Furthermore, although these materials are derived from renewable sources, some fabrication methods such as high-temperature pyrolysis or the use of strong chemical activators can introduce environmental and energy trade-offs that diminish their overall sustainability footprint.

To address these challenges, future research should focus on developing modular, continuous pyrolysis systems and standardized activation protocols to improve scalability and consistency. The adoption of pre-treatment techniques such as torrefaction, and the integration of machine learning algorithms to predict and optimize output quality based on feedstock variability, could significantly enhance production efficiency. In terms of material stability, innovations in surface functionalization and interface engineering, including the use of self-healing binders and flexible conductive networks, offer promising pathways to extend device lifespan and mechanical integrity. Moreover, the exploration of green synthesis approaches such as hydrothermal carbonization and low-temperature activation combined with comprehensive life-cycle assessments (LCA), will be essential to ensuring that these materials fulfill both performance and sustainability criteria in future energy storage technologies.

5.3.3 Polymer Nanocomposites

Polymer nanocomposites have emerged as a versatile and high-potential class of materials in the development of next-generation energy storage and conversion systems. These materials consist of a polymer matrix embedded with nanoscale fillers such as metal oxides, carbon nanotubes (CNTs), graphene or bio-derived nanoparticles that impart enhanced electrical, thermal, and mechanical properties. When sourced from renewable feedstocks and combined with biodegradable or bio-based polymers, such nanocomposites align well with the global push for greener and more sustainable energy technologies.

In the context of electrochemical energy applications, polymer nanocomposites play critical roles in a variety of device components, including solid or gel polymer electrolytes, flexible electrodes, and dielectric layers in capacitors. The integration of nanofillers into the polymer matrix facilitates the creation of continuous ion or electron transport pathways, enhances interfacial interactions, and improves mechanical resilience factors that are particularly important for flexible, stretchable, and wearable energy devices. For instance, nanocomposites based on polyvinyl alcohol (PVA) reinforced with biochar, cellulose nanocrystals or polyaniline-modified nanofillers have shown promising results in solid-state supercapacitors and proton exchange membranes.

Moreover, polymer nanocomposites offer the unique advantage of processability, allowing them to be fabricated into thin films, coatings, or 3D architectures via scalable techniques such as electrospinning, solution casting, or additive manufacturing. These fabrication advantages are especially relevant for energy applications requiring lightweight, conformable, and customizable structures. Importantly, the synergistic effect between the organic polymer matrix and the inorganic or carbon-based nanofillers results in improved ionic conductivity, dielectric constant, and overall electrochemical performance, thereby expanding the design flexibility and functional integration of energy storage systems.

Given their multifunctionality, polymer nanocomposites hold tremendous promise in a wide range of energy applications, including lithium-ion and sodium-ion battery electrolytes, supercapacitor electrodes, fuel cell membranes, and solar cell encapsulants. As research continues to optimize filler dispersion, interfacial bonding, and long-term stability, these materials are poised to play a pivotal role in the realization of high-performance, low-impact, and sustainable energy devices.

Polymer nanocomposites utilized in energy applications represent a diverse category of hybrid materials, where a polymeric matrix often bio-based or biodegradable is reinforced with functional nanoscale fillers. These fillers, which may be derived from inorganic, carbonaceous, or bio-based sources, enhance the electrical, ionic, thermal, and mechanical properties of the base polymer, thus making them highly suitable for use in energy storage and conversion systems. The choice of both the polymer matrix and the nanofiller is critical, as their synergistic interaction governs the electrochemical performance, stability and processing compatibility of the final composite.

One widely studied system involves biopolymer-carbon nanocomposites, where natural or biodegradable polymers such as polylactic acid (PLA), chitosan, cellulose derivatives, or starch are reinforced with graphene, carbon nanotubes (CNTs), or biochar nanoparticles. These materials have shown promise in flexible supercapacitors and solid-state batteries, owing to their high specific surface area, improved mechanical flexibility, and electrical conductivity. For instance, chitosan reinforced with graphene oxide or CNTs has been demonstrated to function as a proton-conductive membrane or a flexible electrode in pseudo-capacitive devices, with enhanced electrochemical stability and ionic transport pathways [35].

Another prominent category includes polymer-metal oxide nanocomposites, such as polyvinyl alcohol (PVA) or polyethylene oxide (PEO) matrices embedded with nanosized TiO_2, ZnO, Fe_3O_4, or MnO_2 particles. These are particularly relevant in battery separators, dielectric capacitors, and proton exchange membranes, where the metal oxide filler enhances dielectric performance, mechanical strength, or ion exchange capacity. For example, MnO_2/PVA nanocomposites have been reported to exhibit superior pseudocapacitance behavior when employed as electrodes in hybrid supercapacitors, leveraging both double-layer and Faradaic charge storage mechanisms.

Emerging systems also incorporate bio-derived nanofillers, such as cellulose nanocrystals (CNCs), bacterial nanocellulose (BNC), and lignin nanoparticles, into biopolymers to develop fully sustainable and biodegradable nanocomposites. These systems are gaining attention in gel polymer electrolytes and dielectric films, where the polar surface groups of nanofillers improve ion conduction and interfacial compatibility with electrodes. Notably, CNC-reinforced PEO nanocomposites have demonstrated improved ionic conductivity and dimensional stability in solid polymer electrolytes for lithium-ion and sodium-ion batteries [9].

Additionally, conducting polymer-based nanocomposites, such as polypyrrole (PPy), polyaniline (PANI), or poly (3,4-ethylenedioxythiophene) (PEDOT) blended with nanofillers, are widely applied in electrodes for supercapacitors and electrochemical sensors. These systems benefit from the pseudocapacitive properties of the conducting polymer and the structural support of the nanofillers, resulting in enhanced cycling stability and charge transfer kinetics. For instance, PANI/BNC nanocomposites have been reported to achieve high specific capacitance, flexibility, and long-term durability in wearable energy devices. The selection and design of polymer nanocomposites for energy applications are highly dependent on the interplay between the polymer matrix and the type of nanofiller used. From carbon-based and metal oxide-reinforced composites to bio-derived nanofiller systems, these materials offer a customizable platform to engineer device-specific properties such as conductivity, flexibility, and environmental compatibility. As advancements in nanotechnology and sustainable materials science continue to evolve, polymer nanocomposites will remain at the forefront of eco-efficient, multifunctional energy storage and conversion technologies.

Table 5.4 shows types of polymer nanocomposites. The integration of nanoscale fillers into bio-based or biodegradable polymers has given rise to a new class of functional materials polymer nanocomposites tailored for advanced energy applications. As shown in Table 5.4, a wide range of polymer matrices, including chitosan, polyvinyl alcohol (PVA), polyethylene oxide (PEO), polyaniline (PANI), and starch, have been successfully reinforced with carbonaceous, metallic, or bio-derived nanofillers to enhance their physicochemical and electrochemical properties. These systems have been strategically engineered to address limitations in conductivity, flexibility, ionic transport, and mechanical durability, making them highly suitable for next-generation energy storage devices such as supercapacitors, solid-state

Table 5.4 Types of polymer nanocomposites in energy applications

Polymer nanocomposite type	Nanofiller function	Application	References
Chitosan-graphene oxide (GO)	Enhances electrical conductivity, mechanical strength, and ion transport	Flexible supercapacitors, proton exchange membranes	[35]
PVA-MnO_2	Provides pseudocapacitance and structural support for hybrid capacitors	Electrodes in hybrid supercapacitors	[31]
PEO-cellulose nanocrystals (CNC)	Improves ionic conductivity and mechanical stability in electrolytes	Solid polymer electrolytes for LIBs and SIBs	[9]
Polyaniline (PANI)-bacterial nanocellulose (BNC)	Enables flexibility and charge storage in wearable supercapacitors	Flexible energy storage for wearable electronics	[11]
Starch-carbon nanotubes (CNTs)	Improves charge mobility and dispersion in biodegradable matrices	Eco-friendly energy storage films and biodegradable batteries	[2]

batteries, and wearable power systems. For instance, the chitosan-graphene oxide (GO) composite exploits the synergistic interaction between the proton-conducting nature of chitosan and the high electrical conductivity and aspect ratio of GO. This results in a flexible, highly conductive membrane applicable in proton exchange and supercapacitor devices. Similarly, PVA-MnO_2 nanocomposites harness the redox-active properties of MnO_2 to impart pseudocapacitive behavior, while the PVA matrix maintains structural integrity and processability, making the system well-suited for hybrid supercapacitor electrodes.

The PEO-CNC system exemplifies the role of bio-derived nanofillers, where cellulose nanocrystals not only reinforce mechanical strength but also enhance ionic conductivity via polar surface groups. These features are particularly beneficial for solid polymer electrolytes in lithium- and sodium-ion batteries, where both mechanical stability and ion transport are critical. In contrast, PANI-BNC nanocomposites target flexible energy storage, combining the pseudocapacitance of polyaniline with the nanofibrillar network of bacterial nanocellulose to produce conductive and mechanically resilient films suitable for wearable electronics.

Furthermore, the starch-CNT composite underscores the importance of combining biodegradability with high conductivity. Carbon nanotubes, when dispersed within starch matrices, form conductive pathways that improve charge mobility, making the system attractive for eco-friendly energy storage films and disposable or biodegradable batteries. While these nanocomposites vary in their constituent materials and specific energy applications, they all share a common advantage: multifunctionality. They simultaneously address multiple performance criteria conductivity, flexibility, mechanical integrity, and sustainability within a single material system. These examples collectively highlight the growing trend toward sustainable nanocomposite engineering, where renewable polymers and advanced nanoscale additives are co-optimized for energy devices. The continued evolution of these systems will rely on improving filler dispersion, interface compatibility, and long-term cycling performance. Moreover, scalable fabrication methods, such as in situ polymerization, solution casting, or 3D printing, will be crucial in translating lab-scale formulations into commercial energy storage technologies.

The functional performance of polymer nanocomposites in energy storage and conversion systems is intricately linked to their fabrication techniques and underlying charge storage mechanisms. The integration of nanoscale fillers such as carbon nanotubes (CNTs), graphene oxide (GO), metal oxides, or bio-derived nanoparticles into a polymer matrix alters the material's morphology, interfacial interactions, and transport properties. These enhancements directly influence critical electrochemical characteristics such as ionic conductivity, charge storage capacity, flexibility and long-term cycling stability.

A variety of scalable fabrication methods are employed to synthesize polymer nanocomposites for energy devices, with the goal of achieving uniform filler dispersion and optimal interfacial bonding. Solution casting remains one of the most widely used methods, particularly for producing thin-film electrodes or solid polymer electrolytes. In this approach, the polymer (e.g., PVA, PEO, chitosan) and nanofiller (e.g., CNTs, MnO_2, CNCs) are dissolved or dispersed in a suitable solvent and cast into a mold or substrate, followed by solvent evaporation under controlled conditions. For example, Xue et al. [77] fabricated a PEO-CNC solid electrolyte via solution casting, achieving high ionic conductivity and dimensional stability suitable for lithium-ion battery applications.

In situ polymerization is another commonly employed strategy, particularly for conducting polymer-based composites. Here, nanofillers are introduced into a monomer solution (e.g., aniline or pyrrole), and polymerization is initiated either chemically or electrochemically. This technique enables the formation of a continuous conductive network while anchoring the filler within the polymer backbone. Sivadas et al. [62] successfully applied in situ polymerization to synthesize PANI-BNC nanocomposites for flexible supercapacitors, demonstrating enhanced charge transport and mechanical resilience. Electrospinning, a high-voltage technique used to produce fibrous mats, is increasingly utilized for polymer nanocomposites in energy devices. The resultant nanofibrous membranes offer high surface-to-volume ratios and porosity, ideal for electrolytes and separators in batteries and capacitors. Additionally, hydrothermal synthesis, layer-by-layer assembly and 3D printing of composite inks are emerging as advanced methods to tailor nanocomposite architectures for device-specific functions.

The mechanism of energy storage in polymer nanocomposites depends on the nature of both the polymer and the embedded nanofillers. In most supercapacitor systems, nanocomposites function via electric double-layer capacitance (EDLC) and pseudocapacitance. EDLC arises from the electrostatic accumulation of charges at the electrode-electrolyte interface, typically facilitated by carbon-based fillers such as CNTs or GO. These fillers provide a high surface area and conductive pathways for charge accumulation. For instance, starch-CNT nanocomposites exhibit EDLC behavior, with CNTs creating conductive bridges that promote rapid electron mobility and low equivalent series resistance [10]. In contrast, nanocomposites incorporating redox-active fillers (e.g., MnO_2, polyaniline) exhibit pseudocapacitive behavior, wherein Faradaic redox reactions at the filler-polymer interface contribute to enhanced charge storage. In the case of PVA-MnO_2 nanocomposites, MnO_2 nanoparticles undergo reversible redox transitions while the PVA matrix provides mechanical flexibility and ionic conduction pathways, resulting in improved energy density and cycling performance [33].

In solid polymer electrolyte systems, polymer nanocomposites serve as ion-conducting matrices that facilitate ionic transport via hopping or segmental motion mechanisms. The presence of polar functional groups in the polymer (e.g., ether groups in PEO) and the high surface area of nanofillers (e.g., CNCs, GO) aid in coordinating with lithium or sodium ions, thus enhancing conductivity. CNCs in particular, due to their hydroxyl-rich surface, can act as ion conduction facilitators while simultaneously improving the mechanical stiffness of the electrolyte. The successful application of polymer nanocomposites in energy devices relies not only on the intrinsic properties of the materials but also on rational design of the composite structure, optimized fabrication routes, and a deep understanding of the underlying charge storage and transport mechanisms. As material science continues to evolve, the integration of bio-derived polymers and nanofillers with advanced processing techniques will remain pivotal in delivering safe, flexible, and high-performance energy systems.

Table 5.5 shows the advantages, limitations and future prospects of polymer nanocomposites in energy applications. Five types of polymer nanocomposites used in energy applications. These materials combine renewable polymers with nanofillers to enhance performance such as conductivity, flexibility, and energy storage capacity. Chitosan-Graphene Oxide (GO) is biodegradable and flexible, with improved conductivity due to GO. However, it absorbs moisture and has limited stability over time. Future improvements can focus on chemical modifications to resist moisture, making it suitable for use in proton exchange membranes. PVA-MnO_2 provides high energy storage through redox reactions and is easy to scale up. Its main weakness is the degradation of PVA at high temperatures or voltages. Combining it with more stable conductive polymers may improve its lifespan and energy output.

PEO-Cellulose Nanocrystals (CNC) shows good ionic conductivity and strength, making it useful in solid-state batteries. However, it does not perform well at room temperature without added plasticizers. Future research could include adding ionic liquids to boost performance in normal conditions. Polyaniline (PANI)-Bacterial Nanocellulose (BNC) is highly flexible and stable during charging cycles, making it ideal for wearable electronics. Still, its performance depends on how well the

Table 5.5 Advantages, limitations and future prospects of polymer nanocomposites in energy applications

Polymer nanocomposite type	Advantages	Limitations	Future prospects
Chitosan-graphene oxide (GO)	Biodegradable, flexible, and improved ionic/electronic conductivity [59]	Moisture sensitivity, poor long-term stability	Functionalization for moisture resistance; integration into proton exchange membranes
PVA-MnO_2	High pseudocapacitance and electrochemical performance; scalable [51]	Polymer degradation at high voltage or temperature	Hybridization with conductive polymers to extend cycle life and energy density
PEO-cellulose nanocrystals (CNC)	Enhanced ionic conductivity and mechanical strength in electrolytes [85]	Low ionic conductivity at room temperature without plasticizer	Incorporation of plasticizers or ionic liquids for higher conductivity at ambient conditions
Polyaniline (PANI)-bacterial Nanocellulose (BNC)	Excellent flexibility and cycle stability in wearable devices [15]	Conductivity dependent on uniform dispersion of nanofillers	3D printing for wearable and structural energy storage applications
Starch-carbon nanotubes (CNTs)	Eco-friendly; good dispersion of CNTs improves charge mobility [18]	Poor mechanical integrity; difficulty in scaling fabrication	Development of biodegradable, high-performance batteries for transient electronics

nanofillers are spread. Advanced printing or coating methods can help achieve more consistent structures. Starch-Carbon Nanotubes (CNTs) offers an eco-friendly option with good charge mobility. But its mechanical strength is low, and scaling up is difficult. It holds promise for biodegradable batteries in short-term or disposable electronics.

5.3.4 *Metal-Organic Frameworks (MOFs) and Hybrid Materials*

Metal-Organic Frameworks (MOFs) and bio-based hybrid materials have emerged as transformative classes of materials at the intersection of nanotechnology, materials science, and green energy. MOFs are crystalline porous materials consisting of metal ions or clusters coordinated to organic ligands, forming multidimensional networks with ultra-high surface areas and tunable chemical environments. When integrated with bio-based precursors such as polysaccharides, lignin, chitosan, or plant-derived polyphenols. MOFs evolve into hybrid systems that exhibit not only structural flexibility and chemical robustness but also environmental compatibility. The transition to renewable and sustainable energy systems requires materials that can perform multifunctional roles from energy storage and conversion to catalysis and environmental remediation while minimizing ecological footprints. Bio-derived MOF hybrids are well-suited for these demands due to their abundant active sites, biodegradability, and the possibility of sustainable synthesis routes. These materials are increasingly utilized in applications such as supercapacitors, fuel cells, photocatalysis, hydrogen storage, and electrochemical sensors. Their tunable pore structures and redox-active surfaces make them especially suitable for catalyzing energy-relevant reactions and facilitating efficient ion/electron transport.

The operational mechanisms of Metal-Organic Frameworks (MOFs) and their hybrid bio-based counterparts in energy applications are fundamentally governed by their unique structural features, redox behavior, and the synergistic interactions between their organic and inorganic constituents. These materials exhibit high surface areas, tunable porosity, and the ability to incorporate functional groups, making them highly versatile in energy-related systems. In gas storage and adsorption applications, for instance, the microporous architecture of MOFs allows for effective physisorption of gases like hydrogen or methane. When modified with bio-based components such as lignin, cellulose, or chitosan the resulting hybrid materials offer additional polar functional groups that enhance gas-binding affinity and improve storage capacity under ambient conditions.

In electrochemical energy storage devices, such as batteries and supercapacitors, MOFs contribute through their redox-active metal centers and conjugated organic linkers, which facilitate ion diffusion and electron transport. The incorporation of conductive bio-based materials, like biochar or carbonized polysaccharides, further enhances charge transfer, structural stability, and cyclic performance. Additionally, in photocatalytic systems, MOFs functionalized with semiconducting or light-absorbing components can generate electron-hole pairs under light irradiation, enabling photodegradation of pollutants or water splitting for hydrogen production. Bio-based ligands or coatings can modulate bandgap energies, stabilize reactive sites, and improve photocatalytic efficiency by suppressing recombination of charge carriers.

Moreover, the hybridization of MOFs with bio derived substrates introduces mechanical flexibility, biocompatibility, and hierarchical porosity, all of which are beneficial for emerging applications in flexible and wearable energy devices. These synergistic effects not only enhance electrochemical and catalytic performance but also contribute to the environmental sustainability of the materials. Importantly, bio-functionalization can address the inherent limitations of pristine MOFs, such as poor water stability and limited recyclability, by reinforcing their structural integrity and providing additional crosslinking. The operational mechanisms of MOFs and their bio-hybrid counterparts in energy systems depend on their structural features, redox behavior and synergistic interactions between organic and inorganic components.

(i) Adsorption and storage mechanism

In hydrogen storage and gas adsorption, MOFs exhibit physisorption-dominated mechanisms due to their large surface area and uniform microporosity. The inclusion of bio-based ligands or scaffolds can enhance specific binding interactions via polar functional groups (e.g., -OH, $-NH_2$), increasing storage capacity at ambient conditions.

(ii) Ion and electron transport

In electrochemical applications (e.g., supercapacitors and batteries), the metal nodes act as redox-active centers, while the conjugated organic linkers facilitate electron delocalization. The incorporation of conductive biopolymers or carbonized bio-based matter (such as biochar or lignin derivatives) improves electronic conductivity and structural integrity under repeated cycling.

(iii) Catalytic and photocatalytic activity

MOFs functionalized with photoactive or catalytic moieties can absorb visible or UV light, generating electron hole pairs that drive photocatalytic reactions (e.g., water splitting or pollutant degradation). Bio-based functional groups can modulate the bandgap, enhance charge separation, and provide anchoring sites for nanoparticles (e.g., TiO_2 or metal NPs), further increasing photocatalytic efficiency.

(iv) Synergistic hybrid functionality

Hybrid systems combining MOFs with cellulose nanocrystals, chitosan, or plant-based carbon substrates benefit from improved mechanical strength, biodegradability, and moisture tolerance. These attributes are critical for real-world deployment in flexible energy devices or bio-integrated electronics. The bio-components may also introduce hierarchical porosity, improving ion diffusion and charge storage kinetics.

(v) Stability and recyclability

One of the challenges in MOF application is structural instability in aqueous or acidic environments. Bio-based modification such as crosslinking with polyphenolic compounds can enhance water resistance and mechanical resilience, allowing for repeated use and long-term operation in energy systems.

In recent years, bio-based hybrid materials incorporating metal-organic frameworks (MOFs) have garnered significant attention due to their exceptional structural tunability, high surface area, and potential for green synthesis. These materials integrate the advantages of organic biopolymers such as biodegradability, environmental compatibility, and renewable origin with the structural and functional properties of MOFs. The resulting composites offer multifunctionality, including enhanced ionic conductivity, redox activity, catalytic performance, and mechanical strength, making them suitable for diverse energy-related applications such as energy storage, gas adsorption, and photocatalysis. The following Table 5.6 presents representative examples of MOF-biopolymer hybrid systems, their fabrication strategies and applications.

Table 5.6 shows a variety of bio-based MOF hybrid materials, illustrating the diversity of fabrication techniques and their adaptability to different application domains in energy systems. For instance, chitosan-based MOFs (Zn-MOF or Cu-MOF) demonstrate notable biocompatibility and electrochemical stability, suitable for sensors and membranes. Their fabrication involves straightforward solution casting or film formation, which is cost-effective and scalable. Cellulose-ZIF-8 hybrids, produced via in situ hydrothermal synthesis, are particularly efficient for gas adsorption and hydrogen storage due to their rigid porous structure and hydrogen-friendly framework. Similarly, lignin-MOF-derived carbon materials, processed through high-temperature carbonization, serve as promising candidates

Table 5.6 Types of micros to nano bio-based MOF hybrid materials in energy applications

Material	Fabrication process	Application	References
Chitosan-Zn-based MOF	Solvothermal synthesis of Zn-MOF followed by incorporation into chitosan matrix	Electrochemical biosensors and supercapacitor electrodes	[4]
Cellulose-ZIF-8 (Zeolitic imidazolate framework)	In situ growth of ZIF-8 crystals on cellulose nanofiber substrate under hydrothermal conditions	Gas adsorption and hydrogen storage	[82]
Lignin-MOF-derived porous carbon	Carbonization of lignin/MOF composite under inert atmosphere at high temperature	Anode materials for lithium-ion batteries	[37]
Chitosan-Cu-MOF hybrid	Mixing of Cu-MOF with chitosan solution and casting into films, followed by drying	Photocatalytic degradation of dyes and antimicrobial membranes	[19]
Silk fibroin-MOF nanocomposite	Blending of silk fibroin protein with MOF precursor solution followed by self-assembly	Flexible energy storage and bioelectronic devices	[60]

for lithium-ion battery anodes, offering excellent conductivity and structural integrity. The integration of silk fibroin with MOFs stands out for its utility in flexible devices, benefiting from the protein's natural mechanical strength and conformability. This diversity in biopolymer selection allows tuning of both physical and chemical properties of the hybrids to meet specific application requirements. In summary, these MOF-bio-based hybrids not only leverage the functional features of MOFs (e.g., porosity, catalysis) but also introduce eco-sustainability, biodegradability, and enhanced mechanical properties via the biopolymer components. Future development should focus on improving synthesis reproducibility, scalability, and long-term stability to enable industrial-scale deployment of these materials in next-generation energy systems.

These hybrids leverage the functional versatility of MOFs such as tunable porosity, redox activity and catalytic behavior alongside the sustainability, biodegradability, and mechanical adaptability of natural biopolymers. While promising, each material system exhibits its own balance of advantages and limitations that influence its feasibility in specific energy domains. The following Table 5.7 summarizes the key strengths and challenges associated with several representative MOF-biopolymer hybrids offering insights into their design and application constraints.

The materials reviewed highlight the tailored functionalities introduced by incorporating MOFs into various biopolymers. Chitosan-Zn-MOF hybrids are especially advantageous in biosensing and supercapacitors due to their natural electroactivity and compatibility, although they are hindered by moisture sensitivity and limited intrinsic conductivity. Cellulose-ZIF-8 combinations offer structural robustness and gas storage capacity, but stability under acidic environments remains a concern. Lignin-MOF-derived carbon materials showcase significant promise for battery electrodes with improved conductivity and stability, yet the high-temperature carbonization needed for synthesis poses scalability and energy concerns. In environmental applications, Chitosan-Cu-MOF hybrids provide antimicrobial and

Table 5.7 Advantages and limitations of MOF-bio-based hybrid materials in energy applications

Material	Advantages	Limitations
Chitosan-Zn-based MOF	Biodegradable, electroactive, good film-forming ability, enhanced surface area and porosity	Moisture sensitivity, limited electrical conductivity, potential metal leaching
Cellulose-ZIF-8	High mechanical strength, renewable, excellent gas adsorption capacity, environmentally benign	ZIF-8 may degrade under acidic conditions, moderate thermal stability
Lignin-MOF-derived porous carbon	High electrical conductivity, stable framework, sustainable precursor, good cycling stability	Complex carbonization process, potential structural collapse at high temperature
Chitosan-Cu-MOF hybrid	Antimicrobial, photocatalytic activity, biodegradable, tunable surface chemistry	Photocorrosion of Cu-based MOF, low long-term stability in aqueous conditions
Silk fibroin-MOF nanocomposite	Biocompatible, flexible, suitable for wearable electronics, supports uniform MOF dispersion	Mechanical integrity may degrade under humid conditions, processing challenges

photocatalytic benefits, making them suitable for membranes and water purification systems. However, photo-corrosion and stability in aqueous media remain technical barriers. Silk fibroin-MOF nanocomposites are well-suited for wearable and flexible energy devices due to their biocompatibility and flexibility, though mechanical stability under humid conditions needs further optimization. MOF-bio-based hybrid materials stand at the forefront of sustainable innovation in energy technology, offering unique advantages such as renewability, biocompatibility, and tunable physicochemical properties. However, material-specific limitations including environmental instability, fabrication complexity and mechanical weaknesses must be addressed through targeted material engineering and hybridization strategies. Future research should prioritize the development of scalable synthesis routes, structural stabilization techniques, and multifunctional architectures to fully realize the potential of these materials in real-world energy systems.

5.4 Applications in Renewable Energy Systems

The integration of renewable energy technologies into mainstream energy systems is a critical step toward achieving global sustainability goals and mitigating climate change. In this context, bio-based micro- and nanomaterials have emerged as transformative components due to their eco-friendly nature, resource abundance, and exceptional functional properties. These materials have shown significant potential in enhancing the performance, efficiency, and environmental compatibility of various renewable energy systems, including solar cells, biofuel production, fuel cells, and energy storage devices. Their structural tunability, high surface activity, and biocompatibility allow for improved energy conversion, catalytic activity, and charge transport. As renewable energy systems evolve to meet the increasing demand for clean and decentralized power, the incorporation of bio-derived materials presents a viable and innovative pathway to greener, more sustainable energy infrastructure.

5.4.1 Energy Storage Devices

Energy storage plays a critical role in enabling stable and efficient renewable energy systems by mitigating the intermittency of solar and wind power. Micro- to nanoscale bio-based materials have attracted growing attention for next-generation storage technologies such as lithium-ion batteries (LIBs), sodium-ion batteries (SIBs), and supercapacitors (SCs). Their inherent renewability, tunable chemistry, and structural versatility position them as sustainable alternatives to synthetic materials. This section provides a focused and critical review of the principal classes of bio-based materials used in energy storage devices, highlighting their functional roles as electrodes, electrolytes, and separators. The materials include bio-derived carbons from lignin and chitosan, nanocellulose-based separators, and hybrid biopolymer composites for flexible energy systems. Performance characteristics, synthesis approaches, and application contexts are summarized in Table 5.7, while Table 5.8 compares their advantages and limitations.

Carbonized biomaterials such as lignin, gelatin, chitosan, and starch offer high surface areas and functional group availability that support efficient charge storage. Through pyrolysis and chemical activation, these precursors yield porous carbons that serve as active electrode materials in LIBs and SCs. For example, nitrogen-doped carbon from lignocellulosic waste can achieve specific capacitances exceeding 300 F/g, with notable cycling durability. Chitosan- and gelatin-derived carbons, often enhanced by heteroatom doping, also demonstrate promising capacity and rate performance. However, electrical conductivity and structural uniformity remain limiting factors unless advanced templating or graphitization methods are employed. Meanwhile, nanocellulose derived from wood pulp or bacterial sources, is emerging as a biodegradable, mechanically robust alternative to conventional polymer separators. Its nanofibrous architecture facilitates ion transport while offering thermal and dimensional stability, thereby enhancing battery safety and performance. Yet, its moisture sensitivity and scalability must be addressed to transition from laboratory-scale demonstrations to commercial systems. Furthermore, flexible and wearable energy devices benefit from bio-nano hybrid materials that combine conductive polymers (e.g., polyaniline) with biochar, starch, or natural fibers. These composites are engineered to balance electrochemical activity with mechanical flexibility. For instance, starch-based biopolymer electrolytes exhibit promising ionic conductivity in sodium-ion battery configurations. Rice husk-derived biochar combined with polyaniline has enabled flexible supercapacitors with capacitance above 300 F/g. Nevertheless, issues of durability and uniformity under mechanical stress remain. Whilst bio-based materials are gaining traction as environmentally sustainable alternatives in energy storage devices such as lithium-ion batteries (LIBs), sodium-ion batteries (SIBs), and supercapacitors (SCs). Despite their eco-friendliness and abundant availability, they face limitations such as inconsistent conductivity, moisture sensitivity, and structural variability. To overcome these constraints, advanced processing techniques including heteroatom doping, 3D printing, and bio-templating are being explored to optimize pore structures and improve electrochemical properties. Moreover, lifecycle assessments and cost-efficiency evaluations are critical to ensure the scalability of these materials. Table 5.8 provides a comprehensive overview of various bio-derived materials, their processing parameters, performance metrics, and specific applications in energy storage technologies.

The table highlights six classifications of bio-based materials evaluated for use in energy storage systems. Among carbon-based electrodes, lignin-derived carbon demonstrated the highest surface area (up to 3742 m^2/g) and strong energy storage performance, with a specific capacitance of 504.7 F/g and energy density of 18.1 Wh/kg [80]. Chitosan- and gelatin-derived carbons showed slightly lower performance, but their renewable origin and processing ease make them viable for supercapacitor applications. Nanocellulose, sourced from bacterial cultures, was shown to be a promising separator or electrode material with surface areas ranging

Table 5.8 Comprehensive summary of bio-based material in energy storing devices

Classification	Material type	Source/precursor	Study/parameter	Yield/performance	Application	References
Bio-derived carbon for supercapacitors and batteries	Lignin-derived carbon	Kraft lignin (extracted from black liquor)	Sol-gel method to form aerogel Pre-carbonization at 400 °C for 3 h. (2 °C/min) under N_2 KOH activation at 700–900 °C for 1 h (1.5 °C/min). Mass ratio (carbon/KOH) = 1:3 Washed with HCl and DI water, dried at 105 °C	Surface area: up to 3742 m^2/g Specific capacitance: 504.7 F/g (at 0.2 A/g) Energy density: 18.1 Wh/kg Power density: 16,000 W/kg Cycle retention: 87.7% after 10,000 cycles	Supercapacitor electrode	[72]
	Chitosan-derived carbon	Shrimp shell chitosan	Carbonization at 300 °C, aerogel formation, NiO and Ni $(OH)_2$ nanoparticle deposition	Specific capacitance: 316 mAh/g at 1 A/g Energy density: 65 Wh/kg at power density 1500 W/kg Retention: 84% after 10,000 cycles	Supercapacitor electrode	[53]
	Gelatin-derived carbon	Gelatin + glutaraldehyde	Pre-carbonized at 300 °C (Ar, 1 h, 5 °C/min) KOH activation at 700 °C (1:1 ratio, Ar)	Specific capacitance: 392 F/g at 1 A/g Retention: 64% at 10 A/g Energy density: 6.39 Wh/kg at 50.03 W/kg; 5.00 Wh/kg at 502.79 W/kg	Supercapacitor	[70, 71]

(continued)

Table 5.8 (continued)

Classification	Material type	Source/precursor	Study/parameter	Yield/performance	Application	References
Nanocellulose in battery separators	Nanocellulose	Bacterial nanocellulose	Cultured in fique juice Chemical activation using H_3PO_4 and KOH Pyrolysis at 500–800 °C	Specific surface area: 680–893 m^2/g (KOH) Specific surface area: ~780 m^2/g (H_3PO_4) capacitance: 97.5–220 F/g (EIS in 1 M H_2SO_4) Best performance at 700–800 °C (KOH) 11 wt% P on surface (H_3PO_4, XPS)	Supercapacitor electrode	[69]
Hybrid composites for flexible storage	Biochar-polymer composite	Durian rind	Functionalization with HNO_3 and $KMnO_4$ Dried at 70 °C for 5 days Pyrolysis at 800 °C for 25 min under vacuum in muffle furnace Composite formation with PANI using in-situ polymerization	BET surface area: 835 m^2/g Specific capacitance: 615 F/g (MBCA) at 10 mV/s Energy density: 76.88 Wh/kg	Supercapacitor electrode	[66]
	Starch-based biopolymer electrolyte	Corn starch	Boric acid-PVA crosslinking method	Ionic conductivity: 9.2×10^{-3} (solid CMS), 1.13×10^{-2} S/cm (gel SA) Capacity:>210 mAh/g CE: 100% before 10 cycles	Lithium-ion battery electrolyte	[24]

from 680 to 893 m^2/g and capacitance up to 220 F/g, particularly after activation using KOH or H_3PO_4. In the hybrid category, durian rind-based biochar composites functionalized with polyaniline exhibited high capacitance (615 F/g) and notable energy density (76.88 Wh/kg), indicating excellent potential for flexible supercapacitor designs [46]. The studies demonstrate the versatility and promise of bio-based materials across various components of energy storage systems, though optimization in conductivity and long-term stability remains essential for commercialization. Table 5.9 summarizes the key advantages and challenges of selected bio-based materials commonly employed in energy storage devices such as supercapacitors and lithium-ion batteries. This comparative overview provides insights into their functional potential and identifies areas requiring further research and material engineering.

Table 5.8 categorizes six types of bio-based materials based on their strengths and inherent limitations in energy storage contexts. Lignin-derived carbon offers high carbon content, tunable porosity, and scalability, but suffers from irregular pore size and poor conductivity without enhancement. Nanocellulose stands out for its mechanical strength, flexibility, and biodegradability, yet it requires conductive additives and is sensitive to moisture due to its hygroscopic nature. Biochar polymer

Table 5.9 Advantages and limitations of selected bio-based materials in energy storage applications

Material type	Advantages	Limitations
Lignin-derived carbon	High carbon content; tunable porosity; low cost High carbon content; scalable; electroactive	Low structural uniformity; low initial efficiency Irregular pore size distribution; low electrical conductivity
Nanocellulose	Excellent mechanical strength; biodegradable; stable High surface area; flexible; sustainable	Requires conductive additives; moderate capacitance Highly hygroscopic; potential swelling; low thermal stability
Biochar-polymer composite	Lightweight; scalable; tunable surface chemistry Synergistic conductivity; scalable; customizable	Aging of polymer; reproducibility issues Poor polymer-filler interaction; low energy density
Chitosan-derived carbon	Abundant; N-doped conductivity enhancement Heteroatom doping improves activity; biodegradable	Poor electronic conductivity without doping Moisture-sensitive; batch variability
Gelatin-derived carbon	Flexible; eco-friendly; good electrochemical response Low-cost; porous; eco-friendly	Fragile structure; low capacitance compared to activated C Weak structural rigidity; lower performance at high rates
Starch-based biopolymer electrolyte	Non-toxic; cost-effective; compatible with Li salts Renewable; biodegradable; decent ionic transport	Poor mechanical integrity; lower voltage window Low ionic conductivity; mechanical fragility

composites, often derived from agricultural waste, combine lightweight design with customizable surface chemistry, but face issues like polymer aging, low energy density, and weak filler interaction. Chitosan-derived carbon, a nitrogen-rich biomaterial, enhances electrochemical activity through heteroatom doping, though its moisture sensitivity and low conductivity (in undoped forms) present obstacles. Gelatin-derived carbon is praised for its porosity and eco-friendliness, making it suitable for moderate-performance electrodes; however, it tends to have lower structural rigidity and inferior high-rate performance. Lastly, starch-based biopolymer electrolytes offer safe and biodegradable ionic conduction, but lack mechanical integrity and have a narrow voltage window, which limits their function in high-performance lithium-based batteries. The table underscores the eco-friendly promise of bio-based materials while pointing to the need for strategic engineering solutions to overcome their current technical constraints.

5.4.2 *Photovoltaic Systems*

Bio-based materials are gaining increasing attention in photovoltaic (PV) technologies as environmentally responsible alternatives to synthetic and inorganic components. Their renewable origin, low toxicity, and structural adaptability make them attractive for a range of photovoltaic applications, including light absorption, charge transport, and structural support. Micro- to nano-scale bio-based materials offer not only sustainability advantages but also opportunities for low-energy processing, material circularity, and novel device architectures [40]. This section outlines the major classes of bio-based materials in photovoltaic systems, including natural dyes, biopolymer electrolytes, bio-templated nanostructures, and flexible bio-derived substrates. Table 5.4 highlights the comparative benefits and limitations of each category.

Natural dyes, primarily derived from plant-based pigments such as anthocyanins, chlorophylls, and carotenoids, function as light-harvesting agents in dye-sensitized solar cells. These molecules can bind to semiconductor surfaces and initiate photo-induced charge transfer. Their molecular structures often include conjugated systems and functional groups like hydroxyl or carbonyl moieties, which facilitate electron injection into materials such as TiO_2. One of the main advantages of natural dyes lies in their biodegradability and accessibility from common biomass residues. Moreover, they can be extracted using low-impact solvents and mild processing conditions. However, their limited absorption range and susceptibility to photodegradation remain challenges for high-efficiency and long-term applications [34]. Therefore, structural modification and natural compound selection are being explored to enhance their photoactive properties and anchoring performance.

Biopolymer-based electrolytes represent a promising approach to replacing conventional liquid or gel electrolytes, which often involve hazardous solvents [25]. Polymers such as chitosan, gelatin, agar, starch, and cellulose derivatives can form quasi-solid or gel matrices that host ionic redox mediators. These materials are valued for their biodegradability, film-forming capability, and compatibility with low-temperature processing. Furthermore, biopolymer electrolytes improve the safety of photovoltaic devices by reducing leakage and evaporation. The ionic conductivity of these systems can be enhanced by adding plasticizers, salts, or ionic liquids, which increase ion mobility. However, maintaining a balance between mechanical integrity and ionic transport remains a significant design consideration. Crosslinking techniques, nanofiller inclusion, and advanced polymer blending are being investigated to address these limitations and improve the overall performance.

Bio-templated nanostructures utilize biological molecules or organisms as templates for controlling the morphology of semiconducting materials. DNA, proteins, viruses, and polysaccharides offer structurally organized frameworks that can guide the formation of nanostructured films or crystallites [64]. These biological templates can influence grain boundaries, porosity, and surface area, which in turn affect charge transport and light absorption within the photovoltaic device. One of the key advantages of bio-templating is the ability to generate complex structures under ambient conditions, aligning with principles of green fabrication [26]. Additionally, it enables bottom-up assembly processes that require minimal energy inputs. Nevertheless, reproducibility, material compatibility, and integration into device fabrication workflows remain barriers to commercialization. Ensuring thermal and chemical stability of biological templates under operational environments is essential for further development.

Flexible bio-derived substrates are emerging as key components in lightweight and portable PV systems [27] Materials such as regenerated cellulose, gelatin films, and cellulose acetate provide transparent and flexible supports for active layers in dye-sensitized or organic solar cells These substrates are obtained from renewable biomass and offer biodegradability, optical transparency, and compatibility with roll-to-roll manufacturing. Their mechanical flexibility makes them ideal for wearable electronics, foldable solar panels, and disposable sensors. However, they are typically sensitive to moisture and temperature, which may limit their stability in outdoor or high-load applications. Surface coatings, lamination techniques, and structural reinforcements are therefore applied to enhance water resistance, adhesion, and thermal tolerance, enabling broader functionality.

In recent years, the integration of bio-based and natural materials into solar cell components has gained attention for promoting greener and more sustainable photovoltaic technologies [45]. These materials offer promising benefits such as tunable light absorption, biodegradability, flexibility, and compatibility with environmentally friendly processing. However, their implementation still faces several limitations, including issues of stability, conductivity, and durability. Table 5.10 summarizes various bio-based materials currently explored for solar applications, highlighting their key advantages and the primary challenges limiting their widespread adoption. The table presents a comparison of various bio-based materials used in solar cell development, emphasizing their functional advantages and limitations. Natural dyes and biopolymer electrolytes offer eco-friendly alternatives due to their renewability, non-toxicity, and flexibility, though they suffer from drawbacks like poor stability and lower conductivity. Bio-templated films and gelatin substrates enable precise nanostructuring and flexible support, respectively, but are

Table 5.10 Advantages and limitations of bio-based materials for solar cell applications

Material type	Advantages	Limitations
Natural dyes	Renewable; non-toxic; biodegradable Light absorption, electron transfer	Poor light stability; narrow absorption range
Biopolymer electrolytes	Eco-friendly; flexible; easily processed Ion conductivity, flexibility	Lower conductivity than liquid electrolytes; aging effects
Bio-templated films	Precise nanostructuring; bio-recognition elements Nanostructure formation, morphology control	Processing complexity; limited scalability
Gelatin substrates	Flexible; biocompatible; low cost Flexible support layers	Limited durability; water sensitivity Moisture sensitivity
Cellulose acetate films	High transparency; mechanically robust	Difficult surface modification

limited by processing complexity and sensitivity to moisture. Cellulose acetate films, while mechanically robust and transparent, pose challenges in surface modification. Overall, while these materials show promise for sustainable solar technologies, their practical use is constrained by performance and durability issues.

5.4.3 Bioenergy Systems

Biochemical or thermochemical pathways. These systems form a critical part of the renewable energy portfolio, particularly in supporting waste-to-energy conversion, carbon neutrality, and rural energy access. Micro- and nano-scale bio-based materials have gained prominence in this sector, offering enhanced catalytic activity, improved stability, and better compatibility with biological and chemical processes [41]. Their high surface area, customizable surface chemistry, and tunable porosity enable precise control over reaction environments, supporting both microbial and enzymatic processes across a wide range of bioenergy platforms. Notably, materials such as biochar, chitosan nanoparticles, and nano-engineered carbon structures are being applied to various system types including microbial fuel cells (MFCs), nano-enabled biomass conversion, and algae-derived fuel production [22]. The integration of these bio-derived nanomaterials plays a pivotal role in making bioenergy systems more efficient, scalable, and environmentally benign.

One major application of nano-enabled bio-materials in bioenergy is nano-catalysts in biomass conversion. These catalysts, often synthesized from biochar, lignin, or plant-derived metals, serve as active agents in processes like pyrolysis, hydrothermal liquefaction, and fermentation. Their nanostructured morphology provides abundant active sites, improving catalytic selectivity and reducing energy input requirements. For example, iron- or nickel-loaded biochar can facilitate the cracking of lignocellulosic biomass into syngas or bio-oils under milder conditions. Furthermore, nano-catalysts enable the selective removal of oxygen-containing compounds from bio-oil, enhancing its energy density and compatibility with existing fuel infrastructure. The use of such catalysts also promotes valorization of agricultural waste, aligning with circular economy principles. For instance, a study by Yang et al. [79] investigated various engineered biochar-based catalysts including physically activated, nitrogen-doped, and nickel-doped biochars for use in catalytic biomass pyrolysis without additional steam input. Among these, nickel-doped biochar demonstrated superior performance, producing high syngas yields (78.2 wt%) and hydrogen yields (26 mmol H_2/g biomass) with improved gas energy conversion efficiency (73%). The catalyst's high surface area and porous structure were key to its performance. Notably, it retained catalytic activity across three consecutive cycles, highlighting its potential for sustainable, high-efficiency biomass conversion.

Microbial fuel cells (MFCs) represent another bioenergy system where nano-bio materials demonstrate critical functionality. In MFCs, microbes oxidize organic substrates and transfer electrons to an external circuit, generating electricity. To enhance electron transfer and microbial adhesion, nanostructured carbon materials such as those derived from bamboo, coconut shells, or algae are used as electrodes [54]. Their hierarchical porosity and high conductivity support dense biofilm formation and efficient charge transport. Additionally, surface modifications with heteroatoms (e.g., nitrogen or sulphur) can further improve electrocatalytic activity. These bio-based electrodes not only improve power density and operational lifespan but also offer a sustainable and cost-effective alternative to synthetic materials. For example, a study by Guo et al. [23] fabricated nitrogen-doped 3D expanded graphite foam (NEGF) through liquid nitrogen expansion and hydrothermal treatment of commercial graphite foil. The resulting NEGF electrodes exhibited significantly enhanced surface area, nitrogen-doped active sites, and improved electron transfer properties. When used as anodes in MFCs, NEGF achieved a maximum power density of 0.739 W/m^2, which was 17.4 times higher than the unmodified graphite foil anode (0.0451 W/m^2). This improvement was attributed to the synergistic effect of the 3D graphene-like structure and nitrogen doping, which promoted dense biofilm formation and effective microbial-electrode interaction. In the context of nano-encapsulation for biofuels, micro- and nano-materials serve to stabilize enzymes and microbial communities involved in biomass breakdown. For instance, chitosan nanoparticles can encapsulate cellulase enzymes, preserving their activity in harsh industrial environments. This encapsulation prevents enzyme denaturation, reduces leaching, and enhances substrate accessibility, particularly in the production of bioethanol or biobutanol from lignocellulosic feedstocks. Moreover, encapsulation techniques can be extended to whole-cell biocatalysts, improving their tolerance to toxic intermediates and increasing overall process efficiency. These advances contribute to lowering the cost of enzymatic biofuel production and expanding the usability of diverse biomass sources. For example, chitosan-functionalized magnetic nanocomposites (Ch-MNCs) have been developed to immobilize cellulases for biomass hydrolysis, offering high reusability, thermal and chemical stability, and ease of enzyme recovery through magnetic separation. These nanobiocatalytic systems provide a cost-effective and sustainable platform for the enzymatic conversion of lignocellulosic agro-residues into fermentable sugars, addressing the economic viability of second-generation biofuels [30].

Algae-based biofuels represent a promising avenue for third-generation bioenergy, with microalgae offering high lipid productivity and rapid growth under non-arable conditions. Nano-materials are used to enhance photosynthetic efficiency, harvesting, and oil extraction processes. For instance, a study demonstrated the application of biochar and bio-derived iron nanoparticles as cost-effective heterogeneous nano-biocatalysts in the transesterification process for biodiesel production from the marine microalgae *Phaeodactylum tricornutum* and *Isochrysis* sp. [6]. The use of these catalysts resulted in higher fatty acid methyl ester (FAME) yields compared to conventional acid catalysts confirming their suitability and enhanced activity. Moreover, biomass and chlorophyll productivity were improved through optimized nutrient supplementation (urea and rice husk), showing significant gains over control conditions. This study highlights how nano-enabled bio-catalysts not only enhance conversion efficiency but also contribute to eco-friendly and scalable algal biodiesel production. Carbon dots or iron oxide nanoparticles can be used to increase light absorption or stimulate metabolic pathways, thereby boosting lipid accumulation. Nano-additives may also serve as micronutrient carriers or growth promoters in cultivation systems, reducing the need for intensive chemical inputs. In downstream processing, nanostructured membranes and flocculants enable more energy-efficient dewatering and oil recovery. These approaches help address key limitations in algal biofuel scalability and economic feasibility. Table 5.11 shown

Table 5.11 Comparison of nano- and bio-based materials in bioenergy systems

Application area	Materials used	Strengths	Limitations
Biomass conversion	Biochar, lignin-based nano-catalysts (e.g., Ni-, Fe-loaded biochar)	High catalytic activity due to large surface area and porosity. Operates under milder conditions- improves syngas/bio-oil yield and quality	Catalyst deactivation over time Material performance may vary with feedstock. Recovery and reuse challenges
Microbial fuel cells (MFCs)	Nanostructured carbon (e.g., N-doped 3D expanded graphite foam)	Enhanced electron transfer Dense biofilm formation High power density Sustainable electrode alternative	Material degradation over long term use Fabrication cost for advanced structures
Enzyme/ microbial nano-encapsulation	Chitosan nanoparticles, magnetic nanocomposites (Ch-MNCs)	Protects enzyme activity under industrial stress Reduces enzyme loss Allows enzyme recovery via magnetism	Complexity in scale-up Potential leaching of nanoparticles into the environment
Algal biofuel production	Biochar, Fe nanoparticles, nano-additives (e.g., carbon dots)	Enhances lipid yield and photosynthesis Improves nutrient utilization Eco-friendly catalyst for transesterification	Economic feasibility still under development Potential nanotoxicity and material regulation issues

below summarize the comparison of nano- and bio-based materials in bioenergy systems. The application of bio-based micro- and nano-materials across these bioenergy systems demonstrates significant potential for improving energy conversion efficiency, material sustainability, and integration with circular bioeconomy models. Nevertheless, technical challenges remain particularly around material consistency, long-term biocompatibility, and the environmental fate of nano-additives. Future efforts should focus on scalable fabrication methods, hybrid bio-inorganic systems, and lifecycle assessments to guide sustainable deployment.

5.4.4 Hydrogen Production and Storage

Hydrogen is increasingly regarded as a key energy carrier for decarbonizing various sectors, from transportation and industrial processes to power generation. Its clean combustion profile, producing only water as a by-product, makes it especially attractive for sustainable energy systems. However, current hydrogen technologies often depend on fossil-based inputs and energy-intensive methods for production and storage. In response, bio-based micro- and nano-materials are being actively explored to enhance the sustainability, efficiency, and practicality of hydrogen generation and containment [87]. These materials, derived from renewable biomass or inspired by biological systems, offer advantages in photocatalysis, enzymatic activity, thermal stability, and adsorption properties.

Photocatalytic water splitting using bio-derived materials is one of the most promising methods for clean hydrogen production. In this approach, light energy typically from sunlight is used to drive the decomposition of water molecules into hydrogen and oxygen. Bio-based carbon materials derived from lignin, cellulose, or chitosan can serve as photocatalyst supports or even active components when modified through heteroatom doping or hybridization with semiconductors [44]. These materials are valued for their high surface area, structural defects, and functional groups that facilitate light absorption, charge separation, and catalytic activity. When integrated with metal oxides or quantum dots, their performance in visible light conditions improves considerably. This method presents an attractive alternative to traditional metal-based catalysts, while also addressing issues of sustainability, material abundance, and environmental impact.

In parallel, bio-inspired enzymatic catalysts offer a novel and efficient route to hydrogen production by mimicking the function of natural enzymes such as hydrogenases and nitrogenases. These biological catalysts operate under mild conditions and exhibit high selectivity for hydrogen evolution. Researchers have developed synthetic analogs of these enzymes, often using iron, nickel, or cobalt complexes supported on bio-derived matrices [36]. Carbon frameworks made from natural polymers, for instance, can stabilize these metal centers and offer a conducive environment for electron transfer. Such catalysts bridge the gap between biological efficiency and engineering scalability, enabling low-cost, biocompatible systems for hydrogen production. Furthermore, the low toxicity and biodegradability of the supporting materials align well with the goals of green chemistry and life cycle sustainability.

Biomass-derived hydrogen production is another approach that capitalizes on the chemical energy stored in organic waste. Using thermochemical processes such as gasification, pyrolysis, or steam reforming, biomass is converted into hydrogen-rich gases [3]. In these systems, micro- and nano-structured bio-materials often act as catalysts or catalyst supports. Biochar and carbon nanotubes derived from agricultural residues can be functionalized or impregnated with metals like nickel or cobalt to enhance catalytic activity. Their porous architecture allows for efficient heat and mass transfer, while their surface chemistry can be tuned to favor specific reaction pathways. Moreover, these systems can make use of non-food biomass and industrial by-products, contributing to resource efficiency and waste valorization. The main challenges lie in managing tar formation, catalyst fouling, and process scalability.

Hydrogen storage remains a critical aspect of the hydrogen economy, and bio-nanocomposites are emerging as potential alternatives to high-pressure or cryogenic systems. These storage materials typically rely on adsorption mechanisms and are composed of bio-based polymers or carbons integrated with metal nanoparticles or hydrides [48]. Porous carbons derived from sources like cellulose, starch, or alginate exhibit high surface areas and tunable pore structures, making them suitable for hydrogen physisorption at moderate pressures. Additionally, incorporating metal-organic frameworks (MOFs) or nanoparticles can further improve storage density and desorption kinetics. Bio-nanocomposites are particularly attractive due to their light weight, structural flexibility, and low environmental footprint. However, challenges such as low hydrogen capacity and thermal management remain, requiring further optimization in material design and processing. Collectively, bio-based micro- and nano-materials present valuable opportunities for advancing hydrogen production and storage technologies. Their roles in photocatalysis, enzymatic processes, biomass reforming, and hydrogen adsorption contribute to cleaner and more efficient systems [42]. Continued progress in material design and integration will be essential for practical deployment.

5.4.5 *Wind and Wave Energy Systems*

Marine and wind energy systems are increasingly central to sustainable energy transitions, offering clean and renewable alternatives to fossil fuels. However, both systems face harsh environmental conditions that demand advanced materials with high mechanical strength, durability, and resistance to environmental degradation. Bio-based micro- and nano-materials are gaining attention for their potential to meet these demands while also aligning with sustainability goals. These materials offer benefits such as biodegradability, tunable mechanical properties, and multi-functionality, which are valuable in enhancing the performance and longevity of energy infrastructure [5]. As these systems scale, integrating bio-derived materials into their design presents both an opportunity and a challenge, requiring balance between environmental compatibility and technical performance.

In the context of turbine blade design, the use of bio-composites presents a promising route toward more sustainable manufacturing [50]. Natural fibers such as flax, hemp, and bamboo, when combined with bio-based or partially bio-based resins, can yield lightweight materials with commendable tensile strength and fatigue resistance. These bio-composites are particularly appealing for small- to medium-scale wind turbines, though recent developments in nanofiber reinforcement such as with nanocellulose have opened possibilities for their use in larger structural components. Moreover, advances in resin chemistry and fabrication methods such as vacuum-assisted resin transfer molding have enabled better dispersion of fibers and improved bonding, which are crucial for load-bearing reliability. Their low embodied energy and reduced end-of-life environmental impact further contribute to their growing relevance in turbine applications.

Protective coatings represent another critical area where bio-based nanomaterials offer significant benefits. In marine environments, where corrosion from saltwater and biofouling by microorganisms can rapidly degrade metal and composite surfaces, nano-engineered coatings can extend equipment lifespan and reduce maintenance needs. Coatings derived from bio-based compounds such as chitosan or tannin, especially when combined with nanoparticles like zinc oxide or silver, exhibit both hydrophobic and antimicrobial properties [67] These functional coatings can be applied in thin layers, ensuring minimal weight addition while offering resistance to UV degradation, salt spray, and microbial adhesion. Furthermore, the ability to customize surface properties via chemical modification or nanoparticle incorporation enhances their adaptability across components ranging from rotor housings to submerged structural elements.

The conversion of ambient mechanical energy into electrical output using piezoelectric materials has long been of interest in renewable systems. Within marine and wind applications, bio-based piezoelectric materials offer a sustainable solution for harvesting vibrational energy from dynamic components. Materials such as cellulose nanocrystals and chitin-based films have shown piezoelectric responses suitable for powering low-energy devices [76]. Their flexibility, combined with environmental compatibility, makes them well-suited for embedding into moving parts of turbines or floating platforms, where continuous oscillations or mechanical stress occur. These systems can power small sensors or transmitters autonomously, contributing to the development of self-powered monitoring and control networks. Moreover, as piezoelectric responses in bio-based materials can be enhanced through orientation and hybrid structuring, their utility is expected to expand in more demanding operational contexts.

Monitoring the structural health of energy infrastructure is essential for ensuring safety, optimizing maintenance schedules, and preventing unexpected failures. Bio-based materials are increasingly used in the development of embedded or surface-mounted sensors capable of detecting strain, temperature fluctuations, and chemical exposure [39]. Nanostructured films composed of nanocellulose or gelatin, often infused with conductive or responsive nanoparticles, can function as flexible sensors. These sensors can be integrated into composite blades or anchored onto submerged structures, providing real-time data on material degradation, mechanical loading, or environmental stress. Their ability to conform to complex surfaces and their biocompatibility support long-term, in-situ monitoring without requiring frequent replacement. Furthermore, when combined with wireless transmission and energy-harvesting technologies, these materials support more autonomous and efficient maintenance regimes. Overall, the application of bio-based micro- and nano-materials in marine and wind energy systems reflects a growing emphasis on multifunctionality and sustainability. These materials are not only replacing synthetic alternatives but are also introducing new capabilities in terms of energy harvesting, corrosion resistance, and real-time monitoring. While there remain challenges related to long-term stability and scalability, continued advancements in fabrication methods and material integration are likely to accelerate their adoption.

5.5 Summary

The increasing demand for clean and efficient energy solutions has driven significant interest in the development of micro- to nano-scale bio-based materials as key enablers of next-generation renewable energy technologies. This chapter offers a comprehensive overview of the transformative role that these materials play in advancing sustainable energy systems. Emphasis is placed on the utilization of renewable biological resources such as lignocellulosic biomass, algal matter, and agricultural residues in thermochemical conversion processes like pyrolysis, gasification, and hydrothermal liquefaction to produce clean fuels such as bio-oil and syngas. Additionally, it highlights the integration of bio-derived nanomaterials such as nanocellulose, biochar, and carbon hybrids into energy storage and catalytic systems, where their high surface area, structural tunability, and redox activity improve performance in batteries, supercapacitors, and capacitive deionization applications. The text also explores the growing use of polymer nanocomposites and metal-organic frameworks (MOFs) hybridized with biopolymers for multifunctional energy devices. These innovations demonstrate strong potential in areas ranging from electrochemical sensing to flexible solid-state energy storage. Despite their advantages in biodegradability and environmental compatibility, challenges such as low conductivity, moisture sensitivity, and life-cycle impacts remain. To ensure broader deployment and industrial scalability, continued research is needed to optimize synthesis methods, minimize ecological trade-offs, and enhance techno-economic feasibility for next-generation renewable energy systems.

References

1. Ahmad A, Gondal MA, Hassan M, Iqbal R, Ullah S, Alzahrani AS, Memon WA, Mabood F, Melhi S (2023) Preparation and characterization of physically activated carbon and its energetic application for all-solid-state supercapacitors: a case study. ACS Omega 8(24):21653–21663. https://doi.org/10.1021/acsomega.3c01065
2. Alva MS, Nazareth RA, Sudhakar YN, Desai N (2025) Carbon quantum dot incorporated Zanthan gum based gel polymer electrolytes for high performance supercapacitors, pp 1–14
3. Alvarado-flores JJ, Alcaraz-vera JV, Guzm E, Rutiaga-quiñones JG (2024) Thermochemical production of hydrogen from biomass
4. Alzard MH, Alzard RH, Abdellah M (2025) Elaborating the knowledge structure and emerging research in the utilization of metal - organic formwork in engineering applications: a scientometric analysis from 2004 to 2024. Clean Mat 16(January):100314. https://doi.org/10.1016/j.clema.2025.100314
5. Ang TZ, Salem M, Kamarol M, Das HS, Nazari MA, Prabaharan N (2022) A comprehensive study of renewable energy sources: classifications, challenges and suggestions. Energ Strat Rev 43(November 2021):100939. https://doi.org/10.1016/j.esr.2022.100939
6. Anto S, Karpagam R, Renukadevi P, Jawaharraj K, Varalakshmi P (2019) Biomass enhancement and bioconversion of brown marine microalgal lipid using heterogeneous catalysts mediated transesterification from biowaste derived biochar and bionanoparticle. Fuel 255(July):115789. https://doi.org/10.1016/j.fuel.2019.115789
7. Armelin E, Pérez-Madrigal MM, Alemán C, Díaz DD (2016) Current status and challenges of biohydrogels for applications as supercapacitors and secondary batteries. J Mater Chem A 4(23):8952–8968. https://doi.org/10.1039/c6ta01846g
8. Baisch JS, Djadjo CL, Alves Nunes ÉJ, de Oliveira Carneiro L, Tres MV, Zabot GL (2025) Promising products based on hydrothermal liquefaction of agricultural biomass: an overview. Discover Appl Sci 7(5). https://doi.org/10.1007/s42452-025-06898-2
9. Baranov A, Leonova L, Belmesov A, Domashnev D, Levchenko A, Shmygleva L, Karelin A, Dremova N (2022) Acidic cesium salts of phosphotungstic acid: morphology, water content and ionic conductivity. Solid State Ionic 379(November 2021):115902. https://doi.org/10.1016/j.ssi.2022.115902
10. Bu F, Zhou W, Xu Y, Du Y, Guan C (2020) Recent developments of advanced micro-supercapacitors: design, fabrication and applications. NPJ Flex Electron. https://doi.org/10.1038/s41528-020-00093-6
11. Chaudhary MS, Seth D, Hasan S, Sultana S, Haider MA, Agarwal M, Khan MZ (2025) Hydrothermally synthesized PANI-decorated $CuMnO_2$ nanostructured electrode: powering flexible solid-state supercapacitors for wearable and portable electronics. https://doi.org/10.1021/acsaem.4c02311
12. Chénais B (2021) Algae and microalgae and their bioactive molecules for human health. Molecules 26(4):2–5. https://doi.org/10.3390/molecules26041185
13. Chin SX, Lau KS, Ginting RT, Tan ST, Khiew PS, Chia CH, Wongchoosuk C (2023) Facile preparation of carbon nanotubes/cellulose nanofibrils/manganese dioxide nanowires electrode for improved solid-sate supercapacitor performances. Polymers 15(18). https://doi.org/10.3390/polym15183758
14. Choudhary N, Singh S, Bhardwaj S, Gupta S, Nandi U, Chandra R, Maji PK (2024) Recent advancements in nanocellulose-based supercapacitors for energy storage devices: a review. Carbohydr Polym Technol Appl 7(December 2023):100416. https://doi.org/10.1016/j.carpta.2023.100416

15. de Assis SC, Morgado DL, Scheidt DT, de Souza SS, Cavallari MR, Ando Junior OH, Carrilho E (2023) Review of Bacterial Nanocellulose-Based Electrochemical Biosensors: Functionalization, Challenges, and Future Perspectives. Biosensors, 13(1), 142. https://doi.org/10.3390/bios13010142
16. Ding J, Yang Y, Poisson J, He Y, Zhang H, Zhang Y, Bao Y, Chen S, Chen YM, Zhang K (2024) Recent advances in biopolymer-based hydrogel electrolytes for flexible supercapacitors. ACS Energy Letts 9(4):1803–1825. https://doi.org/10.1021/acsenergylett.3c02567
17. Elisadiki J, Kibona TE, Machunda RL, Saleem MW, Kim W-S, Jande YAC (2020) Electrode materials for capacitive deionization: a review. Biomass Convers Biorefin 10(4). https://doi.org/10.1016/j.jelechem.2020.114416
18. Gamage A, Thiviya P, Mani S, Ponnusamy PG, Manamperi A, Evon P, Merah O, Madhujith T (2022) Environmental Properties and Applications of Biodegradable Starch-Based Nanocomposites. Polymers 14(21), 4578. https://doi.org/10.3390/polym14214578
19. Garg, R., Sabouni, R., Alaamer, A., Alali, A., Al-muqbel, D., Alqassem, H., & Almazrooei, K. (2023). Environmental Technology & Innovation Recent development in metal-organic framework-based hybrid nanocomposites for pollutants remediation from wastewater: challenges and opportunities. Environ Technol Innov, 32(October), 103446. https://doi.org/10.1016/j.eti.2023.103446
20. Geng Y, Wang J, Chen X, Wang Q, Zhang S, Tian Y, Liu C, Wang L, Wei Z, Cao L, Zhang J, Zhang S (2023) In situ N, O-dually doped Nanoporous biochar derived from waste eutrophic spirulina for high-performance supercapacitors. Nano 13(17). https://doi.org/10.3390/nano13172431
21. Ghazali I, Abdul-Rashid SH, Dawal SZM, Huda N, Shariff AHM, Herawan SG, Ho FH, Sakundarini N (2021) Guidelines for designing green products considering customers' cultural preferences. Sustainability (Switzerland) 13(2):1–27. https://doi.org/10.3390/su13020673
22. Guchhait SK, Khatana S, Saini RK, Pranay, Singh AD, Sarma AK (2024) Recent advances in biomass derived nano-structured carbon materials for low-temperature fuel cell application. Appl Cataly O Open 189(January):206924. https://doi.org/10.1016/j.apcato.2024.206924
23. Guo W, Chao S, Chen Q (2020) Improved power generation using nitrogen-doped 3D graphite foam anodes in microbial fuel cells. Bioprocess Biosyst Eng 43(1):143–151. https://doi.org/10.1007/s00449-019-02212-8
24. Hadad S, Hamrahjoo M, Dehghani E, Salami-Kalajahi M, Eliseeva SN, Moghaddam AR, Roghani-Mamaqani H (2022) Starch acetate and carboxymethyl starch as green and sustainable polymer electrolytes for high performance lithium ion batteries. Appl Energy 324(July). https://doi.org/10.1016/j.apenergy.2022.119767
25. Hasan MM, Islam MD, Rashid TU (2020) Biopolymer-based electrolytes for dye-sensitized solar cells: a critical review. Energy Fuel 34(12):15634–15671. https://doi.org/10.1021/acs.energyfuels.0c03396
26. Herrera-Beurnio MC, Hidalgo-Carrillo J, López-Tenllado FJ, Martin-Gómez J, Estévez RC, Urbano FJ, Marinas A (2021) Bio-templating: an emerging synthetic technique for catalysts. A review. Catalysts 11(11). https://doi.org/10.3390/catal11111364
27. Huang L, Li Y, Zheng Z, Bai Y, Russell TP, He C (2024) Flexible, transparent, and sustainable cellulose-based films for organic solar cell substrates. Mat Horiz 11(6):1560–1566. https://doi.org/10.1039/D3MH01998E
28. Jha S, Nanda S, Acharya B, Dalai AK (2022) A review of thermochemical conversion of waste biomass to biofuels. Energies 15(17):1–23. https://doi.org/10.3390/en15176352
29. Jia H, Shahi S, Shrestha LK, Ariga K, Michinobu T (2023) Improved supercapacitor performances by adding carbonized C60-based nanospheres to PVA/TEMPO-cellulose hydrogel-based electrolyte. RSC Adv 13(31):21502–21509. https://doi.org/10.1039/d3ra03349j
30. Kaur G, Taggar MS, Kalia A (2023) Cellulase-immobilized chitosan-coated magnetic nanoparticles for saccharification of lignocellulosic biomass. Environ Sci Pollut Res 30(52):111627–111647. https://doi.org/10.1007/s11356-023-27919-w

31. Kim M, Shin Y, Ahn K, Haeng C, Kang B (2022) New stepwise potentiostatic charge protocol for preventing dendrite formations in Li metal batteries. J Power Sources 532(April):231348. https://doi.org/10.1016/j.jpowsour.2022.231348
32. Klemm D, Kramer F, Moritz S, Lindström T, Ankerfors M, Gray D, Dorris A (2011) Nanocelluloses: a new family of nature-based materials. Angewandte Chemie Int Edn 50(24):5438–5466. https://doi.org/10.1002/anie.201001273
33. Kumar S, Muhammad S, Mehdi Z, Taunk M, Kumar S, Aherwar A, Singh S, Singh T (2025) Properties, stability, and applications of MXenes. https://doi.org/10.1039/d4ta08094g
34. Kusumawati Y, Hutama AS, Wellia DV, Subagyo R (2021) Natural resources for dye-sensitized solar cells. Heliyon 7(12):e08436. https://doi.org/10.1016/j.heliyon.2021.e08436
35. Lan J, Wu Y, Lin C, Chen J, Zhu R, Ma X, Cao S (2023) Totally-green cellulosic fiber with prominent sustained antibacterial and antiviral properties for potential use in spunlaced nonwoven fabric production. Chem Eng J 464(February):142588. https://doi.org/10.1016/j.cej.2023.142588
36. Leone L, Sgueglia G, La Gatta S, Chino M, Nastri F, Lombardi A (2023) Enzymatic and bioinspired systems for hydrogen production. Int J Mol Sci 24(10). https://doi.org/10.3390/ijms24108605
37. Li W, Shi J (2023, March) Lignin-derived carbon material for electrochemical energy storage applications: insight into the process-structure-properties- performance correlations. Front Bioeng Biotechnol:1–19. https://doi.org/10.3389/fbioe.2023.1121027
38. Lin X, Dong Y, Tao S, Feng X, Wang X, Song T, Liu J, Zhong Z, Wang Y, Qi H (2023) Temperature-gated nanocellulose membrane for enhanced and controllable osmotic energy harvesting. Nano Energy 107(December 2022):108156. https://doi.org/10.1016/j.nanoen.2022.108156
39. Mardanshahi A, Sreekumar A, Yang X, Barman SK, Chronopoulos D (2025) Sensing techniques for structural health monitoring: a state-of-the-art review on performance criteria and new-generation technologies. Sensors 25(5). https://doi.org/10.3390/s25051424
40. Miettunen K, Hadadian M, García JV, Lawrynowicz A, Akulenko E, Rojas OJ, Hummel M, Vapaavuori J (2024) Bio-based materials for solar cells. Wiley Interdisciplin Rev Energy Environ 13(1):1–16. https://doi.org/10.1002/wene.508
41. Mishra K, Siwal SS, Nayaka SC, Guan Z, Thakur VK (2023) Waste-to-chemicals: green solutions for bioeconomy markets. Sci Total Environ 887(December 2022):164006. https://doi.org/10.1016/j.scitotenv.2023.164006
42. Miskan SN, Abdulkadir BA, Setiabudi HD (2025) Materials on the frontier: a review on groundbreaking solutions for hydrogen storage applications. Chem Phys Impact 10(March):100862. https://doi.org/10.1016/j.chphi.2025.100862
43. Mutuma BK, Sylla NF, Bubu A, Ndiaye NM, Santoro C, Brilloni A, Poli F, Manyala N, Soavi F (2021) Valorization of biodigestor plant waste in electrodes for supercapacitors and microbial fuel cells. Electrochim Acta 391:138960. https://doi.org/10.1016/j.electacta.2021.138960
44. Nisar M, Khan N, Qadir MI, Shah Z (2025) Recent advances in TiO 2-based photocatalysts for efficient water splitting to hydrogen. pp 1–27. https://creativecommons.org/licenses/by/4.0/
45. Nkinyam CM, Ujah CO, Nnakwo KC, Kallon DVV (2025) Insight into organic photovoltaic cell: prospect and challenges. Unconven Resour 5(April 2024):100121. https://doi.org/10.1016/j.uncres.2024.100121
46. Norouzi O, Di Maria F, Dutta A (2020) Biochar-based composites as electrode active materials in hybrid supercapacitors with particular focus on surface topography and morphology. J Energy Storage 29(October 2019):101291. https://doi.org/10.1016/j.est.2020.101291
47. Ntunka MG, Khumalo SM, Makhathini TP, Mtsweni S, Tshibangu MM, Bwapwa JK (2025) Valorization of lignocellulosic biomass to biofuel: a systematic review. Chem Eng 9 (3):1–22
48. Osman AI, Nasr M, Eltaweil AS, Hosny M, Farghali M, Al-Fatesh AS, Rooney DW, Abd El-Monaem EM (2024) Advances in hydrogen storage materials: harnessing innovative technology, from machine learning to computational chemistry, for energy storage solutions. Int J Hydrog Energy 67:1270–1294. https://doi.org/10.1016/j.ijhydene.2024.03.223

49. Penchev H, Ivanova G, Hubenov V, Boyadzieva I, Budurova D, Ublekov F, Gigova A, Stoyanova A (2025) Supercapacitor cell performance with bacterial nanocellulose and bacterial nanocellulose/polybenzimidazole impregnated membranes as separator. Membranes 15(1). https://doi.org/10.3390/membranes15010012
50. Pender K, Bacharoudis K, Romoli F, Greaves P, Fuller J (2024) Feasibility of natural fibre usage for wind turbine blade components: a structural and environmental assessment. Sustainability (Switzerland) 16(13). https://doi.org/10.3390/su16135533
51. Periyasamy T, Asrafali SP, Islam M, Bari GAKMR, Lee J (2024) Polymer composite hydrogel based on polyvinyl alcohol/polyacrylamide/polybenzoxazine carbon for use in flexible supercapacitors. Polymers 16(11), 1463. https://doi.org/10.3390/polym16111463
52. Prabakar P, Mustafa Mert K, Muruganandam L, Sivagami K (2024) A comprehensive review on biochar for electrochemical energy storage applications: an emerging sustainable technology. Front Energy Res 12(November):1–17. https://doi.org/10.3389/fenrg.2024.1448520
53. Quan LH, Dieu Thuy UT, Nam PV, Van Chi N, Duong TX, Van Hoa N (2023) Chitosan-derived carbon aerogel nanocomposite as an active electrode material for high-performance supercapacitors. J Sci Adv Mat Dev 8(3). https://doi.org/10.1016/j.jsamd.2023.100586
54. Ramesh P, Gupta R, Koventhan C, Muralitharan G, Lo AY, Huang YJ, Ramasamy S (2025) Recent trends in the use of electrode materials for microbial fuel cells accentuating the potential of photosynthetic cyanobacteria and microalgae: a review. PRO 13(5). https://doi.org/10.3390/pr13051348
55. Rana AK, Hart P, Thakur VK (2024) Nanocellulose/zero, one- and two-dimensional inorganic additive based electrodes for advanced supercapacitors. Mater Today Electron 8(January):100103. https://doi.org/10.1016/j.mtelec.2024.100103
56. Sayed ET, Olabi AG, Shehata N, Al Radi M, Majdy Muhaisen O, Rodriguez C, Ali Atieh M, Abdelkareem MA (2023) Application of bio-based electrodes in emerging capacitive deionization technology for desalination and wastewater treatment. Ain Shams Eng J 14(8):102030. https://doi.org/10.1016/j.asej.2022.102030
57. Sezali NAA, Ong HL, Villagracia AR, Hoang TD (2024) Bio-based nanomaterials for energy application: a review. Vietnam J Chem 62(1):1–12. https://doi.org/10.1002/vjch.202300158
58. Shafizadeh A, Rastegari H, Shahbeik H, Mobli H, Pan J, Peng W, Li G, Tabatabaei M, Aghbashlo M (2023) A critical review of the use of nanomaterials in the biomass pyrolysis process. J Clean Prod 400(December 2022):136705. https://doi.org/10.1016/j.jclepro.2023.136705
59. Sharif M, Tavakoli S (2023) Biodegradable chitosan-graphene oxide as an affective green filler for improvingof properties in epoxy nanocomposites. International Journal of Biological Macromolecules 233, 123550. https://doi.org/10.1016/j.ijbiomac.2023.123550
60. Shen X, Shi H, Wei H, Wu B, Xia Q, Yeo J (2022) Engineering natural and recombinant silks for sustainable biodevices. Front Chem 10(May):1–16. https://doi.org/10.3389/fchem.2022.881028
61. Silva AP, Argondizo A, Juchen PT, Ruotolo LAM (2021) Ultrafast capacitive deionization using rice husk activated carbon electrodes. Sep Purif Technol 271(January):118872. https://doi.org/10.1016/j.seppur.2021.118872
62. Sivadas BO, Ashcroft I, Khlobystov AN, Goodridge RD (2021) Advanced industrial and engineering polymer research laser sintering of polymer nanocomposites. Adv Ind Eng Polym Res 4(4):277–300. https://doi.org/10.1016/j.aiepr.2021.07.003
63. Siwal SS, Sheoran K, Saini AK, Vo DVN, Wang Q, Thakur VK (2022) Advanced thermochemical conversion technologies used for energy generation: advancement and prospects. Fuel 321(April):124107. https://doi.org/10.1016/j.fuel.2022.124107
64. Sotiropoulou S, Sierra-Sastre Y, Mark SS, Batt CA (2008) Biotemplated nanostructured materials. Chem Mater 20(3):821–834. https://doi.org/10.1021/cm702152a
65. Tadesse MG, Lübben JF (2023) Review on hydrogel-based flexible supercapacitors for wearable applications. Gels 9(2):1–20. https://doi.org/10.3390/gels9020106

66. Thines KR, Abdullah EC, Ruthiraan M, Mubarak NM, Tripathi M (2016) A new route of magnetic biochar based polyaniline composites for supercapacitor electrode materials. J Anal Appl Pyrolysis 121:240–257. https://doi.org/10.1016/j.jaap.2016.08.004
67. Thomas P, Sahoo BN, Thomas PJ, Greve MM (2024) Recent advances in emerging integrated anticorrosion and antifouling nanomaterial-based coating solutions. Environ Sci Pollut Res 31(60):67550–67576. https://doi.org/10.1007/s11356-024-33825-6
68. Urdl K, Weiss S, Christöfl P, Kandelbauer A, Müller U, Kern W (2020) Diels-Alder modified self-healing melamine resin. Eur Polym J 127(February). https://doi.org/10.1016/j.eurpolymj.2020.109601
69. Villarreal-Rueda J, Zapata-Benabithe Z, Posada L, Martínez E, Herrera S, López S, Sobrido ABJ, Castro CI (2023) Bacterial nanocellulose from Komagataeibacter Medellinensis in fique juice for activated carbons production and its application for supercapacitor electrodes. Polymers 15(7). https://doi.org/10.3390/polym15071760
70. Wang G, Lin Z, Jin S, Li M, Jing L (2022a) Gelatin-derived honeycomb like porous carbon for high mass loading supercapacitors. J Energy Storage 45(September 2021):103525. https://doi.org/10.1016/j.est.2021.103525
71. Wang Z, Ge X, Li H, Chen S, Yang Y, Zhu Y, & Yan J. (2022). Gelatin-derived hierarchical porous carbon for high-performance supercapacitors. Carbon Trends, 8, 100183. https://doi.org/10.1016/j.cartre.2021.100183
72. Wang T, Liu Z, Li P, Wei H, Wei K, Chen X (2023) Lignin-derived carbon aerogels with high surface area for supercapacitor applications. Chem Eng J 466(February). https://doi.org/10.1016/j.cej.2023.143118
73. Wu Z, Wang X, Hu D, He N, Zhao L, Yin C, Pan H, Chen Z, Lin Q (2025) Sawdust-based N, O and P co-doped porous biochar for high-performance supercapacitors. New J Chem 49(26):11451–11461. https://doi.org/10.1039/D5NJ01662B
74. Xiaorui L, Haiping Y (2023) A state-of-the-art review of N self-doped biochar development in supercapacitor applications. Front Energy Res 11(March):1–22. https://doi.org/10.3389/fenrg.2023.1135093
75. Xu T, Du H, Liu H, Liu W, Zhang X, Si C, Liu P, Zhang K (2021) Advanced nanocellulose-based composites for flexible functional energy storage devices. Adv Mater 33(48). https://doi.org/10.1002/adma.202101368
76. Xu M, Wen Y, Shi Z, Xiong C, Zhu F, Yang Q (2024) Piezoelectric biopolymers: advancements in energy harvesting and biomedical applications. Polymers 16(23). https://doi.org/10.3390/polym16233314
77. Xue W, Ahangaran F, Wang H, Theato P, Cheng YJ (2025) Gel polymer electrolytes for lithium batteries: advantages, challenges, and perspectives. Macromol Rapid Commun:2500207. https://doi.org/10.1002/MARC.202500207;JOURNAL:JOURNAL:01732803;WGROUP:STRING:PUBLICATION
78. Yan Y, Sun W, Wei Y, Liu K, Ma J, Hu G (2025) Review of biomass-derived carbon nanomaterials—from 0D to 3D—for supercapacitor applications. Nano 15(4):1–24. https://doi.org/10.3390/nano15040315
79. Yang H, Cui Y, Jin Y, Lu X, Han T, Sandström L, Jönsson PG, Yang W (2023) Evaluation of engineered biochar-based catalysts for syngas production in a biomass pyrolysis and catalytic reforming process. Energy Fuel 37(8):5942–5952. https://doi.org/10.1021/acs.energyfuels.3c00410
80. Younes K, Amine S, El Sawda C, El-Zahab S, Arayro J, Mezher R, Halwani J, Ouddane B, Gazo-Hanna E (2025) Principal component analysis of biomass-derived carbon aerogels: unveiling key performance factors for supercapacitor applications. Sustainability 17(10):4530. https://doi.org/10.3390/SU17104530
81. Zhan Y, Hu Y, Chen Y, Yang Q, Shi Z, Xiong C (2021) In-situ synthesis of flexible nanocellulose/carbon nanotube/polypyrrole hydrogels for high-performance solid-state supercapacitors. https://doi.org/10.21203/RS.3.RS-184104/V1

82. Zhang X, Liu Y, Li S, Kong L, Liu H, Li Y, Han W, Yeung KL, Zhu W, Yang W, Qiu J (2014) New membrane architecture with high performance: ZIF – 8 membrane supported on vertically aligned ZnO nanorods for gas permeation and separation. Chem Mater 26:1975–1981
83. Zhang XY, Zhu YR, Chen Y, Dou SY, Chen XY, Dong B, Guo BY, Liu DP, Liu CG, Chai YM (2020) Hydrogen evolution under large-current density based on fluorine-doped cobalt-iron phosphides. Chem Eng J 399(April):125831. https://doi.org/10.1016/j.cej.2020.125831
84. Zheng Q, Cai Z, Ma Z, Gong S (2015) Cellulose nanofibril/reduced graphene oxide/carbon nanotube hybrid aerogels for highly flexible and all-solid-state supercapacitors. ACS Appl Mater Interfaces 7(5):3263–3271. https://doi.org/10.1021/am507999s
85. Zhou J, Wang H, Du C, Zhang D, Lin H, Chen Y, Xiong J (2022) Cellulose for sustainable triboelectric nanogenerators. Advanced Energy and Sustainability Research 3, 2100161. https://doi.org/10.1002/aesr.202100161
86. Zhu Y, Xu X, Guo Q, Han Y, Jiang H, Jiang H, Wang H, Evdokimov PV, Kapitanova OO, Volkov VS, Wang Y, Xiong S, Zheng C, Xie K, Jiao X, Liu Y (2023) Reinforced interface endows the lithium anode with stable cycle at high-temperature of 80 °C. J Energy Chem 78:325–332. https://doi.org/10.1016/j.jechem.2022.12.004
87. Żukowska G, Durczyńska Z, Roszkowski S, Myszura-Dymek M, Bik-Małodzińska M (2024) Possibilities of using bio-based nanomaterials in sustainable agriculture. J Ecol Eng 25(3):313–322. https://doi.org/10.12911/22998993/182966

Chapter 6
Nanostructured Bio-aerogels for Energy Conversion and Storage Applications

6.1 Overview of Bio-aerogels

Bio-aerogel is a highly porous, lightweight material derived from biological or renewable sources. It retains the ultra-light, high-surface-area characteristics of traditional aerogels while being more environmentally friendly and often biodegradable. These materials are typically synthesized from biopolymers such as cellulose, chitosan, alginate, starch, or proteins, and they can be used in various applications in biomedicine, thermal insulation, air purification, and energy conversion and storage. Because of their biocompatibility and sustainability, bio-aerogels are gaining attention as alternatives to synthetic aerogels derived from fossil fuels. Although there has long been interest in aerogels, the last few decades have seen a significant surge in the production of goods based on these compounds because of their unique qualities. The porous network-like nanostructure of bio-aerogels, which are solid materials, has unique properties like low density and thermal conductivity, high surface area and porosity, good impact damping, resistance to flame and moisture, low optical index of refraction, and low dielectric constant. Their aforementioned qualities make them suitable for a wide range of uses in the food industry, wastewater treatment, environmental contamination, pharmaceutical, cosmetics, and medical sectors. Bio-aerogels are created by adding air to the gel network meshes in place of the solvent. Supercritical drying is one method of carrying out this replacement during the drying process. Because aerogels contain around 95% air or gas by volume, they are extremely light in weight and exhibit great porosity found in mixes of mesopores, micropores (less than 2 nm), and macropores (more than 50 nm). Based on the Nita et al. [42], the aerogel depends on the reaction circumstances and they also have an outstanding interior surface area.

Bio-based aerogels are typically made from renewable resources such plant oils, sugar cane, proteins, and starches. Furthermore, aerogels with particular properties appropriate for biomedical applications like tissue engineering, regenerative

M. I. Ahmad, A. K. H. P. S., *Environmental and Energy Technology*, Green Energy and Technology, https://doi.org/10.1007/978-3-032-10706-0_6

medicine, and drug delivery systems have already been successfully prepared using biopolymers (such as chitosan, alginate, pectin, lignin, cellulose, and protein). Ionic liquids, sometimes referred to as "green solvents," are widely employed in the production of aerogels based on polysaccharides as an alternative to the traditional organic solvents used as templates for porous materials. Organic salts that are liquids at room temperature, have strong electric conductivity and thermostability, and are easily recyclable are known as ionic liquids. Ionic liquids can offer the most necessary processing platforms for these biopolymers as high added-value materials, as the main drawback that can prevent their use in creating aerogels for medical applications is the dissolution of certain polysaccharides in organic solvent or water. Furthermore, the development of a variety of porous bio-based aerogels has been made possible by the combination of green solvents, like ionic liquids, which are used to dissolve polysaccharides, and an environmentally friendly manufacturing technique, like supercritical fluid technology, which can be readily modified to extract the ionic liquids. Thus, utilizing ionic liquids, polysaccharides like chitin, cellulose, starch, or lignin have been employed to create aerogels with varying levels of porosity. Furthermore, the gelation of cellulose is facilitated by ionic liquids like 1-allyl-3-methylimidazolium chloride, which results in the creation of nanoporous materials with a consistent structure and a high degree of elasticity. However, the melting point is one of the physical characteristics of ionic liquids that might affect the materials' surface area. Investigation emphasized the impact of this characteristic on the creation of porous cellulose-based aerogels. They found that aerogels with larger surface areas were produced when ILS with lower melting points, like 1-ethyl-3-methylimidzolium acetate or 1-ethyl-3-methylimidazolium diethyl phosphate, were used [11, 38].

6.2 Importance of Nanostructuring in Energy Applications

Bio-based aerogels have gained significant attention in energy applications due to their exceptional thermal insulation, low density, and environmentally friendly composition. Derived from renewable resources such as cellulose, chitosan, and lignin, these bio-aerogels provide a sustainable alternative to synthetic insulation materials commonly used in construction and industrial sectors. Their highly porous nanostructure enables extremely low thermal conductivity, reducing energy loss in buildings, pipelines, and thermal energy storage systems. By minimizing heat dissipation, bio-based aerogels enhance energy efficiency, making them ideal for use in green building designs and passive heating solutions. Additionally, these materials contribute to renewable energy technologies by optimizing thermal management in concentrated solar power plants, where efficient heat retention is essential for maintaining energy output. As highlighted by Mandić et al. [39], the lightweight nature of bio-based aerogels further reduces material transportation costs and environmental footprints, aligning with global efforts to transition toward sustainable energy solutions.

Beyond insulation, bio-based aerogels are revolutionizing energy storage and conversion technologies, particularly in batteries, supercapacitors, and fuel cells. Their ultralow density and large surface area facilitate faster ion and electron transport, enhancing charge storage capacity and cycling stability in lithium-ion and sodium-ion batteries. Additionally, bio-based aerogels function as effective electrode scaffolds and separators, improving battery safety and longevity while reducing reliance on environmentally harmful synthetic polymers. These aerogels also demonstrate promising applications in hydrogen energy systems, where they serve as catalysts and support structures for fuel cells, optimizing reaction efficiency and fuel utilization. Recent research has explored the functionalization of bio-based aerogels with conductive nanomaterials, such as graphene or carbon nanotubes, further improving their electrochemical performance. Given the increasing demand for high-performance energy storage solutions, the development of bio-based aerogels represents a crucial step toward more sustainable and biodegradable alternatives in the energy sector [39].

According to Mandić et al. [39], bio-based aerogels are playing an essential role in energy harvesting and environmental remediation. Their highly porous network allows for efficient adsorption of pollutants, including carbon dioxide and volatile organic compounds, contributing to greenhouse gas reduction and cleaner energy production. In solar-driven applications, bio-based aerogels are used in photocatalysis for hydrogen generation and wastewater treatment, leveraging their ability to support catalytic nanoparticles. Furthermore, they enhance thermal insulation in electronic devices and aerospace components, preventing energy loss and overheating in high-performance applications. Their potential extends to smart grid technologies, where they improve insulation in power transmission lines, reducing energy dissipation and enhancing grid efficiency. As industries continue to explore eco-friendly materials for advanced energy applications, bio-based aerogels emerge as a transformative solution, offering sustainability, efficiency, and multifunctionality across various sectors. With ongoing research and development, their role in shaping the future of energy technologies is expected to expand further.

6.3 Properties and Characteristics of Nanostructured Bio-aerogels

Nanostructured bio-aerogels exhibit highly intricate and interconnected porous structures at the nanoscale, making them exceptional materials for a wide range of applications. **According to** Mandić et al. [39], these aerogels are derived from renewable bio-based precursors such as cellulose, chitosan, lignin, and alginate, processed using advanced techniques like sol-gel polymerization and freeze-drying or supercritical drying. Their low density, which can be as low as 0.01 g/cm^3, and porosity exceeding 90% allow them to be incredibly lightweight yet structurally robust. Nita et al. [42] further reported that the nano-sized pores contribute to superior mechanical strength, high surface area, and excellent thermal insulation.

Furthermore, the 3D network of these materials plays a crucial role in their adsorption and diffusion capabilities, making them ideal for applications requiring controlled release, such as drug delivery and environmental remediation. Unlike conventional aerogels, bio-aerogels offer sustainability advantages, as they are biodegradable and do not pose environmental hazards upon disposal. Their tunable porosity and surface chemistry enable customization for specific applications, including biomedical scaffolding, catalytic supports, and energy storage materials. Additionally, the nanoscale design of bio-aerogels provides enhanced interaction with other materials, which improves their integration into composite systems for multifunctional uses. The combination of high surface-area-to-volume ratio and flexible mechanical properties makes them highly desirable in fields where lightweight yet strong materials are essential. Their ability to retain their structural integrity under extreme conditions, such as high humidity or varying pH environments, further enhances their potential for real-world applications in healthcare, energy, and environmental industries.

The thermal and mechanical properties of nanostructured bio-aerogels are among their most distinguishing features, making them suitable for applications in insulation, flexible electronics, and energy storage. **As described by** Nita et al. [42]**,** bio-aerogels possess extremely low thermal conductivity often below 0.02 W/m·K due to their nanoscale porosity and ability to trap air. This makes them one of the most efficient thermal insulation materials available, surpassing traditional insulators such as fiberglass and polystyrene foams. These properties are particularly valuable in aerospace, construction, and cryogenic applications where minimizing heat transfer is crucial. Additionally, their lightweight nature makes them highly attractive for portable and wearable thermal management systems. Mandić et al. [39] highlighted that bio-aerogels can be engineered to show remarkable flexibility and compressibility, depending on composition and crosslinking density. Some bio-aerogels, like cellulose- or silk-based variants, demonstrate exceptional elasticity and can recover their original shape after deformation, making them ideal for use in strain sensors and flexible electronics. However, their mechanical robustness can be a challenge, as pure bio-aerogels tend to be brittle. To overcome this limitation, researchers often employ composite reinforcement strategies, such as integrating bio-polymers with carbon nanotubes or graphene, to enhance strength without compromising porosity. Furthermore, their resistance to high temperatures and flame-retardant capabilities have expanded their use in fireproofing materials. These features, combined with their biocompatibility, also allow bio-aerogels to serve as excellent candidates for biomedical applications, such as tissue engineering, where a balance between mechanical strength and biodegradability is essential for successful implementation [39, 42].

One of the most advantageous aspects of nanostructured bio-aerogels is their ability to undergo chemical functionalization, which enhances their physicochemical properties and expands their application potential. The surface of bio-aerogels typically contains abundant functional groups depending on the bio-based precursor used. These reactive sites allow for post-synthetic modifications, including grafting with nanoparticles, metal ions, or hydrophobic/hydrophilic coatings, to achieve specific functionalities. For example, in biomedical applications, bio-aerogels can be functionalized with drug molecules, proteins, or growth factors to serve as efficient carriers for controlled drug delivery or tissue regeneration scaffolds. In environmental science, bio-aerogels modified with functional groups like sulfonates or amines exhibit enhanced adsorption capacity for heavy metals, dyes, and organic pollutants, making them effective for wastewater treatment. Their surface chemistry also enables them to act as catalysts or catalyst supports in chemical reactions, particularly in green chemistry applications where sustainability is a priority. Hydrophilic modifications allow bio-aerogels to absorb large amounts of water, making them useful for wound dressings and moisture retention in agricultural applications. On the other hand, hydrophobic modifications can make them effective oil-absorbent materials for environmental spill cleanups. The ability to precisely modify the surface chemistry of bio-aerogels significantly enhances their performance, allowing them to function efficiently in diverse environments, from biological systems to extreme industrial applications. This adaptability enables their use in cutting-edge fields such as nanomedicine, where they facilitate targeted drug delivery, as well as in environmental engineering for pollution control and sustainable materials science for next-generation eco-friendly innovations [39, 42].

Nanostructured bio-aerogels have demonstrated remarkable potential in energy storage and biomedical applications due to their unique structural, thermal, and surface characteristics. Mandić et al. [39] emphasized that their high porosity and low density make them suitable for supercapacitors, lithium-ion batteries, and fuel cells. Their large surface area promotes efficient ion diffusion and charge transfer, thus improving storage capacity, conductivity, and cycling stability. Nita et al. [42] noted that carbonized bio-aerogels derived from lignin, chitin, or cellulose show high electrical conductivity, making them ideal for flexible and lightweight energy storage devices. Additionally, their mechanical flexibility allows seamless integration into wearable electronics. In biomedical applications, bio-aerogels have gained attention for drug delivery, wound healing, and tissue engineering. Their porous matrix enables controlled release in response to stimuli like pH or temperature especially useful in cancer treatment where precise dosing is crucial. According to Mandić et al. [39], they also mimic the extracellular matrix, supporting cell adhesion and proliferation for regenerative therapies. The ability to encapsulate bioactive molecules further supports their use as wound dressings, accelerating healing while preventing infections. With continued research, these multifunctional materials are expected to revolutionize both energy and medical technologies.

6.4 Synthesis and Fabrication of Nanostructured Bio-aerogels for Energy Conversion and Storage Applications

The synthesis of bio-based aerogels generally follows a multi-step process that involves the careful combination of precursor materials, subsequent gelation, and, most critically, the removal of solvents that occupy the porous structure of the wet gel as illustrated in Fig. 6.1. This final step is particularly crucial, as it must be executed in a way that preserves the gel original volume and prevents structural collapse or significant shrinkage. To achieve this, the solvent within the pores is transformed into a supercritical fluid, which is then gradually released in a controlled manner as a gas. This approach ensures that the aerogel maintains the intricate network structure formed during the gelation stage, thereby retaining the properties necessary for its intended applications. Given the wide range of potential applications for aerogels, researchers and material scientists have increasingly focused on customizing their properties to suit specific functional requirements. One of the key aspects of this customization involves the precise modulation of the internal pore structure, which directly influences the material thermal insulation capabilities, mechanical strength, and overall performance. Additionally, surface modifications, various types of coatings, and specialized post-treatment techniques have been explored as effective strategies for enhancing aerogel functionality. These modifications can improve properties such as hydrophobicity, thermal stability, and adsorption efficiency, thereby expanding the potential uses of aerogels in diverse fields ranging from biomedical engineering to environmental remediation [42].

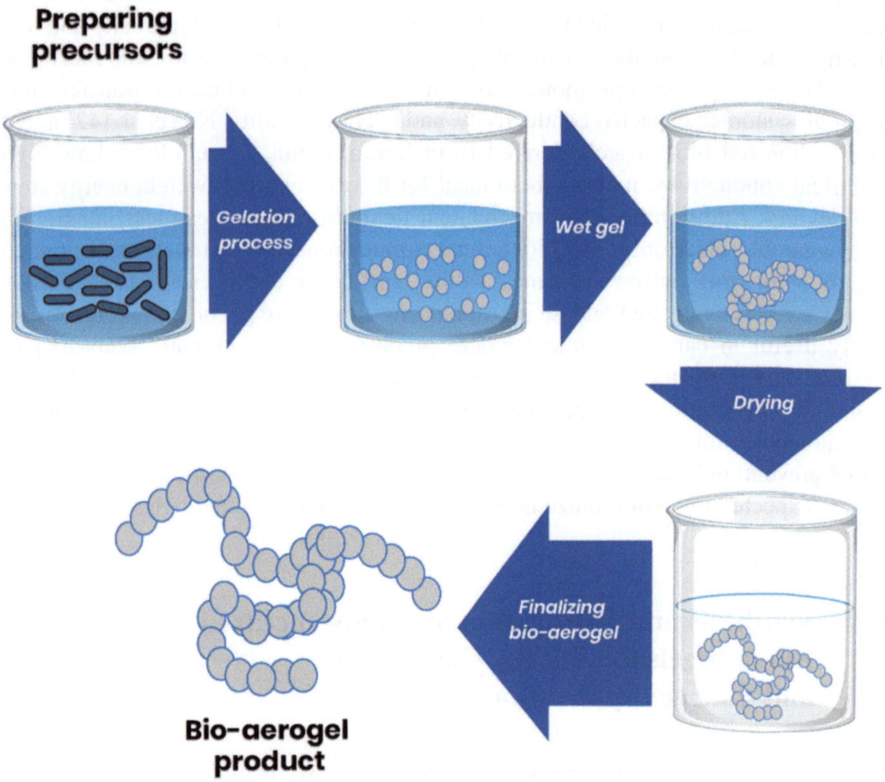

Fig. 6.1 Synthesis of nanostructured bio-aerogels

Another important aspect of recent research efforts in aerogel development is the search for sustainable and alternative raw materials for their synthesis. Traditional aerogel production has relied heavily on synthetic or petroleum-derived precursors, but increasing emphasis is being placed on the utilization of renewable and biodegradable sources. Biomass-derived polymers, such as cellulose, chitosan, alginate, and lignin, are being investigated for their potential to serve as environmentally friendly precursors. The incorporation of such bio-based materials not only contributes to sustainability but also enhances the biodegradability of the final aerogel products, making them more suitable for green applications. In addition to modifying aerogel composition and structure, researchers have also explored the development of organic-inorganic hybrid networks to enhance their mechanical properties. Purely organic aerogels often exhibit limitations in terms of mechanical stability and flexibility, which can restrict their applicability in certain fields. To address this challenge, hybrid materials combining organic polymeric networks with inorganic components, such as siloxanes, have been successfully developed. The integration of inorganic elements imparts improved mechanical robustness and flexibility, making the aerogels more resistant to mechanical stress and deformation while retaining their ultralight nature and high porosity. Overall, advancements in aerogel research continue to focus on optimizing their structural integrity, enhancing functional properties, and ensuring sustainability through the exploration of novel materials and synthesis techniques. These efforts aim to expand the utility of aerogels across multiple industrial and scientific domains, including energy storage, catalysis, drug delivery, and advanced insulation technologies [49, 50].

A critical stage in the fabrication of aerogels is the drying process, as it directly influences the final material's morphology, porosity, and overall structural stability. The effectiveness of this step determines whether the aerogel retains its highly porous network or undergoes deformation, shrinkage, or fracturing. Traditional drying techniques, such as simple solvent evaporation, often generate capillary forces within the gel's intricate pore system. These forces create internal stress, which can lead to pore collapse or even cracks in the material. **According to** Zuo et al. [74], to prevent such structural damage, researchers have explored alternative drying techniques that minimize or eliminate capillary stress. Among these methods, supercritical drying (using alcohol, acetone, or CO_2) is highly effective, as it avoids liquid-gas phase transitions. Additionally, ambient pressure drying, freeze-drying, microwave drying, and vacuum drying have been utilized, each offering distinct advantages depending on the required aerogel properties and the intended application.

Supercritical drying is a specialized process used to remove liquid from wet gels while preserving their intricate porous structures. The technique involves placing the wet gel inside a sealed chamber and gradually increasing both the temperature and pressure until they surpass the critical temperature and pressure of the liquid present in the gel pores. Under these extreme conditions, the distinction between the liquid and vapor phases disappears, effectively eliminating capillary forces that would otherwise cause structural collapse due to surface tension effects [10]. Once the gel reaches a supercritical state, the gas is released from the chamber, followed by controlled cooling, ultimately yielding an aerogel that retains its unique structural properties. One of the defining characteristics of supercritical drying is that under supercritical conditions, the surface tension between liquid and gas is reduced to zero. This absence of interfacial tension is crucial in preserving the nano-scale architecture of the gel, leading to materials with minimal shrinkage, finer pore structures, and significantly higher specific surface areas. When drying is conducted using supercritical carbon dioxide, the gel structure is particularly well-protected, ensuring that key nanostructural features remain intact. This method is particularly advantageous in applications requiring extremely low thermal conductivities, even lower than that of air, making aerogels excellent insulators.

Despite its effectiveness in producing high-quality aerogels, supercritical drying presents several notable challenges. The process often requires extended durations spanning several days making it impractical for large-scale or industrial manufacturing. Moreover, it relies on substantial solvent usage and necessitates access to expensive high-pressure equipment, which significantly elevates production costs and raises concerns about solvent disposal and environmental impact [52]. The critical point conditions vary by solvent: for water, the critical temperature is approximately 374 °C at around 22 MPa, whereas ethanol has a lower critical temperature (~243 °C) and critical pressure (~6.3 MPa) [9]. Supercritical carbon dioxide ($scCO_2$), by contrast, offers milder operating conditions (~31 °C at 7.4 MPa) and has been widely adopted in aerogel research for biopolymers like carrageenan, chitin, alginate, cellulose, and agar due to its low toxicity and recyclability [1]. Researchers have explored alternative drying methods such as freeze-drying and ambient pressure drying, which, while potentially less effective at preserving nanoscale structures, offer more cost-effective and environmentally sustainable solutions to aerogel production [19, 36, 42].

Ambient pressure drying is one of the most commonly employed techniques in industrial applications due to its simplicity and energy efficiency. As discussed by Zuo et al. [74], unlike supercritical drying, which requires high temperatures and pressures, ambient pressure drying operates under normal atmospheric conditions, making it a cost-effective and accessible method for producing aerogels. In this process, organic aerogels are first synthesized through a sol-gel method, forming a wet gel structure. To prepare the material for drying, the gel undergoes a solvent exchange process, typically using acetone or ethanol, which replaces the original solvent present in the gel network. This step is crucial as it helps in minimizing the capillary stresses that could lead to structural collapse. Once the solvent exchange is complete, the gel is dried under ambient pressure without the need for specialized equipment. However, this method has its drawbacks. During drying, the evaporation of liquid from the gel can result in significant shrinkage, causing the material to lose its porous structure. In some cases, rather than forming a lightweight, highly porous aerogel, the material may collapse into a dense film with minimal porosity. Despite these challenges, ambient pressure drying remains a viable alternative for aerogel production due to its lower cost and reduced energy consumption.

Freeze-drying, also known as lyophilization, is a technique that involves the removal of frozen water from the pores of a wet precursor material through a process called sublimation. According to Long et al. [36], this technique operates under low pressure, enabling the frozen liquid to transition directly from solid to gas without entering the liquid phase. The initial step involves freezing the wet gel, which solidifies the liquid phase within the porous structure. Once frozen, the gel is subjected to sublimation, where ice crystals formed during freezing evaporate under controlled vacuum conditions. The resulting dried material, referred to as a cryogel, differs from traditional aerogels in its properties. Cryogels tend to have a higher density and a lower surface area compared to aerogels, with porosities reaching up to 80% but only about half of the internal surface area of aerogels. This occurs due to the growth of large ice crystals within the gel matrix during freezing, which expands the number of macropores and contributes to volume shrinkage. Despite these differences, cryogels exhibit less shrinkage and a more uniform pore size distribution when compared to materials dried through hot-drying or vacuum-drying techniques. Freeze-drying is widely regarded as a relatively simple, cost-effective, and environmentally sustainable method, as it primarily utilizes water as a solvent. Furthermore, it is particularly suitable for bio-based polymers such as casein, pectin, alginate, gelatin, hyaluronic acid, and cellulose. However, there are notable drawbacks, including extended processing times, high energy consumption, and the risk of structural collapse due to volume changes during the freezing stage. Despite these challenges, cryogels offer superior thermal and mechanical properties compared to conventional polymeric foams, making them valuable in applications where durability and insulation are essential.

Microwave drying is another innovative method that has been explored for the production of aerogels with desirable properties, including high surface area and well-developed porosity. This technique involves the application of microwave energy to rapidly remove moisture from the gel structure, facilitating the drying process in a significantly shorter time compared to conventional methods. Aerogels produced through microwave drying exhibit structural characteristics similar to those obtained via freeze-drying. However, they tend to have a more intricate network of small-sized, interconnected macropores, which can enhance their mechanical stability and functional properties. Additionally, microwave drying is recognized for its efficiency, as it drastically reduces processing time while still yielding aerogels with promising porosity and textural properties. This method presents a valuable alternative to more time-consuming drying techniques, making it a promising option for large-scale aerogel production [74].

Another innovative technique for producing spherical biopolymer particles involves jet cutting, a method that enables precise control over particle size and shape. As reported by Preibisch et al. [45] solutions of biopolymers such as amidated pectin, sodium alginate, and chitosan (with concentrations ranging from 1 to 3 wt.%) were used as precursor materials in the jet-cutting process to form uniform gel particles. The jet-cutting process allowed for the formation of uniform gel particles, which were then subjected to further processing to enhance their structural and functional properties. Following their formation, these gel particles underwent a solvent exchange process using ethanol to replace the water content within their porous network. This step was essential for preparing the particles for subsequent supercritical drying, which preserved their high porosity and minimized shrinkage. The final aerogel particles exhibited an exceptionally large surface area of approximately 500 m^2/g, making them highly suitable for applications requiring strong adsorption properties and stability. This method offers great potential for producing high-performance aerogel materials.

6.5 Applications in Energy Conversion

Bio-aerogels, derived from renewable biopolymers, offer exceptional properties like high porosity, low density, and large surface area, making them ideal for energy conversion applications. These eco-friendly materials enhance performance in supercapacitors, batteries, and fuel cells by improving charge storage, ion transport, and thermal insulation, supporting sustainable and efficient energy solutions. Figure 6.2 presents the applications of bio-aerogels in energy conversion.

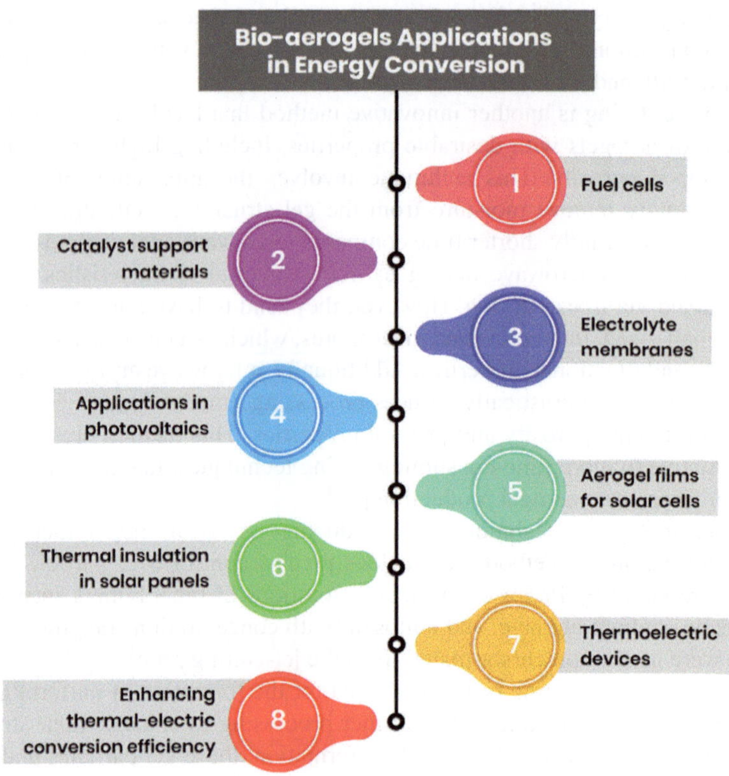

Fig. 6.2 Applications of bio-aerogels in energy conversion

6.5.1 *Bio-aerogels in Fuel Cells*

Bio-based aerogels have garnered significant attention in recent years as sustainable and multifunctional materials for energy applications, particularly in fuel cells. Their unique properties such as high porosity, large surface area, low density, and tunable surface chemistry make them excellent candidates for supporting catalytic activity, enhancing ion transport, and improving overall system efficiency. When derived from natural polymers like chitosan and cellulose, these aerogels not only offer environmental compatibility but also open new pathways for cost-effective and efficient fuel cell designs. Recent advancements have highlighted their potential in developing next-generation components, including catalysts and membranes, underscoring their pivotal role in advancing green and renewable energy technologies. The limitations associated with using powdered materials in catalytic applications particularly issues related to handling, recovery, and reusability can be effectively addressed by transitioning to monolithic materials. Among these, polymer-based aerogels have emerged as highly promising alternatives due to their distinctive structural and functional properties. Notably, polymer aerogels are characterized by their exceptionally high specific surface area, which plays a critical role in facilitating the diffusion and transport of gases such as hydrogen. This makes them particularly suitable for catalytic processes where efficient mass transfer is essential. To further enhance the functional capabilities of these materials, researchers have explored the integration of inorganic matrices with bio-derived substances. The rationale behind this approach lies in the complementary nature of these materials. While inorganic matrices often possess favorable mechanical properties, bio-based materials contribute unique physicochemical characteristics such as hydrophilicity, biodegradability, and tunable porosity. By combining these two classes of materials, it becomes possible to fabricate hybrid aerogels with tailored properties that meet the demands of specific catalytic applications. In a notable study, it was demonstrated the effectiveness of utilizing bio-based materials in hydrogen generation systems. The work focused on the hydrolysis of sodium borohydride ($NaBH_4$) to produce hydrogen gas—a reaction that can be controlled simply by adjusting the position of the catalyst within the system [65]. This level of control offers a practical method for achieving on-demand hydrogen generation, which is a significant step toward the development of portable and efficient hydrogen-based energy solutions.

Additionally, the aerogel catalyst used in their experiments exhibited excellent reusability. After each reaction cycle, the catalyst could be regenerated through simple mechanical compression, which removed the residual reactants and restored the aerogel's original 3D porous structure. This recyclability greatly enhances the cost-efficiency and sustainability of the system. Beyond hydrogen generation, the application of bio-based aerogels is expanding into other energy-related domains, including wastewater treatment and energy harvesting. For instance, an investigation was conducted on the use of chitosan-derived aerogels as supporting materials for Fe-N-C catalysts in microbial fuel cells [29]. Chitosan, a nitrogen-rich biopolymer, served as a nitrogen precursor that improved the catalytic activity of the system. The findings revealed a notable increase in power density, thereby confirming the feasibility of using economically sustainable and eco-friendly materials in microbial fuel cell technologies. Further advancements in fuel cell technology were made by researchers who explored the combination of cellulose-based aerogels with a polymeric matrix. Specifically, they incorporated quaternized branched polyethyleneimine (q-bPEI) into a cellulose aerogel matrix, producing a composite material with outstanding chemical stability and elevated ionic conductivity. This hybrid material was employed in the fabrication of an anion exchange membrane for fuel cell applications. Although the maximum power density achieved was 46.3 mW cm^{-2}—a relatively modest output—it demonstrated sufficient potential to warrant further investigation. Electrochemical analyses revealed an activation energy of approximately 16 kJ mol^{-1}, corroborating the promising ionic transport characteristics of the system. While this discussion highlights several novel applications of bio-based aerogels in catalytic and energy systems, it is important to note that a considerable portion of current research focuses on their use in energy storage devices, such as batteries and supercapacitors. Given the depth and breadth of research in this specific area, a comprehensive review dedicated solely to the role of aerogels in energy storage may be more appropriate for a separate publication [68].

6.5.2 *Catalyst Support Materials*

Bio-aerogels serve as excellent scaffolds for dispersing metal or metal oxide nanoparticles, which act as active catalytic sites. Their porous structure allows for enhanced diffusion of reactants and products, while their large surface area provides ample anchoring sites for catalysts, minimizing aggregation and maintaining high activity over prolonged use. One of the key advantages of bio-aerogels is their compatibility with sustainable energy systems. Unlike conventional catalyst supports such as carbon black or silica, bio-aerogels are biodegradable and derived from low-cost, abundant natural resources. This aligns with the growing need for green chemistry approaches in energy technology. For instance, cellulose-based aerogels have been used to support platinum and palladium nanoparticles for fuel cell applications, showing improved electrocatalytic performance and durability. As noted by Jayan et al. [25], cellulose-based aerogels have been used to support platinum and palladium nanoparticles for fuel cell applications, showing improved electrocatalytic performance and durability.

In hydrogen production, chitosan-derived aerogels loaded with metal catalysts have demonstrated efficient catalytic activity in water splitting and reforming reactions. These materials not only enhance catalytic efficiency but also offer thermal and chemical stability, which is critical for practical energy conversion systems. Additionally, bio-aerogels offer opportunities for functionalization and hybridization with other materials, allowing for the fine-tuning of properties to match specific catalytic requirements. For example, doping with heteroatoms (e.g., nitrogen or sulfur) or integrating carbonaceous materials can improve electrical conductivity and active site availability, further optimizing the performance of the catalyst-support system. As a result, researchers are exploring various composite bio-aerogels to develop next-generation energy devices with higher efficiency and lower environmental impact.

6.5.3 Electrolyte Membranes

Bio-aerogels, due to their exceptional physicochemical characteristics, have emerged as promising materials in the development of next-generation electrolyte membranes for energy storage and conversion devices. These lightweight, highly porous materials, derived from renewable biopolymers such as cellulose, chitosan, alginate, and lignin, offer unique advantages in ionic conductivity, mechanical stability, and environmental sustainability. As highlighted by researchers, the mesoporous structure of bio-aerogels enhances ion transport by facilitating the efficient diffusion of charge carriers through interconnected pores [5]. This characteristic is particularly beneficial in applications such as lithium-ion batteries and fuel cells, where ion conductivity and stability at elevated temperatures are crucial. Furthermore, bio-aerogels can be engineered to exhibit high surface area and tunable pore size, enabling their incorporation as scaffolds or separators within solid-state electrolytes. Studies have demonstrated that when functionalized or doped with ionic liquids, inorganic fillers, or cross-linkers, bio-aerogels exhibit enhanced electrochemical performance, thermal stability, and reduced swelling in aqueous environments. In polymer electrolyte membranes, the integration of cellulose-based aerogels has shown to improve water retention, proton conductivity, and mechanical durability, making them suitable candidates for proton exchange membrane fuel cells [39]. Meanwhile, a study emphasizes that the abundant hydroxyl and amino groups on bio-based aerogel surfaces enable strong interactions with ionic species, improving charge mobility and compatibility with polymer matrices. Additionally, their biodegradability and low toxicity contribute to a more sustainable lifecycle for electrochemical devices. However, challenges such as scalability, uniform pore distribution, and the need for controlled functionalization remain areas for further research. Addressing these limitations through advanced fabrication techniques like freeze-drying, supercritical drying, and sol-gel synthesis could significantly enhance the consistency and commercial viability of bio-aerogel-based electrolyte membranes. Overall, the incorporation of bio-aerogels into electrolyte systems represents a synergistic approach that merges green chemistry principles with cutting-edge material science, offering a sustainable pathway toward high-performance, eco-friendly energy devices [12].

6.5.4 Applications in Photovoltaics

Bio-aerogels, with their inherently high porosity, tunable surface chemistry, and lightweight nature, are emerging as innovative materials in the field of photovoltaics, offering promising enhancements to both the efficiency and sustainability of solar energy devices. Derived from renewable biopolymers such as cellulose, chitosan, and lignin, bio-aerogels present a unique combination of optical transparency, thermal insulation, and structural flexibility that makes them ideal for integration into various photovoltaic architectures. Alwin and Shajan [5] highlighted how aerogels, including bio-based variants, can be engineered with specific refractive indices and scattering properties essential for optimizing light management in solar cells. These characteristics enable bio-aerogels to function effectively as anti-reflective coatings or light-scattering layers, thereby increasing the light absorption within the active layers of photovoltaic devices without adding significant weight or compromising transparency. Significantly, the incorporation of bio aerogels as encapsulants or substrates in organic and dye-sensitized solar cells has been shown to improve device durability and energy conversion efficiency. Baia et al. [7], reported that using porous materials with controlled nanostructures such as aerogels have facilitates better diffusion of photoactive materials and electrolytes, contributing to enhanced charge separation and reduced recombination losses. Bio-aerogels also offer excellent thermal insulation properties, which are advantageous in protecting photovoltaic devices from heat-induced degradation a critical factor in maintaining long-term performance under outdoor conditions. Moreover, they can be doped or functionalized with conductive or photoactive nanoparticles (e.g., TiO_2, ZnO, or graphene derivatives) to enhance electron transport and support multifunctional roles within the device structure. Arachchige and Brock [6], emphasized the importance of mesoporous structures in improving the performance of dye-sensitized solar cells, where bio-aerogels can serve as scaffolding materials for anchoring light-absorbing dyes and facilitating efficient electrolyte penetration. Additionally, Mohanan et al. [40] explored the potential of integrating bio-aerogels into hybrid photovoltaic systems, such as perovskite solar cells, where their compatibility with organic-inorganic interfaces supports improved stability and enables flexible device fabrication. Notably, the environmental advantages of using bio-aerogels biodegradability, non-toxicity, and the use of renewable feedstocks address the ecological concerns surrounding traditional photovoltaic materials, making them attractive candidates for sustainable solar technologies. Despite their promise, challenges remain in optimizing the mechanical robustness and long-term environmental stability of bio-aerogels under operational conditions. Furthermore, achieving consistent material quality at scale requires further refinement of processing techniques such as sol-gel synthesis and supercritical drying. Nevertheless, the interdisciplinary convergence of green materials science, nanotechnology, and photovoltaic engineering positions bio-aerogels as versatile components that could revolutionize the design of high-performance, eco-friendly solar cells. As research advances, the strategic use of bio-aerogels could not only enhance the efficiency of current photovoltaic technologies but also pave the way for innovative solar harvesting systems with reduced environmental footprints and improved adaptability for flexible or wearable applications.

6.5.5 Transparent Aerogel Films for Solar Cells

Transparent aerogel films derived from bio-based materials are emerging as innovative components in solar cell technology, particularly due to their multifunctionality in light management, thermal insulation, and environmental sustainability.

Bio-aerogels, ultra-lightweight, porous materials synthesized from renewable biopolymers such as cellulose, chitosan, and alginate offer a combination of high optical transparency, low thermal conductivity, and mechanical flexibility, making them ideal for use as transparent coatings, substrates, or encapsulants in solar energy devices. According to Wu et al. [59], transparent bio-aerogel films exhibit outstanding optical transmittance while maintaining a high surface area and tunable porosity, allowing them to serve as both anti-reflective layers and light-scattering interfaces in solar cells. These properties are essential for enhancing light trapping and absorption, especially in thin-film and flexible solar technologies, where maximizing photon capture within limited thicknesses is critical. Moreover, transparent bio-aerogel films contribute significantly to the thermal and environmental stability of photovoltaic systems. As highlighted by some researchers Kim and Kim [28], incorporating transparent aerogels into the upper layers of solar cells can suppress thermal buildup caused by prolonged solar exposure, thereby mitigating thermal degradation of active materials and enhancing overall device lifespan. The use of bio-aerogels in this capacity also aligns with green manufacturing practices, as they are derived from abundant, biodegradable feedstocks and processed under relatively mild conditions, reducing the environmental footprint associated with conventional synthetic aerogels. Functionalization of these films. For example, by embedding nanoparticles such as TiO_2, SiO_2, or conductive polymers can further improve their light management properties and mechanical resilience while maintaining transparency. Hamann et al. [21] demonstrated that such hybrid structures can enhance power conversion efficiency by increasing light harvesting without introducing additional optical losses. Additionally, the inherent flexibility and formability of bio-aerogel films make them particularly attractive for next-generation solar applications, including building-integrated photovoltaics, wearable solar textiles, and foldable solar modules.

Their capacity to conform to irregular surfaces without compromising performance opens new avenues for device integration in non-traditional settings. However, achieving uniform thickness, consistent refractive index, and long-term durability under environmental stressors such as UV radiation, moisture, and mechanical strain remain challenges that require further research and development. Optimizing sol-gel chemistry, cross-linking techniques, and drying methods will be essential to scaling up transparent bio-aerogel film production with reproducible properties suitable for industrial photovoltaic applications. In conclusion, the application of transparent bio-aerogel films in solar cell technology represents a promising advancement that combines high-performance functionality with ecological responsibility. These films not only improve light transmission and thermal management but also contribute to the creation of sustainable, adaptable, and aesthetically pleasing solar solutions. As fabrication technologies evolve and interdisciplinary efforts continue to address current material limitations, bio-aerogel-based transparent films are poised to play a transformative role in the next generation of solar energy devices.

6.5.6 Thermal Insulation in Solar Panels

Bio-aerogels have attracted significant attention as innovative thermal insulating materials for solar panels, offering a sustainable and high-performance alternative to conventional insulators. Composed of renewable biopolymers such as cellulose, starch, chitosan, and alginate, bio-aerogels exhibit ultra-low thermal conductivity, high porosity, and exceptional lightness, making them ideal candidates for minimizing heat loss in solar energy systems. As highlighted by some studies, the unique nano-porous network structure of bio-aerogels limits solid and gas phase conduction, enabling them to serve as super-insulators with thermal conductivities as low as 0.02 W/m·K. This property is critical in solar panels, especially in photovoltaic-thermal hybrid systems, where excessive heat accumulation can reduce electrical efficiency and degrade materials over time. Integrating bio-aerogels into the panel structure, either as a back-sheet insulating layer or as part of the encapsulation helps maintain optimal operating temperatures and improves long-term energy yield. In PV/T systems, efficient thermal insulation is necessary not only to enhance electrical output but also to support effective heat recovery. Thermal management directly influences the overall system performance, and materials like bio-aerogels can significantly reduce thermal losses while maintaining the structural and optical integrity of the panel. Additionally, the low density and flexibility of bio-aerogels simplify integration into diverse solar technologies, including flat-plate collectors, concentrated solar power units, and building-integrated photovoltaics. The studies further emphasize that aerogels, when engineered with transparency or semi-transparency, can also serve dual roles in thermal insulation and optical management, an important factor for preserving light transmission while minimizing infrared heat gain.

The environmental benefits of using bio-aerogels are equally compelling. Traditional insulation materials like polyurethane foams or mineral wool are often non-renewable, energy-intensive to produce, and difficult to recycle. In contrast, bio-aerogels are derived from natural, biodegradable sources and can be processed using relatively low-energy methods such as freeze-drying or supercritical CO_2 drying [35]. Their green credentials align with the sustainability goals of modern solar infrastructure, reducing the carbon footprint of the panel manufacturing process. Furthermore, bio-aerogels can be functionalized with additives such as infrared reflectors, fire retardants, or hydrophobic agents to enhance durability and performance under real-world conditions, making them more viable for outdoor solar applications. Despite their advantages, there are still challenges to address in scaling up bio-aerogel use in solar insulation. Mechanical fragility, moisture sensitivity, and the high cost of fabrication remain barriers to widespread adoption [46, 47]. However, ongoing advancements in composite aerogels and cross-linking techniques are steadily improving their mechanical and environmental stability. Overall, the application of bio-aerogels in solar panel insulation represents a powerful synergy of nanotechnology, renewable materials, and clean energy engineering. With continued innovation, bio-aerogels are poised to play a vital role in improving the efficiency, sustainability, and lifespan of solar energy systems worldwide [67].

6.5.7 *Thermoelectric Devices*

Bio-aerogels, derived from natural biopolymers like bacterial cellulose, have emerged as promising materials in the development of thermoelectric devices due to their unique combination of low thermal conductivity, flexibility, and environmental friendliness. These porous structures facilitate efficient heat-to-electricity conversion by maintaining significant temperature gradients essential for thermoelectric performance. Its thermoelectric performance was tunable through adjustments in temperature differences and applied compressive strain. Specifically, the aerogel-based thermoelectric generator achieved an optimal output power of 84 nW under a 50 K temperature difference and 30% strain, highlighting its potential in flexible and wearable energy-harvesting applications. Another advancement reported by Lin et al. [33], involved the creation of a composite aerogel integrating Bi_2Te_3 nanosheets with a three-dimensional carbon network (CN@Bi_2Te_3). This material effectively combined electromagnetic wave absorption with thermoelectric properties. The CN@Bi_2Te_3 aerogel demonstrated ultra-low thermal conductivity (0.07 W m^{-1} K^{-1}) and achieved a minimum reflection loss of 51.30 dB with an effective absorption bandwidth of 6.20 GHz at an optimal thickness of 2.8 mm. Under an 80 K temperature gradient, the flexible thermoelectric generator incorporating this aerogel produced a maximum output power of 42.2 μW.

Furthermore, by harnessing 2.45 GHz microwaves to establish the temperature difference, the EMA-TE-coupled device attained an open-circuit voltage of 38.4 mV, a short-circuit current of 1.03 mA, and an output power of 9.87 μW after 50 s of radiation. This approach not only addresses electromagnetic pollution but also contributes to energy recycling, paving the way for large-area flexible electromagnetic wave-to-electric energy conversion devices [37]. The integration of organic porous materials, such as bio-aerogels, with nanostructured thermoelectric materials has been extensively reviewed by Chen et al. [14], emphasizing their potential in next-generation thermoelectric applications. These hybrid materials leverage the synergistic effects of organic matrices and inorganic nanostructures to enhance thermoelectric performance while maintaining mechanical flexibility and environmental sustainability. The unique structural properties of bio-aerogels, including high porosity and surface area, contribute to their effectiveness in thermoelectric devices by facilitating efficient charge transport and minimizing thermal conductivity. bio-aerogels represent a significant advancement in thermoelectric technology, offering a sustainable and efficient solution for energy harvesting. Their integration into thermoelectric devices not only enhances performance but also contributes to environmental sustainability by utilizing natural materials and addressing issues like electromagnetic pollution. Continued research and development in this field are expected to further optimize these materials, expanding their applicability in wearable electronics, environmental energy harvesting, and beyond.

6.5.8 *Enhancing Thermal-Electric Conversion Efficiency*

Bio-aerogels, derived from natural biopolymers such as cellulose, have garnered significant attention for their potential to enhance thermoelectric conversion efficiency. Their intrinsic properties which is lightweight nature, high porosity, and exceptional thermal insulation make them ideal candidates for improving the performance of thermoelectric devices. By integrating bio-aerogels with conductive materials, researchers have developed composite structures that effectively bridge the gap between thermal insulation and electrical conductivity, thereby optimizing thermoelectric efficiency. One notable advancement involves the development of a three-dimensional directional cellulose-based carbon aerogel integrated with epoxy resin. This composite not only exhibits excellent photothermal conversion efficiency but also delivers an impressive output voltage of 160.29 mV in solar-thermal-electric applications. The hierarchical porous structure of the cellulose-based carbon aerogel facilitates efficient light absorption and heat transfer, while the epoxy resin enhances the mechanical stability of the aerogel, making it a robust candidate for sustainable energy harvesting. In another study, researchers fabricated a composite material combining carboxymethylcellulose sodium-derived carbon aerogel with an organic phase change material. This hybrid structure demonstrated a remarkable light-thermal conversion efficiency of up to 90.3%. Furthermore, the composite achieved a maximum light-thermal-electric energy conversion output, underscoring its potential in efficient solar energy utilization. The integration of bio-aerogels with phase change materials has also been explored to enhance thermal management in thermoelectric devices. A bio-based MXene hybrid aerogel/paraffin composite exhibited ideal heat storage performance, excellent cycle stability, and thermal stability.

The aerogel's porous structure allows for effective encapsulation of the phase change material, preventing leakage and improving the thermal response rate. Such composites not only improve the thermal regulation of thermoelectric devices but also contribute to their overall efficiency and longevity. Moreover, the development of bio-inspired core-shell structural aerogels has introduced programmable water transport and thermal management capabilities. These aerogels, designed with hydrophobic shells covering hydrophilic cores, can effectively manage moisture and heat within thermoelectric systems, thereby enhancing performance and durability. In addition to structural innovations, the incorporation of conductive polymers into bio-aerogels has been investigated. For instance, omnidirectional 3D printing of Poly(3,4-ethylenedioxythiophene) polystyrene sulfonate aerogels has enabled the creation of stretchable interconnects and thermoelectric components with tunable electromechanical properties. These aerogels exhibit a high power factor and ultra-low thermal conductivity, contributing to an improved figure of merit in thermoelectric applications. Furthermore, bio-aerogels have been utilized in passive daytime radiative cooling applications [8]. Materials such as superhydrophobic waste paper-based aerogels have demonstrated high solar reflectance and thermal emittance, leading to significant temperature reductions. When integrated into building materials, these aerogels can reduce cooling energy consumption by an average of 43.4%, showcasing their potential in energy-efficient building designs. The application of bio-aerogels in enhancing thermoelectric conversion efficiency is a burgeoning field that leverages the unique properties of these materials. Through innovative composite designs and integration with conductive elements, bio-aerogels offer promising avenues for developing efficient, sustainable, and versatile thermoelectric devices [24, 55, 64, 69].

6.6 Applications in Energy Storage

Bio-aerogels, derived from sustainable biopolymers, have emerged as promising materials in energy storage applications due to their ultra-lightweight structure, high porosity, and excellent surface area. These properties enable efficient ion transport and charge storage, making bio-aerogels ideal candidates for use in supercapacitors, batteries, and other advanced energy storage systems. Figure 6.3 explores the most important applications of bio-aerogels on energy storage applications.

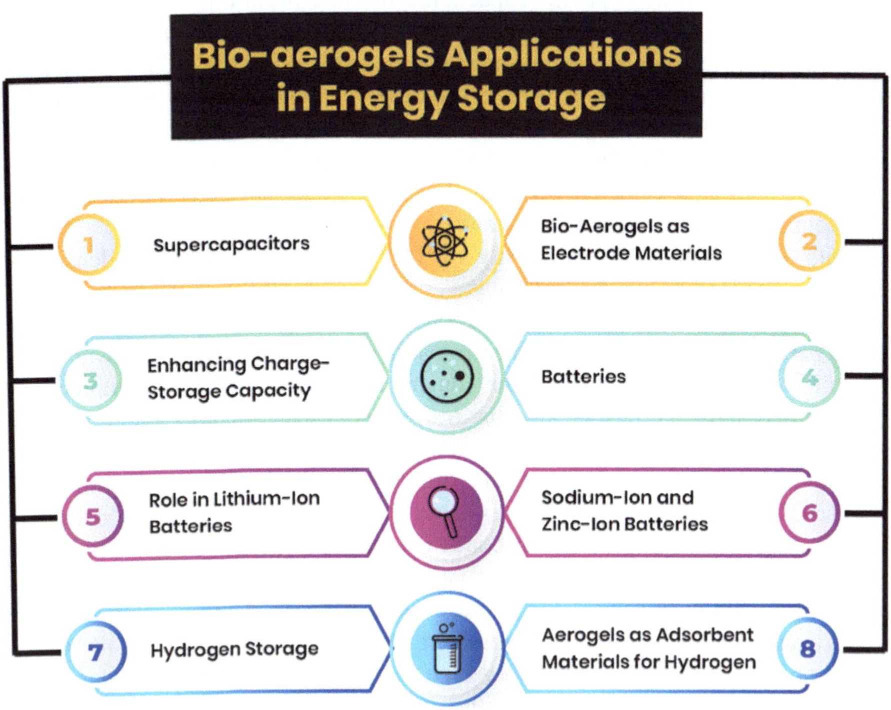

Fig. 6.3 Applications of bio-aerogels in energy storage

6.6.1 Supercapacitors

Supercapacitors are energy storage devices that bridge the gap between conventional capacitors and batteries, offering high power density, rapid charge/discharge rates, and long cycle life. Bio-aerogels, derived from renewable biomass, are used in supercapacitors as lightweight, porous electrode materials that enhance ion transport and improve overall electrochemical performance. It is commonly known that hierarchical porous architectures including micropores, mesopores, and macropores are desirable in electrode materials. Ion diffusion lengths can be greatly shortened by macropores (>50 nm), which act as sizable ion buffer storage tanks. Micropores (<2 nm) offer various ion diffusion channels and a large number of ion storage active sites, whereas mesopores (2 ~ 50 nm) give low-resistance pathways for quick ion transport to the electrode's interior regions. According to Wang et al. [56, 57], carbon aerogels are widely used in the production of supercapacitor electrodes due to their exceptional conductivity, cost-effectiveness, and distinctive three-dimensional network structure. A study shown that a highly effective electrode material may be created by substituting lignin for resorcinol. By sol-gel polymerizing lignin and resorcinol with formaldehyde, the scientists created lignin-based carbon aerogel. The resorcinol was partially substituted with lignin in varying mass ratios. The sample with the best electrical and morphological characteristics was found to be the 20-weight percent replacement. In terms of electrical characteristics, the produced electrode's specific capacitance reached 142.8 F g^{-1} at a current density of 0.5 A g^{-1}, and the specific surface area obtained was 590 m^2 g^{-1}. The produced aerogel electrode's specific capacitance stayed at 112.5 F g^{-1} even when the current density increased to 10 A g^{-1}. The observed data show that the sample has good rate capability after optimal capacitive behavior. As highlighted by Xu et al. [61], the lignin-based carbon aerogel also demonstrated superior durability, maintaining 96% of its specific capacitance after 2000 cycles attributed to the aerogel's interconnected porous network structure.

With a maximum energy density of 28.2 Wh kg^{-1} at a power density of 262.6 W kg^{-1} and a specific capacitance of 193 F g^{-1} at 0.5 A g^{-1}, a symmetric supercapacitor constructed with two carbon aerogel electrodes demonstrated exceptional performance. In order to create the distinctive and controllable porosity structure of carbon aerogels, crosslinker content is essential to the development of lignin-based aerogels. In this way, the researchers used formaldehyde as a crosslinker and showed that the ratio of lignin to crosslinker had a considerable impact on a number of the generated aerogels' properties, including their appropriateness for uses like supercapacitor electrodes. They discovered that improved ion accessibility results from a higher lignin content during the carbon aerogel production process. When compared to commercial activated carbon and RF-based carbon aerogel evaluated using the same two-electrode method, superior electrochemical performance and cycle stability were noted. Because surface activation altered the interior pore structure of carbon aerogels, it also improved specific capacitance and capacitive rate retention, which further improved the performance of electric double-layer capacitors. At a power density of 209.1 W kg^{-1}, symmetric supercapacitor cells made of activated carbon aerogel with 87 weight percent lignin had an excellent energy density of 3.2 Wh kg^{-1}. These findings demonstrate the potential of lignin-epoxy-based activated carbon aerogels as renewable, affordable, and promising materials for a range of energy applications, such as carbon electrodes for other hybrid battery systems including Li-ion capacitors and supercapacitors [13, 27, 43].

Ice-templating, lyophilization, and carbonization were used in a green procedure to create hierarchically porous carbon aerogels. According to Dong et al. [17], carbon aerogels made at a cooling rate of 7.5 K min^{-1} are claimed to have achieved an unusually high specific surface area of 1260 m^2 g^{-1} without any physical or chemical activation for the first time. With a cyclic stability of 94% following 4500 charge and discharge cycles and a specific capacitance of 410 F g^{-1} at 2 mV s^{-1}, the electrode fabricated from this aerogel demonstrated exceptional electrochemical performance. It was investigated how the structure and electrochemical performance of the carbon aerogels were affected by the ice-templating cooling rate as well as the solid concentration of lignin and cellulose nanofibers in the suspension. It was discovered that while the solid content of the lignin-nanocellulose suspension had no effect on the nanoporous structure and specific surface area of the carbon aerogels, the ice-templating procedure and cooling pace had a substantial impact. The carbon aerogel electrode attained a specific capacitance of 240 F g^{-1} at 0.1 A g^{-1} when assembled as a supercapacitor. The supercapacitor demonstrated a quick reaction with a relaxation time constant of 1.3 s and an energy density of 4.3 Wh kg^{-1} at a power density of 500 W kg^{-1}. The development of high-performance, eco-friendly carbon aerogel electrodes for energy storage applications is thus made possible by this study.

A study Thomas et al. [51], demonstrated the superior qualities of the bidirectional freezing method for creating a composite aerogel made of graphene oxide and cellulose nanofibrils. The formation of multilayer ice crystals repels the uniformly distributed composite suspensions due to the temperature difference created by bidirectional freezing. Carbon aerogels with regular layered structures are created through carbonization. Regarding the mechanical characteristics, carbon aerogels made using a bidirectional freezing technique demonstrated exceptional mechanical qualities as well as excellent electrochemical performance. With a capacitance retention of 93% after 5000 cycles at 2 A g^{-1}, the supercapacitor electrodes made from these aerogels demonstrated a high specific capacity of 140 F g^{-1} at 0.1 A g^{-1}. Al Haj et al. [2] used various salt anions, chlorides, citrates, and acetates to create a sophisticated carbon aerogel based on cellulose nanocrystals for adaptive supercapacitor performance. The authors demonstrated that the carbon aerogel containing nitrates had a higher specific capacitance than carbon aerogels containing other salts at a current density of 1 A g^{-1}. At a current density of 1 A g^{-1}, the nitrate- carbon aerogel electrode's specific capacitance was 413.3 F g^{-1}.

Graphene-based or graphene-modified materials are receiving a lot of interest these days because of their important electrical conductivity and capacitance characteristics. Since layer-stacking aggregation is the primary issue with graphene-based materials, it should be stabilized with other materials for effective applications. In order to examine tetramethylpiperidine-1-oxyl radical-oxidized nanocellulose/pristine graphene aerogels as electrode materials in supercapacitors, researchers described its preparation and possible uses. They looked into the effects of the carbonization temperature on the material's characteristics and found that raising the temperature to 1100 °C significantly improved the electrode's electrochemical performance; for example, the specific capacitance gained was 361.74 F g^{-1} at a current density of 0.5 A g^{-1}, and the capacitance retention was 99.3% after 5000 cycles. The authors examined the contact interface using electrochemical impedance spectroscopy and found that the sample with the best supercapacitor performance had an impedance of 4 ohms, indicating the constituents' best interface adhesion [56, 57].

However, [60] conducted an examination into the performance of electrode supercapacitors using porous biochar produced from cellulose and aerogel through heteroatomic doping. After presenting the synthesis of N,B-co-doped porous biochar with thermal activation at 650 °C, the authors came to the conclusion that the XRD examination supported their conclusion that the N,B-co-doped porous biochar preparation had no effect on the cellulose (or, for that matter, carbonic) structure. The produced samples had a minimal average pore size of 6.6 nm and high specific surface areas starting at 891 m^2 g^{-1}. As anticipated given these characteristics, the samples' specific capacitance (220.9 F g^{-1} at 1 A g^{-1}) and capacitance retention (98.6% after 3000 cycles) were both high. The authors verified low resistance values brought on by the nitrogen and boron doping, as well as well-connected structures inside the of N,B-co-doped porous biochar framework, using the electrochemical analysis. The scientists successfully synthesized a 3D hierarchical porous and conductive carbon aerogel material from CNF/rGO/Zn/Co ZIF using a novel and straightforward method that involved high-temperature carbonization and directional freeze drying. The good capacitance retention rate (83.4% at 10 A g^{-1}) and high specific capacitance (364.6 F g^{-1} at 1 A g^{-1}) of the CNF/rGO/Zn/Co ZIF electrode were made possible by the combination of several materials and the special 3D hierarchical porous structure with micropores, mesopores, and macropores. Furthermore, the CNF/rGO/Zn symmetric supercapacitor had a maximum energy density of 18.9 Wh kg^{-1} at 288.4 W kg^{-1}, demonstrating its wide range of energy storage potential. Interestingly, it also demonstrated outstanding cycling performance, maintaining capacitance at 78.9% after 10,000 cycles. Thus, the CNF/rGO/Zn/Co ZIF hybrid carbon aerogel exhibits great potential as a supercapacitor electrode material [63]. Table 6.1 explores bio-aerogel separators and their charge/discharge properties.

Table 6.1 Some bio-aerogels as separators in energy storage systems

Bio-aerogel material	Porosity %	Charge capacity (mAh/g)	Number of cycles	References
Celgard 2400	41	138.2	100	[54]
Lignin	74	131.8	50	[72]
Cellulose	61.5	143.6	400	[20]
Cellulose/polyimide	78	166.2	200	[15]
Chitosan	98.4	92.2	500	[66]

6.6.2 Bio-aerogels as Electrode Materials

Bio-aerogels, derived from renewable biomass sources such as lignin, cellulose, and chitosan, have garnered significant attention as sustainable electrode materials in energy storage systems. Their unique properties including high porosity, large specific surface area, and excellent electrical conductivity make them ideal candidates for applications in supercapacitors, lithium-ion batteries, and fuel cells. In supercapacitors, the hierarchical porous structure of bio-based carbon aerogels facilitates efficient ion transport and storage. Macropores act as ion reservoirs, mesopores provide pathways for rapid ion movement, and micropores offer abundant active sites for ion adsorption. For instance, Alagarsamy et al. [3] reported lignin-based carbon aerogels synthesized through sol-gel polymerization have demonstrated specific surface areas up to 590 m^2/g and specific capacitances reaching 142.8 F/g at a current density of 0.5 A/g. These aerogels also exhibit remarkable durability, retaining 96% of their capacitance after 2000 cycles, attributed to their interconnected porous networks. The incorporation of crosslinkers plays a pivotal role in tailoring the porous architecture of lignin-based aerogels. Adjusting the lignin-to-crosslinker ratio influences ion accessibility and electrochemical performance.

Studies from Li et al. [32] have shown that higher lignin content enhances ion transport pathways, leading to improved capacitive behavior. Surface activation further augments the performance of electric double-layer capacitors (EDLCs) by modifying the internal pore structure, thereby increasing specific capacitance and rate retention. Symmetric supercapacitor cells assembled with activated carbon aerogels containing 87 wt% lignin have achieved energy densities of 3.2 Wh/kg at power densities of 209.1 W/kg, underscoring their potential as cost-effective and renewable materials for energy applications. Beyond supercapacitors, bio-based aerogels have shown promise as anode materials in LIBs. Their mesoporous structures facilitate lithium-ion diffusion, leading to enhanced charge-discharge rates and cycling stability. For example, Mandić et al. [39] highlighted lignin-SiO_2 aerogels have exhibited excellent cycling performance and rate capabilities, positioning them as viable alternatives to traditional non-renewable carbon materials. The sustainable nature of biomass-derived carbon aerogels aligns with the global push towards environmentally friendly energy solutions. Their low density, mechanical stability, and chemical robustness further contribute to their suitability as electrode materials. However, challenges such as scalability, standardization, and cost-effectiveness remain. Ongoing research focuses on refining synthesis methods and exploring novel applications to overcome these hurdles. Bio-aerogels represent a promising frontier in the development of sustainable electrode materials for energy storage systems. Their tunable properties and renewable origins offer a pathway towards greener and more efficient energy storage solutions [3, 32, 39].

6.6.3 Enhancing Charge-Storage Capacity

Bio-based aerogels have garnered significant attention in energy storage applications due to their exceptional properties, including high porosity, large specific surface area, and environmental sustainability. Derived from renewable resources such as cellulose, lignin, and chitosan, these aerogels serve as advanced materials for electrodes and separators in devices like lithium-ion batteries, supercapacitors, and fuel cells. In the realm of lithium-ion batteries, bio-aerogels have been explored as potential anode materials. Researchers developed a composite of silica encapsulated in honeycomb-like porous carbon derived from lignin, achieving a remarkable reversible capacity of 1109 mAh g^{-1}, superior rate capability, and excellent cycling performance. The material high specific surface area, ranging from 882 to 1107 m^2 g^{-1}, and dominant mesoporous structure facilitate efficient lithium storage. Similarly, a study synthesized N-doped porous carbon nanofiber aerogels from marine biomass waste, exhibiting a high surface area of 2290 m^2 g^{-1}. This structure enhances contact between the active material and the electrolyte, resulting in a specific capacity of 572 mAh g^{-1} at 1 A g^{-1} after 600 cycles. These studies underscore the potential of bio-aerogels to replace non-renewable carbon materials in lithium-ion batteries anodes, offering sustainable solutions with high performance. Beyond lithium-ion batteries, bio-aerogels have demonstrated efficacy in supercapacitor applications. A group of researchers investigated cellulose-lignin aerogels, which exhibited specific capacitances around 120 F g^{-1} and exceptional cycle stability, retaining 98% capacity after 10,000 charge-discharge cycles. The synergistic combination of cellulose and lignin contributes to the aerogel enhanced electrochemical properties. Additionally, a previous study explored lignin-resorcinol-formaldehyde aerogels prepared using green deep eutectic solvents, achieving specific capacitances of 284 F g^{-1} and cycle stability of 97% over 5000 cycles at 10 A g^{-1}. These findings highlight the role of bio-aerogels in developing high-performance supercapacitors with sustainable materials. The application of bio-aerogels extends to battery separators as well. Cellulose-nanofiber aerogel separators were evaluated, noting their high porosity of 99.5% and strong mechanical stability. These separators demonstrated an ionic conductivity of 2.64 mS cm^{-1} and effective diffusion coefficients comparable to bulk electrolytes, leading to excellent electrochemical performance with good cycling stability over 200 cycles. Similarly, chitin-based aerogels prepared via freeze-drying exhibited high electrolyte uptake and ionic conductivity, enhancing the rate performance and cycle life of the batteries.

These studies illustrate the potential of bio-aerogels to improve battery efficiency and longevity through innovative separator designs. In lithium-sulfur batteries, bio-aerogels have been employed to address challenges related to sulfur volume expansion and polysulfide migration. Researchers have developed 3D aerogel frameworks incorporating materials like tetramethylpiperidine-1-oxyl -oxidized cellulose nanofibers and reduced graphene oxide, which serve as hybrid sulfur host materials. These structures minimize volume expansion effects and enhance rate capability, achieving capacities of 744 mAh g^{-1} at 5C and maintaining high capacity over

Table 6.2 Bio-aerogels performance in energy storage devices

Bio-aerogel material	Specific capacity mAh/g	Efficiency (%)	References
Lignin	1109	99	[23]
Cellulose	572	–	[31]
Resorcinol-formaldehyde	288	97	[4]
TEMPO-cellulose	869	99.7	[34]

extended cycles. Such advancements demonstrate the versatility of bio-aerogels in accommodating different battery chemistries and improving overall performance. Despite these promising developments, challenges remain in the widespread adoption of bio-aerogels in energy storage systems. Issues such as scalability, standardization, and cost-effectiveness need to be addressed to facilitate commercial applications. Future research aims to refine synthesis methods, explore novel bio-based materials, and develop composite aerogels that combine multiple functionalities. By overcoming these challenges, bio-aerogels can play a pivotal role in advancing sustainable energy storage technologies, offering environmentally friendly alternatives without compromising performance [39]. Table 6.2 shows performances of bio-aerogels in energy storage devices.

6.6.4 *Batteries*

Lithium-ion batteries are attractive options for energy storage research and development due to their high energy densities and long lifespans. Graphite is a common anode material for LIBs because of its hierarchical structure. Its poor intercalation capacity, however, is a major disadvantage that is driving the search for new anode materials. Although the electric characteristics of Li metal anodes make them an excellent electrode, problems with dendritic growth still exist. Scientists have concentrated on carbon-derived anodes made from sustainable precursors because of their benefits in terms of cost-effectiveness, longer cycle life, and safety, as well as the urgent sustainability issues, as noted by Muddasar et al. [41]. Low electrical conductivity, shrinkage during the lithiation and delithiation processes, and uncontrolled lithium dendritic growth are the main obstacles to the use of bio-based materials for active electrodes. Pyrolysis could improve electron transport in these materials by speeding up electrochemical reactions and encouraging the creation of conductive carbon compounds. However, the main reason silicon-based materials have drawn a lot of interest as anodes is because of their enormous theoretical capacity (about 4200 mAh g^{-1}), which is more than ten times greater than graphite's (approximately 370 mAh g^{-1}) as highlighted by Zhang et al. [70]. For the use of electrical cars, this increased theoretical capacity is essential since it allows for power and energy density while reducing battery size.

Hard carbon, or non-graphite carbon, has disordered internal crystallites that cause reduced reversibility and capacity loss, despite the fact that carbon-based materials may be easily produced from many bio-based sources. However, hard carbon is worth looking into because it is described as a very stable material. Reversibility is greatly influenced by elements including particle size, high porosity, and additives. Lithium-ion diffusion length can be significantly shortened by decreasing particle size, improving structural stability and specific capacity. Depending on the pyrolysis settings, several aerogels could be produced by pyrolyzing bio-based aerogels. However, as conductivity is a crucial factor for energy storage applications, it is crucial for electrical and mechanical qualities to optimize pyrolysis conditions in order to obtain graphite-like structures rather than amorphous non-conductive carbon aerogels [39].

Although lithium-ion batteries are the subject of most study, certain studies focus on the performance of lithium-sulfur batteries. In particular, one of the most recent studies by Liu et al. [34] focused on creating a rigid framework using 1D TEMPO-oxidized cellulose nanofibers (TEMPO-CNF) that had been altered with 2D rGO. The constructed framework was then evenly incorporated into the MXene monolayer. By acting as a hybrid sulfur host material, composite 3D aerogel may reduce the impact of sulfur volume expansion during charge/discharge cycles. When evaluated as cathode materials in lithium-sulfur batteries, the produced materials demonstrated a significant boost in rate capability, measuring 744 mAh g^{-1} at 5C with an internal resistance of 2.2 ohms. Furthermore, after 200 cycles, electrodes such as this one demonstrated a high capacity of 869 with a low capacity-fading rate. The flexible 3D aerogel material serves as a carbon support for sulfur since, in the past, bacterial cellulose was utilized as the template to create cathode material with a high sulfur content [73]. The composite strong framework to tolerate the strain brought on by the volume fluctuations of active components and its superior electrical conductivity were made possible by the special interconnected networks of bacterial-cellulose-based carbon aerogel. Lithium ions were able to travel quickly through the organic electrolyte thanks to the composite's macroporous structure. The total electrode resistance may also be decreased by the ultralight BC interlayer that is placed between the separator and the sulfur cathode. Above all, the excess sulphur buildup on the cathode surface was successfully removed by the CBC interlayer. The new CBC interlayer is where migratory polysulfides that typically settle on the cathode surface now land. For the practical use of high-performance lithium-sulfur batteries in the future, the combined effects of this CBC structure on the cathode side result in an inexpensive, environmentally benign electrode material design.

Researchers demonstrated that carbon-based aerogels made with sweet potatoes as the source may be utilized as an effective separator material for lithium-sulfur batteries since bio-based aerogels could be made directly from natural sources [22]. In particular, the generated carbon-aerogel was employed as a modifier for the commercially used separator in order to improve characteristics like low utilization rate and poor cycle life. In addition to lowering battery resistance, coated carbon aerogel increased the rate at which sulfur was used, acting as an upper current collector. Furthermore, the reversible discharge capacity of 431 mAh g^{-1} after 1000 cycles at 1 C, where Coulombic efficiency was above 95%, followed by the initial discharge capacity of 1216 mAh g^{-1} at 0.1 C was significantly higher than in the pure commercial separators.

A high-strength bamboo cellulose-based membrane specifically made for zinc-based batteries was proposed by researchers. Interconnected cellulose nanofibers and 2D nanosheets with organic nanopores are features of cellulose-based membrane hierarchical structure. Compared to glass fiber separators, this structure, which is controlled by van der Waals forces and H-bonding, offers remarkable mechanical strength and successfully inhibits the growth of zinc dendrites. Additionally, glass fiber is quickly destroyed, but BCM retains its structural integrity even after being submerged in neutral and alkaline electrolytes. Cellulose-based membrane mesopores provide for homogenous deposition by facilitating a consistent Zn2+ flow. Zn/BCM/Zn symmetric cells, thus, exhibit a longer cycle life of up to 5000 h, whereas the Zn/GF/Zn symmetric cell only exhibits 30 h. Cellulose-based membrane provides an environmentally friendly and pollution-free preparation method since it is made from bamboo, a plentiful and renewable biomass [18].

6.6.5 Role in Lithium-Ion Batteries

Bio-aerogels, particularly those derived from renewable biopolymers such as cellulose, lignin, alginate, and chitosan, have emerged as promising candidates for enhancing the performance and sustainability of lithium-ion batteries. Their unique structural and physicochemical properties—such as high porosity, large surface area, low density, and tunable functional groups—make them ideal materials for various components within lithium-ion batteries, including electrodes, separators, and binders. In the context of anodes, bio-based aerogels can serve as excellent carbon precursors. Through controlled carbonization processes, these aerogels yield porous carbon structures with high electrical conductivity and a stable framework that accommodates lithium-ion intercalation. For instance, carbonized cellulose or lignin aerogels exhibit hierarchical porosity, which facilitates rapid ion transport and efficient electrolyte penetration both critical for improving charge/discharge rates and cycling stability. Furthermore, their sustainable origin offers a compelling alternative to conventional fossil-based carbon materials, aligning with the broader goals of green energy storage technologies. In addition to their application in anodes, bio-aerogels are also being explored as advanced separators and electrolytic scaffolds. Their interconnected pore networks can be tailored to control ionic conductivity and minimize dendritic lithium formation, thereby enhancing battery safety. Some studies have incorporated functional dopants such as nitrogen, phosphorus, or sulfur into bio-aerogel matrices to further improve electrochemical activity, surface wettability, and lithium-ion diffusion kinetics. Moreover, the mechanical flexibility and thermal stability of these aerogels contribute to the structural integrity of lithium-ion batteries under various operating conditions [39]. A major challenge in lithium-ion batteries is volume expansion and structural degradation of electrodes during repeated cycling, which often leads to capacity fade. Bio-derived aerogels, due to their elasticity and resilience, can mitigate these effects by buffering volume changes and maintaining electrode cohesion. Additionally, their eco-friendly decomposition and potential recyclability provide long-term environmental advantages. Despite these benefits, challenges remain in the scalability, cost-effectiveness, and standardization of bio-aerogel production. Techniques such as supercritical drying and freeze-drying, which are essential for aerogel synthesis, are energy-intensive and not yet industrially optimized. However, ongoing research focuses on streamlining these methods and developing hybrid bio-aerogels that incorporate both natural and synthetic components for enhanced performance. The paper emphasizes that further work is needed to establish a clear structure-property-performance relationship, especially in real-world battery architectures. Nonetheless, preliminary results from various studies suggest that bio-aerogels not only match but can surpass traditional materials in key performance metrics such as capacity retention, rate capability, and cycling life. In conclusion, the incorporation of bio-aerogels into lithium-ion batteries represents a promising intersection of sustainable materials science and advanced energy storage. Their versatility, environmental compatibility, and tunable electrochemical properties position them as strategic materials for the next generation of high-performance, green batteries.

6.6.6 *Sodium-Ion and Zinc-Ion Batteries*

Bio-aerogels, derived from renewable biopolymers like cellulose, chitosan, and alginate, have garnered significant attention for their applications in energy storage systems, notably in sodium-ion and zinc-ion batteries [39]. Their distinctive properties, including high porosity, large surface area, mechanical flexibility, and environmental friendliness, make them ideal candidates for enhancing battery performance and sustainability. Sodium-ion batteries are considered a promising alternative to lithium-ion batteries due to the abundance and low cost of sodium resources. However, challenges such as the larger ionic radius of Na^+ compared to Li^+ lead to slower diffusion kinetics and structural instability in electrode materials. Bio-aerogels offer potential solutions to these issues. Their interconnected porous networks facilitate efficient ion transport and accommodate the volumetric expansion of electrode materials during cycling [14]. For instance, carbonized bio-aerogels derived from cellulose or lignin can serve as anode materials, providing a conductive matrix that enhances electron transport and structural integrity. Additionally, the tunable surface chemistry of bio-aerogels allows for the introduction of functional groups that can improve the electrochemical reactivity and stability of sodium-ion storage. While specific studies focusing solely on bio-aerogels in sodium-ion batteries are limited, the inherent characteristics of bio-aerogels suggest their potential in addressing the current limitations of sodium-ion batteries technology [48].

Zinc-ion batteries have emerged as a viable option for large-scale energy storage due to zinc's high theoretical capacity, safety, and cost-effectiveness. However, issues such as dendrite formation, hydrogen evolution, and electrode corrosion hinder their practical application. Recent research has demonstrated that bio-aerogels can play a crucial role in overcoming these challenges. For example, a study introduced a biomass-based hydrogel interface composed of carboxymethyl chitosan and sodium alginate (CCS/SA) onto the zinc anode ([62]). This bio-inspired hydrogel layer exhibited an ion-selective sieving mechanism that blocked detrimental polyiodide ions while facilitating Zn^{2+} ion transport. The presence of functional groups such as carboxyl and amino groups in the hydrogel interacted with Zn^{2+} ions, promoting uniform deposition and suppressing dendrite growth. As a result, the modified Zn-I_2 battery achieved an ultra-long cycle life of 60,000 cycles at 5 A g^{-1}, showcasing the effectiveness of bio-aerogels in enhancing Zinc-ion batteries performance. In another approach studies from Zhang et al. [71], silica aerogel layers with ultra-high specific surface areas and abundant polar functional groups were applied to zinc anodes. These layers provided numerous active sites for uniform zinc ion deposition and acted as desolvation layers and zinc ion flux regulators. The modified zinc anodes exhibited significantly improved cycling stability, with symmetric cells demonstrating stable performance for over 1100 h at 0.5 mA cm^{-2}. Furthermore, metal oxide aerogels, such as oxygen vacancy-rich CeO_2 aerogels, have been employed as interface layers for zinc anodes. These aerogels acted as ion sieves, redistributing Zn^{2+} at the anode surface and leading to uniform zinc deposition. The presence of oxygen vacancies facilitated strong interactions with sulfate ions, forming a negatively charged layer that attracted Zn^{2+} ions and accelerated their migration kinetics.

This strategy effectively suppressed dendrite growth and side reactions, resulting in a long-term lifespan of over 4000 h at 4 mA cm^{-2}. Additionally, the development of environmentally friendly, degradable zinc-ion batteries has been explored using guar gum-cellulose aerogel electrolytes. These natural biomaterial-based electrolytes exhibited high ionic conductivity and mechanical strength, effectively restraining zinc dendrite formation. The resulting batteries demonstrated excellent electrochemical performance, with specific capacities of 309.1 mAh g^{-1} after 100 cycles and complete degradation in phosphate-buffered saline after approximately 40 days, highlighting the potential of bio-aerogels in creating sustainable and transient energy storage devices. Bio-aerogels offer significant advantages in the development of sodium-ion and zinc-ion batteries. Their structural versatility, combined with the ability to tailor surface functionalities, enables improvements in ion transport, electrode stability, and overall battery performance. While research in this area is still evolving, the promising results underscore the potential of bio-aerogels as sustainable and effective components in next-generation energy storage systems.

6.6.7 Hydrogen Storage

Bio-aerogels, derived from renewable biopolymers such as cellulose, lignin, and chitosan, have garnered significant attention for their potential in sustainable energy applications, notably in hydrogen storage [32]. Their unique properties including high porosity, large surface area, low density, and tunable surface functionalities-make them promising candidates for efficient hydrogen adsorption and desorption processes. Hydrogen storage remains a critical challenge in the advancement of hydrogen energy technologies, particularly for applications requiring safe, efficient, and compact storage solutions. Traditional storage methods, such as high-pressure gas cylinders and liquid hydrogen tanks, pose safety risks and are often impractical for mobile applications. In this context, solid-state hydrogen storage materials, especially those based on carbon structures, have been extensively explored [53]. Among these, carbon aerogels have emerged as a class of materials with notable potential due to their structural characteristics. Carbon aerogels are synthesized through the pyrolysis of organic aerogels, resulting in a highly porous, amorphous carbon network. Their high surface area and tunable pore sizes are advantageous for hydrogen storage, as they provide ample sites for hydrogen adsorption. Studies have demonstrated that the hydrogen storage capacity of carbon aerogels is closely related to their surface area and pore volume [26, 58]. For instance, research has shown that carbon aerogels with surface areas exceeding 3000 m^2/g can achieve hydrogen storage capacities of up to 5 wt% at cryogenic temperatures (77 K) and moderate pressures.

The synthesis process of carbon aerogels allows for the incorporation of various dopants or the creation of composites to enhance hydrogen storage performance. For example, the introduction of metal nanoparticles or metal-organic frameworks (MOFs) into the aerogel matrix can create additional active sites for hydrogen adsorption, thereby increasing storage capacity. Moreover, the surface chemistry of bio-aerogels can be tailored to improve interactions with hydrogen molecules. Functional groups such as hydroxyls, carboxyls, or amines can be introduced to enhance the binding energy between the hydrogen molecules and the aerogel framework, facilitating better adsorption at ambient conditions. Despite these promising attributes, challenges remain in optimizing bio-aerogels for practical hydrogen storage applications. One significant issue is the relatively low hydrogen adsorption capacity at room temperature and ambient pressure, which limits their applicability in real-world scenarios. Efforts are ongoing to enhance the room-temperature storage capacity through strategies such as pore size optimization, surface functionalization, and the development of hybrid materials that combine bio-aerogels with other hydrogen-absorbing materials. Bio-aerogels represent a promising avenue for hydrogen storage solutions, offering a sustainable and potentially efficient alternative to conventional materials. Their structural versatility, combined with the ability to tailor surface properties and incorporate functional additives, positions them as key candidates in the quest for effective hydrogen storage systems. Continued research and development in this field are essential to overcome current limitations and fully realize the potential of bio-aerogels in hydrogen energy applications.

6.6.8 Aerogels as Adsorbent Materials for Hydrogen

Aerogels, renowned for their exceptional porosity, low density, and high surface area, have emerged as promising adsorbent materials for hydrogen storage applications [30]. These unique properties facilitate enhanced hydrogen adsorption capacities, positioning aerogels as potential candidates to address the challenges associated with efficient hydrogen storage. Carbon-based aerogels, in particular, have been extensively studied for hydrogen storage due to their tunable pore structures and high surface areas. Research indicates that the hydrogen storage capacity of carbon aerogels is closely related to their surface area and pore volume [16]. For instance, carbon aerogels with surface areas exceeding 3000 m^2/g have demonstrated hydrogen storage capacities approaching 5 wt% at cryogenic temperatures (77 K) and moderate pressures [44].

The synthesis process of carbon aerogels allows for the incorporation of various dopants or the creation of composites to enhance hydrogen storage performance. The introduction of metal nanoparticles or metal-organic frameworks into the aerogel matrix can create additional active sites for hydrogen adsorption, thereby increasing storage capacity. Silica aerogels have also been explored for hydrogen storage applications. Studies have shown that silica aerogel blankets can adsorb significant quantities of hydrogen at cryogenic temperatures [44]. Unlike traditional adsorbents, these aerogel blankets are robust composite materials that can be shaped into various forms, enhancing the efficiency of storage system designs. This versatility makes them particularly attractive for developing modular hydrogen storage systems. Despite these promising attributes, challenges remain in optimizing aerogels for practical hydrogen storage applications. One significant issue is the relatively low hydrogen adsorption capacity at room temperature and ambient pressure, which limits their applicability in real-world scenarios. Efforts are ongoing to enhance the room-temperature storage capacity through strategies such as pore size optimization, surface functionalization, and the development of hybrid materials that combine aerogels with other hydrogen-absorbing materials. Aerogels represent a promising avenue for hydrogen storage solutions, offering a sustainable and potentially efficient alternative to conventional materials. Their structural versatility, combined with the ability to tailor surface properties and incorporate functional additives, positions them as key candidates in the quest for effective hydrogen storage systems. Continued research and development in this field are essential to overcome current limitations and fully realize the potential of aerogels in hydrogen energy applications [30, 44].

6.7 Summary

Nanostructured bio-aerogels emerged as promising materials for energy conversion and storage applications, offering a sustainable alternative to conventional materials. Derived from renewable sources such as cellulose, lignin, and chitosan, these bio-based aerogels exhibit high porosity, large specific surface area, biocompatibility, and biodegradability, making them highly advantageous for environmental sustainability. They are synthesized through a process involving the careful combination of precursor materials, gelation, and subsequent solvent removal. Due to their exceptional thermal insulation, charge capacity, electrical conductivity, low density, and eco-friendly composition, bio-aerogels garnered significant attention in the energy sector. They are used in a wide range of energy conversion and storage applications, including fuel cells, electrolyte membranes, photovoltaics, solar cell films, thermal insulators, thermoelectric devices, supercapacitors, electrode materials, batteries, and hydrogen storage or adsorption. Recent advancements demonstrated that bio-aerogels offer excellent cycling stability and rate performance compared to conventional materials. Future research is focused on improving synthesis methods and exploring novel applications, positioning bio-aerogels as sustainable solutions for advancing energy technologies.

References

1. Abdul Khalil HPS, Adnan AS, Yahya EB, Olaiya NG, Safrida S, Hossain MS, Balakrishnan V, Gopakumar DA, Abdullah CK, Oyekanmi AA, Pasquini D (2020) A review on plant cellulose nanofibre-based aerogels for biomedical applications. Polymers 12(8). https://doi.org/10.3390/polym12081759
2. Al Haj Y, Soliman AB, Vapaavuori J, Elbahri M (2024) Carbon aerogels derived from anion-modified nanocellulose for adaptive supercapacitor performance. Adv Funct Mater 34(28). https://doi.org/10.1002/adfm.202313117
3. Alagarsamy S, Jasmine Vincent John A, Chen S-M, Sakthinathan S, Chang Y-H, Tang J-Y (2024) Novel method of phosphorous doped bismuth oxide with carbon aerogel for a selective voltammetry quantification of anti-cancer drug: Regorafenib. Chem Eng J 494:153054. https://doi.org/10.1016/j.cej.2024.153054
4. Alex AS, M.S., A. L., V., S., John B, C., G., S.A. I (2017) Microporous carbon aerogel prepared through ambient pressure drying route as anode material for lithium ion cells. Polym Adv Technol 28(12):1945–1950. https://doi.org/10.1002/pat.4085
5. Alwin S, Sahaya Shajan X (2020) Aerogels: promising nanostructured materials for energy conversion and storage applications. Mater Renew Sustain Energy 9(2):7. https://doi.org/10.1007/s40243-020-00168-4
6. Arachchige IU, Brock SL (2006) Sol–Gel assembly of CdSe nanoparticles to form porous aerogel networks. J Am Chem Soc 128(24):7964–7971. https://doi.org/10.1021/ja061561e
7. Baia L, Peter A, Cosoveanu V, Indrea E, Baia M, Popp J, Danciu V (2006) Synthesis and nanostructural characterization of TiO2 aerogels for photovoltaic devices. Thin Solid Films 511-512:512–516. https://doi.org/10.1016/j.tsf.2005.12.024
8. Baysal HE, Yu T, Naenen V, De Smedt S, Hiz D, Zhang B, Xia H, Florenciano I, Rosenthal M, Cardinaels R, Molina-Lopez F (2025) Omnidirectional 3D printing of PEDOT: PSS aerogels with tunable electromechanical performance: a playground for unconventional stretchable interconnects and Thermoelectrics. Adv Sci 12(11). https://doi.org/10.1002/advs.202412491
9. Beitl KN, Reimhult E (2024) Effect of solvent properties on the critical solution temperature of thermoresponsive polymers. Int J Mol Sci 25(14). https://doi.org/10.3390/ijms25147734
10. Błaszczyński T, Ślosarczyk A, Morawski M (2013) Synthesis of silica aerogel by supercritical drying method. Proc Eng 57:200–206. https://doi.org/10.1016/j.proeng.2013.04.028
11. Budtova T (2019) Cellulose II aerogels: a review. Cellulose 26(1):81–121. https://doi.org/10.1007/s10570-018-2189-1
12. Budtova T, Aguilera DA, Beluns S, Berglund L, Chartier C, Espinosa E, Gaidukovs S, Klimek-kopyra A, Kmita A, Lachowicz D, Liebner F, Platnieks O, Rodríguez A, Navarro LKT, Zou F, Buwalda SJ (2020) Biorefinery approach for aerogels. Polymers 12(12):1–63. https://doi.org/10.3390/polym12122779

13. Cao Q, Lou R, Dong L, Niu T, Wei G, Lyu G (2023) Evaluation of gelation time affecting the self-assembled framework of lignin nanoparticle-based carbon aerogels and their electrochemical performances. ACS Appl Energy Mater 6(21):10874–10882. https://doi.org/10.1021/acsaem.3c01574
14. Chen H, Zhou L, Sun Y, Zhang T, Wang H, Huang H, Ning J, Hu Y (2025) Bio-inspired biomass hydrogel interface with ion-selective responsive sieving mechanism for corrosion-resistant and dendrite-free zinc-iodine batteries. Energy Storage Mater 76:104113. https://doi.org/10.1016/j.ensm.2025.104113
15. Deng J, Cao D, Yang X, Zhang G (2022) Cross-linked cellulose/carboxylated polyimide nanofiber separator for lithium-ion battery application. Chem Eng J 433:133934. https://doi.org/10.1016/j.cej.2021.133934
16. Dolbin AV, Khlistyuck MV, Esel'son VB, Gavrilko VG, Vinnikov NA, Basnukaeva RM, Martsenuk VE, Veselova NV, Kaliuzhnyi IA, Storozhko AV (2018) Sorption of hydrogen by silica aerogel at low-temperatures. Low Temp Phys 44(2):144–147. https://doi.org/10.1063/1.5020910
17. Dong Z, Zhong L, Zhang Y, Liu F, Huang H, Xu Z (2023) Construction of multifaunctional cellulose nanofibers/reduced graphene oxide carbon aerogels by bidirectional freezing for supercapacitor electrodes and strain sensors. Diam Relat Mater 140:110555. https://doi.org/10.1016/j.diamond.2023.110555
18. Fu J, Wang H, Xiao P, Zeng C, Sun Q, Li H (2022) A high strength, anti-corrosion and sustainable separator for aqueous zinc-based battery by natural bamboo cellulose. Energy Storage Mater 48:191–191.f6. https://doi.org/10.1016/j.ensm.2022.02.052
19. Gawryla M, Arndt E, Sánchez-Soto M, Schiraldi D (2018) Poly(amide-imide) aerogel materials produced via an ice templating process. Materials 11(2):233. https://doi.org/10.3390/ma11020233
20. Gou J, Liu W, Tang A, Wu L (2022) Interfacially stable and high-safety lithium batteries enabled by porosity engineering toward cellulose separators. J Membr Sci 659:120807. https://doi.org/10.1016/j.memsci.2022.120807
21. Hamann TW, Martinson ABF, Elam JW, Pellin MJ, Hupp JT (2008) Aerogel templated ZnO dye-sensitized solar cells. Adv Mater 20(8):1560–1564. https://doi.org/10.1002/adma.200702781
22. Huang Y, Zheng M, Lin Z, Zhao B, Zhang S, Yang J, Zhu C, Zhang H, Sun D, Shi Y (2015) Flexible cathodes and multifunctional interlayers based on carbonized bacterial cellulose for high-performance lithium-sulfur batteries. J Mater Chem A 3(20):10910–10918. https://doi.org/10.1039/C5TA01515D
23. Huang S, Yang D, Zhang W, Qiu X, Li Q, Li C (2021) Dual-templated synthesis of mesoporous lignin-derived honeycomb-like porous carbon/SiO2 composites for high-performance Li-ion battery. Microporous Mesoporous Mater 317:111004. https://doi.org/10.1016/j.micromeso.2021.111004
24. Huang D, Wang Z, Sheng X, Chen Y (2023) Bio-based MXene hybrid aerogel/paraffin composite phase change materials with superior photo and electrical responses toward solar thermal energy storage. Sol Energy Mater Sol Cells 251:112124. https://doi.org/10.1016/j.solmat.2022.112124
25. Jayan SS, Jayan JS, Saritha A (2023) A review on recent advances towards sustainable development of bio-inspired agri-waste based cellulose aerogels. Int J Biol Macromol 248:125928. https://doi.org/10.1016/j.ijbiomac.2023.125928
26. Kabbour H, Baumann TF, Satcher JH, Saulnier A, Ahn CC (2006) Toward new candidates for hydrogen storage: high-surface-area carbon aerogels. Chem Mater 18(26):6085–6087. https://doi.org/10.1021/cm062329a
27. Karaaslan MA, Lin L-T, Ko F, Renneckar S (2022) Carbon aerogels from softwood Kraft lignin for high performance supercapacitor electrodes. Front Mat 9. https://doi.org/10.3389/fmats.2022.894061

28. Kim C, Kim K (2021) Enhancement of solar thermoelectric power generation by optical and thermal management with highly transparent aerogel window. Sol Energy Mater Sol Cells 230:111224. https://doi.org/10.1016/j.solmat.2021.111224
29. Lan W, Dong D, Zhang M, Xiao Y, Zhao Z, Yang Z, Cao Y, Fan M (2023) Fabrication of dense anion exchange membranes by filling polymer matrix into quaternized branched polyethyleneimine @cellulose aerogel through in-situ polymerization. J Macromol Sci A 60(11):761–777. https://doi.org/10.1080/10601325.2023.2257758
30. Lee S-Y, Lee J-H, Kim Y-H, Kim J-W, Lee K-J, Park S-J (2022) Recent Progress using solid-state materials for hydrogen storage: a short review. PRO 10(2):304. https://doi.org/10.3390/pr10020304
31. Li D, Wang Y, Sun Y, Lu Y, Chen S, Wang B, Zhang H, Xia Y, Yang D (2018) Turning gelidium amansii residue into nitrogen-doped carbon nanofiber aerogel for enhanced multiple energy storage. Carbon 137:31–40. https://doi.org/10.1016/j.carbon.2018.05.011
32. Li M, Pang B, Dai S, Cui Y, Wu Y, Li H, Luo B (2024) Sustainable biomass-derived carbon aerogels for energy storage applications. Chem Eng J 499:156693. https://doi.org/10.1016/j.cej.2024.156693
33. Lin M-H, Hong S-H, Ding J-F, Liu C-L (2024) Organic porous materials and their nanohybrids for next-generation thermoelectric application. ACS Appl Mater Interfaces 16(49):67116–67133. https://doi.org/10.1021/acsami.4c12729
34. Liu Y, Zhang M, Gao Y, Guo J (2022) Regulate the reaction kinetic rate of lithium-sulfur battery by rational designing of TEMPO-oxidized cellulose nanofibers/rGO porous aerogel with monolayer MXene coating. J Alloys Compd 898:162821. https://doi.org/10.1016/j.jallcom.2021.162821
35. Liu H, Tian Y, Liu J, Zhang D, Wu X, Li Z (2023) Performance analysis of solar drying system with sunlight transparent thermally insulating aerogels. Energy 269:126698. https://doi.org/10.1016/j.energy.2023.126698
36. Long L-Y, Weng Y-X, Wang Y-Z (2018) Cellulose aerogels: synthesis, applications, and prospects. Polymers 10(6):623. https://doi.org/10.3390/polym10060623
37. Long H, Qian Y, Gang S, Zhang W, Yang B, Wei Y, Wang B, Li W, Jiang Q (2025) High-performance thermoelectric composite of bi 2 Te 3 nanosheets and carbon aerogel for harvesting of environmental electromagnetic energy. ACS Nano 19(1):1819–1831. https://doi.org/10.1021/acsnano.4c16834
38. Lopes JM, Mustapa AN, Pantić M, Bermejo MD, Martín Á, Novak Z, Knez Ž, Cocero MJ (2017) Preparation of cellulose aerogels from ionic liquid solutions for supercritical impregnation of phytol. J Supercrit Fluids 130:17–22. https://doi.org/10.1016/j.supflu.2017.07.018
39. Mandić V, Bafti A, Panžić I, Radovanović-Perić F (2024) Bio-based aerogels in energy storage systems. Gels 10(7). https://doi.org/10.3390/gels10070438
40. Mohanan JL, Arachchige IU, Brock SL (2005) Porous semiconductor chalcogenide aerogels. Science 307(5708):397–400. https://doi.org/10.1126/science.1104226
41. Muddasar M, Beaucamp A, Culebras M, Collins MN (2022) Cellulose: characteristics and applications for rechargeable batteries. Int J Biol Macromol 219:788–803. https://doi.org/10.1016/j.ijbiomac.2022.08.026
42. Nita LE, Ghilan A, Rusu AG, Neamtu I, Chiriac AP (2020) New trends in bio-based aerogels. Pharmaceutics 12
43. Niu J, Shao R, Liu M, Liang J, Zhang Z, Dou M, Huang Y, Wang F (2018) Porous carbon electrodes with battery-capacitive storage features for high performance Li-ion capacitors. Energy Storage Mater 12:145–152. https://doi.org/10.1016/j.ensm.2017.12.012
44. Otto M, Khlyapov Y, Fernandez E, Kapat J, Blair RG, Swanger AM, Schmitt J (2024) Cryo-adsorbed hydrogen storage using aerogel blankets: optimizing modular storage systems. In: AIAA SCITECH 2024 Forum. https://doi.org/10.2514/6.2024-0101
45. Preibisch I, Niemeyer P, Yusufoglu Y, Gurikov P, Milow B, Smirnova I (2018) Polysaccharide-based aerogel bead production via jet cutting method. Materials 11(8):1287. https://doi.org/10.3390/ma11081287

46. Reim M, Beck A, Körner W, Petricevic R, Glora M, Weth M, Schliermann T, Fricke J, Schmidt C, Pötter F (2002) Highly insulating aerogel glazing for solar energy usage. Sol Energy 72(1):21–29. https://doi.org/10.1016/S0038-092X(01)00086-X
47. Reim M, Körner W, Manara J, Korder S, Arduini-Schuster M, Ebert H-P, Fricke J (2005) Silica aerogel granulate material for thermal insulation and daylighting. Sol Energy 79(2):131–139. https://doi.org/10.1016/j.solener.2004.08.032
48. Shi Z, Chen S, Xu Z, Liu Z, Guo J, Yin J, Xu P, Zhang N, Zhang W, Alshareef HN, Liu T (2023) Metal oxide aerogels: a new horizon for stabilizing anodes in rechargeable zinc metal batteries. Adv Energy Mater 13(20). https://doi.org/10.1002/aenm.202300331
49. Shimizu T, Kanamori K, Nakanishi K (2017) Silicone-based organic-inorganic hybrid aerogels and xerogels. Chem Eur J 23(22):5176–5187. https://doi.org/10.1002/chem.201603680
50. Smirnova I, Gurikov P (2017) Aerogels in chemical engineering: strategies toward tailor-made aerogels. Ann Rev Chem Biomol Eng 8(1):307–334. https://doi.org/10.1146/annurev-chembioeng-060816-101458
51. Thomas B, Geng S, Wei J, Lycksam H, Sain M, Oksman K (2022) Ice-templating of lignin and cellulose nanofiber-based carbon aerogels: implications for energy storage applications. ACS Appl Nano Mat 5(6):7954–7966. https://doi.org/10.1021/acsanm.2c01033
52. Turhan Kara I, Kiyak B, Colak Gunes N, Yucel S (2024) Life cycle assessment of aerogels: a critical review. J Sol-Gel Sci Technol 111(2):618–649. https://doi.org/10.1007/S10971-024-06455-0/METRICS
53. Usman MR (2022) Hydrogen storage methods: review and current status. Renew Sust Energ Rev 167:112743. https://doi.org/10.1016/j.rser.2022.112743
54. Wan J, Zhang J, Yu J, Zhang J (2017) Cellulose aerogel membranes with a tunable Nanoporous network as a matrix of gel polymer electrolytes for safer lithium-ion batteries. ACS Appl Mater Interfaces 9(29):24591–24599. https://doi.org/10.1021/acsami.7b06271
55. Wang C, Huang Z, Wang T, Liu X, Zhu P (2022) Light-thermal-electric energy conversion based on polyethylene glycol infiltrated carboxymethylcellulose sodium-derived carbon aerogel. Energy Convers Manag 267:115948. https://doi.org/10.1016/j.enconman.2022.115948
56. Wang Y, Yang J, Song Y, Yang Q, Xiong C, Shi Z (2023a) Porous and three-dimensional carbon aerogels from nanocellulose/pristine graphene for high-performance supercapacitor electrodes. Diam Relat Mater 132:109626. https://doi.org/10.1016/j.diamond.2022.109626
57. Wang T, Liu Z, Li P, Wei H, Wei K, Chen X (2023b) Lignin-derived carbon aerogels with high surface area for supercapacitor applications. Chem Eng J 466(February). https://doi.org/10.1016/j.cej.2023.143118
58. Worsley MA, Satcher JH, Read C (2010) IV. C. 1h carbon aerogels for hydrogen storage accomplishments, pp 502–505
59. Wu L, Zhao B, Gao D, Jiao D, Hu M, Pei G (2023) Solar transparent and thermally insulated silica aerogel for efficiency improvement of photovoltaic/thermal collectors. Carbon Neutrality 2(1):6. https://doi.org/10.1007/s43979-023-00046-8
60. Xiao J, Yuan X, Li W, Zhang TC, He G, Yuan S (2024) Cellulose-based aerogel derived N, B-co-doped porous biochar for high-performance CO_2 capture and supercapacitor. Int J Biol Macromol 269:132078. https://doi.org/10.1016/j.ijbiomac.2024.132078
61. Xu J, Zhou X, Chen M, Shi S, Cao Y (2018) Preparing hierarchical porous carbon aerogels based on enzymatic hydrolysis lignin through ambient drying for supercapacitor electrodes. Microporous Mesoporous Mater 265:258–265. https://doi.org/10.1016/j.micromeso.2018.02.024
62. Xu R, Zhou J, Gong H, Qiao L, Li Y, Li D, Gao M, Xu G, Wang M, Liang X, Zhang X, Luo M, Qiu H, Liang K, Li Y (2022) Environment-friendly degradable zinc-ion battery based on guar gum-cellulose aerogel electrolyte. Biomater Sci 10(6):1476–1485. https://doi.org/10.1039/D1BM01747K
63. Yan C, Jia S, Wei J, Shao Z (2022) Construction of hierarchical porous derived from the cellulose nanofiber/graphene/Zn/Co ZIF 3D conductive carbon aerogels for high-performance supercapacitors. J Alloys Compd 920:165868. https://doi.org/10.1016/j.jallcom.2022.165868

64. Yan J, Sun Y, Jia T, Tao B, Hong M, Song P, Xu M (2025) Vertically aligned cellulose nanofiber/carbon nanotube aerogel-infused epoxy nanocomposites for highly efficient solar-thermal-electric conversion. J Mat Sci Technol 214:313–321. https://doi.org/10.1016/j.jmst.2024.07.018
65. Yang W, Wang X, Rossi R, Logan BE (2020) Low-cost Fe-N-C catalyst derived from Fe (III)-chitosan hydrogel to enhance power production in microbial fuel cells. Chem Eng J 380:122522. https://doi.org/10.1016/j.cej.2019.122522
66. Yang X, Zhang X, Li D, Zhou Y, Ren X, Zhang Q, Jiang T, Shi D, You J (2023) Nanofibrillar aerogels produced via one-pot chitin nanofibrils liquid exfoliation and self-crosslinking in the DMSO/KOH pseudosolvent system for high-performance lithium-ion batteries. J Power Sources 580:233398. https://doi.org/10.1016/j.jpowsour.2023.233398
67. Yashim MM, Sainorudin MH, Mohammad M, Fudholi A, Asim N, Razali H, Sopian K (2021) Recent advances on lightweight aerogel as a porous receiver layer for solar thermal technology application. Sol Energy Mater Sol Cells 228:111131. https://doi.org/10.1016/j.solmat.2021.111131
68. Ye S, Wang Y, Du B, Cheng L, Sun L, Yan P (2023) Recoverable sepiolite coated B-CoP/cellulose hybrid aerogel as monolithic catalysts for hydrogen generation via NaBH4 hydrolysis. Chem Eng J 474:145772. https://doi.org/10.1016/j.cej.2023.145772
69. Yu Z, Li S, Su J, Zhang J, Zhang D, Zhou Z, Qin Z, Liu X, Lai Y, Fu S (2023) Bio-inspired core-shell structural aerogel with programmable water release capacity for efficient solar thermoelectricity-freshwater cogeneration. Matter 6(10):3509–3525. https://doi.org/10.1016/j.matt.2023.07.015
70. Zhang Y, Wu B, Mu G, Ma C, Mu D, Wu F (2022) Recent progress and perspectives on silicon anode: synthesis and prelithiation for LIBs energy storage. J Ener Chem 64:615–650. https://doi.org/10.1016/j.jechem.2021.04.013
71. Zhang J, Jiang B, Ling L, Lu Y, Zou T, Cao B, Zhang T (2024) Using silica aerogel layers for the stabilization of Zn metal anodes in Zn-ion batteries. Chem Sel 9(18). https://doi.org/10.1002/slct.202400620
72. Zhao M, Wang J, Chong C, Yu X, Wang L, Shi Z (2015) An electrospun lignin/polyacrylonitrile nonwoven composite separator with high porosity and thermal stability for lithium-ion batteries. RSC Adv 5(122):101115–101120. https://doi.org/10.1039/C5RA19371K
73. Zhu L, You L, Zhu P, Shen X, Yang L, Xiao K (2018) High performance lithium-sulfur batteries with a sustainable and environmentally friendly carbon aerogel modified separator. ACS Sustain Chem Eng 6(1):248–257. https://doi.org/10.1021/acssuschemeng.7b02322
74. Zuo L, Zhang Y, Zhang L, Miao Y-E, Fan W, Liu T (2015) Polymer/carbon-based hybrid aerogels: preparation. Prop Appl Mat 8(10):6806–6848. https://doi.org/10.3390/ma8105343

Chapter 7
Biodiesel Production from Waste Frying Oils: The Role of Nanomaterials, Sustainable Development Goals Linkage and Economics Perspective

7.1 Overview of Renewable Energy

The unyielding pursuit of sustainable energy solutions is driven by two main factors: the increasing global energy demand and the urgent need to mitigate the environmental harm caused by fossil fuel combustion The combustion of fossil fuels continues to be a significant source of greenhouse gas emissions, air pollution, and global climate change, thereby increasing the necessity for alternative energy sources that can mitigate environmental effects while fostering sustainable economic development [58]. The development, exploration, and widespread adoption of renewable energy sources such as solar, hydroelectric, tidal, wind, biomass, and geothermal energy, have become fundamental and sustainable approaches to mitigating the adverse impacts of environmental degradation. These renewable technologies present a viable alternative to fossil fuels by significantly reducing greenhouse gas emissions, conserving natural resources, and lowering pollution levels [64]. Furthermore, the shift towards renewable energy not only aligns with global efforts to combat climate change but also contributes to enhancing energy security, fostering economic growth, and driving technological innovation. As the global demand for clean and sustainable energy continues to grow, advancing renewable energy technologies remains essential in addressing the complex challenges of environmental conservation and achieving sustainable development goals [2].

These sources are not only plentiful but also present markedly lower environmental impacts compared to conventional fossil fuels [30]. Solar and wind energy generate no direct emissions during operation, and advancements in energy storage technologies have significantly improved their feasibility in contemporary energy systems [32]. Simultaneously, biomass, especially as biofuel has garnered considerable interest for its capacity to deliver energy and valuable by-products, facilitating a circular economy model [37].

M. I. Ahmad, A. K. H. P. S., *Environmental and Energy Technology*, Green Energy and Technology, https://doi.org/10.1007/978-3-032-10706-0_7

According to recent statistics from the International Energy Agency (IEA), global energy consumption experienced a significant increase of 4.6% in 2023 [35]. This surge highlights the ongoing and persistent rise in worldwide energy demand, driven by multiple factors. A key driver of the increased energy consumption is the ongoing economic recovery in the aftermath of the COVID-19 pandemic. As nations strive to rejuvenate their economies, efforts are focused on restoring industrial and commercial activities to pre-pandemic levels. Moreover, the rapid pace of industrialization and urbanization, particularly in emerging economies, has significantly amplified the demand for energy, further accelerating consumption patterns globally. The escalating demand underscores the urgent need for sustainable energy strategies to balance economic expansion with environmental and resource conservation goals. These findings emphasize the critical importance of transitioning towards renewable energy sources and implementing energy-efficient technologies to address the challenges posed by growing global energy needs. The increase in demand highlights the necessity of shifting to cleaner, more sustainable energy sources to alleviate the environmental consequences of fossil fuel reliance. As the international community progresses towards fulfilling the climate objectives established in the Paris Agreement, the adoption of renewable energy has intensified in various regions.

In reaction to this escalating demand, international energy policies are progressively supporting renewable energy. The European Union (EU) has led renewable energy transitions by establishing ambitious targets for its member states. The EU's Renewable Energy Directive stipulated that a minimum of 10% of final energy consumption in the transport sector must originate from renewable sources by 2020, with additional targets established for subsequent years [27]. This directive forms a component of a comprehensive EU strategy aimed at attaining climate neutrality by 2050, encompassing a shift to 100% renewable energy in electricity generation and substantial reductions in greenhouse gas emissions across various sectors. Other regions, in addition to the EU, are implementing analogous policies. The Economic Research Institute for ASEAN and East Asia (ERIA) anticipates a significant rise in biofuel utilisation within the global renewable energy portfolio from 2015 to 2035, underscoring biofuel's contribution to fulfilling regional and global energy requirements [22, 47, 69].

7.1.1 Biodiesel as Renewable Energy Option

Biodiesel occupies a significant role among renewable energy options, as it is derived from vegetable oils and animal fats, rendering it a renewable and biodegradable alternative to petroleum-based diesel [11, 40]. Biodiesel is widely recognized for its environmentally favorable attributes, making it a sustainable alternative to conventional fossil fuels. Its low toxicity and inherent biodegradability contribute significantly to reducing the environmental burden associated with energy production and usage. Furthermore, biodiesel combustion produces substantially lower emissions of harmful pollutants such as sulfur compounds, hydrocarbons, and carbon monoxide, which are major contributors to air pollution and health-related concerns. These reductions play a critical role in mitigating the adverse impacts of fossil fuel use on air quality and public health. The cleaner-burning characteristics of biodiesel are essential in advancing worldwide efforts to mitigate greenhouse gas emissions. These features not only help reduce environmental impacts but also promote the transition to renewable and sustainable energy systems, in accordance with worldwide sustainability objectives [48].

It has been identified as a crucial factor in diminishing greenhouse gas (GHG) emissions and facilitating sustainable energy transitions, particularly within the transport sector, a major source of global CO_2 emissions. The IEA Renewables 2022 report forecasts a more than 60% increase in global renewable energy capacity by 2026, propelled by substantial investments in solar, wind, and bioenergy sectors [35]. Biodiesel is expected to play a substantial role in this growth, providing a sustainable substitute for traditional diesel, characterised by reduced lifecycle emissions and the possibility of circular economy integration via waste feedstocks.

A diverse range of feedstocks, encompassing both consumable and non-consumable sources, has been investigated for biodiesel production. Edible oils, including palm, peanut, and sunflower oils, are effective yet are criticised for competing with food resources and potentially jeopardising food security. The utilisation of consumable feedstocks for biofuel production has been associated with variations in food prices, especially in developing countries [45]. In response to these issues, researchers have utilised non-edible feedstocks like jatropha, linseed, and other inedible oils, which do not compete with food crops; however, they pose challenges related to cultivation efficiency, yield variability, and land use change [69].

Furthermore, current research is investigating third-generation biodiesel feedstocks sourced from algal oils and waste products. These comprise waste cooking oil, animal fats, and microalgae, which provide the dual advantage of repurposing waste streams while generating high-quality biodiesel [17, 47]. Microalgae have garnered significant interest owing to their rapid growth rates, elevated lipid content, and capacity to thrive on non-arable land, rendering them a viable candidate for large-scale biodiesel production [4]. Fourth-generation biodiesel, presently in development, entails genetically modified crops engineered to yield greater quantities of biofuels, presenting a progressive avenue for sustainable energy [70].

Biodiesel manufacturing has historically depended on several feedstocks, including vegetable oils (edible and non-edible), animal fats, and waste frying oils, which are accessible and extensively employed [47]. Feedstocks are categorized into four distinct generations based on their origin, sustainability, and environmental impact [69]. The first generation primarily includes oils derived from edible agricultural products such as palm oil, soybean oil, maize oil, coconut oil, rapeseed oil, sunflower oil, mustard oil, and olive oil. This classification reflects the dependence on food-based resources for biodiesel production and highlights the associated challenges related to sustainability and food security. Although these feedstocks are essential for biodiesel synthesis, their dependence on food-grade resources poses issues regarding food security, land utilisation, and ecological equilibrium, rendering sustainability a crucial factor for future biodiesel advancement. While these first-generation feedstocks have proven to be highly effective for biodiesel production, their utilization has sparked significant concerns related to market pricing volatility and food security. This is primarily due to their reliance on edible oils, which places biodiesel production in direct competition with food supply chains, potentially driving up the cost of essential food commodities and exacerbating food scarcity in vulnerable regions. To address these challenges, the second generation of biodiesel feedstocks focuses on non-edible oils, including karanja, rubber seed, castor, tall oil, jojoba, mahua, neem, and radish oils. These feedstocks provide a promising alternative by circumventing the conflict with food resources, thereby helping to mitigate food security concerns. However, despite their advantages, second-generation feedstocks face notable limitations in terms of production efficiency and agricultural productivity, as these oils are often derived from low-yielding or underutilized plant species. Consequently, optimizing the cultivation, harvesting, and processing of these non-edible oils remains a critical area of research to enhance their viability and scalability for sustainable biodiesel production.

The third generation encompasses oils sourced from algae and diverse waste materials, including used cooking oil, chicken fat, poultry byproducts, beef tallow, fish remnants, and algae feedstocks such as Azolla filiculoides. This generation is deemed more sustainable due to its dependence on waste materials and non-food sources, thereby alleviating the food-versus-fuel controversy. The fourth generation of biodiesel feedstocks, currently under active investigation, focuses on the utilization of genetically modified organisms (GMOs) to overcome the limitations of earlier generations and further enhance biodiesel production. This approach employs advanced genetic engineering techniques to optimize microorganisms, such as *Chlamydomonas reinhardtii*, *Phaeodactylum tricornutum*, and *Thalassiosira pseudonana*, for higher lipid accumulation and improved efficiency in biodiesel synthesis [69, 70]. These genetically modified species are designed to maximize their ability to convert carbon dioxide into valuable lipids, offering a sustainable and scalable solution to biodiesel production without competing with food supplies or relying on arable land. As a cutting-edge innovation, fourth-generation feedstocks hold immense potential to revolutionize the biofuel industry by integrating biotechnology and renewable energy development, aligning with global sustainability goals.

Genetically engineered feedstocks possess the capacity to transform biodiesel production by providing enhanced yields and improved sustainability, although they also elicit ethical and environmental apprehensions. Biodiesel offers a multifaceted and eco-friendly substitute for fossil fuels, with considerable potential to diminish greenhouse gas emissions. Nonetheless, its development must meticulously evaluate the trade-offs linked to various feedstock generations. Figure 7.1 presents a comprehensive comparison of the advantages and disadvantages of each generation of biodiesel feedstocks, elucidating their environmental, economic, and technical feasibility.

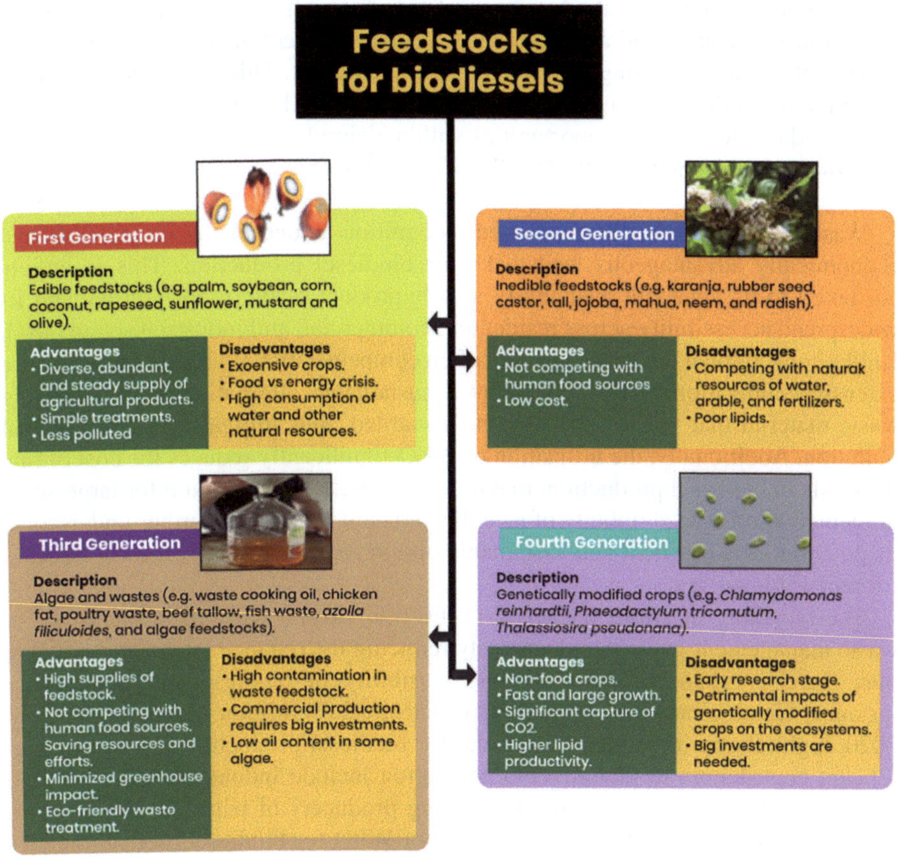

Fig. 7.1 Biodiesel production of different four generations of feedstocks

7.2 Overview of Waste Frying Oils

Waste frying oil (WFO), a by-product of the food industry frequently disposed of improperly, represents a highly promising source for biodiesel production and contributes to environmental degradation. This not only mitigates the disposal challenges linked to WFO but also diminishes dependence on edible oils, which have traditionally fuelled the "food vs fuel" controversy. The use of WFO in biodiesel production promotes circular economy principles by converting an environmental pollutant into a renewable fuel source, thereby closing the waste lifecycle loop. Notwithstanding its potential, the extensive implementation of WFO as a biodiesel feedstock encounters numerous challenges, such as variability in oil quality, the existence of contaminants, and the economic feasibility of large-scale production.

Furthermore, the ecological advantages of transforming waste frying oil into biodiesel must be considered alongside the logistical and technical challenges related to the collection, processing, and refinement of waste oil. This paper seeks to investigate the feasibility of converting waste frying oils into biodiesel, focussing on the associated challenges and opportunities within a circular economy framework, and assessing how this method advances the overarching objective of sustainable energy production.

Waste frying oil (WFO) has gained recognition as one of the most viable and economically advantageous feedstocks for biodiesel production. This is largely attributed to its abundant availability as a byproduct of the food industry and its widespread accessibility across residential, commercial, and industrial sectors [84]. Unlike conventional feedstocks, which often compete with food resources or require extensive cultivation, WFO represents a sustainable alternative by repurposing waste material that would otherwise be discarded, thus mitigating environmental pollution. Additionally, the utilization of WFO significantly reduces the cost of raw materials in biodiesel production, making it a cost-effective solution for large-scale implementation. The simplicity of its collection and processing further underscores its potential as a practical and sustainable option for meeting the growing demand for renewable energy.

The recycling of waste frying oil (WFO) provides a sustainable approach to waste management while simultaneously reducing the reliance on virgin vegetable oils, thereby enhancing the economic feasibility of biodiesel production. For instance, China alone generated approximately 32.89 million tonnes of WFO in 2020, highlighting its immense potential for biodiesel production [21]. Other significant contributors to global WFO generation include Indonesia, Malaysia, and Thailand, which are also among the leading producers of palm oil [18, 63, 75]. Despite its advantages, the widespread use of WFO for biodiesel production does not come without environmental challenges. A significant portion of WFO generated in households is often improperly disposed of through drainage systems, leading to substantial environmental damage, including water contamination and disruption of aquatic ecosystems [68]. The environmental hazards posed by improper disposal of WFO highlight the importance of implementing more effective waste collection systems and raising public awareness about the environmental impacts of oil waste.

The chemical composition of biodiesel is highly dependent on the feedstock used, particularly the type of waste frying oil. Biodiesel derived from waste palm oil, for instance, is characterized by a higher concentration of saturated fatty acids, which improves both its calorific value and oxidation stability [50]. This makes palm oil-derived biodiesel particularly suitable for regions with high-temperature climates, where oxidation stability is crucial. Biodiesel, on the other hand, is generated from waste sunflower oil and has a high concentration of unsaturated fatty acids. This results in a reduced viscosity and enhanced flow characteristics, both of which might be useful in cold regions [68]. Additionally, waste corn oil has been identified as a promising feedstock for biodiesel production, primarily due to its favorable density and viscosity characteristics, which enhance its combustion efficiency [53]. As research into the various types of waste frying oils continues, the potential to optimize biodiesel properties through selective feedstock utilization becomes increasingly apparent. The variability in the chemical composition of biodiesel derived from different types of waste oils suggests that targeted feedstock selection could be used to meet specific biodiesel performance requirements for various climates and engine types [53, 68].

Moreover, the global push toward a circular economy has further highlighted the importance of reusing waste frying oil in biodiesel production. In this context, WFO serves as a critical element in reducing dependency on fossil fuels and minimizing the environmental impact of waste oil disposal. Countries with significant waste oil generation should adopt comprehensive strategies to facilitate the collection and conversion of WFO into biodiesel, fostering a sustainable energy future while addressing the pressing need for waste management solutions [84].

There is a significant gap in the literature regarding the challenges and opportunities related to the incorporation of circular economy principles in the production of biodiesel from waste frying oils, despite the fact that extensive research has been conducted on the methods of producing biodiesel, including the evaluation of a variety of feedstocks and comparisons with conventional diesel fuels. The environmental and economic advantages of employing WFO as a biodiesel feedstock are significant; however, they are frequently eclipsed by challenges associated with feedstock variability, collection logistics, and contamination issues that complicate production.

7.3 Sources and Types of Waste Frying Oils

Waste frying oil (WFO) has garnered significant interest as a feasible feedstock for biodiesel production, supporting sustainable energy objectives by providing an eco-friendly approach to waste management. In contrast to virgin vegetable oils that compete with food resources, waste frying oil (WFO) is a byproduct of culinary processes, generally disposed of by households, restaurants, industrial food manufacturers, non-food sectors and households [3]. Employing WFO for biodiesel offers a substitute for petroleum-based fuels while alleviating waste disposal challenges, thus diminishing environmental pollution and promoting a circular economy.

7.3.1 Waste Frying Oil Sources

(i) **Household waste frying oil**

Households constitute a substantial source of WFO, especially in areas with elevated cooking oil consumption. Frequently utilised oils comprise sunflower, canola, palm, and soybean oil. While individual households may generate minimal amounts of WFO, their collective output constitutes a significant volume. Gathering WFO from residences can be problematic due to logistical complications and variable quality. Nevertheless, certain municipal programs and waste collection agencies have implemented systems for households to responsibly dispose of used cooking oil, thereby promoting its utilisation in biodiesel production Raising consumer awareness and establishing community collection points can improve the supply chain for biodiesel production from household waste frying oil.

(ii) **Restaurant and food service industry**

The restaurant and food service sector is a significant contributor to WFO, generating substantial quantities of used oil daily as a result of the regular frying of foods. Waste frying oil from restaurants generally comprises oils such as vegetable, soybean, and palm oil, which are utilised for prolonged durations at elevated temperatures. Extended usage modifies the oil's characteristics, introducing free fatty acids and impurities that can hinder biodiesel production. Notwithstanding these challenges, WFO from restaurants is a preferred feedstock due to its accessibility and comparatively straightforward collection process. Many biodiesel producers partner directly with restaurants to secure a steady supply of WFO, thereby creating a sustainable feedstock supply chain.

(iii) **Industrial food processing plants**

Extensive food processing facilities, especially those engaged in snack food, meat, and fast-food manufacturing, produce significant quantities of WFO. These establishments frequently employ substantial amounts of oil for deep frying, resulting in a continuous production of WFO. Industrial food processors typically handle their WFO disposal more systematically than smaller sources, facilitating more efficient collection and processing. Waste feedstock from industrial sources typically exhibits a more uniform composition, rendering it more predictable and simpler to process for biodiesel production. Furthermore, numerous industrial facilities are progressively prioritising sustainable practices, rendering WFO-to-biodiesel conversion a compelling alternative for waste management.

(iv) **Hospitality sector and institutional kitchens**

Major institutions, including hospitals, educational facilities, hotels, and military kitchens, substantially contribute to WFO production. These facilities frequently produce meals for hundreds of individuals each day, resulting in significant waste food output. Institutional kitchens utilise a diverse array of oils, akin to restaurants, but on a scale comparable to industrial food processing. Collection systems in these environments can be structured to facilitate the efficient retrieval of WFO. This sector presents a substantial opportunity for biodiesel producers, as collaborations with institutions can ensure a steady supply of feedstock.

(v) **Fast-food chains and franchise outlets**

Fast-food chains and franchise establishments focussing on fried foods constitute a significant source of WFO. The international expansion of fast-food chains has resulted in a uniform procedure for oil disposal, with numerous companies actively participating in recycling initiatives. Fast-food establishments typically uphold stringent standards for oil quality and replacement frequency, leading to waste frying oil that is comparatively simple to convert into biodiesel. Several prominent fast-food chains have engaged in partnerships with biodiesel manufacturers to transform their waste frying oil into fuel, thereby advancing their corporate social responsibility (CSR) initiatives and environmental sustainability objectives.

7.3.2 *Waste Frying Oil Types and Categories*

Various types and categories of waste frying oils exist, including vegetable oils, animal fats, mixed oils from commercial frying, hydrogenated oils, and waste oils from specialised cooking methods. Vegetable-derived waste frying oil is among the most prevalent types for biodiesel production, owing to its accessibility and advantageous characteristics. While animal-derived oils are infrequently utilised for frying, they are frequently present in the waste streams of certain food processing industries, especially in the manufacture of fast foods and meat products. Numerous commercial food establishments utilise a combination of vegetable oils or a mixture of vegetable and animal oils for frying. The formulation of these blended oils can differ markedly due to the amalgamation of various oil types. Hydrogenated oils, frequently utilised in the food industry to improve stability and prolong shelf life, additionally function as a source of waste frying oil (WFO). Hydrogenation alters the chemical composition of the oil, increasing its saturation. Moreover, certain niche applications utilise specialised oils that can be repurposed for biodiesel production following frying. Table 7.1 presents a summary of the different types and categories of waste frying oils utilised for biodiesel production.

Table 7.1 Different types and categories of waste frying oils utilised for biodiesel production

Vegetable-based waste frying oil	
Soybean oil	Widely utilised in restaurants and industrial food processing, soybean oil constitutes one of the primary sources of waste frying oil (WFO). Its composition of fatty acids is relatively balanced, rendering it an appropriate feedstock for biodiesel with favourable stability and combustion properties
Sunflower oil	Sunflower oil, commonly utilised in domestic and commercial environments, possesses a high concentration of unsaturated fatty acids, enabling the production of biodiesel with superior cold flow characteristics. Nonetheless, its elevated oxidation rate necessitates meticulous processing and stabilisation during biodiesel production
Palm oil	Predominantly utilised in areas rich in affordable palm oil, like Southeast Asia, waste palm oil possesses a high saturated fat content, resulting in biodiesel characterised by elevated viscosity and cloud points. Nonetheless, it provides exceptional combustion efficiency, rendering it a favoured option for tropical climates
Canola (rapeseed) oil	Renowned for its low saturated fat content and high monounsaturated fat levels, canola oil generates biodiesel with excellent oxidative stability. It is widely utilised in Europe and Canada, where rapeseed oil is predominant

(continued)

Table 7.1 (continued)

Animal-based waste frying oil	
Tallow (beef fat)	Tallow is a byproduct of beef cooking and is commonly utilised in fast-food establishments. The elevated saturated fat content enhances the biodiesel's stability at elevated temperatures, yet it also increases the cloud point, potentially restricting its applicability in colder regions
Lard (pork fat)	Lard is utilised in specific fried and baked products. It serves as a suitable feedstock for biodiesel; however, its elevated melting point requires pre-treatment and particular processing conditions to yield biodiesel with optimal flow characteristics
Fish oil	Fish oils possess elevated concentrations of polyunsaturated fatty acids, resulting in biodiesel with advantageous cold flow characteristics, yet necessitating additives or antioxidants to enhance oxidative stability
Mixed oil types from commercial frying	
Blended vegetable oils	Restaurants may utilise a mixture of soybean, canola, and corn oil for frying, owing to economic and performance advantages. The composition of these oils alters over time with repeated use, impacting the quality of the biodiesel. Mixed oils necessitate meticulous oversight during biodiesel production to guarantee uniformity and quality in the final product
Oil mixtures from deep frying of meats and vegetables	Restaurants may utilise a mixture of soybean, canola, and corn oil for frying owing to economic and performance advantages. The composition of these oils alters over time with repeated use, impacting the quality of the biodiesel. Mixed oils necessitate vigilant oversight during biodiesel production to guarantee uniformity and quality in the final product
Hydrogenated oils	
Hydrogenated soybean oil	Hydrogenation enhances the stability of soybean oil, which is frequently utilised in industrial frying for snack foods. Nevertheless, the procedure elevates the melting point, necessitating further processing in biodiesel production to attain satisfactory fluidity
Hydrogenated palm oil	Hydrogenated palm oil, recognised for its stability at elevated temperatures, is utilised in specific frying applications. The saturated composition improves the oxidative stability of biodiesel but may result in elevated cold flow characteristics, necessitating mitigation through blending with alternative biodiesel sources or additives
Waste oil from specialty cooking applications	
Coconut oil	Frequently utilised in tropical areas for frying because of its distinctive flavour and high-temperature stability, waste coconut oil serves as a significant biodiesel feedstock. The elevated lauric acid concentration results in biodiesel with reduced viscosity, appropriate for tropical applications
Peanut oil	Frequently utilised in specific frying applications because of its elevated smoke point, waste peanut oil serves as an effective biodiesel feedstock. The composition facilitates a relatively simple transesterification process, producing biodiesel with acceptable characteristics

7.4 Fatty Acid Composition of Waste Frying Oils

The plant-based source of waste frying oils is essential in determining their fatty acid profiles and, therefore, their overall chemical composition. The frequent utilisation of these oils for frying, coupled with processes like oxidation, hydrolysis, transfer of food residues, and polymerisation, can substantially modify their inherent properties [20]. Moreover, waste frying oils typically possess a greater proportion of free fatty acids (FFA) than their virgin vegetable oil equivalents. The high FFA content, combined with the intricate chemical composition from repeated thermal processing, directly affects the quality and conversion efficiency of biodiesel derived from these oils [53]. The chemical properties of waste frying oils require suitable pretreatment to enhance the transesterification process, particularly for adherence to international biodiesel standards.

During frying operations, vegetable oil experiences substantial chemical and physical changes. These alterations are chiefly ascribed to elevated temperature exposure, oxygen interaction, and the degradation of triglycerides. The volatility of oil diminishes, while its molecular weight escalates as a result of polymerisation and the formation of oligomeric compounds during frying [66, 67]. WFO, a byproduct of frying processes, primarily comprises oleic acid (43.67%) and linoleic acid (11.39%), as documented in various studies [72]. The free fatty acid (FFA) composition of WFO demonstrates significant variability, which directly influences the physicochemical characteristics of the resultant biodiesel. The composition of fatty acids, especially the saturation level and carbon chain length, is essential in determining the overall fatty acid profile [80]. Saturated fatty acids, like stearic acid, are completely hydrogenated, with carbon atoms bonded to the maximum number of hydrogen atoms possible. Conversely, unsaturated fatty acids, like linoleic acid, possess one or more double bonds that diminish the saturation level, leading to carbon atoms being attached to fewer hydrogen atoms [66, 67]. The existence of double bonds in unsaturated fatty acids affects the reactivity and thermal stability of WFO During frying, these double bonds may undergo oxidation or polymerisation, thereby modifying the free fatty acid profile and introducing polar compounds. The differences in chemical structure, molecular weight, and average composition of typical fatty acids in WFO, as outlined in Table 7.2, are crucial in influencing the performance characteristics of biodiesel. The composition may vary depending on the type of oil used, frying duration, and food content.

(i) Variation in composition

- The composition depends on the type of cooking oil (e.g., soybean oil, palm oil, sunflower oil) initially used.
- Extended frying increases free fatty acids (FFA) due to oxidation, hydrolysis, and thermal degradation.

Table 7.2 Fatty acids composition of WFO

Palmitic acid (C16:0)	Saturated fatty acid	$C_{16}H_{32}O_2$ $\{CH_3(CH_2)_{14}COOH\}$	10–30
Oleic acid (C18:1)	Monounsaturated fatty acid	$C_{18}H_{34}O_2$ $\{CH_3(CH_2)_7CH_5CH(CH_2)_7COOH\}$	40–60
Linoleic acid (C18:2)	Polyunsaturated fatty acid	$C_{18}H_{32}O_2$ $\{CH_3(CH_2)_4CH_5CH\text{-}CH_2\text{-}CH_5CH\text{-}(CH_2)_7COOH\}$	20–40
Linolenic acid (C18:3)	Polyunsaturated fatty acid	$C_{18}H_{30}O_2$ $\{CH_3(CH_2)_4CH_5CH\text{-}CH_2\text{-}CH_5CH\text{-}CH_2\text{-}CH_5CH(CH_2)_4COOH\}$	1–5
Stearic acid (C18:0)	Saturated fatty acid	$C_{18}H_{38}O_2$ $\{CH_3(CH_2)_{16}COOH\}$	2–6
Myristic acid (C14:0)	Saturated fatty acid	$C_{14}H_{28}O_2$	0–3
Lauric acid (C12:0)	Saturated fatty acid	$C_{12}H_{24}O_2$	Trace
Other fatty acids	Various	–	1–5

(ii) Significance for biodiesel production

- The high content of unsaturated fatty acids (oleic and linoleic acids) makes WFO a suitable feedstock for biodiesel production.
- Saturated fatty acids (palmitic and stearic acids) contribute to biodiesel stability but may affect cold-flow properties.

(iii) Processing considerations

- Before biodiesel conversion, waste frying oil often requires pre-treatment to remove impurities, reduce free fatty acids, and improve feedstock quality.

7.5 Characteristics of Biodiesel from Waste Frying Oils

For the most part, vegetable oils are used as the primary feedstock in the production of biodiesel. This is owing to the fact that vegetable oils are renewable and readily available. Using these oils as fuels directly provides a number of significant technical issues due to the fact that they have a high density, a low volatility, and an enhanced viscosity. This can lead to troublesome particle deposition in engines [39]. Furthermore, these properties lead to incomplete combustion, negatively affecting engine performance and emissions [29]. The transesterification process is extensively utilised to transform triglycerides present in natural oils, including animal fats and vegetable oils, into biodiesel, thereby mitigating these limitations. This procedure entails a chemical reaction between triglycerides and a short-chain alcohol, usually methanol or ethanol, facilitated by a catalyst which typically sodium hydroxide or potassium hydroxide [3]. This reaction yields biodiesel (fatty acid methyl esters) and glycerine as a co-product. Biodiesel generated via transesterification addresses the shortcomings of unrefined vegetable oils and adheres to the fuel quality standards established by ASTM D6751 and EN 14214, ensuring compatibility with standard diesel engines [42]. A principal advantage of biodiesel is its beneficial environmental characteristics.

Biodiesel is distinguished by its markedly lower sulphur content and minimal presence of aromatic hydrocarbons, both of which facilitate decreased greenhouse gas emissions relative to petroleum-derived diesel [13, 29]. Moreover, elevated cetane number of biodiesels enhances combustion efficiency and diminishes engine knock, rendering it an appropriate alternative or blending agent with conventional diesel [42]. The physicochemical characteristics of biodiesel, notably its lower density and viscosity compared to raw vegetable oils, improve its applicability for direct utilisation or blending in diesel engines, thereby fostering a cleaner and more sustainable fuel alternative [13]. Biodiesel presents a viable avenue for sustainable energy, corresponding with international initiatives to decrease reliance on fossil fuels and alleviate the detrimental environmental effects of traditional diesel emissions.

According to the American Society for Testing and Materials (ASTM D6751), the European Standard (EN 14214), and the Malaysian Standard (MS2008:2008), biodiesel that is produced from waste frying oils is able to consistently fulfil the quality criteria that have been established by significant standardisation organisations. In order to guarantee that biodiesel is compatible with diesel engines, these guidelines delineate essential requirements for fuel performance. These parameters include viscosity, density, acid value, and oxidation stability. According to Atadashi et al. [9], nations such as the United States of America, the European Union, India, Brazil, Japan, South Africa, and China have all created their own national criteria for biodiesel fuel. The EN 14214 standard from the European Union and the ASTM D6751 standard from the United States are widely acknowledged as international standards for the protection of biodiesel quality [82]. Biodiesel that has not been mixed with petroleum fuel is referred to as pure biodiesel, which is denoted by the marking B100. The term "BX" refers to a combination in which "X" indicates the proportion of biodiesel that is present in the mixture, which is always less than 100. Biodiesel blends, on the other hand, are different. The requirements established by ASTM D6751 are only applicable to pure biodiesel (B100), and they do not include biodiesel mixes.

These standards delineate essential parameters for fuel performance, such as viscosity, density, acid value, and oxidation stability, thereby ensuring the compatibility of biodiesel with diesel engines. These criteria in the standard serve as a foundation, providing essential principles that support quality control protocols. Moreover, these standards are essential for guaranteeing that biodiesel integrates effectively with diesel engines, regarding both compatibility and performance. Table 7.3 briefly describes the physicochemical properties of WFO biodiesel and diesel fuel.

Table 7.3 Physicochemical properties of typical WFO biodiesel and biodiesel

Physicochemical properties	Units	WFO biodiesel	Diesel	Description
Density @ 15 °C	kg/m^3	860–900	820–845	WFO biodiesel typically has a higher density than diesel fuel, which can affect fuel injection timing and combustion efficiency
Viscosity @ 40 °C	mm^2/s (cSt)	4.0–6.0	2.0–4.5	Higher viscosity in WFO biodiesel can influence fuel atomization in engines. It remains within acceptable limits but is higher than conventional diesel
Flash point	°C	160–170	>60	The higher flash point of biodiesel enhances safety during storage and handling, reducing the risk of fire
Cetane number	–	50–65	45–55	A higher cetane number indicates better ignition properties. WFO biodiesel's higher cetane number can lead to smoother engine operation
Heating value (calorific value)	MJ/kg	37–40	42–45	WFO biodiesel has a slightly lower calorific value due to its oxygen content, resulting in a minor reduction in engine power output
Cloud point	°C	0 to 15	−15 to 5	Higher values for biodiesel may cause cold flow issues in colder climates, necessitating the use of additives or blending with diesel
Pour point	°C	−5 to 10	−35 to −15	
Acid value	mg KOH/g	≤0.50	≤0.05	Indicates the presence of free fatty acids; a low acid value is essential to prevent corrosion and engine deposits
Sulfur content	% (wt)	<0.001	≤0.05	WFO biodiesel has negligible sulfur content, contributing to lower sulfur oxide (SOx) emissions compared to diesel fuel
Oxygen content	% (wt)	10–12	~0	The inherent oxygen in biodiesel improves combustion efficiency but reduces the energy content per unit mass
Lubricity (HFRR @ 60 °C)	μm	<460	>460	Biodiesel enhances lubricity, which can reduce wear on engine components such as fuel pumps and injectors
Ash content	% (wt)	≤0.02	≤0.01	Lower values in biodiesel reduce the likelihood of particulate emissions and engine deposits
Carbon residue	% (wt)	≤0.05	≤0.35	
Distillation range	°C	340–355 (approximate)	180–360	Biodiesel has a narrower and higher distillation range, affecting its volatility and combustion characteristics
Water content	mg/kg	≤500	≤200	Biodiesel can absorb more water than diesel fuel, which may promote microbial growth and fuel degradation if not properly managed

7.6 Production Techniques of Biodiesel from Waste Frying Oils

The production of biodiesel encompasses various methodologies, each tailored to accommodate the diversity of feedstock types, operational conditions, and the specific quality standards of the final biodiesel product. Four techniques are particularly notable for their extensive use and practical applicability: pyrolysis, transesterification, micro-emulsification, and dilution as shown in Fig. 7.2. These methods, while differing in principles and processes, are acknowledged for their efficacy in transforming raw materials into biodiesel, simultaneously addressing economic, environmental, and technological limitations.

7.6.1 *Pyrolysis*

Pyrolysis is a thermochemical conversion process that involves the thermal decomposition of biomass in an oxygen-free environment. This technique harnesses high temperatures, typically exceeding 500 °C, to break down complex biopolymers present in the feedstock, resulting in the formation of three primary products: bio-oil, char, and combustible gases [1]. The absence of oxygen prevents combustion and ensures the selective decomposition of organic materials into these valuable by-products. The process is fundamentally driven by heat, which facilitates the depolymerization and rearrangement of molecular structures in the biomass. The bio-oil produced can serve as a precursor for biodiesel production through further refining processes, making it a critical component in renewable energy systems. Char, another by-product, has applications as a soil amendment (biochar) or as a solid fuel, while the combustible gases generated can be utilized as a source of thermal energy in industrial processes. Pyrolysis is a process that is renowned for its versatility, since it enables the processing of a broad variety of biomass sources. These materials include agricultural wastes, waste from forestry, and organic industrial by-products. Charring wood to form charcoal is a typical example of pyrolysis. In this process, partial combustion happens due to the restricted supply of oxygen [34]. Another example of pyrolysis is the production of charcoal. As a result, this proves the practicability of the technique as well as its applicability to both conventional and contemporary applications.

Pyrolysis is an essential thermochemical method that is used to transform organic resources into energy products that are used in many applications. Due to the fact that this process includes the breakdown of biomass under situations where oxygen is limited, it is a vital approach for the creation of additional energy. There are three basic forms of pyrolysis that may be distinguished from one another: slow pyrolysis, fast pyrolysis, and flash pyrolysis [1]. The capability of this technology to dramatically change its behaviour in response to changes in operating parameters is the source of its adaptability. There are a few key criteria that are utilised to differentiate

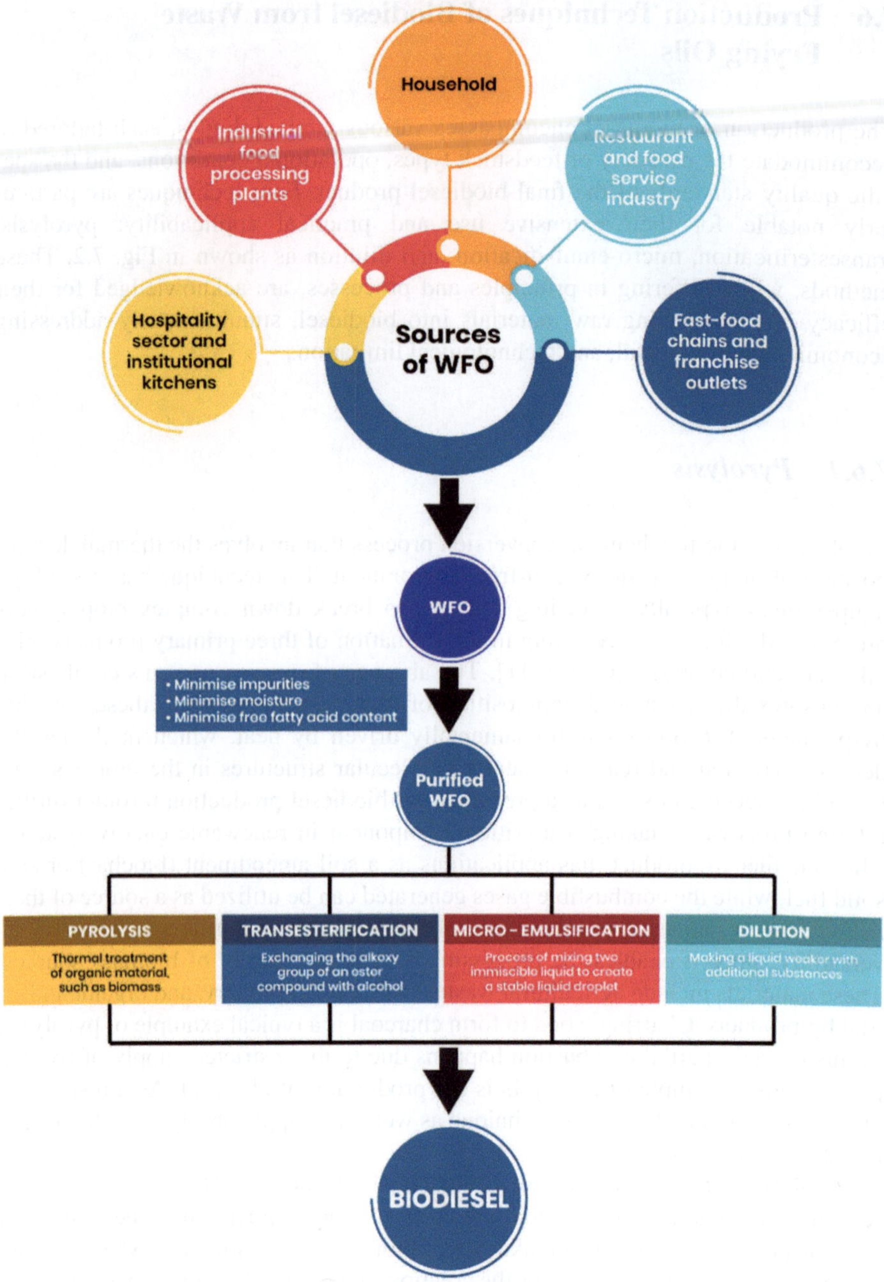

Fig. 7.2 Processes for the production of biodiesel from WFO

between these different forms of pyrolysis [33]. These characteristics include the temperature range, the heating rate, the residence duration, and the size of the biomass particles. It is because of this flexibility that pyrolysis is a viable strategy for the generation of sustainable energy and the management of resources. Individual types can be separated from one another. This flexibility allows for the process to be optimised to suit a variety of energy and material requirements, as these parameters determine the efficiency of the pyrolysis process as well as the distribution of the products that are produced.

In the process of slow pyrolysis, which is also commonly known as traditional pyrolysis, the heating durations are protracted, and the temperatures are moderate. Solid char, liquid bio-oil, and non-condensable gas are the three primary products that are generated by this process. All of this takes place in an oxygen-deficient (anoxic) environment [23]. The thermal breakdown of the organic elements in biomass begins at temperatures ranging from 350 °C to 550 °C and reaches completion at higher temperatures between 700 °C and 800 °C. Pyrolysis is an essential process for the production of sustainable energy and the creation of bio-based materials because they are controlled circumstances that ensure the selective destruction of the components of biomass.

Pyrolysis is a thermochemical process that involves the breakdown of complex carbon, hydrogen, and oxygen chains present in biomass into smaller, more manageable molecules. This decomposition occurs under controlled high-temperature conditions in the absence of oxygen, resulting in the formation of a mixture of syngas (comprising hydrogen (H_2), carbon monoxide (CO), and carbon dioxide (CO_2)), solid char, and condensable vapors such as tars and oils [59]. The liquid fraction, commonly referred to as bio-oil, holds significant potential as an alternative fuel source, providing a renewable and sustainable energy option. Pyrolysis is considered an efficient and relatively straightforward process that generates less waste compared to other biomass conversion methods, making it a desirable option for energy recovery [72].

Despite its advantages, the application of pyrolysis comes with notable challenges. It requires specialized equipment capable of withstanding and operating at high temperatures, often leading to substantial initial financial investments. Furthermore, the biodiesel produced through pyrolysis often exhibits lower purity levels, which can limit its suitability for large-scale production and use in commercial applications [69]. This limitation poses a significant drawback in efforts to scale the process for widespread adoption. To address these challenges, ongoing research seeks to improve the efficiency and quality of biodiesel obtained through pyrolysis.

The conversion of used frying oil into biodiesel is accomplished by the use of pyrolysis, which is a thermochemical process. The creation of bio-oil is accomplished by the utilisation of this method, which entails putting leftover frying oil through a series of processes that take place at high temperatures and do not require oxygen. A mixture of hydrocarbons, oxygenated molecules, and a variety of organic ingredients are included in the bio-oil that is produced following the process. There are major differences between bio-oil and waste frying oil in terms of their composition, origin, and features. Fats, contaminants, and degradation byproducts are all components of frying oil, which is a waste product from food preparation. The production of bio-oil can be achieved through the application of pyrolysis and other thermochemical processes to biomass, such as wood or agricultural leftovers. There are differences in composition as a result of different sources. In contrast to bio-oil, which is a more complex mixture of hydrocarbons, oxygenated compounds, and organic elements originating from biomass, waste frying oil is predominantly composed of triglycerides and may contain pollutants that are produced during the cooking process. Because of the different processing methods, their chemical compositions and features are different, which in turn affects the industrial applications of these substances.

The method of converting waste frying oil into biodiesel by pyrolysis is a complicated procedure that requires the oil to be processed at high temperatures in an atmosphere that is devoid of oxygen. A complex mixture of hydrocarbons, oxygenated compounds, and organic materials is produced by subjecting waste frying oil to regulated heating during the preparatory step. This results in the development of bio-oil, which is a complicated amalgamation of these components. There are a number of necessary stages involved in the subsequent refining of this bio-oil into biodiesel. Through the process of hydrodeoxygenation (HDO), oxygen is removed from bio-oil, which ultimately results in a reduction in the amount of oxygenated molecules present in the oil. After that, stabilisation operations are carried out, with the primary focus being on the removal of lighter components in order to enhance the overall stability of the hydrotreated bio-oil being treated. A subsequent step involves separating the bio-oil that has been stabilised into light and heavy fractions. Furthermore, the heavy fraction, which has a boiling point that is higher than 350 °C, is put through extra processing in a hydrocracker, which ultimately results in components that are suitable for use in petrol and diesel engines [65]. Pyrolysis is the process that facilitates the transformation of waste frying oil into biodiesel and other important fuel ingredients. This intricate method demonstrates the process.

7.6.2 Micro-emulsification

Micro-emulsification is an innovative and promising method in the realm of fuel production, particularly for biofuels. Fundamentally, a microemulsion represents a thermodynamically stable fluid, in contrast to kinetically stable systems that eventually separate into distinct water and oil phases. This inherent stability enables the creation of a distinctive approach for fuel production with significantly reduced by-products. Structurally, microemulsions are optically isotropic liquid microstructures with particle dimensions typically ranging from 1 to 150 nanometers. These systems form spontaneously when two generally immiscible liquids, such as low molecular weight alcohol and vegetable oil, are combined in the presence of a surfactant, resulting in a stable and homogenous mixture [54]. This technique offers both energy efficiency and economic viability for biofuel production, with the absence of a chemical reaction being a key advantage. It simplifies the biodiesel synthesis process while producing biofuels with superior atomization properties, thereby reducing nitrogen oxide emissions and avoiding the generation of by-products [31].

One of the most noteworthy benefits of the micro-emulsification technique is its ability to significantly lower the viscosity of vegetable oils. Conventional vegetable oils, such as soybean and sunflower oil, are characterized by high viscosity, which can present challenges in combustion engine performance. Through the inclusion of surfactants, microemulsions create diminutive, stable oil droplets dispersed in water, effectively reducing the overall viscosity of the mixture. This reduction facilitates improved fuel management, better atomization during combustion, and enhanced performance of vegetable oil-based fuels in diesel engines. Additionally, microemulsions, being thermodynamically stable mixtures, form spontaneously under specific conditions, eliminating the need for complex chemical processing. Micro-emulsification technology has emerged as a promising approach to improve the practicality and efficiency of utilizing vegetable oils as alternative fuels. This method offers a sustainable and effective solution for biofuel applications by addressing key challenges associated with the direct use of vegetable oils, such as viscosity and combustion efficiency. As a result, micro-emulsification presents a viable pathway for advancing renewable energy technologies in the context of sustainable fuel production.

The increased viscosity of biodiesel generated via the micro-emulsification method poses considerable operational difficulties, especially in its use in combustion engines. A significant concern is the possibility of injector needle obstructions, which can hinder engine performance and result in expensive maintenance needs. Additionally, the diminished combustion efficiency of high-viscosity biodiesel leads to incomplete fuel combustion, causing significant carbon deposits to accumulate in engine components. These deposits jeopardise engine durability and elevate emissions, so undermining the environmental advantages of biodiesel as a sustainable fuel option.

Moreover, the stability of biodiesel generated using the micro-emulsification process has been documented as inadequate, hence constraining its practical use [38, 52]. Instability in biodiesel may result in phase separation, diminished shelf life, and erratic fuel performance, rendering it less appropriate for extensive utilisation. These constraints highlight the necessity for more refinement and optimisation of the micro-emulsification process to improve biodiesel quality and consistency. Table 6 presents a thorough review of findings from multiple studies examining the emulsification of waste frying oils, offering significant insights into the obstacles and progress related to this method.

7.6.3 Dilution

When it comes to the manufacturing of biodiesel, dilution is a unique technology that is recognised for its non-chemical nature and uncomplicated methodology. Adding a solvent like diesel or ethanol to a solution is the first step in the process, which essentially involves lowering the concentration of the solute, which is normally biodiesel, in the solution. This method is particularly useful for reducing the viscosity of biodiesel, which is an essential factor in determining how well it performs in combustion engines. It is usual practice to employ sunflower and cottonseed oils as feedstocks in the dilution process. These oils are well-known for the high quantities of unsaturation that they contain. The generation of biodiesel with substantially reduced viscosity, which makes it appropriate for use in engines with few or no modifications, is one of the advantages of dilution. Other advantages include its simplicity and cost-effectiveness while also producing biodiesel. As a result of these characteristics, dilution is a viable method for the manufacture of biodiesel in areas that have restricted access to sophisticated refining technologies.

Dilution, on the other hand, contains its own set of intrinsic limitations and difficulties, just like any other approach. Insufficient volatility, decreased stability, and a tendency towards incomplete combustion are common characteristics of biodiesel that is created by the process of dilution. According to Ambat et al. [6], these concerns can have a negative impact on the performance of the engine, which can result in decreased durability, increased maintenance requirements, and complications such as the obstruction of the injector nozzle. According to Ogunkunle and Ahmed's research from 2020 [57], biodiesel with large dilution ratios frequently fails to achieve the severe ASTM fuel quality standards, which restricts its ability to participate in commercial applications. According to research conducted by Phan and Phan [60], the 20 vol% biodiesel dilution ratio is the most generally recommended dilution ratio since it provides the best possible engine performance with the least amount of changes required. On the other hand, surpassing this ratio may result in the diluted biodiesel exhibiting qualities that are comparable to those of pure biodiesel, which might potentially render the benefits of dilution as ineffective [55]. As a consequence of this, the choice of the dilution ratio is an extremely important factor in defining the physicochemical qualities of the biodiesel that is produced. It may be essential to perform additional processing processes or refine the biodiesel further in order to generate biodiesel of a high quality that is in accordance with industry requirements. This highlights the importance of continuing research and making technological developments in order to improve the practicability and dependability of the dilution process in the manufacture of biodiesel because it is essential.

7.6.4 Transesterification

Transesterification is an essential chemical process that transforms triglycerides in feedstocks into biodiesel, a renewable energy source that diminishes reliance on fossil fuels [10]. Triglycerides, the principal constituents of lipids, react chemically with alcohol, usually methanol or ethanol, in the presence of a catalyst to provide two principal products: methyl esters (biodiesel) and glycerol as a by-product. In the process, alcohol interacts with fatty acids in triglycerides, producing fatty acid alkyl esters while glycerol is generated as a byproduct. The efficacy of transesterification is significantly influenced by the catalyst employed, which may be homogeneous (acidic or basic) or heterogeneous, as it enhances the reaction rate and facilitates the successful conversion of fatty acids into biodiesel. The stoichiometry of the reaction and the operating parameters, including temperature, pressure, and the molar ratio of alcohol to oil, are crucial in optimising the production and quality of biodiesel. For specific feedstocks, like waste frying oils, an esterification phase is necessary prior to transesterification to decrease the free fatty acid (FFA) content to below 2 mg KOH/g, which is crucial for preventing soap production during biodiesel synthesis [14]. This two-step reaction process improves the efficiency of biodiesel generation from waste oils by mitigating the high free fatty acid levels that could impede transesterification. Esterification is the chemical transformation of free fatty acids into alkyl esters and water by their interaction with alcohol, facilitated by acidic catalysts like sulphuric acid. This procedure not only decreases the FFA content but also inhibits side reactions, such as saponification, which could decrease the biodiesel production. Subsequent to esterification, the transesterification process is executed, transforming triglycerides into biodiesel and glycerol with the aid of basic catalysts like sodium hydroxide or potassium hydroxide. The biodiesel and its byproducts are further separated by methods such as decantation and washing to guarantee fuel purity. This combined method of esterification and transesterification enables the conversion of waste frying oils, a prevalent byproduct of the food industry, into a sustainable, economical, and eco-friendly substitute for fossil-derived diesel fuels [41]. This approach tackles waste management challenges while enhancing global energy sustainability through the provision of a renewable and cleaner fuel source.

7.7 Role of Nanomaterials as Catalysts in the Transesterification Process of Biodiesel

Catalysts used in chemical processes are broadly categorized into two types: homogeneous and heterogeneous catalysts, both of which are fundamental in driving catalytic reactions by significantly lowering the activation energy and facilitating chemical transformations. Homogeneous catalysts, in particular, possess distinct features that make them highly suitable for biodiesel synthesis. Their ability to dissolve completely in the reaction medium ensures uniform interaction with reactants, leading to efficient catalytic activity [26]. However, this solubility also introduces challenges, such as difficulty in separating and recovering the catalyst after the reaction, which can complicate large-scale industrial applications. The management, handling, and potential reusability of homogeneous catalysts in biodiesel manufacturing facilities require meticulous oversight and sophisticated techniques, increasing the overall operational complexity. Despite these drawbacks, homogeneous catalysts play a pivotal role in biodiesel production, particularly in addressing the solubility limitations of alcohol in the oil phase. Alcohol and oil are naturally immiscible, and without a catalyst, the reaction between them proceeds at an extremely slow rate [8]. Homogeneous catalysts act as powerful solubilizers, enhancing the miscibility of alcohol with the oil, which accelerates the transesterification process, a critical step in biodiesel synthesis. By improving alcohol solubility, these catalysts not only boost the reaction rate but also increase the overall efficiency and yield of biodiesel production. This crucial functionality underscores their importance in biodiesel synthesis, even as advancements in catalyst technology strive to address their inherent limitations in scalability and recovery [49].

However, the removal of homogeneous catalysts from the reaction mixture presents a considerable obstacle when it comes to the utilisation of these catalysts. Following the conclusion of the reaction, the catalysts will require neutralisation or extraction, which will often call for a substantial amount of hot water. Significant volumes of wastewater from industrial processes are generated as a result of this process, which not only causes environmental problems but also drives up operating costs. Figure 7.3 is a diagrammatic illustration of the process of producing biodiesel, with a particular emphasis on the transesterification method. This figure illustrates the transesterification reaction, a fundamental process in biodiesel production. In this reaction, triglycerides (fats or oils) react with alcohol (commonly methanol or ethanol) in the presence of a catalyst to produce fatty acid esters (biodiesel) and glycerol as a byproduct. The chemical structure of triglycerides is broken down, with each ester group (R_1COO, R_2COO, and R_3COO) reacting with an alcohol molecule (ROH) to form three separate ester molecules, which constitute biodiesel. Simultaneously, the glycerol backbone is liberated, forming glycerol (CH_2-COH, CH-COH, CH_2-COH) as a secondary product. The catalyst, which can be acidic, basic, or enzymatic, plays a critical role in accelerating the reaction and improving the efficiency of biodiesel production. This reaction demonstrates the molecular transformation of fats and oils into a renewable energy source, emphasizing the chemical basis of biodiesel synthesis.

Table 7.4 provides information on the types of catalysts that are utilised in the transesterification process. In the process of producing biodiesel, one of the most important factors to consider is the specific surface area of the catalyst. The catalytic efficacy of nanocatalysts is significantly improved when the surface-to-volume ratio is increased. This is because the nanocatalysts obtain a greater concentration of active sites on their surfaces, which in turn speeds up reaction rates and increases efficiency. When it comes to stimulating transesterification processes, the increased specific surface area of nanocatalysts is particularly useful. Calcium oxide, often known as CaO, is a nanocatalyst that is widely used for the production of biodiesel. This is mostly due to the fact that it is both cost-effective and possesses excellent catalytic activity. Enhancing the synthesis of biodiesel may be accomplished by the use of this material either as a standalone catalyst or in conjunction with other compounds. Furthermore, magnesium oxide (MgO), strontium oxide (SrO), and

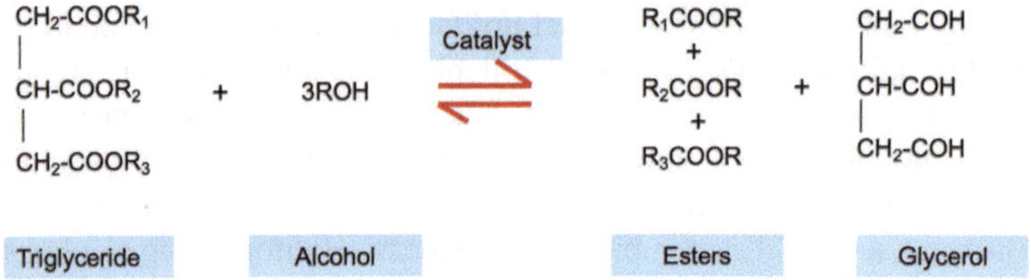

Fig. 7.3 A diagrammatic illustration of the process of producing biodiesel

Table 7.4 Catalysts used in the transesterification process

Types of catalysts	Description
Homogeneous acid catalysts	These catalysts are energy-efficient and well-suited for low-quality oils with high free fatty acid (FFA) content. They integrate esterification and transesterification, resulting in high biodiesel yields. However, they have slower reaction rates, require elevated temperatures, are difficult to recover, and may cause engine corrosion. They are compatible with oils and fats containing both high and low FFA levels. Examples include sulfuric acid (H_2SO_4), hydrochloric acid (HCl), phosphoric acid (H_3PO_4), and sulfonated organic acids
Homogeneous base catalysts	These are inexpensive, easily available, and require less energy. They also minimize engine corrosion and enable faster reactions compared to acid catalysts. Nonetheless, they generate alkaline wastewater, are ineffective for oils with high FFA content, require additional purification, and pose challenges in glycerol recovery. They are limited to oils and fats with low FFA content. Examples include potassium hydroxide (KOH), sodium hydroxide (NaOH), sodium methoxide (CH_3ONa), and potassium methoxide (CH_3OK).
Heterogeneous acid catalysts	These catalysts handle high FFA oils effectively, work well with low-grade feedstocks, produce minimal waste, and are easy to recover and reuse. However, they involve expensive synthesis, demand more energy, have slower reaction rates, and require higher alcohol volumes. They are suitable for oils and fats with both high and low FFA levels. Examples include tungstated zirconia, sulfated zirconia, and chloroaluminate
Heterogeneous base catalysts	These catalysts generate less waste, are easy to separate, offer high selectivity, are recyclable, and need less energy. However, they produce lower biodiesel yields from high FFA oils, necessitate purification, and are prone to soap formation. They are ideal for oils and fats with low FFA levels. Examples include calcium oxide (CaO), magnesium oxide (MgO), zinc oxide (ZnO), and mixed oxides like CaO-ZnO and CaO-La_2O_3
Enzyme catalysts	Enzymes operate at low temperatures, produce biodiesel of high purity, and can handle feedstocks with high water or FFA content. Despite these advantages, they are expensive, take longer to react, and are sensitive to methanol and glycerol, which can deactivate them. They are ideal for oils and fats with high water and FFA levels. Examples include Novozym 435, Lipozyme TL IM, and Bacillus subtilis
Nano-catalysts	These catalysts feature a high surface area, accelerating reaction rates while using smaller amounts of catalyst. They allow faster reactions and are easy to recover, but their production is costly, requires more alcohol, and involves operational complexities. They are suitable for feedstocks with high water and FFA content. Examples include Nano-CaO, smoke-deposited MgO, and MgO/Al_2O_3

zinc oxide (ZnO) are examples of other nanocatalysts that have been shown to be effective in the transesterification process. It is possible to combine materials such as iron oxide (Fe_3O_4), copper ferrite ($CuFe_2O_4$), activated carbon, and other sophisticated compounds into these catalysts in order to further improve their characteristics and functioning. The development of new catalysts and their applications is absolutely necessary in order to improve the effectiveness and long-term viability of the biodiesel manufacturing process.

7.7.1 Biodiesel Production from Waste Frying Oil Using Different Nanocatalysts

The development of nanocatalysts has been revolutionised by nanotechnology and nanomaterials, which has resulted in a significant increase in the production of biodiesel. According to Polshettiwar and Varma [61], these nanocatalysts are made up of nanoparticles, which may be identified by their diverse forms, morphologies, and diameters that are predominantly less than 100 nanometres. The nanoscale size of these materials result in a high surface area-to-volume ratio, which leads to an increase in their catalytic activity and efficiency. The creation of pseudo-homogeneous systems that take use of the advantages offered by both heterogeneous and homogeneous catalysis is accomplished by nanocatalysts through the combination of the principles of both types of catalysis [24]. This one-of-a-kind integration makes it possible to achieve remarkable catalytic performance, which includes increased activity, greater selectivity, improved efficiency, and increased reusability. In contrast to traditional catalysts, which are typically restricted to either homogeneous or heterogeneous forms, nanocatalysts provide enhanced versatility and adaptability. Their capacity to function efficiently under varied conditions renders them especially appropriate for the intricate transesterification processes associated with biodiesel production [71]. Moreover, nanocatalysts demonstrate remarkable recyclability and longevity, thereby decreasing overall production expenses and minimising waste. Their sophisticated design facilitates accurate regulation of catalytic reactions, thereby enhancing the transformation of diverse feedstocks, including waste frying oil, into premium biodiesel. Consequently, nanocatalysts not only improve the sustainability of biodiesel production but also correspond with global initiatives to advance renewable energy solutions and mitigate the environmental effects of fossil fuels. Figure 7.4 depicts the biodiesel production from various feedstocks using different nanocatalysts.

Because of their positively charged acid sites, nanocatalysts are essential in the manufacture of biodiesel. These sites significantly increase the adsorption of oil molecules on their surface, which is a vital step in the process. The adsorption of triglycerides, which are the primary component of oil, onto the acid-active sites of the nanocatalyst is facilitated by this surface contact. Immediately following the adsorption step, the catalytic reaction with methanol begins, which in turn drives the transesterification process forward. Triglycerides are converted into biodiesel, more precisely fatty acid methyl esters (FAMEs), using this process, which takes place under circumstances that are less severe than those that are required by conventional catalysts [83]. The greater efficiency of nanocatalysts in comparison to traditional catalytic systems is highlighted by the enhanced reaction rate and decreased energy demands that are exhibited by this mechanism.

One of the unique capabilities of nanocatalysts is their capacity to adsorb methanol at particular active sites, which ultimately leads to the production of methoxide anions and hydrogen cations. Methoxide anions play an important role in the process by interacting with triglyceride molecules to break down their structure and

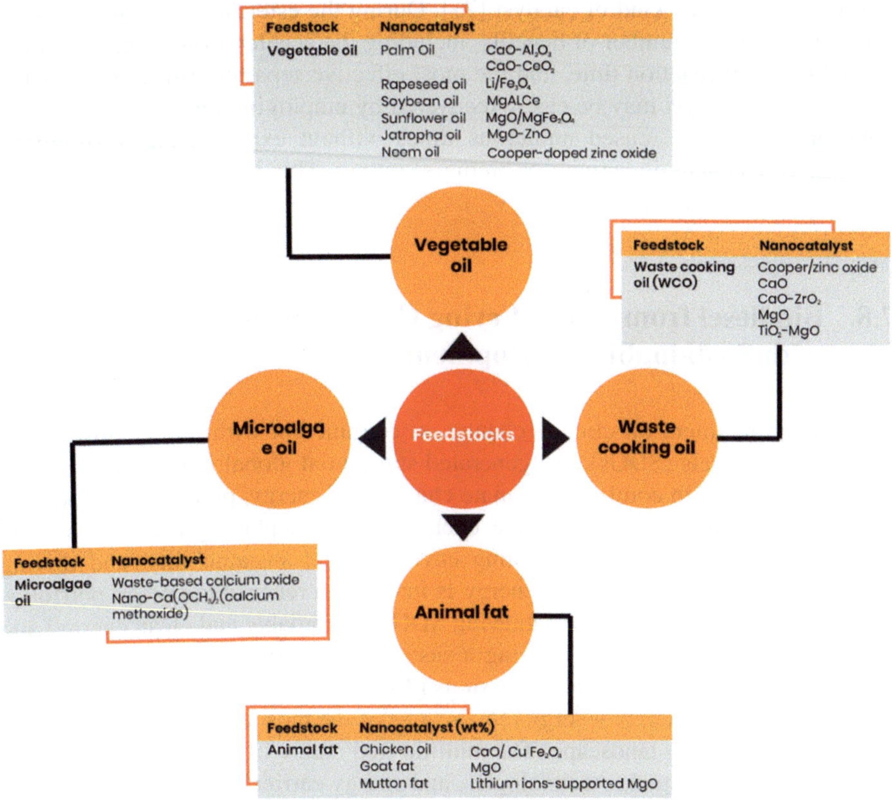

Fig. 7.4 Production of biodiesel from various feedstocks utilising multiple nanocatalysts

encourage the synthesis of FAMEs. Additionally, these anions serve as critical intermediates. This highly developed catalytic process not only ensures higher biodiesel yields, but it also reduces the amount of byproducts produced, the amount of energy consumed, and the overall costs. Nanocatalysts are able to overcome the limitations of typical catalysts, such as decreased reaction speeds, higher energy needs, and restricted recyclability [81]. This is made possible by the nanoscale accuracy and adaptability of nanocatalysts. In line with worldwide attempts to embrace renewable energy alternatives, this breakthrough underscores the crucial role that nanotechnology plays in allowing the manufacture of biodiesel in a manner that is efficient, sustainable, and scalable.

During the process of producing biodiesel, the reusability and recovery of heterogeneous nanocatalysts are the key advantages that they offer. When the biodiesel synthesis process is complete, the nanocatalyst is collected and reused after each cycle. The recovery of nanocatalysts is frequently accomplished by the use of chemical techniques. Utilising heterogeneous catalysts allows for the recovery of both the primary product and the byproduct that are produced by the reaction mixture in a manner that is both efficient and quick. It is not necessary to perform a washing step when using this kind of catalyst [77]. During the esterification process, nanocatalysts provide a number of benefits, including the acceleration of reaction rates, the reduction of reaction time, and the most effective separation from the reaction mixture. The catalyst may be easily recovered by employing an external magnetic field, and it can be reused numerous times without experiencing a significant decrease in its catalytic activity or biodiesel output. This is according to tests that were conducted in the past.

7.8 Biodiesel from Waste Frying Oil Linkages with Sustainable Development Goals

The Paris Agreement on climate change, in conjunction with the 17 Sustainable Development Goals (SDGs), has generated substantial global impetus for international cooperation in combating climate change. Bioenergy, particularly biodiesel, is a vital element among renewable energy solutions, playing a critical role in enhancing energy security, promoting environmental sustainability, and driving economic development [79]. Bioenergy is integral to renewable energy portfolios and significantly contributes to achieving SDG 7 (affordable and clean energy) and SDG 13 (climate action), by offering a sustainable alternative to fossil fuels and reducing greenhouse gas (GHG) emissions [36].

Despite its potential, bioenergy remains one of the most complex sectors within the renewable energy landscape. The multifaceted nature of bioenergy stems from the diversity of feedstocks, technologies, and energy carriers necessary for producing biofuels such as biodiesel. Biodiesel production, in particular, relies on a wide range of feedstocks, including forest products, agricultural crops, crop residues, and municipal waste. These diverse feedstocks require different conversion technologies to effectively transform them into usable energy. Biodiesel's primary advantage lies in the widespread availability of biomass, making it a readily accessible resource in various regions across the globe [25]. Biomass feedstocks, such as forest residues, agricultural by-products, and municipal waste, are abundant in almost every country, further supporting biodiesel's role in facilitating energy transitions at both the national and global levels [56].

Presently, bioenergy constitutes the predominant segment of the global renewable energy portfolio, accounting for over two-thirds of renewable energy consumption worldwide. However, a significant portion of this contribution arises from the traditional use of biomass, particularly for cooking and heating in rural regions of Asia and Africa. Although traditional biomass meets immediate energy needs, it also presents considerable health and environmental risks due to inefficiencies and pollution associated with its combustion. Consequently, there is an urgent need to transition from traditional biomass use to modern bioenergy alternatives, such as wood pellets, biogas, and liquid biofuels, which offer cleaner, more efficient energy solutions.

Biodiesel, as a modern bioenergy source, has the potential to play a pivotal role in future renewable energy frameworks, particularly in the decarbonization of the transport sector. Biofuels, including biodiesel, provide a practical solution for reducing carbon emissions in sectors where electrification is not yet feasible, such as aviation, marine transport, and heavy-duty road transportation. Accelerating the deployment of sustainable bioenergy solutions, particularly biodiesel, is crucial for advancing the transition to a low-carbon energy system, while also supporting global initiatives to meet the SDGs and the climate targets set by the Paris Agreement.

The SDGs provide a comprehensive framework for directing global development toward a more sustainable and equitable future. While the SDGs do not explicitly mention biomass utilization for food, feed, or energy systems, biomass is increasingly recognized as essential for achieving several SDGs. The growing demand for biomass, driven by the need to meet SDG targets, must be managed sustainably to ensure alignment with the principles of other SDGs [15]. In particular, biomass utilization aligns with SDG 2, which aims to eradicate hunger and promote sustainable agriculture. By incorporating bioenergy crops into agricultural systems through agroforestry and intercropping practices, land productivity can be enhanced, thereby supporting both food security and energy production. Biodiesel production from non-food biomass, such as crop residues and municipal waste, offers a dual-purpose solution that promotes sustainable rural development and energy security.

Bioenergy, including biodiesel, plays a vital role in achieving SDG 7, which aims to ensure universal access to affordable, reliable, and sustainable energy. Increasing the share of renewable energy in the global energy mix, particularly through bioenergy, significantly reduces dependence on fossil fuels, fostering a cleaner and more sustainable energy future [36]. Additionally, biodiesel and other modern biofuels offer a practical solution for enhancing decentralized energy access in off-grid areas, particularly in developing countries. This advancement contributes to improving energy equity and security, addressing critical disparities in energy availability and sustainability. Bioenergy's contributions extend beyond energy and environmental benefits, serving as a driver for SDG 8, which focuses on promoting sustained, inclusive, and sustainable economic growth. The expansion of biodiesel production, especially in rural areas, can create employment opportunities across the bioenergy supply chain, from feedstock cultivation to production and distribution [62]. The promotion of green jobs within bioenergy sectors supports economic diversification, enhances rural livelihoods, and contributes to poverty reduction in biomass-rich regions.

Bioenergy technologies also contribute to achieving SDG 9, which emphasizes industrial innovation and infrastructure development. The adoption of biorefineries, which convert biomass into biodiesel and other high-value products, demonstrates how bioenergy systems can modernize industries and drive innovation [36]. These innovations enhance industrial sustainability and promote technological progress in the renewable energy sector, fostering a cleaner, more efficient global economy. One of the most significant contributions of bioenergy is its role in addressing climate change, as outlined in SDG 13. The sustainable production and use of biodiesel can significantly reduce GHG emissions by replacing fossil fuels in key sectors, thus helping to meet global climate targets. By integrating bioenergy into national energy strategies, countries can advance climate mitigation efforts, supporting the overall goal of limiting global temperature rise as stipulated by the Paris Agreement [78].

Moreover, bioenergy aligns with SDG 15, which emphasizes the sustainable management of terrestrial ecosystems. Biodiesel production offers an opportunity to restore degraded lands by cultivating energy crops, thereby enhancing ecosystem resilience and biodiversity [51]. Sustainable biomass cultivation also promotes soil health and helps mitigate land degradation, contributing to long-term environmental sustainability. In a nutshell, biodiesel as a key component of bioenergy, supports and advances a range of Sustainable Development Goals. From promoting sustainable agriculture and modern energy access to driving economic growth, industrial innovation, and climate action, biodiesel plays a critical role in fostering a sustainable, low-carbon future. As the demand for bioenergy increases, it is essential to integrate sustainability principles throughout the biomass supply chain to ensure that bioenergy continues to contribute meaningfully to global sustainable development.

7.9 Economics Perspective of Biodiesel

7.9.1 Investment Cost for Biodiesel Production

The overall investment cost for biodiesel production from waste frying oils (WFO) is affected by various factors, including the production technology used, the production plant's scale, the quality and availability of raw materials, and market volatility in feedstock prices [28]. The production of biodiesel from waste frying oil exhibits distinct cost dynamics stemming from the variability in waste oil quality and the processing demands for purifying and transforming these oils into usable fuel. This requires meticulous evaluation of both capital and operational expenses to guarantee the economic feasibility of biodiesel production from WFO.

The costs associated with biodiesel production are generally classified into two primary categories: fixed capital investment and operating (or working capital) expenses. Fixed capital investment encompasses the initial capital necessary for the installation of production equipment, including reactors, separators, and other vital auxiliary systems that facilitate continuous and efficient processing. When producing biodiesel from waste frying oil (WFO), supplementary equipment may be required for pre-treatment processes, including filtration and degumming, to eliminate impurities from waste oils, potentially elevating fixed capital costs relative to traditional biodiesel production from virgin oils [46].

Operating costs refer to the continuous expenses necessary to sustain production at a specified capacity. This encompasses the expenses associated with raw materials specifically WFO procurement utilities (energy and water), labour, facility maintenance, and other variable costs directly linked to production output [73]. Waste frying oils, while typically regarded as inexpensive feedstocks, can still involve expenses related to collection, transportation, and processing, especially when obtained from various sources or necessitating significant purification to comply with production standards [44]. Numerous studies have sought to quantify the overall investment costs associated with biodiesel production, especially from waste frying oil (WFO), revealing considerable discrepancies influenced by geographic location, production scale, and feedstock availability [7]. Smaller production facilities may incur elevated per-unit production costs due to inefficiencies and the lack of economies of scale. In contrast, large-scale facilities, although requiring greater fixed investment, can attain reduced operational expenses by optimising production processes and utilising resources more efficiently [74].

Cost estimation models for biodiesel production frequently differ in methodology, with some emphasising capital expenditures and others prioritising operational costs. The precision of these models is significantly influenced by the assumptions concerning feedstock availability, market dynamics, and technological efficacy [19]. A thorough comprehension of both fixed and variable cost elements is crucial for decision-makers to enhance production strategies and evaluate the long-term viability of biodiesel production from waste frying oil (WFO). The production of biodiesel from waste frying oils presents a sustainable alternative to conventional fossil fuels; however, the related investment costs necessitate meticulous evaluation of both capital and operational expenditures. As various regions and sectors investigate biodiesel as a renewable energy source, especially through the use of waste materials, precise cost evaluations and technological innovations will be essential for increasing production in a cost-efficient and environmentally sustainable way [5].

7.9.2 Circular Economy Integration in the Conversion of Waste Frying Oil into Biodiesel

The circular economy is characterised as a regenerative economic system aimed at reducing resource consumption, waste production, emissions, and energy losses across the lifecycle of products and processes [76]. It accomplishes this through the strategic application of three fundamental principles: deceleration, consolidation, and constriction of material and energy cycles.

- Decelerating material and energy cycles emphasises prolonging product longevity through practices including maintenance, reuse, refurbishment, and remanufacturing. This method diminishes the necessity for continual manufacturing of new products and preserves natural resources.
- Closing material and energy cycle seeks to establish closed-loop systems in which waste and byproducts are reintroduced as raw materials or energy sources for subsequent production processes. This eradicates the notion of waste, guaranteeing that resources persist in circulation for an extended duration.

- Narrowing material and energy cycles entails enhancing resource utilisation efficiency by refining product designs and manufacturing processes to minimise material and energy consumption. This diminishes the total ecological impact of production and consumption systems.

Figure 7.5 shows circular economy model. This circular economy model illustrates the continuous flow of materials and energy through a system designed to minimize waste and optimize resource efficiency. The process involves several interconnected stages: raw material, design, production and remanufacturing; distribution; consumption; repair/reuse; collection; recycling and residual waste.

This model can be applied to the lifecycle of waste frying oil in the production of biodiesel. Waste frying oil, a byproduct of food preparation, is the raw material in this system. Instead of being discarded, it is repurposed for energy production [43]. The lifecycle of waste frying oil (WFO) in biodiesel production exemplifies the principles of a circular economy, wherein waste is converted into a valuable resource. Rather than being disposed of, WFO, a byproduct of culinary processes, is reutilised as a feedstock for biodiesel synthesis [16]. This process initiates with efficient collection systems that acquire used cooking oils from residences, eateries,

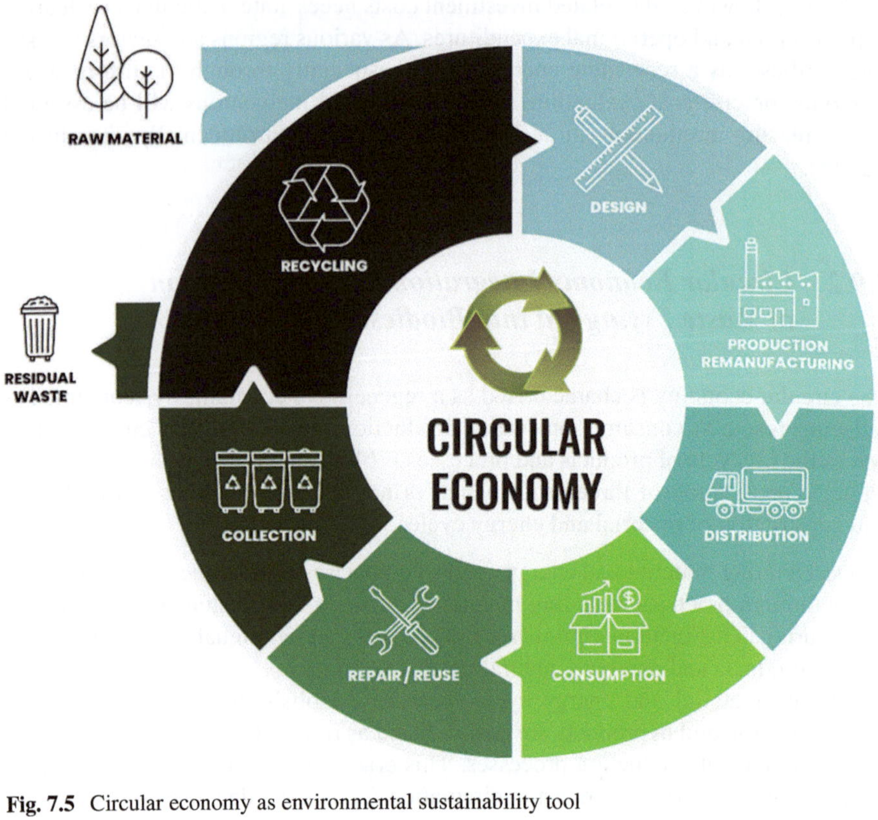

Fig. 7.5 Circular economy as environmental sustainability tool

and industries, thereby averting their improper disposal in landfills or drainage systems. Structured programs guarantee the redirection of these oils into the production cycle, minimising environmental damage and optimising resource utilisation.

The biodiesel production process is meticulously engineered to enhance the transformation of waste frying oil into superior fuel. Advanced technologies enhance and process oils, reducing energy consumption and emissions while generating a renewable energy source that lessens reliance on fossil fuels. Biodiesel is subsequently distributed to energy-intensive sectors, such as transportation, where it substitutes traditional diesel [12]. Vehicles and machinery utilising biodiesel generate markedly reduced emissions, thereby enhancing air quality and alleviating the effects of climate change. This seamless integration of waste conversion into energy exemplifies a closed-loop system in which resources are perpetually reused.

The process produces negligible residual waste, which is managed sustainably to enhance alignment with circular economy principles. Recycling WFO into biodiesel completes the resource cycle, exemplifying an innovative waste management approach that advances sustainable development objectives. This method not only minimises waste and greenhouse gas emissions but also encourages the utilisation of renewable energy, cultivating a more sustainable and environmentally friendly future. The WFO lifecycle demonstrates the capacity of circular economy models to promote environmental and economic sustainability through the conversion of waste into an energy resource.

The circular economy encourages the maintenance of a system in which materials and energy are recycled and reused within a closed loop, thereby reducing the amount of waste that is produced. The incorporation of renewable energy sources, the adoption of cyclical material flows, and the implementation of hierarchical energy recovery methods are the principles that underpin this paradigm shift [14]. Through the enhancement of resource efficiency, the reduction of greenhouse gas emissions, and the repurposing of waste into valuable bioproducts, a robust circular economy contributes to the enhancement of the three pillars of sustainability: economic, environmental, and social. An example of one of the most compelling applications of the circular economy is the transformation of waste frying oils into biodiesel. This process is an example of sustainable resource utilisation, and it addresses significant waste challenges that come from urban residential, commercial, and industrial sources. Furthermore, the production of biodiesel from waste oils preserves the inherent value of the raw materials, thereby extending the lifecycle of the materials and conforming to the principles of a circular economy.

7.10 Summary

Biodiesel production from waste frying oils (WFO) represents a promising avenue within the realms of renewable energy, addressing critical technological, environmental, and economic dimensions. With the increasing global demand for sustainable energy solutions driven by environmental challenges and Sustainable

Development Goals (SDGs), biodiesel emerges as a renewable and biodegradable alternative that reduces greenhouse gas emissions and enhances air quality. Its alignment with SDG 7 (Affordable and Clean Energy) and SDG 13 (Climate Action) further emphasizes its relevance. The chapter categorizes biodiesel feedstocks, highlighting the advantages of third generation options like WFO, which circumvent the "food versus fuel" debate by utilizing non-food waste. Various production methods, such as transesterification, pyrolysis, and micro-emulsification, are analyzed for their efficiency, challenges, and advancements, with a specific focus on nanocatalysts that improve synthesis processes. Additionally, the recycling of WFO is examined as a pathway to promote a circular economy, reducing waste while contributing to energy sustainability. The economic analysis underscores the integration of circular economy principles to optimize resource efficiency, reduce emissions, and advance sustainable energy initiatives. This multifaceted discussion demonstrates the critical role of biodiesel in addressing the pressing environmental and energy needs of today.

References

1. Adelawon BO, Latinwo GK, Eboibi BE, Agbede OO, Agarry SE (2022) Comparison of the slow, fast, and flash pyrolysis of recycled maize-cob biomass waste, box-benhken process optimization and characterization studies for the thermal fast pyrolysis production of bio- energy. Chem Eng Commun:1–31. https://doi.org/10.1080/00986445.2021.1957851
2. Ahmad MI, Jarimi H and Riffat S (2019) Springer briefs in applied sciences and technology nocturnal cooling technology for building applications. http://www.springer.com/series/8884
3. Alade A, Olasesan I, Odofin O, Ajibade O, Kayode A, Olu-Arotiowa O, Odesanmi A (2022) Review on environmental impact assessment of waste textiles and clothing recycling. Adv Tex Technol 30(6):28–36. https://doi.org/10.19398/j.att.202206029
4. Ali SS, El-Sheekh M, Manni A, Ruiz HA, Elsamahy T, Sun J, Schagerl M (2022) Microalgae-mediated wastewater treatment for biofuels production: a comprehensive review. Microbiol Res. 265:127187. https://doi.org/10.1016/j.micres.2022.127187
5. Al-Muhtaseb AH, Osman AI, Murphin KPS, Jamil F, Al-Haj L, Al Nabhani A, Kyaw HH, Myint MTZ, Mehta N, Rooney DW (2021) Circular economy approach of enhanced bifunctional catalytic system of CaO/CeO2 for biodiesel production from waste loquat seed oil with life cycle assessment study. Energ Convers Manage 236:114040. https://doi.org/10.1016/j.enconman.2021.114040
6. Ambat I, Srivastava V, Sillanpää M (2018) Recent advancement in biodiesel production methodologies using various feedstock: a review. Renew Sust Energ Rev 90:356–369. https://doi.org/10.1016/j.rser.2018.03.069
7. Ambaye TG, Vaccari M, Bonilla-Petriciolet A, Prasad S, van Hullebusch ED, Rtimi S (2021) Emerging technologies for biofuel production: a critical review on recent progress, challenges and perspectives. J Environ Manage 290:112627. https://doi.org/10.1016/j.jenvman.2021.112627
8. Ameen M, Ahmad M, Zafar M, Munir M, Mujtaba MM, Sultana S, Rozina E-K, S. E., Soudagar, M. E. M., & Kalam, M. A. (2022) Prospects of catalysis for process sustainability of eco-green biodiesel synthesis via transesterification: a state-of-the-art review. Sustainability 14(12):7032. https://doi.org/10.3390/su14127032

9. Atadashi IM, Aroua MK, Aziz AA (2010) High quality biodiesel and its diesel engine application: a review. Renew Sust Energ Rev 14(7):1999–2008. https://doi.org/10.1016/j.rser.2010.03.020
10. Athar M, Zaidi S (2020) A review of the feedstocks, catalysts, and intensification techniques for sustainable biodiesel production. J Environ Chem Eng 8(6):104523. https://doi.org/10.1016/j.jece.2020.104523
11. Ayoola AA, Hymore FK, Omonhinmin CA, Olawole OC, Fayomi OSI, Babatunde D, Fagbiele O (2019) Analysis of waste groundnut oil biodiesel production using response surface methodology and artificial neural network. Chem Data Collect 22:100238. https://doi.org/10.1016/j.cdc.2019.100238
12. Bhatia SK, Gurav R, Choi TR, Kim HJ, Yang SY, Song HS, Park JY, Park YL, Han YH, Choi YK, Kim SH, Yoon JJ, Yang YH (2020) Conversion of waste cooking oil into biodiesel using heterogenous catalyst derived from cork biochar. Bioresour Technol 302(2019):122872. https://doi.org/10.1016/j.biortech.2020.122872
13. Binhweel F, Hossain MS, Ahmad MI (2022) Recent trends, potentials, and challenges of biodiesel production from discarded animal fats: a comprehensive review. Bioenerg Res 16(2):778–800. https://doi.org/10.1007/s12155-022-10527-w
14. Binhweel F, Pyar H, Senusi W, Shaah MA, Hossain MS, Ahmad MI (2023) Utilization of marine ulva lactuca seaweed and freshwater azolla filiculoides macroalgae feedstocks toward biodiesel production: kinetics, thermodynamics, and optimization studies. Renew Energ 205:717–730. https://doi.org/10.1016/j.renene.2023.01.114
15. Blair J, Gagnon B and Klain A (2021) Biomass supply and the sustainable development goals. International Case Studies. IEA Bioenergy, September, 94
16. Brahma S, Nath B, Basumatary B, Das B, Saikia P, Patir K, Basumatary S (2022) Biodiesel production from mixed oils: a sustainable approach towards industrial biofuel production. Chem Eng J Adv 10(March):100284. https://doi.org/10.1016/j.ceja.2022.100284
17. Cavelius P, Engelhart-Straub S, Mehlmer N, Lercher J, Awad D, Brück T (2023) The potential of biofuels from first to fourth generation. PLoS Biol 21(3):e3002063. https://doi.org/10.1371/journal.pbio.3002063
18. Cerón Ferrusca M, Romero R, Martínez SL, Ramírez-Serrano A, Natividad R (2023) Biodiesel production from waste cooking oil: a perspective on catalytic processes. Processes 11(7). https://doi.org/10.3390/pr11071952
19. Changmai B, Vanlalveni C, Ingle AP, Bhagat R, Rokhum L (2020) Widely used catalysts in biodiesel production: a review. RSC Adv 10(68):41625–41679. https://doi.org/10.1039/d0ra07931f
20. Chen J, Zhang L, Li Y, Zhang N, Gao Y, Yu X (2021) The formation, determination and health implications of polar compounds in edible oils: current status, challenges and perspectives. Food Chem 364:130451. https://doi.org/10.1016/j.foodchem.2021.130451
21. Cheng G, Zhang M, Zhang Y, Lin B, Zhan H, Zhang H (2022) A novel renewable collector from waste fried oil and its application in coal combustion residuals decarbonization. Fuel 323:124388. https://doi.org/10.1016/j.fuel.2022.124388
22. Chia SR, Nomanbhay S, Ong MY, Shamsuddin AHB, Chew KW, Show PL (2022) Renewable diesel as fossil fuel substitution in Malaysia: a review. Fuel 314:123137. https://doi.org/10.1016/j.fuel.2022.123137
23. Chutia S, Gohain M, Borah MJ, Kakoty NM, Deka D (2020) Thermal degradation of Scenedesmus obliquus for biofuel production. Biomass Convers Bior 12(1):5031–5043. https://doi.org/10.1007/s13399-020-00982-w
24. Degfie TA, Mamo TT, Mekonnen YS (2019) Optimized biodiesel production from waste cooking oil (WCO) using calcium oxide (CaO) nano-catalyst. Sci Rep 9(1):1–8. https://doi.org/10.1038/s41598-019-55403-4
25. Demirbas A (2009) Progress and recent trends in biodiesel fuels. Energy Convers Manag 50(1):14–34. https://doi.org/10.1016/j.enconman.2008.09.001

26. Ehsan M, Chowdhury MTH (2015) Production of biodiesel using alkaline based catalysts from waste cooking oil: a case study. Procedia Eng 105(Icte 2014):638–645. https://doi.org/10.1016/j.proeng.2015.05.042
27. European Commission (2020) EUROPE 2020 a European strategy for smart, sustainable and inclusive growth
28. Farid MAA, Roslan AM, Hassan MA, Hasan MY, Othman MR, Shirai Y (2020) Net energy and techno-economic assessment of biodiesel production from waste cooking oil using a semi-industrial plant: a Malaysia perspective. Sustain Energy Technol Assess 39(September 2019):100700. https://doi.org/10.1016/j.seta.2020.100700
29. Fonseca JM, Teleken JG, de Cinque Almeida V, da Silva C (2019) Biodiesel from waste frying oils: methods of production and purification. Energy Convers Manag 184(February):205–218. https://doi.org/10.1016/j.enconman.2019.01.061
30. Gasser M, Pezzutto S, Sparber W, Wilczynski E (2022) Public research and development funding for renewable energy technologies in Europe: a cross-country analysis. Sustainability (Switzerland) 14(9). https://doi.org/10.3390/su14095557
31. Gul M, Mujtaba MA, Masjuki HH, Kalam MA, Zulkifli NWM (2023) Methods for biodiesel production. In: Inamuddin, Ahamed MI, Boddula R, Rezakazemi M (eds) Biodiesel technology and applications. Wiley, pp 267–283. https://doi.org/10.1002/9781119724957.ch9
32. Habibullah MS, Din BH, Tan SH, Zahid H (2022) Impact of climate change on biodiversity loss: global evidence. Environ Sci Pollut Res 29(1):1073–1086. https://doi.org/10.1007/s11356-021-15702-8
33. Heidari M, Salaudeen S, Dutta A, Acharya B (2018) Effects of process water recycling and particle sizes on hydrothermal carbonization of biomass. Energ Fuel 32:11576–11586. https://doi.org/10.1021/acs.energyfuels.8b02684
34. Huang Y, Li B, Liu D, Xie X, Zhang H, Sun H et al (2020) Fundamental advances in biomass autothermal/oxidative pyrolysis: a review. ACS Sustain Chem Eng 8(32):11888–11905. https://doi.org/10.1021/acssuschemeng.0c04196
35. International Energy Agency (2023) World energy outlook 2023 | enhanced reader. pp 23–28. https://www.iea.org/news/the-energy-world-is-set-to-change-significantly-by-2030-based-on-today-s-policy-settings-alone
36. International Renewable Energy Agency (2020) Renewable capacity statistics 2020 international renewable energy agency. In Irena https://tinyurl.com/2p8ws7dm
37. Jaiswal KK, Chowdhury CR, Yadav D, Verma R, Dutta S, Jaiswal KS, SangmeshB, Karuppasamy KSK (2022) Renewable and sustainable clean energy development and impact on social, economic, and environmental health. Energy Nexus 7(July):100118. https://doi.org/10.1016/j.nexus.2022.100118
38. Jin C, Sun T, Ampah JD, Liu X, Geng Z, Afrane S, Yusuf AA, Liu H (2022) Comparative study on synthetic and biological surfactants' role in phase behavior and fuel properties of marine heavy fuel oil-low carbon alcohol blends under different temperatures. Renew Energy 195:841–852. https://doi.org/10.1016/j.renene.2022.06.088
39. Kayode B, Hart A (2019) An overview of transesterification methods for producing biodiesel from waste vegetable oils. Biofuels 10(3):419–437. https://doi.org/10.1080/17597269.2017.1306683
40. Khan HM, Ali CH, Iqbal T, Yasin S, Sulaiman M, Mahmood H, Raashid M, Pasha M, Mu B (2019) Current scenario and potential of biodiesel production from waste cooking oil in Pakistan: an overview. Chin J Chem Eng 27(10):2238–2250. https://doi.org/10.1016/j.cjche.2018.12.010
41. Khan Z, Javed F, Shamair Z, Hafeez A, Fazal T, Aslam A, Rehman F (2021) Current developments in esterification reaction: a review on process and parameters. J Ind Eng Chem 103:80–101. https://doi.org/10.1016/j.jiec.2021.07.018
42. Kumar SB, Sushma US, Chandrasagar L, Raju VB, Devi V (2017) Use of waste frying oil as C.I. engine fuel—a review. OALib 04(11):1–33. https://doi.org/10.4236/oalib.1103958

43. Mahapatra S, Kumar D, Singh B, Sachan PK (2021) Biofuels and their sources of production: a review on cleaner sustainable alternative against conventional fuel, in the framework of the food and energy nexus. Energy Nexus 4(August 2021):100036. https://doi.org/10.1016/j.nexus.2021.100036
44. Mahlia TMI, Syazmi ZAHS, Mofijur M, Abas AEP, Bilad MR, Ong HC, Silitonga AS (2020) Patent landscape review on biodiesel production: technology updates. Renew Sust Energ Rev 118:109526. https://doi.org/10.1016/j.rser.2019.109526
45. Malik K, Capareda SC, Kamboj BR, Malik S, Singh K, Arya S, Bishnoi DK (2024) Biofuels production: a review on sustainable alternatives to traditional fuels and energy sources. Fuels 5(2):157–175. https://doi.org/10.3390/fuels5020010
46. Manaf ISA, Embong NH, Khazaai SNM, Rahim MHA, Yusoff MM, Lee KT, Maniam GP (2019) A review for key challenges of the development of biodiesel industry. Energ Convers Manage 185:508–517. https://doi.org/10.1016/j.enconman.2019.02.019
47. Mazaheri H, Ong HC, Amini Z, Masjuki HH, Mofijur M, Su CH, Badruddin IA, Yunus KTM (2021) An overview of biodiesel production via calcium oxide based catalysts: current state and perspective. Energies 14(13):1–23. https://doi.org/10.3390/en14133950
48. Mishra VK, Goswami R (2018) A review of production, properties and advantages of biodiesel. Biofuels 9(2):273–289. https://doi.org/10.1080/17597269.2017.1336350
49. Montero G and Stoytcheva M (2011) And processing edited by margarita Stoytcheva. https://books.google.com/books/about/Biodiesel.html?hl=fr&id=moqfDwAAQBAJ
50. Muhammed NM, Shaija A (2022) Performance evaluation of diesel engine using biodiesels from waste coconut, sunflower, and palm cooking oils, and their hybrids. Sust Ener Techn 53:102681. https://doi.org/10.1016/j.seta.2022.102681
51. Müller J, Mitesser O, Cadotte MW, van der Plas F, Mori AS, Ammer C, Chao A, Scherer-Lorenzen M, Baldrian P, Bässler C, Biedermann P, Cesarz S, Claßen A, Delory BM, Feldhaar H, Fichtner A, Hothorn T, Kuenzer C, Peters MK et al (2023) Enhancing the structural diversity between forest patches—a concept and real-world experiment to study biodiversity, multifunctionality and forest resilience across spatial scales. Glob Chang Biol 29(6):1437–1450. https://doi.org/10.1111/gcb.16564
52. Najjar R, Heidari S (2018) Modified diesel prepared by stabilization of water as nanodroplets in diesel/colza oil blend: study of phase behavior and affecting parameters. Fuel 214:497–504. https://doi.org/10.1016/j.fuel.2017.11.052
53. Narowska B, Kułażyński M, Łukaszewicz M, Burchacka E (2019) Use of activated carbons as catalyst supports for biodiesel production. Renew Energ 135:176–185. https://doi.org/10.1016/j.renene.2018.11.006
54. Nayab R, Imran M, Ramzan M, Tariq M, Taj MB, Akhtar MN, Iqbal HMN (2022) Sustainable biodiesel production via catalytic and non-catalytic transesterification of feedstock materials – a review. Fuel 328:125254. https://doi.org/10.1016/j.fuel.2022.125254
55. Ndiaye M, Arhaliass A, Legrand J, Roelens G, Kerihuel A (2020) Reuse of waste animal fat in biodiesel: biore fi ning heavily-degraded contaminant-rich waste animal fat and formulation as diesel fuel additive. Renew Energ 145:1073–1079. https://doi.org/10.1016/j.renene.2019.06.030
56. Nikkhah A, El Haj Assad M, Rosentrater KA, Ghnimi S, Van Haute S (2020) Comparative review of three approaches to biofuel production from energy crops as feedstock in a developing country. Bioresour Technol Rep 10(May 2019):100412. https://doi.org/10.1016/j.biteb.2020.100412
57. Ogunkunle O, Ahmed NA (2020) Exhaust emissions and engine performance analysis of a marine diesel engine fuelledwith Parinari polyandra biodiesel-diesel blends. Energy Rep 6:2999–3007. https://doi.org/10.1016/j.egyr.2020.10.070
58. Osman AI, Chen L, Yang M, Msigwa G, Farghali M, Fawzy S, Rooney DW, Yap PS (2023) Cost, environmental impact, and resilience of renewable energy under a changing climate: a review. Environ Chem Lett 21(2):741–764. https://doi.org/10.1007/s10311-022-01532-8

59. Patel A, Agrawal B, Rawal BR (2020) Environmental effects pyrolysis of biomass for efficient extraction of biofuel. Energ Source Part A 42(13):1649–1661. https://doi.org/10.1080/15567036.2019.1604875
60. Phan AN, Phan TM (2008) Biodiesel production from waste cooking oils. Fuel 87:3490–3496. https://doi.org/10.1016/j.fuel.2008.07.008
61. Polshettiwar V, Varma RS (2010) Green chemistry by nano catalysis. Green Chem 12(5):743–775. https://doi.org/10.1039/b921171c
62. Raman R, Sreenivasan A, Kulkarni NV, Suresh M, Nedungadi P (2025) Analyzing the contributions of biofuels, biomass, and bioenergy to sustainable development goals. iScience 28:112157. https://doi.org/10.1016/j.isci.2025.112157
63. Sajjadi B, Raman AAA, Arandiyan H (2016) A comprehensive review on properties of edible and non-edible vegetable oil-based biodiesel: composition, specifications and prediction models. Renew Sust Energ Rev 63:62–92. https://doi.org/10.1016/j.rser.2016.05.035
64. Salehabadi A (2020) Springer briefs in applied sciences nd technology solid-state hydrogen storage materials. http://www.springer.com/series/8884
65. Sánchez-Borrego FJ, Alvarez-Mateos P, García-Martín JF (2021) Biodiesel and other value-added products from bio-oil obtained from agrifood waste. PRO 9(5):797. https://doi.org/10.3390/pr9050797
66. Senusi W, Ahmad MI, Abdul Khalil HPS, Shakir MA, Binhweel F, Shalfoh E, Alsaadi S (2024a) Comparative assessment for biodiesel production from low-cost feedstocks of third oil generation. Renew Energy 236(January):121369. https://doi.org/10.1016/j.renene.2024.121369
67. Senusi W, Ahmad MI, Binhweel F, Shalfoh E, Alsaedi S, Shakir MA (2024b) Biodiesel production and characteristics from waste frying oils: sources, challenges, and circular economic perspective. Environ Sci Pollut Res 31(23):33239–33258. https://doi.org/10.1007/s11356-024-33533-1
68. Simsek S, Uslu S (2020) Determination of a diesel engine operating parameters powered with canola, safflower and waste vegetable oil based biodiesel combination using response surface methodology (RSM). Fuel 270:117496. https://doi.org/10.1016/j.fuel.2020.117496
69. Singh D, Sharma D, Soni SL, Sharma S, Kumar SP, Jhalani A (2020) A review on feedstocks, production processes, and yield for different generations of biodiesel. Fuel 262:116553. https://doi.org/10.1016/j.fuel.2019.116553
70. Singh A, Prajapati P, Vyas S, Gaur VK, Sindhu R, Binod P, Varjani S (2023) A comprehensive review of feedstocks as sustainable substrates for next-generation biofuels. Bioenerg Res 16(1):105–122. https://doi.org/10.1007/s12155-022-10440-2
71. Sun X, Xu L, Yuan H, Wang G, Maleki B (2024) Boosted conversion of restaurant waste oil into biodiesel using Fe3O4@UiO-66-NH2 magnetic heterogeneous nanocatalyst and its application on the diesel engine: optimization via RSM. Renew Energy 223:120007. https://doi.org/10.1016/j.renene.2024.120007
72. Suzihaque MUH, Alwi H, Kalthum U, Abdullah S, Haron N (2022) Biodiesel production from waste cooking oil: a brief review. Mater Today Proc 63:S490–S495. https://doi.org/10.1016/j.matpr.2022.04.527
73. Syafiuddin A, Chong JH, Yuniarto A, Hadibarata T (2020) The current scenario and challenges of biodiesel production in Asian countries: a review. Bioresour Technol Rep 12:100608. https://doi.org/10.1016/j.biteb.2020.100608
74. Tabatabaei M, Aghbashlo M, Dehhaghi M, Panahi HKS, Mollahosseini A, Hosseini M, Soufiyan MM (2019) Reactor technologies for biodiesel production and processing: a review. Prog Energy Comb 74:239–303. https://doi.org/10.1016/j.pecs.2019.06.001
75. Thushari I, Babel S (2022) Comparative study of the environmental impacts of used cooking oil valorization options in Thailand. J Environ Manage 310:114810. https://doi.org/10.1016/j.jenvman.2022.114810
76. Tulashie SK, Alale EM, Agudah PQ, Osei CA, Munumkum CA, Gah BK, Baidoo EB (2024) A review on the production of biodiesel from waste cooking oil: a circular economy approach. Biofuels 0(0):1–21. https://doi.org/10.1080/17597269.2024.2384277

77. Ulakpa WC, Ulakpa ROE, Eyankware EO, Egwunyenga MC (2022) Statistical optimization of biodiesel synthesis from waste cooking oil using NaOH/bentonite impregnated catalyst. Cleaner Waste Systems 3(July):100049. https://doi.org/10.1016/j.clwas.2022.100049
78. UN Climate Change Conference (2024) The Paris agreement. In: Towards a Climate-Neutral Europe: Curbing the Trend, pp 24–45. https://doi.org/10.4324/9789276082569-2
79. United Nations (2024) Progress towards the sustainable development goals. Report of the Secretary-General. Assembly 64782(May):14
80. Wcisło G, Leśniak A, Kurczyński D, Pracuch B (2024) Experimental investigation of physico-chemical properties of the produced biodiesel from waste frying oil and its blend with diesel fuel. Energies 17(16). https://doi.org/10.3390/en17164175
81. Xie Y, Wang D, Almojil SF, Almohana AI, Alali AF, Zhou Y, Raise A (2023) CaO-MgFe2O4@K2CO3 as a novel and retrievable nanocatalyst for two-step transesterification of used frying oils to biodiesel. Process Saf Environ Prot 172:195–210. https://doi.org/10.1016/j.psep.2023.02.005
82. Yasin MHM, Mamat R, Najafi G, Ali OM, Yusop AF, Ali MH (2017) Potentials of palm oil as new feedstock oil for a global alternative fuel: a review. Renew Sustain Energy Rev 79:1034–1049. https://doi.org/10.1016/j.rser.2017.05.186
83. Zhang Y, Duan L, Esmaeili H (2022) A review on biodiesel production using various heterogeneous nanocatalysts: operation mechanisms and performances. Biomass Bioenergy 158(December 2021):106356. https://doi.org/10.1016/j.biombioe.2022.106356
84. Zulqarnain AM, Yusoff MHM, Nazir MH, Zahid I, Ameen M, Sher F, Floresyona D, Budi NE (2021) A comprehensive review on oil extraction and biodiesel production technologies. Sustainability 13(2):1–28. https://doi.org/10.3390/su13020788

Chapter 8
Enhancing Energy Efficiency in Buildings Using Nanomaterial-Based Insulation

8.1 Overview of Energy Efficiency in Buildings

Energy efficiency in buildings is a pivotal aspect of global strategies aimed at combating climate change, reducing energy consumption, and promoting sustainable development. Buildings account for a significant portion of total energy use and are responsible for a large share of greenhouse gas (GHG) emissions. According to Hafez et al. [22], buildings contribute to nearly 30–40% of global energy consumption and over a quarter of CO_2 emissions. Improving the energy performance of buildings not only addresses environmental concerns but also yields economic and social benefits, such as reduced operational costs, enhanced indoor comfort, and better occupant health. In light of growing urbanization, rising energy demands, and climate concerns, there is an urgent need to focus on innovative solutions, technologies, and policies that enhance energy efficiency in the built environment.

The demand for energy in buildings continues to rise due to population growth, urbanization, and increasing reliance on electricity for heating, cooling, and appliances. A significant portion of building energy consumption is attributed to heating, ventilation, and air conditioning (HVAC) systems. This trend, as highlighted by [44] is exacerbated by the expansion of air conditioning in both developed and developing countries, especially in hot and humid climates. Despite improvements in appliance and lighting efficiency, the net energy consumption of buildings remains high. This is partly due to the rebound effect, where energy savings in one area may lead to increased energy use elsewhere. Improving the thermal performance of building envelopes including walls, roofs, floors, windows, and doors—is one of the most effective strategies for enhancing energy efficiency [44]. High-performance insulation materials reduce heat transfer, which minimizes the need for mechanical heating and cooling. Advanced insulation solutions such as vacuum insulation panels, aerogels, and phase change materials (PCMs) offer superior thermal resistance compared to traditional materials like fiberglass or mineral wool. Moreover, passive

M. I. Ahmad, A. K. H. P. S., *Environmental and Energy Technology*, Green Energy and Technology, https://doi.org/10.1007/978-3-032-10706-0_8

design strategies such as building orientation, shading, and natural ventilation which play a critical role in reducing energy demand. When combined with high-performance glazing and airtight construction, these strategies can lead to zero-energy or nearly zero-energy buildings. Thermal bridges and air leakages are also critical issues in energy performance; addressing them through better construction practices and materials can significantly reduce energy losses.

Existing buildings present a significant opportunity for energy savings, as they often have outdated systems and poor thermal performance. Deep energy retrofits as discussed by Fernandes et al. [18] involve comprehensive upgrades to the building envelope, mechanical systems, and controls to achieve energy savings. Unlike shallow retrofits, which focus on low-cost measures, deep retrofits require a holistic approach and careful planning to avoid performance gaps. The process typically includes energy audits, modeling, and scenario analysis to identify the most effective interventions. Retrofitting may involve adding insulation, replacing windows, upgrading HVAC systems, and integrating renewable energy technologies such as solar photovoltaic (PV) panels. Although deep retrofits can be capital-intensive, they provide long-term savings, increase property value, and improve indoor environmental quality. Financial incentives, performance contracting, and innovative business models such as Energy Service Companies are critical to overcoming the financial barriers associated with retrofitting. Integrating renewable energy sources into buildings is essential for reducing dependency on fossil fuels and achieving carbon neutrality. On-site generation through solar PV, solar thermal systems, and small wind turbines allows buildings to become energy producers rather than just consumers. Net metering policies and grid interconnection standards enable buildings to export surplus energy to the grid, further enhancing sustainability. However, renewable energy integration must be supported by efficient energy storage systems and smart grid technologies to balance supply and demand. Battery storage, thermal storage, and hydrogen systems are emerging as viable solutions for storing excess energy and ensuring reliability. In climates with seasonal variability, hybrid systems that combine multiple energy sources and storage solutions offer resilience and flexibility [18].

Effective policies and regulations are critical for driving energy efficiency improvements in the building sector. Building energy codes and standards set minimum performance requirements for new and existing buildings. While technology plays a crucial role, occupant behavior significantly influences actual energy performance as highlighted by Lee et al. [36]. Studies show that buildings with similar physical characteristics can have vastly different energy use due to variations in user behavior. Engaging occupants through awareness campaigns, feedback mechanisms, and incentives is essential for achieving sustained energy savings. Behavioral interventions include providing real-time energy use information, using social comparisons to encourage competition, and implementing gamification techniques to motivate users. Involving occupants in the design and operation of buildings fosters a sense of ownership and responsibility. Building managers and operators also need training and support to effectively manage energy systems and respond to performance data. Cultural and socio-economic factors influence energy consumption patterns, and equity considerations must be addressed in energy efficiency programs. Low-income households, for example, often face barriers such as lack of access to financing and lower awareness of energy-saving opportunities. Policies must ensure that energy efficiency benefits are distributed equitably and contribute to social inclusion [36].

Despite the clear benefits, several barriers hinder the widespread adoption of energy-efficient practices in buildings. Economic barriers include high upfront costs, split incentives (e.g., between landlords and tenants), and uncertain payback periods. Technical barriers involve lack of skilled labor, insufficient data, and interoperability issues among different technologies. Institutional and regulatory barriers include weak enforcement of building codes, inconsistent standards, and limited coordination among stakeholders. Market barriers such as limited access to financing, information asymmetry, and low energy prices also reduce the attractiveness of energy efficiency investments. As discussed by Matos et al. [40], overcoming these barriers requires a multi-faceted approach that combines regulatory reform, financial innovation, capacity building, and stakeholder engagement. International collaboration, knowledge exchange, and research and development are crucial for scaling up successful solutions and adapting them to local contexts. The future of energy efficiency in buildings lies in integrated, smart, and sustainable systems that combine passive design, high-performance materials, advanced technologies, and renewable energy. According to Matos et al. [40], the transition to zero-energy and positive-energy buildings will be accelerated by continued innovation in digital tools, materials science, and energy storage. Policymakers should prioritize mandatory building energy codes, invest in retrofitting programs, and promote the use of digital technologies. Financial institutions can support the transition by offering green loans and performance-based financing. Education and training programs are needed to build capacity among architects, engineers, builders, and facility managers. Public awareness and community engagement will play a critical role in shifting behaviors and creating demand for energy-efficient buildings. Cities and municipalities can lead by example through green public buildings and community energy projects. International frameworks such as the Paris Agreement and Sustainable Development Goals provide the necessary momentum and direction for collective action. In general, energy efficiency in buildings is a cornerstone of sustainable development and climate action. With buildings contributing significantly to global energy use and emissions, enhancing their energy performance is both a necessity and an opportunity. Advances in materials, digital technologies, and renewable energy integration are transforming the way buildings are designed, constructed, and operated. However, realizing the full potential of energy efficiency requires overcoming economic, technical, and behavioral barriers through coordinated efforts by governments, industries, and communities. By adopting a holistic and inclusive approach, the building sector can play a leading role in achieving a low-carbon, resilient, and equitable future.

8.2 Importance of Thermal Insulation in Energy Conservation

Thermal insulation refers to the application of materials or systems designed to reduce the rate of heat transfer. By slowing down the movement of heat through conduction, convection, and radiation, insulation materials help maintain comfortable indoor environments without overreliance on mechanical systems like heaters, air conditioners, and ventilation units. As a result, buildings require less energy input to remain thermally comfortable, which directly translates into lower utility bills and reduced carbon emissions. In the modern world, energy conservation has become a top priority due to increasing global energy demands, rising utility costs, and growing environmental concerns. Among the various sectors that contribute to high energy consumption, the building sector stands out as one of the largest energy consumers. A significant portion of the energy used in buildings is dedicated to regulating indoor temperatures—primarily through heating in the winter and cooling during the summer. This energy expenditure is largely due to the natural transfer of heat through the building envelope, including walls, roofs, and windows. Ali et al. [7] emphasize that adopting effective thermal insulation strategies plays a pivotal role in minimizing energy use and enhancing overall energy efficiency. One of the most effective approaches to energy conservation in buildings is the implementation of passive design strategies. These strategies aim to regulate internal temperatures naturally, without the need for continuous energy input. According to Ali et al. [7], passive thermal control methods include the strategic use of insulation materials, appropriate orientation of the building to take advantage of solar gain, natural shading through vegetation or architectural elements, and the integration of thermal mass to store and slowly release heat. Walls and roofs, which are the most exposed parts of any structure, should be designed with a primary focus on their role as long-term thermal barriers. Enhancing the thermal resistance of these components significantly reduces heat loss during cold months and heat gain during hot months, thereby decreasing the building's overall energy load.

As the global population continues to grow and urbanization expands, the demand for energy-efficient construction methods and materials has surged. Over the past few decades, there has been a notable increase in the interest and application of thermal insulation materials across residential, commercial, and industrial buildings. This shift is driven by the urgent need to curtail excessive energy use and promote sustainability. Thermal insulation materials are engineered to create a barrier against unwanted heat flow. Commonly used insulation materials include fiberglass, mineral wool, cellulose, polystyrene, polyurethane foam, and more recently, advanced materials like aerogels. Each of these materials has distinct properties tailored for specific environments and performance requirements. For instance, aerogels offer exceptional thermal resistance with minimal thickness and weight, making them ideal for space-constrained applications. However, their high manufacturing cost can be a barrier to widespread adoption, especially in developing regions. In contrast, more affordable alternatives such as cellulose, made from recycled paper, and fiberglass, which is widely available and easy to install, continue to be popular in both new constructions and retrofitting projects [73].

The importance of thermal insulation goes beyond just energy efficiency. Its benefits extend into various domains, including economic savings, enhanced comfort, and environmental sustainability. By decreasing the dependence on heating, ventilation, and air conditioning (HVAC) systems, insulation significantly cuts down on the energy needed to condition indoor spaces. This reduction not only lowers utility costs for occupants but also decreases the strain on national power grids, particularly during peak demand periods. Moreover, thermal insulation contributes to a more stable and comfortable indoor environment. It minimizes temperature fluctuations, reduces drafts, and can even improve indoor air quality by lessening the need for forced air circulation. Some insulation materials also offer added advantages such as noise reduction, fire resistance, and moisture control, making them valuable additions to any energy-conscious building design. Thermal insulation is also increasingly being utilized in other industries beyond construction. Sectors such as food cold storage, industrial refrigeration, petrochemical processing, and the transportation of liquefied natural gas rely heavily on insulation to maintain product integrity, prevent energy losses, and ensure safety. These applications further underline the broad scope of insulation's contribution to energy conservation on a global scale. As environmental awareness increases, there has also been a growing trend toward the development and use of sustainable insulation materials. Samardzioska et al. [50] highlight that these products aim to minimize the embodied energy—the total energy required to produce and transport the material—while offering comparable or even superior performance to conventional insulation options. Recycled and repurposed materials, such as denim, wool, hemp, and recycled glass, are now being converted into eco-friendly insulation products. These materials not only divert waste from landfills but also support circular economy principles by giving new life to discarded resources.

The integration of recycled content in insulation manufacturing aligns with the broader goals of sustainable development and environmental stewardship. Such practices significantly lower greenhouse gas emissions by reducing the need for raw material extraction and limiting the consumption of fossil fuels used in production. For example, using recycled glass wool or paper-based cellulose can cut down the energy required during the manufacturing process, contributing further to the conservation of natural resources. The economic implications of thermal insulation are also significant. According to the Joint Research Centre (JRC) of the European Commission, the global market for thermal insulation was valued at approximately USD 22.73 billion in 2015 and is projected to reach USD 38.69 billion by 2027. Europe remains the largest market for insulation materials, driven by strict energy efficiency regulations and sustainability initiatives, with North America following closely behind. This market growth, as also emphasized by Ali et al. [7], indicates increasing global recognition of insulation's role in energy conservation and climate action. Thermal insulation serves as a foundational element in the pursuit of energy efficiency and environmental sustainability. Whether applied in buildings, industrial facilities, or cold-chain logistics, insulation reduces the need for energy-intensive climate control systems, thereby conserving energy, lowering emissions, and improving indoor comfort. As technology evolves and sustainable materials become more accessible, the importance of thermal insulation in achieving global energy conservation goals will only continue to grow. Therefore, understanding the properties of different insulation materials, their suitability for various climates and structures, and their broader ecological impacts is essential for architects, engineers, and policymakers striving to create a more energy-resilient future.

8.3 Role of Nanomaterials in Advanced Insulation Technologies

The integration of nanomaterials into insulation technologies marks a transformative step in advancing energy-efficient building systems. As the global demand for sustainable and thermally efficient structures rises, the need for next-generation insulation materials becomes more pronounced. Nanotechnology, with its promise of enhanced physical and chemical characteristics at the atomic or molecular scale, is at the forefront of this evolution. Across recent research in current nanomaterials, the contribution of nanomaterials—ranging from silica aerogels and carbon nanostructures to hybrid composites—has been explored with the aim of improving thermal insulation, reducing energy consumption, and supporting greener construction practices. One of the most notable advantages of nanomaterials in insulation lies in their ultra-low thermal conductivity. Traditional insulation materials such as mineral wool, polyurethane foam, and polystyrene, while widely used, often fall short in performance compared to emerging nanomaterials. Silica aerogels, in particular, are highlighted extensively in the literature for their extremely low thermal conductivity—around 0.013 W/m·K—which is significantly lower than conventional insulators. Hybrid aerogels synthesized by aggregating inorganic nanomaterials with polymeric fibers have shown superior performance not only in insulation but also in structural resilience. The unique porous architecture of aerogels, comprising up to 90–99% air by volume, disrupts heat transfer mechanisms such as conduction and convection, making them ideal candidates for energy-critical applications [14, 37].

In building applications, nanomaterial-based insulation has proven effective in reducing cooling energy demand. It was demonstrated that integrating silica aerogel insulation within wall assemblies significantly minimized indoor temperature fluctuations and contributed to improved thermal comfort. The integration of nanomaterials into residential building envelopes significantly reduces indoor heat accumulation, thereby decreasing the demand for air conditioning. This reduction in thermal load helps optimize energy use and extends the operational lifespan of cooling systems by minimizing mechanical strain. This directly aligns with global objectives to reduce greenhouse gas emissions and energy usage in buildings, which currently account for over one-third of global energy consumption. Another key aspect highlighted in the literature is the role of vacuum insulation panels (VIPs) that employ nanoporous materials as core substances. A study by Abdelrady et al. [3]. emphasized the potential of VIPs integrated with nanoparticle-based materials in drastically reducing thermal conductivity, sometimes to values as low as 0.002–0.004 W/m·K. These systems achieve their performance by removing air (thus eliminating convective heat transfer) and leveraging nanomaterial cores to inhibit conductive pathways. Their thin profiles also make them advantageous for applications where space is limited but high insulation is essential—such as in retrofitting older buildings or insulating curved or irregular surfaces. Moreover, recent discussions have expanded to the use of carbon-based nanomaterials like carbon nanotubes (CNTs), graphene, and nanodiamonds, which offer promising characteristics due to their exceptional mechanical, thermal, and chemical stability. Carbon nanomaterials can be incorporated into polymer matrices to produce lightweight, flame-retardant, and structurally robust insulation composites. These composites not only provide thermal insulation but also enhance the fire resistance and durability of building components. This multifunctionality is crucial in modern building design, which increasingly demands materials that serve both passive (insulation) and active (structural integrity, fire safety) roles.

While nanomaterials bring forth compelling advantages, the implementation of such advanced insulation materials is not without challenges. Cost remains a significant barrier. The production of aerogels and other nanomaterials often involves complex synthesis processes, including supercritical drying or sol-gel reactions, which are energy-intensive and expensive. Despite the thermal performance gains, the economic feasibility of these materials in large-scale residential construction must be addressed through subsidies, innovations in manufacturing, or policy incentives. Additionally, there are concerns over long-term performance under real environmental conditions. Issues such as material aging, mechanical degradation, and moisture sensitivity can impair the effectiveness of nanomaterial-based insulators over time. Research is ongoing to overcome these challenges through material hybridization and bio-based alternatives. For example, Liu et al. [37] explored the potential of using bio-derived fibers (e.g., cellulose, chitosan) in conjunction with nanomaterials to create environmentally friendly aerogels. Such combinations not only reduce environmental impacts but also introduce a pathway to lower costs and improve biodegradability. To improve the long-term performance of nanostructured insulation in demanding environments, researchers are developing advanced encapsulation methods and surface treatments. These innovations enhance moisture resistance, thermal stability, and mechanical durability, ensuring reliable functionality even in extreme or fluctuating climate conditions.

Beyond the thermal insulation of walls and roofs, nanomaterials are increasingly being explored for their versatility in other areas of building systems, including energy storage enclosures, HVAC duct linings, and even architectural glazing. Their incorporation into energy storage systems helps minimize heat loss or gain, which is crucial for maintaining efficiency and extending battery life. In HVAC systems, nanomaterial-based coatings and foams reduce thermal bridging and enhance overall energy performance. One particularly innovative application is in transparent nanogel-based window glazings. These systems utilize silica or polymer nanogels to create smart windows that effectively block infrared radiation while maintaining high visible light transmittance. This dual capability reduces unwanted solar heat gain while still allowing natural daylight, leading to enhanced indoor comfort and energy savings. According to Karanafti et al. [31], nanomaterials are thus not only redefining conventional insulation but also enabling the development of multifunctional building elements. Their adaptability whether as panels, aerogel blankets, foams, or sprays, makes them suitable for a wide range of building typologies. Furthermore, their role in supporting net-zero energy designs and adaptive reuse projects highlights their significance in sustainable, future-oriented construction strategies. Nanomaterials are playing a pivotal role in shaping the next generation of insulation technologies. Through their unique thermal, mechanical, and multifunctional properties, they address critical limitations of conventional insulators and enable the design of more sustainable, energy efficient, and resilient buildings. While challenges related to cost, scalability, and long-term durability remain, ongoing research and innovation continue to push the boundaries of what is possible in thermal insulation. The synergistic use of advanced nanomaterials, whether in the form of aerogels, carbon nanocomposites, or nanoporous VIPs is unlocking new potential for achieving global energy efficiency targets and reducing the environmental footprint of the built environment.

8.4 Nanomaterials for Building Insulation

Nanomaterials offer innovative solutions for building insulation due to their exceptional thermal properties and energy efficiency. With high surface area and low thermal conductivity, they significantly reduce heat transfer, enhancing indoor comfort and sustainability. Their application in construction represents a promising step toward greener, smarter, and more energy-efficient buildings. In this part, six nanomaterial types were elaborated in the context of building insulation as displayed in Fig. 8.1.

8.4.1 Nanoporous Materials

Nanoporous materials have emerged as a revolutionary solution in the realm of building insulation, offering unparalleled thermal performance and energy efficiency. Among these, silica stand out due to their exceptional properties. Composed predominantly of air—ranging from 85% to 99.8% by volume—silica exhibit extremely low thermal conductivity, often as low as 0.012 W/(m·K), significantly outperforming traditional insulation materials like expanded polystyrene and mineral wool. This remarkable thermal resistance is attributed to their nanoporous structure, which effectively inhibits heat transfer through conduction, convection, and radiation. The effectiveness of silica aerogels in thermal insulation is primarily

Fig. 8.1 Nanomaterials for building insulation

due to their unique microstructure. The nanopores are typically smaller than the mean free path of air molecules, leading to the Knudsen effect, where gas-phase thermal conductivity is drastically reduced. This phenomenon occurs because the confined air molecules collide more frequently with the solid matrix than with each other, thereby limiting heat transfer. Additionally, the solid component of silica aerogels consists of a sparse network of interconnected nanostructures, which minimizes solid-phase thermal conductivity. Furthermore, the transparent nature of silica aerogels allows them to be used in applications where light transmission is desired, such as in translucent panels and daylighting systems, without compromising thermal insulation. Despite their outstanding thermal properties, the practical application of silica aerogels in building insulation faces several challenges. Their inherent brittleness and mechanical fragility necessitate reinforcement strategies to enhance durability. Recent advancements have focused on developing composite aerogels by incorporating fibers or polymers to improve mechanical strength while retaining low thermal conductivity. Additionally, the hydrophilic nature of silica aerogels can lead to moisture absorption, which degrades their insulating performance. To address this, surface modifications and the incorporation of hydrophobic agents have been employed to render aerogels water-resistant [38].

Beyond silica, other nanoporous materials are being explored for building insulation. Nanocellulose-based aerogels, derived from renewable biomass, have garnered attention due to their sustainability and biodegradability. These materials exhibit low thermal conductivity and can be engineered to possess desirable mechanical properties. However, challenges such as flammability and moisture sensitivity remain, prompting ongoing research into chemical modifications and composite formulations to enhance their applicability in building insulation. As highlighted by Wu et al. [60], nanoporous materials, like silica and alumina, offer transformative potential for building insulation by providing superior thermal performance and contributing to energy efficiency. Ongoing research and development efforts are focused on overcoming current limitations related to mechanical strength and moisture resistance, paving the way for broader adoption of these advanced materials in sustainable building practices.

8.4.2 Nano-Aerogels

Aerogels have garnered significant attention in the construction industry due to their exceptional thermal insulation properties. Characterized by their highly porous nanostructure, aerogels exhibit extremely low thermal conductivity, making them superior insulators compared to traditional materials like fiberglass or mineral wool. This low thermal conductivity is primarily attributed to the Knudsen effect, where the pore sizes in aerogels are smaller than the mean free path of air molecules, effectively minimizing heat transfer through conduction and convection. The application of aerogels in building insulation has been explored in various studies, highlighting their potential in enhancing energy efficiency. For instance, simulations conducted on residential buildings demonstrated that incorporating a half-inch layer of aerogel insulation in walls and windows could lead to an average energy savings of 20.9% compared to conventional insulation materials. Notably, the energy loss through windows was reduced by approximately 39.1%, while wall energy loss decreased by about 13.3%. These findings underscore the effectiveness of aerogels in reducing thermal bridging and improving overall building performance [56].

Despite their impressive insulating capabilities, aerogels face challenges that hinder their widespread adoption in the construction sector. One of the primary concerns is their mechanical fragility; traditional silica aerogels are brittle and can be easily damaged under stress. To address this, researchers have developed composite forms such as aerogel plasters (AP), aerogel fibrous composites (AFC), and aerogel concrete (AC). Among these, AFCs have shown the highest improvement in thermal insulation performance, achieving enhancements between 46.0% and 53.5% over non-aerogel walls. Moreover, AFCs offer approximately 50% cost savings to achieve the same thermal resistance as other forms, making them a more economically viable option. Another limitation is the high production cost of aerogels, which stems from complex manufacturing processes and the need for specialized equipment. This economic barrier has restricted their use to niche applications or high-performance buildings. However, advancements in production techniques and the development of hybrid materials are gradually reducing costs, paving the way for broader implementation. Additionally, the hydrophilic nature of silica aerogels poses challenges in humid environments, as they tend to absorb moisture, leading to a decline in insulating performance. As Yang et al. [64] noted, surface modifications and the incorporation of hydrophobic agents have been employed to enhance moisture resistance. Aerogels represent a promising frontier in building insulation technology, offering unparalleled thermal performance that can significantly contribute to energy conservation and sustainability goals. While challenges related to mechanical strength, moisture sensitivity, and production costs persist, ongoing research and technological advancements are addressing these issues. As solutions continue to emerge, the integration of aerogels into mainstream construction practices is becoming increasingly feasible, heralding a new era of energy-efficient building design.

8.4.3 Vacuum Insulation Panels

Vacuum insulation panels (VIPs) are advanced thermal insulation materials that offer exceptional performance for building applications. Comprising a porous core material, such as fumed silica, encapsulated within a gas-tight envelope and evacuated to create a vacuum, VIPs achieve remarkably low thermal conductivity values typically between 0.002 and 0.004 W/(m·K) in their pristine state. This performance is significantly superior to conventional insulation materials, providing 5–10 times higher thermal resistance for the same thickness. The superior insulating capability of VIPs stems from the elimination of gas molecules within the panel, thereby minimizing heat transfer through conduction and convection. The key to VIPs' extraordinary insulating capacity lies in the nanoporous structure of the core material. Fumed silica and similar nanomaterials provide a high specific surface area and small pore diameters (<100 nm), which minimize radiative heat transfer and reduce the mean free path of air molecules, thereby maintaining a near-perfect vacuum effect. Other nanomaterials under research for VIP applications include aerogels, carbon nanotube-enhanced silica, and nano-clay composites, which offer tailored pore architectures, mechanical reinforcement, and improved thermal stability. These attributes make VIPs particularly advantageous in space-constrained building scenarios, such as retrofitting heritage buildings or insulating thin wall assemblies in high-performance constructions. Nevertheless, challenges remain particularly concerning long-term durability, as even trace amounts of gas ingress over time can degrade vacuum quality and increase thermal conductivity. To counter this, high-barrier laminates (often multi-metalized polymer films) and getter materials (such as finely dispersed reactive metal nanoparticles) are incorporated to scavenge residual gases and moisture, prolonging the panel's lifespan. Additionally, the mechanical fragility of VIPs is a practical concern. Nanomaterials, while offering excellent thermal performance, often possess brittle microstructures susceptible to mechanical damage under impact or compression [27]. Hence, protective casing, careful handling, and hybrid designs such as combining VIPs with rigid foam or structural insulation panels are being developed to improve robustness.

Recent advancements have focused on enhancing the durability, thermal stability, and cost-effectiveness of vacuum insulation panels (VIPs) through the integration of nanomaterials and innovative system designs. Key innovations include the development of more robust multi-layered envelope materials with nanocomposite barrier films that offer superior resistance to gas and moisture permeation. Simultaneously, improvements in core compositions such as the use of nanostructured fumed silica, aerogels, and graphene-enhanced or carbon nanotube (CNT)-reinforced matrices have significantly enhanced both thermal performance and mechanical resilience. These nanomaterial-based cores maintain ultra-low thermal conductivity while offering increased compressive strength and resistance to aging effects caused by gas ingress. Moreover, hybrid insulation systems that combine VIPs with nano-engineered foams or phase change materials (PCMs) are being explored to balance thermal efficiency with impact resistance and cost, especially for large-scale building applications. These developments aim not only to extend the service life of VIPs but also to make them more adaptable across diverse architectural scenarios, including retrofit and modular constructions. According to Mukhopadhyaya [42] VIPs represent a cutting-edge solution for achieving superior thermal insulation in buildings, especially where space constraints limit the use of conventional materials. While challenges related to cost, mechanical vulnerability, and long-term durability persist, the incorporation of advanced nanomaterials is steadily transforming these limitations into opportunities. As the demand for high-performance, energy-efficient building solutions continues to rise, nanomaterial-enhanced VIPs are increasingly positioned to become a cornerstone technology in the future of sustainable and smart construction.

8.4.4 Carbon Nanotube

Carbon nanotubes (CNTs) have emerged as a promising material for building insulation due to their exceptional thermal, mechanical, and environmental properties. Recent advancements have led to the development of CNT-based sponge-like structures exhibiting ultra-low thermal conductivity, high mechanical stability, and excellent fire-retardant characteristics, making them suitable for energy-efficient building applications. The unique structure of CNTs allows for the fabrication of highly porous materials with densities as low as 2.1 mg/cm^3. These materials demonstrate thermal conductivities in the range of 10–14 mW/(m·K), significantly lower than traditional insulation materials like expanded polystyrene or mineral wool. The low thermal conductivity is attributed to the high porosity, which limits the cross-sectional area for heat conduction, and the high contact thermal resistance between CNTs, which suppresses phonon transport [68].

Beyond thermal insulation, CNT-based materials exhibit remarkable mechanical robustness. The deposition of amorphous carbon on CNT surfaces and junctions enhances the structural integrity, allowing the material to withstand compressive strains up to 90% without permanent deformation. This resilience ensures durability under mechanical stress, a critical factor for building materials subjected to various loads and environmental conditions. Fire safety is another crucial aspect of building materials. CNT-based sponges have demonstrated excellent fire-retardant properties, withstanding temperatures up to 535 °C in air and maintaining structural stability even after exposure to −196 °C in liquid nitrogen. Unlike conventional polymer-based insulators that often require additional flame retardants, CNT materials inherently resist ignition and do not propagate flames, enhancing building safety without compromising environmental standards. The integration of CNTs into building insulation aligns with the growing demand for sustainable and high-performance construction materials. Their lightweight nature reduces structural load, while their superior insulating properties contribute to energy efficiency by minimizing heat loss. Moreover, the environmental stability of CNT-based materials ensures longevity, reducing the need for frequent replacements and associated maintenance costs. Carbon nanotube-based materials offer a multifaceted solution for building insulation, combining ultra-low thermal conductivity, mechanical resilience, and fire resistance. As research progresses and manufacturing processes become more cost-effective, CNTs are poised to play a significant role in the development of next-generation, energy-efficient buildings [26, 68].

8.4.5 *Phase Change Materials Enhanced Nanoparticles*

Phase change materials (PCMs) enhanced with nanoparticles, known as nano-enhanced PCMs (NEPCMs), have emerged as a promising solution for improving thermal insulation in buildings. Traditional PCMs are valued for their ability to absorb and release latent heat during phase transitions, aiding in temperature regulation. However, their widespread application is hindered by inherently low thermal conductivity, which limits the rate of heat transfer. Incorporating nanoparticles into PCMs addresses this limitation by significantly enhancing thermal conductivity, thereby improving the efficiency of thermal energy storage and release. Research indicates that adding nanoparticles such as copper oxide (CuO), aluminum oxide (Al_2O_3), silver (Ag), and carbon-based materials like carbon nanotubes (CNTs) and graphene to PCMs can substantially increase thermal conductivity [48]. For instance, integrating 3 wt.% of CuO nanoparticles into palmitic acid has been shown to boost thermal conductivity by up to 60%, enhancing the material's capacity to store and release heat effectively. According to Barreneche et al. [11], incorporating 7.2 wt.% of Ag nanoparticles into polyethylene glycol (PEG) combined with diatomite resulted in a 127% increase in thermal conductivity, reaching 0.82 W/(m·K), while maintaining high latent heat storage and thermal stability over multiple cycles. Carbon-based nanoparticles also demonstrate significant improvements in thermal performance. Studies have shown that adding 3 wt.% of expanded graphite or multi-walled carbon nanotubes (MWCNTs) to paraffin-based PCMs can enhance thermal conductivity by up to 297% and 218%, respectively. These enhancements facilitate faster charging and discharging cycles, making NEPCMs more responsive to temperature fluctuations and thereby more effective in maintaining indoor thermal comfort.

Beyond thermal conductivity, nanoparticles contribute to the overall stability and durability of PCMs. For example, Hayat and Chen [23] found that the inclusion of TiO_2 nanoparticles has been found to improve the thermal stability and reduce supercooling in microencapsulated PCMs, ensuring consistent performance over extended periods. Additionally, the use of bio-based support materials, such as carbonized Cyperus alternifolius, combined with Ag nanoparticles, has led to the development of shape-stable PCMs with high thermal conductivity and excellent photo-thermal conversion efficiency, highlighting the potential for sustainable and efficient thermal energy storage solutions in building applications as demonstrated by Shah et al. [51]. Despite these advancements, challenges remain in the large-scale implementation of NEPCMs. Issues such as nanoparticle agglomeration, potential toxicity, and the high cost of certain nanoparticles like silver and carbon nanotubes need to be addressed. Ongoing research focuses on optimizing nanoparticle dispersion within PCMs, developing cost-effective synthesis methods, and ensuring the environmental safety of these materials. The integration of nanoparticles into phase change materials significantly enhances their thermal properties, making them more effective for building insulation applications. By improving thermal conductivity and stability, NEPCMs offer a viable solution for energy-efficient building design, contributing to reduced energy consumption and enhanced indoor comfort. Continued research and development in this field are essential to overcome existing challenges and facilitate the widespread adoption of NEPCMs in the construction industry.

8.4.6 Nanoclay

Nanoclay-based materials are gaining attention as effective and sustainable options for building insulation, offering improvements in thermal performance, mechanical strength, and environmental impact. A study by Anjomshoa [9] investigated the incorporation of montmorillonite nanoclay into gypsum composites. The findings revealed that adding nanoclay reduced the density of the samples and enhanced their thermal insulation properties. This suggests that nanoclay can contribute to the development of lightweight gypsum-based materials with improved energy efficiency for building applications. Another study by Ghyati et al. [20] explored the thermal behavior of polymer/clay-based composites using natural clay and poly(ethylene glycol) (PEG4000). The research demonstrated that incorporating PEG into the clay matrix improved both thermal stability and thermophysical properties. Notably, the hybrid materials exhibited good thermal insulation characteristics when the polymer content exceeded 2.5%, indicating their potential as eco-friendly bio-insulation composites for buildings. These studies highlight the versatility of nanoclay in enhancing the properties of building materials. By improving thermal insulation and mechanical strength, nanoclay-based composites can contribute to more energy-efficient and sustainable construction practices. Moreover, the use of natural and abundant materials like montmorillonite aligns with environmental goals, offering a promising avenue for the development of green building materials. The integration of nanoclay into building materials presents a viable strategy for improving insulation performance while promoting sustainability. Continued research and development in the field of nanoclay-based composites are crucial for harnessing their full potential in modern construction, enabling advancements in material strength, durability, sustainability, and overall structural performance.

8.5 Synthesis and Fabrication of Nanomaterial-Based Insulation

The synthesis of nanomaterial-based insulation, particularly aerogels, involves various methods that contribute to their unique properties such as low density, high porosity, and exceptional thermal insulation capabilities. These methods are crucial in developing materials suitable for applications requiring efficient thermal management. The sol-gel process, which is used to prepare almost all inorganic aerogels, consists of several essential steps: the hydrolysis of precursor molecules to create a colloidal solution (sol), the sol polycondensation process to create a wet network (gel), the aging of the wet gel, and the gel drying as depicted in Fig. 8.2. In general, silica aerogels are synthesized using three primary sources of silicon: silicon alkoxides such as tetraethyl orthosilicate (TEOS), ion-exchange water glass which provides a silicic acid solution, and sodium silicate solution, commonly referred to as water glass.

These sources serve as the foundational precursors in the sol-gel process, a chemical route that significantly determines the structural and morphological properties of both the wet gel and the final aerogel product. Among the crucial factors in this process, Ganobjak et al. [19] emphasize influencing this process are the type of precursor used and the pH level of the reaction medium. These parameters have a direct impact on the formation, structure, and performance of the aerogel. In a comparative study, three distinct silicon-based precursors tetramethyl orthosilicate (TMOS), tetraethyl orthosilicate (TEO), and polyethoxydisiloxane (PEDS) were evaluated for their effects on the characteristics of colloidal silica spheres. It was found that the type of catalyst used in conjunction with each precursor has a strong correlation with the resulting particle size of the silica spheres. When acid catalysts are applied with TEO and PEDS, the result is monodisperse silica spheres particles

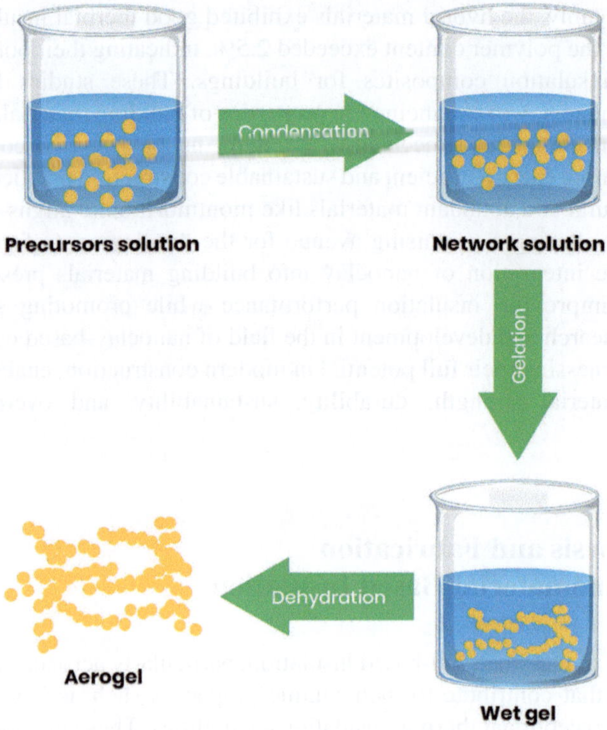

Fig. 8.2 General synthesis of nanoparticle aerogels for insulation

of uniform size and shape. Likewise, TMOS, when used with a basic catalyst, also yields monodisperse colloidal silica spheres. Notably, the silica spheres synthesized using TMOS and PEDS produce particles that are smaller and display a more uniform, spherical pore distribution compared to those produced using TEOS. During the aging step of the sol-gel process, the wet gel undergoes further condensation reactions leading to dehydration and shrinkage of the gel network. This occurs due to the high number of reactive hydroxyl groups present, which continue to polymerize over time. As neighboring hydroxyl groups condense to form M-O-M bonds, the internal liquid within the pores is expelled, making room for further network densification. Drying is the final and most critical stage for preserving the delicate porous structure of aerogels. Common methods include supercritical drying using CO_2, freeze-drying, and ambient pressure drying. Liu et al. [37] also reported that in addition to silica aerogels, alumina (Al_2O_3) aerogels offer excellent thermal stability and dielectric insulation. They are prepared using methods similar to those used for silica, typically employing organic aluminum compounds such as isopropanol aluminum, sec-butyl alcohol aluminum, and various inorganoaluminol salts as precursors.

Beside sol-gel process, there are other pathways of nanomaterial fabrication for thermal insulation of buildings. Hou et al. [25] highlight electrospinning is another method employed in the synthesis of nanomaterial-based insulation. This technique uses a high-voltage electric field to draw charged threads of polymer solutions or melts into fibers with diameters in the nanometer range. The resulting nanofibers can be assembled into mats or composites that exhibit high porosity and low thermal conductivity, making them suitable for insulation applications. Electrospinning allows for the incorporation of various materials, including polymers and inorganic nanoparticles, to tailor the properties of the insulation material. Freeze-drying, or lyophilization, is a drying process used to remove solvents from gels, resulting in the formation of aerogels. In this method, the gel is first frozen, and then the solvent is removed through sublimation under reduced pressure. This process preserves the porous structure of the gel, leading to aerogels with high surface area and low density. Freeze-drying is particularly useful for producing aerogels from heat-sensitive materials, as it avoids high-temperature drying processes that could compromise the material's structure and properties. Chemical vapor deposition (CVD) is a technique used to deposit thin films of materials onto substrates through chemical reactions of gaseous precursors. In the synthesis of nanomaterial-based insulation, CVD can be used to coat substrates with materials that enhance thermal insulation properties. For example, a study by Liu et al. [37] reported depositing thin layers of silica or other insulating materials onto fibers or foams can improve their thermal resistance. CVD allows for precise control over the thickness and composition of the deposited films, enabling the customization of insulation materials for specific applications. Hydrothermal synthesis as discussed by Ji et al. [29], involves the crystallization of substances from high-temperature aqueous solutions at high vapor pressures. This method is used to produce nanostructured materials with controlled morphology and size. In the context of insulation materials, hydrothermal synthesis can be employed to create nanomaterials with specific properties that contribute to thermal insulation, such as low thermal conductivity and high thermal stability. The ability to control the synthesis conditions allows for the tailoring of material properties to meet the requirements of various insulation applications. Self-assembly techniques rely on the spontaneous organization of molecules or nanoparticles into structured arrangements without external guidance. These techniques can be used to create hierarchical structures with nanoscale features that contribute to thermal insulation. For instance, the self-assembly of block copolymers or other amphiphilic molecules can lead to the formation of porous materials with controlled pore sizes and distributions. Such structures can impede heat transfer, enhancing the thermal insulation performance of the material. The synthesis of nanomaterial-based insulation materials involves a variety of methods, each contributing unique features to the final product. The choice of synthesis method depends on the desired properties of the insulation material and the specific application requirements. By leveraging these techniques, researchers can develop advanced insulation materials with superior thermal performance. Table 8.1 explores the fabrication methods of nanomaterials-based insulation.

Table 8.1 Synthesis methods of nanomaterials-based insulation

Insulation Nanomaterials	Synthesis method	References
Silicon nanocrystals (SiNCs)	A silica sol mixture—comprising tetramethyl orthosilicate (TMOS), methanol, deionized water, and ammonium hydroxide—was introduced at a concentration of 5 mg/mL into the functionalized solution	[4]
Al_2O_3-SiO_2	Alumina/silica sols prepared; PCM and PO added; aged, solvent-exchanged with ethanol; supercritically dried	[28]
Graphene oxide (GO)	An aqueous suspension of GO was combined with ethylenediamine (EDA), serving as the reducing agent, while a sodium borate solution (SBS) was simultaneously introduced to enhance the reduction process and structural stability	[61]
Cellulose nanofibers from *Citrullus colocynthis* seeds	Alkali-treated, bleached, acid-hydrolyzed, and mechanically defibrillated Citrullus colocynthis seeds yield cellulose nanofibers	[34]
Cellulose nanofibers from orange bagasse	Orange bagasse was alkali-treated, acid-hydrolyzed, bleached, and ultrasonicated to produce high-purity, crystalline cellulose nanofibers	[39]
Chitin nano-whiskers (CNWs)	Acid hydrolysis, electrospinning, mechanical treatment, ultrasonication, and self-assembly were applied to process and organize the material.	[10]
Nanofibers	Centrifugal spinning at high speed	[69]
Immiscible blends of cellulose acetate butyrate (CAB) and isotactic polypropylenes (iPPs)	Melt extrusion phase separation	[63]

8.6 Mechanisms of Heat Transfer Reduction

Heat transfer reduction is a critical principle in thermal management and energy-efficient material design, especially within the fields of building science, electronics, and industrial systems. The three primary modes of heat transfer: **conduction**, **convection**, and **radiation**, each require targeted strategies to minimize energy loss. Conduction involves heat flow through solids and is typically reduced by using materials with low thermal conductivity or by incorporating nanoporous structures that disrupt phonon pathways. Convection, the transfer of heat through fluid motion, can be mitigated by designing systems that restrict air movement or using encapsulated structures that eliminate convective currents. Radiative heat transfer, which occurs through electromagnetic waves, is minimized by employing reflective surfaces or low-emissivity coatings that inhibit infrared radiation. Advances in material science, particularly the development of nanomaterials and composite systems, have opened new pathways for controlling these heat transfer modes at micro and nano scales. A comprehensive understanding of these mechanisms is essential for engineers and designers aiming to develop effective thermal insulation systems and energy-saving technologies.

8.6.1 Conduction Control

Heat transfer is a fundamental physical process that underpins numerous natural phenomena and technological applications. It occurs via three primary mechanisms: conduction, convection, and radiation. Among these, conduction—defined as the transfer of thermal energy through matter due to a temperature gradient—is particularly significant in solid materials, where it often dominates heat transfer behavior. The control of conduction is, therefore, central to improving the energy efficiency and thermal stability of many systems, from microelectronic devices to metallurgical processes. Conduction control as a mechanism of heat transfer reduction is not only a matter of material choice but also of structural engineering at multiple scales, from the macro to the nanoscale. This essay delves into various strategies for conduction control, emphasizing their underlying mechanisms and practical implications as evidenced in both theoretical and experimental research. One of the foundational strategies in reducing heat transfer via conduction involves the deliberate manipulation of material structures at the micro- and nanoscale. Heat in non-metallic solids is primarily carried by phonons—quantized lattice vibrations. Thus, thermal conductivity in these materials is largely governed by phonon behavior. According to Habibishandiz and Saghir [21], introducing phonon scattering centers such as grain boundaries, dislocations, or embedded nanoparticles, phonon mean free paths can be significantly shortened, reducing the efficiency of thermal transport. This phenomenon has been extensively explored in thermoelectric materials, where the goal is to maintain high electrical conductivity while minimizing thermal conductivity, a balance crucial for high thermoelectric performance. Nanostructured materials, particularly superlattices and nanocomposites, are prominent examples of successful conduction control. Nakamura [43] highlights that in superlattices, periodic layers of dissimilar materials create interfaces that scatter phonons more effectively than electrons, leading to thermal conductivity suppression without compromising electrical flow. Similarly, nanocomposites embed nanoparticles in a host matrix, disrupting phonon pathways through interfacial mismatch and localized resonance effects. These approaches have enabled the development of materials with ultra-low thermal conductivities, approaching the theoretical minimum predicted by the Cahill-Pohl model. Such advances are vital for thermoelectric generators, heat insulators, and electronic devices requiring thermal management.

Beyond engineered nanostructures, natural phase transformations in materials also offer a potent method of conduction control. A striking example is the crystallization of mold fluxes used in the continuous casting of steel. As detailed by researches, the formation of crystalline phases within the mold flux significantly reduces heat transfer between the solidifying steel shell and the mold. This reduction is primarily attributed to two effects. First, the crystallized layer exhibits a higher reflectivity due to enhanced scattering of thermal radiation at the crystal grain boundaries. Second, the microstructure formed during crystallization introduces interfaces that hinder both radiative and conductive heat flow. Interestingly, while the effect of crystallization on radiative heat transfer is well-documented, its role in conduction control is equally critical. As the crystalline network develops, the continuity of conductive pathways is disrupted, effectively lowering the thermal diffusivity of the mold flux layer. This results in a thermally resistive barrier that modulates the cooling rate of the steel, an essential parameter in achieving desirable microstructures in the final product. Thus, by strategically promoting or inhibiting crystallization through flux chemistry and thermal management, engineers can fine-tune the thermal performance of the casting mold. In microscale applications, especially in electronics and microfluidics, conduction control is often achieved through structural design. For example, in microchannel heat exchangers, heat transfer is strongly influenced by axial conduction, heat flowing along the direction of fluid movement. If not properly managed, axial conduction can lead to thermal short-circuiting, where heat bypasses the fluid medium and moves directly through the solid structure. This phenomenon degrades the overall heat exchange efficiency and can cause thermal hotspots. A study investigated the thermal performance of silicon-based microchannels and found that optimizing wall thickness and material properties could suppress undesired axial conduction [21, 66].

Geometric optimization is another key aspect of conduction control. By varying the dimensions of fins, channel walls, or insulating layers, the effective thermal resistance can be modulated. For instance, decreasing the thickness of a high-conductivity material layer while surrounding it with low-conductivity cladding can trap heat or redirect it, much like optical waveguides confine light. The same principle is applied in thermal barrier coatings (TBCs), where ceramic layers with controlled porosity and microstructure serve as insulating shields for turbine components. These materials are designed to exhibit low thermal conductivity while withstanding extreme thermal and mechanical stresses. Moreover, the choice of material inherently influences thermal conduction. Traditional materials like metals exhibit high thermal conductivities due to free electron movement, whereas ceramics and polymers typically have lower thermal conductivities owing to their phonon-dominated transport. However, material innovation has led to the development of hybrid materials, combinations of polymers and fillers like boron nitride or graphene that provide tailored thermal properties [57]. For example, Srikanth et al. [53] demonstrated that increasing the orientation of graphene flakes within a polymer matrix can enhance directional thermal conductivity, enabling anisotropic conduction control.

Another fascinating approach to conduction control involves the manipulation of interfacial thermal resistance, often referred to as Kapitza resistance. This resistance arises at the junction between two dissimilar materials due to mismatch in vibrational spectra. Interfaces can act as bottlenecks for heat flow, particularly in systems with layered structures or composite materials. By increasing the density of interfaces, such as through multilayer coatings or inclusion of nanolaminates, the overall thermal conductivity can be effectively reduced. This principle has been exploited in super-insulative aerogels and phase-change materials for thermal regulation applications. In the context of maritime and coastal infrastructure, Starkov and Starkov [54] examined how heat transfer through concrete structures exposed to dynamic thermal environments can be modulated by embedding PCMs that absorb and release heat through latent transitions. Although this method primarily targets heat storage and regulation, the use of PCMs also introduces localized changes in conduction behavior, particularly by altering the temperature gradient and delaying heat propagation [70].

Recent advances also highlight the potential of bio-inspired materials for conduction control. For instance, structures mimicking the hierarchical organization found in natural materials like nacre or wood can be engineered to have directional thermal properties. These materials exploit a combination of conductive and insulating layers arranged in specific patterns to channel heat along preferred pathways or to block it altogether. Studies by Dey and Saha [16] show how such bio-inspired architectures can yield materials with ultralow thermal conductivities while maintaining mechanical integrity. Conduction control encompasses a multifaceted array of strategies, each leveraging different physical principles and material behaviors to reduce thermal transport. At the atomic scale, phonon scattering and interfacial resistance dominate. At the microscale, geometric design and phase transitions come into play. And at the macroscale, structural engineering and material selection govern the thermal performance of entire systems. The integration of these approaches leads to highly effective thermal management solutions across industries, from steel manufacturing and aerospace engineering to electronics and energy systems. The significance of conduction control continues to grow in the face of global challenges such as climate change and energy conservation. Innovations in this area contribute to more efficient engines, better insulated buildings, and more durable electronic components. Future research will likely focus on smart materials that can adapt their thermal conductivity in real-time, as well as on the integration of machine learning to optimize conduction pathways in complex systems. By deepening our understanding of heat conduction mechanisms and exploring novel materials and structures, we edge closer to a future where thermal energy is managed with unprecedented precision.

8.6.2 Convection Suppression

Convection suppression is a pivotal strategy in reducing heat transfer within buildings. Natural convection, driven by temperature-induced buoyancy forces, can significantly contribute to unwanted heat gain, thereby increasing cooling loads and energy consumption. Implementing convection suppression techniques can enhance thermal comfort and energy efficiency in buildings. One effective method involves the integration of PCMs into building envelopes. PCMs absorb and release latent heat during phase transitions, thereby stabilizing indoor temperatures. Studies have shown that embedding PCMs in multi-layer walls can reduce heat transfer by up to

47.6%, depending on the material properties and placement within the wall structure. The optimal performance is achieved when PCMs with lower thermal conductivity and higher latent heat are positioned closer to the interior surface, aligning the phase change temperature with the desired indoor conditions. Combining PCMs with night ventilation further enhances their effectiveness. In regions like Penang Island, Malaysia, simulations indicate that using PCMs with melting points around 27 °C, in conjunction with night ventilation rates of 8 air changes per hour, can reduce peak indoor temperatures by up to 4.78 °C. This synergy allows the PCM to solidify during cooler nights, ready to absorb heat the following day, thus suppressing convective heat transfer [67].

An effective strategy for suppressing convection in buildings involves designing components that disrupt convective airflows, particularly in attic spaces where heat accumulation is significant. In hot and humid climates, such as Malaysia, attics can reach temperatures between 36 °C and 38 °C in the afternoon, contributing to increased cooling demands and discomfort in occupied spaces below. To mitigate this, integrating structural elements like baffles or winglets has proven beneficial. According to Aimkamon et al. [6] Computational Fluid Dynamics (CFD) studies have demonstrated that installing winglets at specific angles and positions within attic spaces can significantly reduce ceiling surface temperatures. For instance, simulations incorporating two rows of rectangular winglets, each with a length of 30 cm and angled at 45°, showed a decrease in air temperature near surfaces by up to 2 °C for 40 cm downstream from the winglets' rear tips. This configuration enhanced the Nusselt number by three times compared to air gaps without winglets, indicating improved convective heat transfer processes near the winglets. Additionally, the use of exhaust fans in conjunction with winglets increased air velocity in the horizontal channel to 1.0–1.5 m/s and reduced air temperature by 6.5–11.7 °C from an initial 38.0 °C. These structural modifications alter airflow patterns, creating vortices that enhance heat transfer away from the ceiling surface, thereby mitigating the convective heat transfer from the roof to the occupied spaces below. By disrupting the natural convection currents, these interventions effectively lower the thermal load on the building, contributing to improved energy efficiency and occupant comfort. Such passive cooling strategies are particularly advantageous in tropical climates, where reducing reliance on mechanical cooling systems is both economically and environmentally beneficial.

Window design plays a crucial role in suppressing convective heat transfer within buildings, significantly impacting overall thermal performance. Double-glazed windows, consisting of two glass panes separated by an air or gas-filled cavity, are particularly effective in this regard. The dimensions of this cavity, specifically, the air layer thickness and inlet width, are critical factors influencing convective heat transfer within the window assembly. Research indicates that optimizing these parameters can substantially reduce the overall heat transfer coefficient (U-value) of the window. For instance, when the air layer thickness is set to 12 mm and the inlet width is maintained at 1 mm, the total heat transfer coefficient can be reduced to as low as 1.28 $W \cdot m^{-2} \cdot K^{-1}$. This reduction is attributed to the suppression of convective currents within the cavity, leading to decreased thermal transmittance. The mechanism behind this improvement involves the relationship between air layer thickness and convective heat transfer. As the air layer thickens, the average velocity of air within the cavity decreases, diminishing the intensity of convective currents. This reduction in airflow leads to lower convective heat transfer, enhancing the window's insulating properties. Similarly, the inlet width affects the volume of air entering the cavity. A narrower inlet width limits air infiltration, further suppressing convective movements within the air layer. By carefully controlling these dimensions, designers can minimize convective heat transfer, thereby improving the thermal efficiency of the window system. Incorporating these design optimizations into building envelopes is particularly beneficial in hot and humid climates, where managing heat gain is essential for maintaining indoor comfort and reducing energy consumption. By mitigating convective heat transfer through window assemblies, buildings can achieve better thermal performance, contributing to overall energy efficiency and occupant comfort [32, 58].

Reflective insulation materials, such as aluminum foil barriers, play a significant role in suppressing convective heat transfer within buildings. These materials primarily function by reflecting radiant heat, thereby reducing the temperature differential that drives convection currents. When installed in attics or wall cavities, reflective insulation can effectively diminish convective loops that contribute to heat gain. The effectiveness of reflective insulation hinges on its low emissivity surface, which reflects a substantial portion of radiant energy. As noted by Demezhko et al. [15] this reflection reduces the amount of heat absorbed by building components, thereby lowering surface temperatures and minimizing the temperature gradients responsible for initiating convective air movements. By mitigating these gradients, reflective insulation curtails the formation of convective currents within enclosed spaces. Moreover, reflective insulation often incorporates air spaces adjacent to the reflective surface. These air gaps serve as additional barriers to heat transfer, impeding both conductive and convective heat flow. According to Lee et al. [35], the combination of reflective surfaces and air spaces enhances the overall thermal resistance of the building envelope. This dual-action approach not only reflects incoming radiant heat but also disrupts the convective pathways that facilitate heat transfer. Kandeal et al. [30] emphasized that implementing reflective insulation is particularly beneficial in hot and humid climates, where controlling heat gain is essential for maintaining indoor comfort and reducing energy consumption. By suppressing convective heat transfer, reflective insulation contributes to a more stable indoor environment and enhances the energy efficiency of buildings.

Furthermore, controlling air infiltration through proper sealing and insulation can prevent the formation of convective currents within building cavities. Ensuring airtight construction minimizes the entry of warm air, thereby reducing the potential for convective heat transfer. This approach not only suppresses convection but also enhances the overall energy efficiency of the building. Convection suppression is a multifaceted approach to reducing heat transfer in buildings. By integrating PCMs, optimizing structural components, enhancing window designs, utilizing reflective materials, and ensuring airtight construction, buildings can effectively minimize convective heat gains. These strategies are particularly beneficial in hot and humid climates, where controlling indoor temperatures is essential for occupant comfort and energy conservation [5, 12].

8.6.3 Radiation Blocking with Nano-Coatings

Radiation blocking stands as a pivotal mechanism in reducing heat transfer, especially in environments where thermal management is critical. Unlike conduction and convection, which require a medium for heat transfer, radiation involves the emission and absorption of electromagnetic waves, allowing heat to traverse even through a vacuum. This characteristic makes radiative heat transfer particularly significant in applications ranging from building insulation to aerospace engineering. One of the primary methods to mitigate radiative heat transfer is the use of radiation shields. These are thin, reflective barriers placed between radiating surfaces to interrupt the direct path of thermal radiation. Typically made from materials like polished aluminum or copper, these shields possess low absorptivity and high reflectivity. By introducing additional resistance in the heat flow path, radiation shields effectively reduce the net heat flux between surfaces. For instance, inserting a single radiation shield between two parallel plates can halve the heat transfer rate. Adding more shields continues to decrease the heat transfer, with the reduction factor being inversely proportional to the number of shields plus one. This principle is particularly beneficial in applications like cryogenics and high-temperature furnaces, where controlling radiative heat loss is paramount [72–74].

Building upon the concept of radiation shields, Multi-Layer Insulation (MLI) systems employ multiple layers of reflective materials separated by spacers to minimize radiative heat transfer. Commonly used in spacecraft and satellites, MLI systems are designed to maintain thermal stability in the harsh environment of space. Each reflective layer in an MLI system reflects a portion of the incoming radiation, and the intervening spacers prevent conductive heat transfer between layers. The cumulative effect of these layers significantly reduces the overall heat transfer, making MLI systems highly effective in thermal insulation for aerospace applications. In the realm of building construction, radiant barriers serve as a practical solution to reduce heat gain, especially in hot climates. According to Buahom et al. [13], these barriers, often made of aluminum foil laminated onto various substrates, are installed in attics to reflect radiant heat from the sun, thereby lowering cooling costs. The effectiveness of radiant barriers depends on factors such as emissivity, installation angle, and the presence of air gaps. Wang et al. [59] emphasized that properly installed radiant barriers can reflect significant amount of radiant heat, contributing to significant energy savings in residential and commercial buildings..

Beyond physical barriers, advancements in material science have led to the development of reflective coatings that enhance radiation blocking capabilities. These coatings, often applied to surfaces exposed to high temperatures, consist of materials with low emissivity and high reflectivity. For example, ceramic coatings embedded with reflective layers can effectively reduce radiative heat transfer in turbine components, thereby extending their service life and improving efficiency. Such coatings are particularly valuable in aerospace and industrial applications where components are subjected to extreme thermal conditions. Recent research by Xu et al. [62] has explored the use of metamaterials, engineered materials with unique electromagnetic properties for infrared radiation suppression. These materials can be designed to exhibit low emissivity in specific wavelength ranges, effectively reducing thermal radiation. For instance, certain metamaterials have been developed to suppress infrared radiation in the atmospheric window while enhancing radiative cooling in other spectral regions. Such capabilities are promising for applications in thermal camouflage and energy-efficient building materials. In agricultural settings, managing heat within greenhouses is essential for optimal plant growth. As reported by Abdel-Ghany et al. [2], incorporating radiation-blocking techniques, such as applying near-infrared (NIR) reflective coatings to greenhouse covers, can significantly reduce heat gain. These coatings reflect a substantial portion of the solar radiation while allowing photosynthetically active radiation to pass through, maintaining a conducive environment for plants. Studies have shown that such coatings can lower greenhouse temperatures by several degrees Celsius, reducing the need for mechanical cooling and conserving energy. Radiation blocking is a critical aspect of thermal management across various industries and applications. Through the use of radiation shields, MLI systems, radiant barriers, reflective coatings, and advanced materials like metamaterials, it is possible to significantly reduce radiative heat transfer. These technologies not only enhance energy efficiency but also contribute to the longevity and performance of systems operating under thermal stress. Continued research and innovation in this field hold the promise of even more effective solutions for controlling heat transfer in the future..

Table 8.2 presents a comparison of the three fundamental mechanisms used to reduce heat transfer in buildings which are conduction control, convection suppression, and radiation blocking. Each mechanism addresses a different mode of heat transfer and utilizes specific strategies and materials to limit energy loss or gain. Conduction control reduces the transfer of heat through solid materials by using insulators with low thermal conductivity, such as fiberglass, foam, or aerogels.

Table 8.2 Comparison of the three mechanisms of heat transfer reduction

Reduction mechanism	Transfer medium	Work mechanism	Materials used	Examples
Conduction control	Solids	Uses materials with low thermal conductivity to slow down heat flow	Fiberglass, foam, aerogels	Wall insulation, refrigerator linings
Convection suppression	Fluids	Limits or eliminates fluid movement using trapped gases or vacuums	Foam with air pockets, vacuum insulation	Thermos flasks, insulated panels
Radiation blocking	Electromagnetic waves	Reflects or emits less radiant heat using shiny or low-emissivity surfaces	Aluminum foil, reflective coatings, Mylar	Space blankets, radiant barriers in attics

These materials work by slowing down the movement of heat-carrying particles. Convection suppression focuses on limiting heat transfer through fluids (liquids or gases), which typically occurs due to the circulation of air or liquid caused by temperature differences. This is achieved by using materials that trap air in small pockets or by creating a vacuum, both of which prevent the movement of fluid and thus minimize heat flow. Radiation blocking targets the emission or absorption of thermal radiation by incorporating reflective or low-emissivity surfaces, such as aluminum foil or Mylar, to reflect radiant heat and enhance insulation performance.

8.7 Applications in Energy-Efficient Building Design

8.7.1 Building Envelopes: Walls, Roofs, and Windows

The building envelope, comprising the walls, roofs, and windows of a structure, serves as the first line of defense between indoor spaces and the external environment. As the global demand for sustainable and energy-efficient buildings continues to rise, the design and performance of building envelopes have emerged as critical factors influencing thermal comfort, energy consumption, and environmental impact. In particular, in climates characterized by high humidity and solar intensity, such as those found in Southeast Asia, optimizing the building envelope becomes essential not only for reducing reliance on mechanical cooling systems but also for ensuring occupant well-being and long-term cost efficiency. Walls, as a fundamental part of the building envelope, perform dual roles which are providing structural support and acting as a thermal barrier. Their contribution to thermal resistance is largely determined by the materials used and the configuration of insulation layers. In regions where solar radiation is intense, poorly insulated walls can lead to significant heat gain, resulting in increased indoor temperatures and higher cooling energy demand. Incorporating high-performance insulation materials, such as expanded polystyrene, polyurethane foam, or mineral wool, has been shown to significantly enhance the thermal resistance of walls, reducing heat transfer and stabilizing indoor climates [8]. Additionally, wall constructions that integrate thermal mass such as bricks, concrete blocks, or rammed earth can absorb and store heat during the day, releasing it during cooler periods and thus contributing to a more stable indoor thermal environment. This passive thermal regulation reduces the building's dependence on active HVAC systems. Air and moisture control within wall assemblies is another critical consideration. Vapor barriers and waterproof membranes are essential in preventing moisture intrusion, which can lead to structural degradation, mold growth, and a decrease in insulation performance. Proper sealing techniques and the use of moisture-resistant cladding materials can extend the durability of wall systems and preserve energy efficiency over time [47].

Roof structures, due to their large surface area and direct exposure to sunlight, play a pivotal role in a building's energy performance. In tropical and subtropical regions, the roof is often the most significant source of heat gain, especially during midday hours. The use of reflective roofing materials or cool roof coatings can mitigate this issue by reflecting a higher percentage of solar radiation away from the building [17]. These materials, often coated with white or light-colored surfaces or embedded with reflective granules, can reduce roof surface temperatures dramatically, leading to lower interior cooling loads. Moreover, roofs designed with sufficient insulation can prevent conductive heat transfer into living spaces. Innovations such as ventilated roof cavities or radiant barriers further enhance thermal performance by promoting convective cooling and reducing radiant heat gain. Green roofs, which involve planting vegetation over roof membranes, offer a multifunctional solution by providing insulation, reducing urban heat island effects, and managing rainwater runoff [1]. Their ability to lower surface and ambient temperatures around buildings contributes not only to improved energy efficiency but also to enhanced biodiversity and urban aesthetics. In addition, the integration of photovoltaic systems into roofing structures aligns with the global push for net-zero energy buildings. By harnessing solar energy directly from the building's envelope, occupants can offset a significant portion of their electricity demand, making buildings more self-sufficient and environmentally sustainable.

Windows, while essential for natural lighting and ventilation, represent one of the most thermally vulnerable components of the building envelope. Poorly designed or improperly installed windows can account for substantial heat gain or loss, leading to energy inefficiency. Advances in glazing technology have dramatically improved window performance in recent years. Double-glazed and triple-glazed window systems, especially when combined with low-emissivity coatings and inert gas fills like argon or krypton, offer excellent thermal insulation while maintaining high levels of natural light transmittance. These technologies limit both solar heat gain and infrared radiation exchange, allowing for more stable indoor temperatures throughout the day. The strategic orientation and sizing of windows are equally important. In hot climates, minimizing east- and west-facing windows can reduce direct solar exposure, while optimizing openings on the north and south facades can provide ample daylight without excessive heat. Shading devices, including overhangs, louvers, fins, and external blinds, help control the amount of solar radiation entering through windows. These features not only improve energy efficiency but also enhance visual comfort by minimizing glare. Additionally, dynamic glazing systems such as electrochromic or thermochromic windows that change tint in response to temperature or sunlight, offer intelligent solutions that adapt to environmental conditions in real-time, ensuring optimal thermal and visual performance throughout the day [24]. The integration of walls, roofs, and windows into a cohesive and well-orchestrated building envelope design requires a holistic understanding of how each element interacts with the others. It is no longer sufficient to treat these components in isolation; rather, modern energy-efficient design demands an integrated approach that considers the building as a system. Passive design strategies, such as orienting buildings to take advantage of prevailing winds for natural ventilation or designing layouts that optimize cross-ventilation, play a crucial role in reducing energy loads. Similarly, combining thermal mass with operable windows enables night-time cooling, whereby accumulated heat is vented out during cooler periods, thus reducing reliance on air conditioning. Technologies such as Building Information Modeling (BIM) have facilitated the simulation of energy performance during the design stage, enabling architects and engineers to model the effects of various materials and configurations on thermal comfort and energy use. BIM tools allow for performance-based design, where decisions are guided by real-time data and predictive analysis rather than assumptions [71].

Retrofitting existing buildings to improve their envelope performance is a growing area of focus, especially in rapidly urbanizing regions. Many older buildings were constructed without consideration for energy efficiency and thus represent significant opportunities for improvement. Simple measures such as installing additional insulation, sealing leaks, replacing outdated windows, and applying reflective coatings to roofs can yield significant energy savings [52]. These interventions are often cost-effective and have relatively short payback periods, especially in areas with high energy costs or peak cooling demands. Governments and regulatory bodies in many countries are increasingly recognizing the importance of building envelopes in energy policy and have introduced building codes and incentive programs aimed at encouraging high-performance envelope retrofits. In Malaysia, for instance, the Green Building Index (GBI) emphasizes the importance of envelope performance in its certification criteria, encouraging designers and developers to prioritize energy-efficient materials and passive strategies [55]. Ultimately, the building envelope is more than a physical barrier, it is a dynamic interface that mediates environmental forces, facilitates energy exchange, and defines the comfort of the occupants within. As buildings continue to account for a significant portion of global energy consumption and carbon emissions, the pursuit of energy-efficient envelope design remains a cornerstone of sustainable architecture. Advances in materials science, simulation technologies, and design methodologies provide architects and engineers with unprecedented tools to push the boundaries of what is possible. By prioritizing thoughtful envelope design in both new construction and retrofitting projects, the built environment can move toward a future that is not only energy-conscious but also resilient, healthy, and responsive to the challenges of a changing climate.

8.7.2 *Interior Insulation Solutions*

Interior insulation has emerged as a pivotal strategy in enhancing the energy efficiency of buildings, particularly in retrofitting scenarios where external modifications are constrained. This approach involves adding insulation layers to the interior surfaces of walls, ceilings, and floors, thereby improving thermal performance without altering the building's exterior. Such solutions are especially beneficial for heritage buildings, urban dwellings with aesthetic restrictions, and structures in densely populated areas. A comprehensive study conducted in Chengdu, China, demonstrated the efficacy of interior insulation in improving energy performance. The research focused on residences with intermittent cooling patterns, revealing that interior insulation could achieve an average energy-saving rate of 42.09% compared to uninsulated counterparts. By insulating internal opaque structures, including walls, floors, and ceilings, the study highlighted significant reductions in cooling loads, emphasizing the potential of interior insulation in hot-summer and cold-winter climates [65].

In the context of heritage buildings, interior insulation presents a viable solution that respects historical aesthetics while enhancing energy efficiency. Research on using nature-based materials like Typha (cattail) for interior insulation has shown promising results. Typha panels offer thermal insulation properties suitable for tropical climates, contributing to energy savings and preserving the architectural integrity of heritage buildings. Various materials are employed in interior insulation, each offering distinct advantages. Cellulose insulation, made from recycled paper products, provides excellent thermal performance and soundproofing qualities. Its ability to fill cavities and conform to irregular spaces makes it effective in reducing air infiltration. Additionally, cellulose insulation is treated with fire retardants, enhancing its safety profile.

Reflective insulation is an effective interior insulation strategy designed to reduce heat transfer by reflecting radiant heat away from the building interior. Unlike conventional insulation, which primarily resists conductive and convective heat flow, reflective insulation works by using a reflective surface—typically aluminum foil laminated onto various substrate materials, to reflect radiant energy. This makes it particularly advantageous in hot and sunny climates, where solar radiation is a major contributor to indoor heat gain. When installed properly on interior surfaces such as ceilings or walls, it helps maintain cooler indoor temperatures by minimizing the amount of solar heat that penetrates the building envelope. This not only enhances thermal comfort but also reduces reliance on mechanical cooling systems, leading to lower energy consumption and utility costs. In addition to reflective systems, bio-based materials such as straw fiber are gaining attention as sustainable alternatives for interior insulation. Derived from agricultural waste, straw fiber panels are not only renewable and biodegradable but also exhibit thermal insulation properties comparable to conventional materials like expanded polystyrene [41]. Several studies have shown that straw insulation performs exceptionally well in terms of thermal resistance, particularly during hot seasons. Its ability to buffer temperature fluctuations helps reduce indoor overheating and minimizes the need for artificial cooling [49]. Furthermore, using straw-based insulation contributes to environmental sustainability by lowering carbon footprints and promoting circular resource use. As the building industry moves toward greener construction practices, the integration of natural materials like straw in interior insulation applications stands out as both an energy-efficient and ecologically responsible choice.

Implementing interior insulation in buildings offers many benefits, but it also comes with certain challenges that must be addressed to ensure long-term effectiveness and safety. One of the most critical issues is moisture management. When insulation is added to the interior surfaces of a building, it can potentially create a barrier that traps moisture between the insulation layer and the original wall. This can lead to condensation, which in turn may cause mold growth, material degradation, and reduced insulation performance. To avoid these problems, it is essential to use vapor-permeable or moisture-regulating materials, along with proper installation techniques that include air sealing and ventilation strategies [33]. Another consideration is the slight reduction in usable interior space, especially in small rooms or older buildings with already limited dimensions. Although the space taken up by interior insulation is generally minimal, typically a few centimeters—it may affect furniture placement or aesthetics. However, studies and real-world applications consistently show that this small loss of space is a worthwhile trade-off for the significant energy savings and improved indoor comfort that result from better thermal performance [45]. Interior insulation is especially advantageous in cases where external retrofitting is not possible, such as in historic buildings, densely built urban areas, or properties with preserved façades. It allows for energy efficiency upgrades without altering the building's external appearance. When appropriate materials are selected and potential issues are addressed during the design and installation process, interior insulation becomes a highly effective solution for reducing energy consumption, enhancing occupant comfort, and promoting more sustainable and resilient building practices.

8.7.3 Reflective and Transparent Insulation Materials

Reflective and transparent insulation materials (TIMs) are increasingly pivotal in the design of energy-efficient buildings, offering innovative solutions to reduce energy consumption while maintaining aesthetic and functional qualities. These materials address the growing demand for sustainable architecture by enhancing thermal performance and daylight utilization. Reflective insulation systems, including radiant barriers and reflective insulation, operate by reflecting radiant heat rather than absorbing it. Typically composed of low-emissivity materials like aluminum foil, these systems are effective in reducing heat transfer, especially in hot climates. Radiant barriers, when installed in attics, can decrease heat flux by 26–50% and cooling loads by 6–16% during cooling seasons as demonstrated in the study by Lee et al. [35]. Reflective insulation, which includes enclosed air spaces, further enhances thermal resistance by minimizing conductive and convective heat transfer. However, the performance of these systems can degrade over time due to factors like dust accumulation and moisture, emphasizing the need for proper installation and maintenance.

TIMs offer a unique combination of thermal insulation and light transmission, making them ideal for applications where natural lighting is desired without compromising energy efficiency. Materials such as monolithic silica aerogel and capillary structures have been studied for their ability to reduce heat loss while allowing daylight penetration [46]. These materials are particularly beneficial in glazing systems, where they can significantly improve the thermal performance of windows and facades. The integration of TIMs into building designs not only enhances energy savings but also contributes to occupant comfort by providing consistent natural lighting and reducing reliance on artificial lighting. The incorporation of reflective and transparent insulation materials into building designs offers numerous benefits. Reflective insulation is particularly effective in reducing cooling loads in hot climates, making it suitable for regions like Malaysia. TIMs, on the other hand, are advantageous in temperate climates where both insulation and daylighting are essential. Their application in windows, skylights, and facades can lead to significant energy savings by reducing the need for artificial lighting and heating. Moreover, these materials can contribute to the aesthetic appeal of buildings, allowing architects to design structures that are both functional and visually pleasing.

While the benefits of reflective and transparent insulation materials are evident, several challenges must be addressed to maximize their effectiveness. For reflective insulation, maintaining the low-emissivity surface is crucial, as degradation can significantly reduce performance [35]. Proper installation is essential to prevent issues like moisture accumulation and to ensure long-term durability. For TIMs, factors such as material cost, availability, and integration into existing building systems must be considered. Additionally, the long-term performance and durability of TIMs under various environmental conditions require further research to ensure their viability in diverse applications [46]. Reflective and transparent insulation materials represent significant advancements in building technology, offering solutions that enhance energy efficiency without compromising design and comfort. Their integration into building designs can lead to substantial energy savings, reduced environmental impact, and improved occupant well-being. As ongoing research and technological development continue to overcome current limitations and enhance material performance, the widespread adoption of reflective and transparent insulation materials is expected to increase significantly, supporting global initiatives aimed at creating more sustainable, energy-efficient, and environmentally responsible building environments for the future.

The application of nanomaterial-based insulation in energy-efficient building design offers diverse benefits across various architectural elements, each with its own strengths and limitations as explored in Table 8.3. In building envelopes including walls, roofs, and windows, nanomaterials significantly enhance thermal resistance by minimizing heat transfer through conduction, convection, and radiation, leading to substantial energy savings. However, their high initial costs and integration complexity with conventional materials can present challenges. Interior insulation solutions benefit from the compactness and efficiency of nanomaterials, making them ideal for retrofitting existing buildings without compromising indoor space. They also help maintain thermal comfort, though improper installation or poor encapsulation can affect performance and safety. Reflective and transparent insulation materials provide the added advantage of natural daylighting while minimizing heat gain or loss, which reduces reliance on HVAC and lighting systems. Still, their long-term effectiveness can diminish due to surface wear or contamination. Overall, each application presents a promising pathway toward sustainable, energy-efficient building design.

Table 8.3 Benefits and limitations of nanomaterial insulation applications in energy-efficient building design

Application	Advantages	Limitations
Building envelopes (walls, roofs, windows)	Significant reduction in heat transfer (conduction, convection, radiation) Improves overall thermal resistance of building shell Enhances energy savings across seasons	High initial cost of nanomaterials Complex integration with traditional materials Potential moisture trapping if not properly ventilated
Interior insulation	Space-saving due to thin-layer high-performance insulation Easy retrofit for existing buildings Maintains indoor comfort levels	Risk of indoor air quality concerns from poorly encapsulated nanomaterials Reduced effectiveness if not correctly installed Higher cost compared to bulk insulation
Reflective and transparent insulation materials	Dual benefit which are insulation and natural lighting Reduced HVAC and lighting load Improves visual aesthetics and daylighting	Performance degradation over time due to dust or wear High material and maintenance costs Limited thermal mass compared to opaque insulation

8.8 Summary

Improving energy efficiency in buildings is a critical strategy for reducing global energy consumption and minimizing environmental impact. In this context, nanomaterial-based insulation has emerged as a transformative solution, offering superior thermal properties, lightweight characteristics, and multifunctional capabilities. These advanced materials exhibit extremely low thermal conductivity, high surface area, and the ability to integrate with conventional construction components, thereby reducing heat transfer while maintaining structural integrity and design flexibility. Additionally, various types of nanomaterials used in building insulation are discussed, including nanoporous materials, nano-aerogels, vacuum insulation panels, carbon nanotubes, phase change materials enhanced with nanoparticles, and nanoclays. The reduction of heat transfer in buildings is achieved through key mechanisms that are controlling conduction, suppressing convection, and blocking radiation, all of which contribute to improved thermal insulation performance. Enhanced design applications, such as optimized building envelopes, interior insulation, and the use of reflective and transparent insulation materials, play a significant role in promoting energy-efficient buildings. Despite their promising performance, nanomaterial-based insulation systems still face notable challenges including high production costs, concerns over long-term durability, and difficulties in integration with existing building infrastructure. Addressing these issues is essential to advancing the widespread adoption of nanomaterials for energy-efficient building design.

References

1. Abdalazeem ME, Hassan H, Asawa T, Mahmoud H (2022) Review on integrated photovoltaic-green roof solutions on urban and energy-efficient buildings in hot climate. Sustain Cities Soc 82:103919. https://doi.org/10.1016/j.scs.2022.103919
2. Abdel-Ghany AM, Al-Helal IM, Alzahrani SM, Alsadon AA, Ali IM, Elleithy RM (2012) Covering materials incorporating radiation-preventing techniques to meet greenhouse cooling challenges in arid regions: a review. Sci World J 2012:1–11. https://doi.org/10.1100/2012/906360
3. Abdelrady A, Abdelhafez MHH, Ragab A (2021) Use of insulation based on nanomaterials to improve energy efficiency of residential buildings in a Hot Desert climate. Sustainability 13(9):5266. https://doi.org/10.3390/su13095266
4. Aghajamali M, Iqbal M, Purkait TK, Hadidi L, Sinelnikov R, Veinot JGC (2016) Synthesis and properties of luminescent silicon nanocrystal/silica aerogel hybrid materials. Chem Mater 28(11):3877–3886. https://doi.org/10.1021/acs.chemmater.6b01114
5. Agrawal T, Jha NK, Chalamalla VK (2024) Improvement in sealing effectiveness of air curtains using positive buoyancy, pp 137–148. https://doi.org/10.1007/978-981-99-5755-2_17
6. Aimkamon P, Nontakaew U, Sudprasert S (2014) Numerical study of reduction of ceiling heat load by forced convection. Int J Build Urban 4:25–36. Interior and Landscape Technology (BUILT)
7. Ali A, Issa A, Elshaer A (2024) A comprehensive review and recent trends in thermal insulation materials for energy conservation in buildings. Sustainability 16(20):8782. https://doi.org/10.3390/su16208782
8. Altun AF (2022) Determination of optimum building envelope parameters of a room concerning Window-to-wall ratio, orientation, Insulation thickness and window type. Buildings 12(3):383. https://doi.org/10.3390/buildings12030383
9. Anjomshoa E (2024) Investigation of lightweight gypsum based on montmorillonite nanoclay with enhanced insulation properties. Int J Build Pathol Adapt
10. Anwer MAS, Wang J, Guan A(Qi), Naguib HE (2019) Chitin nano-whiskers (CNWs) as a bio-based bio-degradable reinforcement for epoxy: evaluation of the impact of CNWs on the morphological, fracture, mechanical, dynamic mechanical, and thermal characteristics of DGEBA epoxy resin. RSC Adv 9(20):11063–11076. https://doi.org/10.1039/C9RA00769E
11. Barreneche C, Martín M, Calvo-de la Rosa J, Majó M, Fernández AI (2019) Own-synthetize nanoparticles to develop nano-enhanced phase change materials (NEPCM) to improve the energy efficiency in buildings. Molecules 24(7):1232. https://doi.org/10.3390/molecules24071232
12. Ben-Abdelwahed A, Liu S, Sommier A, Reungoat D (2025) Experimental and numerical analysis of heat transfer and thermal deformation in small-dimension liquid mechanical seals. Int J Heat Mass Transf 241:126754. https://doi.org/10.1016/j.ijheatmasstransfer.2025.126754
13. Buahom P, Gong P, Wang C, Yu H, Liu J, Park CB (2022) Carbon as a solution for nanocellular foam superinsulation. Carbon 189:319–338. https://doi.org/10.1016/j.carbon.2021.11.041
14. Chai B, Xu M, Yan J, Ren Z (2018) Remarkably enhanced photocatalytic hydrogen evolution over MoS 2 nanosheets loaded on uniform CdS nanospheres. Appl Surf Sci 430:523–530. https://doi.org/10.1016/j.apsusc.2017.07.292
15. Demezhko DY, Khatskevich BD, Mindubaev MG (2020) Methods of suppressing free thermal convection in water-filled wells during temperature research. Georesursy Georesour 22(1):55–62
16. Dey P, Saha SK (2021) Fluid flow and heat transfer in microchannel with porous bio-inspired roughness. Int J Therm Sci 161:106729. https://doi.org/10.1016/j.ijthermalsci.2020.106729
17. Feng C, Lei Y, Fang J, Lu B, Li X, Xue X (2022) Optimized radiative parameters of building roof surfaces for energy efficiency: case studies in China. J Build Eng 61:105289. https://doi.org/10.1016/j.jobe.2022.105289

18. Fernandes J, Santos MC, Castro R (2021) Introductory review of energy efficiency in buildings retrofits. Energies 14(23):8100. https://doi.org/10.3390/en14238100
19. Ganobjak M, Brunner S, Wernery J (2020) Aerogel materials for heritage buildings: materials, properties and case studies. J Cult Herit 42:81–98. https://doi.org/10.1016/j.culher.2019.09.007
20. Ghyati S, Kassou S, El Jai M, El Kinani EH, Benhamou M, Chang SH (2022) Thermal properties of clay containing polymeric reinforcement as an eco-friendly bio-insulation composite for buildings. J Therm Anal Calorim 147(13):7213–7228. https://doi.org/10.1007/s10973-021-11015-7
21. Habibishandiz M, Saghir MZ (2022) A critical review of heat transfer enhancement methods in the presence of porous media, nanofluids, and microorganisms. Therm Sci Eng Progr 30:101267. https://doi.org/10.1016/j.tsep.2022.101267
22. Hafez FS, Sa'di B, Safa-Gamal M, Taufiq-Yap YH, Alrifaey M, Seyedmahmoudian M, Stojcevski A, Horan B, Mekhilef S (2023) Energy efficiency in sustainable buildings: a systematic review with taxonomy, challenges, motivations, methodological aspects, recommendations, and pathways for future research. Energy Strat Rev 45:101013. https://doi.org/10.1016/j.esr.2022.101013
23. Hayat MA, Chen Y (2021) *A brief review on nano phase change material-based polymer encapsulation for thermal energy storage systems*, pp 19–26. https://doi.org/10.1007/978-3-030-63916-7_3
24. Hosseini SM, Shirmohammadi R, Kasaeian A, Pourfayaz F (2021) Dynamic thermal simulation based on building information modeling: a review. Int J Energy Res 45(10):14221–14244. https://doi.org/10.1002/er.6740
25. Hou P, Zhang F, Zhang L, Liu C, Cheng H (2022) Synthesis of carbon nanotubes by floating catalyst chemical vapor deposition and their applications. Adv Funct Mater 32(11). https://doi.org/10.1002/adfm.202108541
26. Ijjada N, Nayaka RR (2022) Review on properties of some thermal insulating materials providing more comfort in the building. Mater Today Proc 58:1354–1359. https://doi.org/10.1016/j.matpr.2022.02.230
27. Jelle BP, Kalnæs SE (2016) Nanotech based vacuum insulation panels for building applications. In: Nano and biotech based materials for energy building efficiency. Springer, pp 167–214. https://doi.org/10.1007/978-3-319-27505-5_7
28. Ji X, Zhou Q, Qiu G, Yue C, Guo M, Chen F, Zhang M (2018) Preparation of monolithic silica-based aerogels with high thermal stability by ambient pressure drying. Ceram Int 44(11):11923–11931. https://doi.org/10.1016/j.ceramint.2018.03.261
29. Ji D, Lin Y, Guo X, Ramasubramanian B, Wang R, Radacsi N, Jose R, Qin X, Ramakrishna S (2024) Electrospinning of nanofibres. Nat Rev Methods Primers 4(1):1. https://doi.org/10.1038/s43586-023-00278-z
30. Kandeal AW, Ismail M, Basem A, Elsayad MM, Alawee WH, Majdi HS, Abdullah AS, Jang S-H, An M, Omara ZM, Ghazaly NM, Sharshir SW (2024) An overview of the improvement of natural convection heat transfer via surface thermal radiation for different geometries. Res Eng 23:102514. https://doi.org/10.1016/j.rineng.2024.102514
31. Karanafti A, Theodosiou T, Tsikaloudaki K (2022) Assessment of buildings' dynamic thermal insulation technologies-a review. Appl Energy 326:119985. https://doi.org/10.1016/j.apenergy.2022.119985
32. Kathy O (2011) Reduction of heat transfer into building via water and other methods
33. Knarud JI, Geving S, Kvande T (2021) Moisture performance of interior insulated brick wall segments subjected to wetting and drying – a laboratory investigation. Build Environ 188:107488. https://doi.org/10.1016/j.buildenv.2020.107488
34. Kouadri I, Satha H (2018) Extraction and characterization of cellulose and cellulose nanofibers from Citrullus colocynthis seeds. Ind Crop Prod 124:787–796. https://doi.org/10.1016/j.indcrop.2018.08.051

35. Lee, S. W., Lim, C. H., & Salleh, E. @ I. Bin. (2016). Reflective thermal insulation systems in building: a review on radiant barrier and reflective insulation. Renew Sust Energ Rev, 65, 643–661. https://doi.org/10.1016/j.rser.2016.07.002
36. Lee C-C, Wang C, He Z, Xing W, Wang K (2023) How does green finance affect energy efficiency? The role of green technology innovation and energy structure. Renew Energy 219:119417. https://doi.org/10.1016/j.renene.2023.119417
37. Liu Q, Yan K, Chen J, Xia M, Li M, Liu K, Wang D, Wu C, Xie Y (2021) Recent advances in novel aerogels through the hybrid aggregation of inorganic nanomaterials and polymeric fibers for thermal insulation. Aggregate 2(2). https://doi.org/10.1002/agt2.30
38. Lu L, Wang H, Yun S, Hu J, Wang M (2024) A state-of-the-art review of novel aerogel insulation materials for building exterior walls. Energy Sour Part A Recov Util Environ Effect 46(1):16231–16252. https://doi.org/10.1080/15567036.2024.2424915
39. Mariño MA, Rezende CA, Tasic L (2018) A multistep mild process for preparation of nanocellulose from orange bagasse. Cellulose 25(10):5739–5750. https://doi.org/10.1007/s10570-018-1977-y
40. Matos AM, Delgado JMPQ, Guimarães AS (2022) Linking energy poverty with thermal building regulations and energy efficiency policies in Portugal. Energies 15(1):329. https://doi.org/10.3390/en15010329
41. Mohd Ashhar MZ, Haw LC (2022) Recent research and development on the use of reflective technology in buildings – a review. J Build Eng 45:103552. https://doi.org/10.1016/j.jobe.2021.103552
42. Mukhopadhyaya P (2022) High performance thermal insulations—vacuum insulation panels (VIPs), pp 275–288. https://doi.org/10.1007/978-3-030-98693-3_10
43. Nakamura Y (2018) Nanostructure design for drastic reduction of thermal conductivity while preserving high electrical conductivity. Sci Technol Adv Mater 19(1):31–43. https://doi.org/10.1080/14686996.2017.1413918
44. Nam J, Yang S, Choi Y, Kim S (2025) Hybrid membranes incorporating microencapsulated PCM for enhanced hygrothermal performance in enthalpy recovery ventilation systems. Appl Therm Eng 258:124759. https://doi.org/10.1016/j.applthermaleng.2024.124759
45. Năstase G, Doboşi IS, Brezeanu AI, Taus D, Tăbăcaru MB, Vuţoiu BG, Rusu D, Bulmez AM, Iordan NF (2022) Experimental heat transfer, sound insulation and interior comfort parameters assessment on a box double-skin façade. Buildings 12(6):730. https://doi.org/10.3390/buildings12060730
46. Nautiyal A, Duggal P, Tomar RK (2022) An overview of transparent insulation materials and its future developments. In: IRC-Set 2021. Springer Nature Singapore, pp 283–297. https://doi.org/10.1007/978-981-16-9869-9_22
47. Navakrishnan S, Sivakumar B, Senthil R, Kumar RS (2021) Heating and cooling application in energy efficient buildings using Trombe Wall: a review. IOP Conf Ser Mat Sci Eng 1130(1):012015. https://doi.org/10.1088/1757-899X/1130/1/012015
48. Qian T, Li J, Min X, Guan W, Deng Y, Ning L (2015) Enhanced thermal conductivity of PEG/diatomite shape-stabilized phase change materials with Ag nanoparticles for thermal energy storage. J Mater Chem A 3(16):8526–8536. https://doi.org/10.1039/C5TA00309A
49. Saber HH, Yarbrough DW (2023) Assessing the effect of air intrusion on reflective insulations performance with horizontal heat flow. Buildings 13(10):2461. https://doi.org/10.3390/buildings13102461
50. Samardzioska T, Jovanoska M, Grujoska V (2023) Sustainable thermal insulation derived from recycled textile waste. Adv Environ Eng Res 04(01):1–8. https://doi.org/10.21926/aeer.2301004
51. Shah KW, Ong PJ, Chua MH, Toh SHG, Lee JJC, Soo XYD, Png ZM, Ji R, Xu J, Zhu Q (2022) Application of phase change materials in building components and the use of nanotechnology for its improvement. Energ Buildings 262:112018. https://doi.org/10.1016/j.enbuild.2022.112018

52. Sing MCP, Chan VWC, Lai JHK, Matthews J (2021) Energy-efficient retrofitting of multi-storey residential buildings. Facilities 39(11/12):722–736
53. Srikanth R, Nemani P, Balaji C (2015) Multi-objective geometric optimization of a PCM based matrix type composite heat sink. Appl Energy 156:703–714. https://doi.org/10.1016/j.apenergy.2015.07.046
54. Starkov AS, Starkov IA (2024) Effective thermal characteristics of nanostructures in the presence of kapitza resistance. J Appl Mech Tech Phys 65(2):336–344. https://doi.org/10.1134/S0021894424020159
55. Taufan A, Zaki SA, Tuck NW, Singh MK, Rijal HB (2023) Energy-efficient retrofitting strategies in mosque buildings: a review. Renew Sust Energ Rev 183:113479. https://doi.org/10.1016/j.rser.2023.113479
56. Thie C, Quallen S, Ibrahim A, Xing T, Johnson B (2023) Study of energy saving using silica aerogel insulation in a residential building. Gels 9(2):86. https://doi.org/10.3390/gels9020086
57. Vagge ST, Ghogare S (2022) Thermal barrier coatings: review. Mater Today Proc 56:1201–1216. https://doi.org/10.1016/j.matpr.2021.11.170
58. Wang Z, Tian Q, Jia J (2022) The convective heat transfer performance and structural optimization of the cavity in energy-saving thermal insulation windows under cold air penetration condition. Energies 15(7):2481. https://doi.org/10.3390/en15072481
59. Wang Y, Pu L, Leng Y, Wang X, Yu Z (2025) Performance research and optimization of hollow glass microsphere insulation system for large liquid hydrogen sphere. Int J Hydrog Energy 111:555–565. https://doi.org/10.1016/j.ijhydene.2025.02.002
60. Wu Y, Wang X, Yao L, Chang S, Wang X (2023) Thermal insulation mechanism, preparation, and modification of Nanocellulose aerogels: a review. Molecules 28(15):5836. https://doi.org/10.3390/molecules28155836
61. Xu X, Zhang Q, Yu Y, Chen W, Hu H, Li H (2016) Naturally dried graphene aerogels with superelasticity and tunable Poisson's ratio. Adv Mater 28(41):9223–9230. https://doi.org/10.1002/adma.201603079
62. Xu Q, Liu X, Xuan Y, Xu Y, Liu D (2020) High-performance infrared thermal radiation suppression metamaterials enabling inhibited infrared emittance and decreased temperature simultaneously. Int J Heat Mass Transf 161:120318. https://doi.org/10.1016/j.ijheatmasstransfer.2020.120318
63. Xue CH, Wang D, Xiang B, Chiou B, Sun G (2010) Morphology evolution of polypropylene in immiscible polymer blends for fabrication of nanofibers. J Polym Sci B Polym Phys 48(9):921–931. https://doi.org/10.1002/polb.21978
64. Yang J, Wu H, Liang Y, Cen J, Zhang X (2023) A comparative assessment of different aerogel-insulated building walls for enhanced thermal insulation performance. Gels 9(12):943. https://doi.org/10.3390/gels9120943
65. Ye X, Lu J, Zhang T, Wang Y, Fukuda H (2021) Improvements in energy saving and thermal environment after retrofitting with interior insulation in intermittently cooled residences in hot-summer/cold-winter zone of China: A case study in Chengdu. Energies 14(10):2776. https://doi.org/10.3390/en14102776
66. Yu F, Wang T, Zhang C (2018) Effect of axial conduction on heat transfer in a rectangular microchannel with local heat flux. J Therm Sci Technol 13(1):JTST0013-JTST0013. https://doi.org/10.1299/jtst.2018jtst0013
67. Zavrl E, Zupanc G, Stritih U, Dovjak M (2019) Overheating reduction in lightweight framed buildings with application of phase change materials. Strojniški Vestnik J Mech Eng 66(1):3–14. https://doi.org/10.5545/sv-jme.2019.6244
68. Zhan H, Shi QQ, Wu G, Wang JN (2020) A carbon nanotube approach for efficient thermally insulating material with high mechanical stability and fire-retardancy. RSC Adv 10(37):21772–21780. https://doi.org/10.1039/D0RA03472J
69. Zhang X, Lu Y (2014) Centrifugal spinning: an alternative approach to fabricate nanofibers at high speed and low cost. Polym Rev 54(4):677–701. https://doi.org/10.1080/15583724.2014.935858

70. Zhang S, Feng D, Shi L, Wang L, Jin Y, Tian L, Li Z, Wang G, Zhao L, Yan Y (2021) A review of phase change heat transfer in shape-stabilized phase change materials (ss-PCMs) based on porous supports for thermal energy storage. Renew Sust Energ Rev 135:110127. https://doi.org/10.1016/j.rser.2020.110127
71. Zhang Z, Zhang L, Zhou Y, Cui Y, Chen Z, Liu Y, Li J, Long Y, Gao Y (2023) Thermochromic energy efficient windows: fundamentals, recent advances, and perspectives. Chem Rev 123(11):7025–7080. https://doi.org/10.1021/acs.chemrev.2c00762
72. Zhang L, Luo F, Zhou Y, Nan H, Zhang Q, Wang C, Li Y, Qing Y (2024a) Enhanced thermal radiation blocking of YSZ-LaMgAl11O19 thermal barrier coatings consisting of splat interfaces with different refractive indexes. J Eur Ceram Soc 44(15):116776. https://doi.org/10.1016/j.jeurceramsoc.2024.116776
73. Zhang X, Yu J, Zhao C, Si Y (2024b) Elastic SiC aerogel for thermal insulation: a systematic review. Small 20(32). https://doi.org/10.1002/smll.202311464
74. Zhang L, Luo F, Zhou Y, Cao Y, Zhang Q, Wang C, Li Y, Qing Y (2025) Enhanced thermal radiation blocking of plasma spraying thermal barrier coatings with a bimodal splat thickness distribution. J Eur Ceram Soc 45(3):117068. https://doi.org/10.1016/j.jeurceramsoc.2024.117068

Chapter 9
Bio-aerogels: Promising Nanostructured Materials for Air Filtration

9.1 Overview of Air Pollution Challenges

Air pollution remains one of the most critical global environmental issues, significantly contributing to climate change, ecosystem degradation, and public health crises. The primary sources include industrial emissions, vehicular exhaust, biomass burning, and secondary pollutants formed through atmospheric photochemical reactions [154]. Among the various airborne pollutants, fine particulate matter ($PM_{2.5}$ and PM_{10}), nitrogen oxides (NO_x), sulfur dioxide (SO_2), ozone (O_3), and volatile organic compounds (VOCs) are recognized for their adverse health effects, including respiratory and cardiovascular diseases, neurotoxicity, and carcinogenicity [74, 91]. Recent data suggest that over 90% of the world's population is exposed to air quality levels exceeding WHO guideline limits, particularly in urban and densely populated regions [143]. The nanometric size and high surface reactivity of airborne particles facilitate their deep penetration into the alveolar regions of the lungs and even systemic circulation, posing serious long-term health risks [105]. Moreover, indoor air pollution often overlooked accounts for a substantial portion of exposure, especially in developing countries where combustion-based cooking and poor ventilation are prevalent [48].

Traditional filtration systems such as high-efficiency particulate air (HEPA) filters, electrostatic precipitators, and activated carbon beds, while effective, face challenges in terms of energy consumption, maintenance, disposal, and inefficiency against ultrafine particles and gaseous pollutants [122]. These limitations necessitate the development of advanced, sustainable materials that combine high filtration efficiency, low pressure drop, and long operational lifespans. In this context, bio-based nanomaterials particularly bio-aerogels have emerged as transformative candidates due to their exceptional porosity, tunable surface chemistry, low density, and biocompatibility [145]. Their multifunctional capabilities not only enable mechanical filtration but also adsorption and catalytic degradation of airborne toxins.

M. I. Ahmad, A. K. H. P. S., *Environmental and Energy Technology*, Green Energy and Technology, https://doi.org/10.1007/978-3-032-10706-0_9

Therefore, addressing air pollution requires a paradigm shift toward eco-friendly and nanostructured filtration technologies, underpinned by materials science, green engineering, and interdisciplinary collaboration.

9.2 Bio-aerogels: Structural Features and Functional Potential

Bio-aerogels constitute a versatile class of advanced functional materials, distinguished by their ultra-lightweight, highly porous architectures and exceptional physicochemical properties, including elevated specific surface area, low bulk density, and outstanding thermal insulation capabilities [109]. These aerogels are fabricated from renewable biopolymeric sources such as nanocellulose, chitosan, and gelatin, all of which are inherently biodegradable, biocompatible, and environmentally benign [60, 116]. The tunable mechanical strength, surface chemistry, and morphological versatility of biopolymer-based aerogels have enabled their extensive deployment across diverse application domains, notably in environmental remediation, biomedical engineering, energy storage, and advanced filtration technologies [60].

By capitalizing on the intrinsic advantages of natural polymers, biopolymer aerogels present a sustainable alternative to their conventional synthetic counterparts, thereby significantly reducing environmental footprint throughout their life cycle. Moreover, their adaptability to various functionalization techniques allows for customized performance tailored to specific environmental and biomedical challenges. Comparative analyses, such as that presented in Table 9.1, further illustrate the

Table 9.1 General properties of different aerogel prepared from conventional materials

Aerogel type	Synthesis and properties overview	Structural features	Thermal and surface characteristics	References
Silica aerogel	Fabricated via sol-gel processing with supercritical drying; fragile without reinforcement	Porosity: 90–99.8%, density: 100–500 kg/m^3	Surface area: 500–1000 m^2/g; thermal conductivity: 0.01–0.02 W/mK	[25, 89]
Carbon aerogel	Resorcinol-formaldehyde polymerization followed by supercritical drying; high thermal stability	Porosity: 80–99%, density: 200–800 kg/m^3	Surface area: 400–1000 m^2/g; thermal conductivity: 0.01–0.05 W/mK	[24, 84, 121]
Polymeric aerogel	Organic monomer polymerization with supercritical or freeze drying; highly customizable chemistry	Porosity: 85–99.5%, density: 50–300 kg/m^3	Surface area: 200–800 m^2/g; thermal conductivity: 0.02–0.03 W/mK	[119, 138, 148]
Biopolymeric aerogel	Gelation of natural biopolymers; processed by supercritical or freeze drying; eco-friendly applications	Porosity: 80–99%, density: 100–600 kg/m^3	Surface area: 100–800 m^2/g; thermal conductivity: 0.03–0.05 W/mK	[4, 16, 51, 55]

distinctive material properties of bio-aerogels relative to traditional inorganic aerogels, highlighting their superior sustainability profile and functional versatility.

9.3 Historical Evolution of Nanostructured Bio-aerogel Materials

The pioneering development of aerogels can be traced back to 1931 when Samuel Kistler, an American scientist, introduced the term "aerogel" and successfully fabricated the first aerogel structure [104]. Kistler aimed to produce a solid material with minimal density by replacing the liquid phase of a gel with gas without inducing structural collapse [70]. This was accomplished through the innovation of the supercritical drying technique, wherein the solvent within the gel was removed under supercritical conditions, avoiding the detrimental effects associated with a liquid-gas interface transition. The resultant material preserved the original gel network, characterized by an extremely low density and high porosity, thereby establishing the foundation for silica-based aerogels.

For several decades, research predominantly focused on silica aerogels due to their extraordinary properties, such as ultra-low density, exceptionally high surface area, and superior thermal insulation performance [76]. These materials found critical applications in aerospace insulation, such as space suits and spacecraft, as well as in astrophysical studies for capturing cosmic dust particles [65, 78]. Nevertheless, inherent limitations, including their mechanical fragility and brittleness, restricted their broader technological adoption [65]. Consequently, during the mid-twentieth century, researchers directed efforts toward developing aerogels based on alternative inorganic compositions, such as alumina and titania, aiming to address the mechanical weaknesses and reduce production costs [81]. Although these advancements diversified the material base, challenges in structural robustness and economic feasibility remained significant.

A major paradigm shift occurred in the late twentieth century with the emergence of organic and hybrid aerogels, integrating organic polymers with conventional inorganic networks [119]. This innovation sought to enhance the mechanical resilience of aerogels while preserving their hallmark properties of low density and high porosity. Polymeric aerogels, particularly those derived from resorcinol-formaldehyde and polyurethane systems, offered increased flexibility, improved mechanical performance, and greater processability compared to purely inorganic aerogels [102]. As research entered the twenty-first century, attention increasingly turned towards environmental sustainability, culminating in the rise of biopolymeric aerogels [3]. Fabricated from natural, renewable polymers such as cellulose, chitosan, gelatin, alginate, and starch, biopolymer-based aerogels align with the principles of green chemistry by offering biodegradability, resource renewability, and reduced environmental footprint [146]. This transition reflects a growing commitment within materials science to develop high-performance functional materials that harmonize technological advancement with ecological stewardship. An overview of the chronological development of aerogel materials is summarized in Table 9.2.

Table 9.2 An overview of the chronological development of aerogel materials

Year	Aerogel development	Source material	Fabrication technique	References
1930s–1950s				
1931	Pioneering silica aerogel	Sodium silicate	Supercritical solvent extraction	[70]
1932	First organic and metal oxide aerogels	Metal oxides, organic compounds	Supercritical drying technique	[155]
1950	Hydrophobic silica aerogels	Metal oxides combined with silica	Surface silylation using trichloro-methyl silane	[19]
1960s–1980s				
1968	Sol-gel derived silica aerogels	Metal oxides with silica	Sol-gel processing utilizing TMOS and supercritical drying	[19]
1974	Refined sol-gel silica aerogels	Metal oxides with silica	Optimized sol-gel method	[129]
1989	Introduction of organic and carbon aerogels	Organic polymers	Adapted sol-gel process	[103]
1990s–2010s				
1992	Low-density, high-porosity silica aerogels	Metal oxides	Acid-base catalysis with aprotic solvent exchange	[132]
1996	Advanced silica aerogels	Metal oxides and organic polymers	Rapid supercritical extraction (RSCE)	[114]
1997	Ultralight x-aerogels	Metals and organic polymers	Crosslinking via di-isocyanates within silica matrix	[75]
2006	Cellulose-derived aerogels	Cellulose derivatives	Eco-friendly isocyanate gelation with tin catalyst	[40]
2008	Cellulose nanofiber (CNF) aerogels	Cellulose nanofibers	Vacuum-assisted drying of aqueous suspensions	[100]
2009	Metal-organic aerogels	Metals, organic polymers	Hybrid smelting of resorcinol-formaldehyde and iron oxide gels	[75]
2012	Cellulose nanowhisker-based foams	Softwood Kraft pulp	Freeze-casting technology	[32]
2014	Silica-aerogel composite plasters	Silica combined with natural plaster	Granular incorporation approach	[22]
2015	Water-responsive nanocellulose aerogels	Cellulose nanofibrils	Crosslinking maleic acid-modified CNF with hypophosphite	[69]
2016	Superhydrophobic cellulose-based aerogels	Cellulose derivatives	Physical-chemical foaming, plasma modification and silanization	[153]
2018	Cost-effective silica aerogel production	Fly ash and trona ore	Ambient pressure drying technology	[144]
2020s to recent				
2022	Ambient-pressure dried biomass aerogels	Carboxymethylated nanocellulose (CNF) & amyloid nanofibrils	Reversible-gel-assisted ambient pressure drying	[13]

(continued)

Table 9.2 (continued)

Year	Aerogel development	Source material	Fabrication technique	References
2023	Cellulose-based aerogels for air filtration	Lignin, konjac glucomannan, corn protein, nanocellulose	Sol-gel + freeze-drying	[86]
2024	Bio-based composite aerogel for cigarette smoke and airborne toxins removal	Biomass-derived biopolymers + activated carbon	Ice crystal templating/ sol-gel drying	[130]
2025	Transformation into particulate and aerogel forms for adsorption	Cellulose from Durio zibethinus peel	Homogenization, sonication, and freeze drying	[14]
2025	Nanostructured bioaerogels as air filtration	Nanocellulose in the form of nano fibers and chitosan (degree of deacetylation ≥85%)	Synthesize via a solution-gelation and freeze-drying process.	[117]

9.4 Bio-aerogels as Nanostructured Materials for Air Filtration

The intensifying degradation of global environmental quality primarily driven by industrialization, urbanization, and various anthropogenic activities has triggered an urgent demand for advanced materials capable of tackling critical environmental challenges. Among various pollutants, airborne particulate matter (PM), heavy metals, and organic contaminants remain persistent threats to both ecological systems and public health. Conventional mitigation strategies often rely on synthetic materials, many of which pose significant environmental burdens due to their high ecological footprints, limited biodegradability, and risks of secondary pollution. As a result, the development of sustainable, high-performance materials has become a central focus in environmental engineering research.

In this context, nanostructured biopolymer aerogels have emerged as promising eco-friendly alternatives. Synthesized from renewable biomass sources such as cellulose, chitosan, alginate, and gelatin, these materials offer distinct advantages, including ultra-low density, hierarchical porosity, large specific surface area, intrinsic biocompatibility, and full biodegradability. These features support superior adsorption capacities, selective pollutant removal, and integration into multifunctional environmental remediation systems. Moreover, recent advancements in fabrication techniques such as freeze-drying, supercritical drying, and additive manufacturing have enabled precise control over aerogel pore structures, surface chemistries, and mechanical robustness, allowing tailored performance for specific applications including air filtration, heavy metal sequestration, oil-water separation, and organic pollutant removal. The incorporation of hydrophobic modifications, such as those involving nanocellulose and κ-carrageenan composites, has further improved the durability and operational stability of bio-aerogels under harsh and humid conditions [117].

Among the most promising applications of these materials is in air filtration, where bio-aerogels offer a unique combination of functional customizability, low airflow resistance, sustainable lifecycle profiles, and high-efficiency particulate capture. These nanostructured materials, derived from renewable biopolymers like cellulose, chitosan, alginate, proteins, or combinations thereof, are capable of trapping ultrafine particulate matter ($PM_{2.5}$ and $PM_{0.3}$) with remarkable efficiency while maintaining minimal air resistance, which is essential for energy-efficient filtration systems. Recent studies have demonstrated impressive performance outcomes. For example, a polyaryl thioether sulfone nanofiber aerogel featuring grooved fibers and multiscale pores achieved an extraordinary 99.7% capture of $PM_{0.3}$ at a mere 17.2 Pa pressure drop, even under conditions reaching 330 °C highlighting its potential for industrial and high-temperature environments (O-GPPASS aerogel) [128]. In another example, cellulose nanofiber (CNF)/PVA/bamboo charcoal aerogels exhibited over 99.6% efficiency in filtering $PM_{2.5}$, benefitting from the synergistic interaction between the nanofibrous network and hierarchical porous structure [63].

Other composite aerogels have shown similar promise. Silica-reinforced aerogels with super-amphiphobic behavior maintained ~98.5% efficiency for $PM_{0.3}$ removal even after 50 operational cycles, demonstrating durability in high humidity, temperatures up to 250 °C, and corrosive environments [149]. Meanwhile, hybrid composites such as bacterial cellulose embedded with UiO-66-NH_2 metal-organic frameworks reached 96.9% $PM_{2.5}$ removal efficiency, coupled with regenerability and enhanced mechanical strength underscoring how the integration of inorganic nanomaterials can enhance both adsorption performance and structural resilience [30]. Collectively, these advances highlight the transformative potential of bio-aerogels in air filtration. Through nano-architectural design and hybrid functionalization, bio-aerogels are now able to meet or even surpass conventional HEPA standards, while also offering renewable sourcing, biodegradability, and adaptability to extreme operating conditions. Despite remaining challenges such as scaling up ambient drying processes, ensuring long-term mechanical stability, and establishing standardized performance metrics, bio-aerogel-based filters represent a compelling pathway toward the realization of sustainable, high-efficiency air purification technologies.

Bio-aerogels is an ultralight, highly porous materials composed of renewable biopolymers such as cellulose, chitosan, alginate, or proteins have emerged as especially promising candidates for advanced air filtration, combining exceptional filtration efficiency, low pressure drop, sustainability, and biodegradability. Notably, a novel multi-hierarchical nanofiber aerogel recently reported in [128] demonstrated an outstanding 99.7% filtration rate for $PM_{0.3}$ with a mere 17.2 Pa pressure drop that enhance both capture and airflow endurance. Complementary to this, wheat-straw-reinforced konjac glucomannan bio-aerogels achieved ~93.5% filtration of particles ≥0.3 μm with excellent breathability (~271 L s^{-1} m^{-2}) and moderate hydrophobicity, maintaining functionality in humid conditions [115]. All-protein aerogels like the Z1/G1-3 formulation achieved 99.24% $PM_{2.5}$ and 95.54% $PM_{0.3}$ filtration at a low $\Delta P \approx 55$ Pa, illustrating the synergy of biopolymer networks and nanofibrous morphology [80, 149]. Recent work on polyimide cellulose nanofiber composites shows durability under harsh conditions while preserving hierarchical porosity and performance [108]. These advances underscore how controlled pore architecture, surface functionalization, and biopolymer/inorganic hybrids (e.g., MOF-loaded cellulose aerogels) can yield filtration media rivaling or exceeding HEPA-level performance while ensuring recyclability and eco-compatibility. The primary challenges ahead remain in scaling up freeze-drying techniques, enhancing mechanical and moisture resilience through cross-linking or hydrophobic treatments, and standardizing performance benchmarks aligned with HEPA/ULPA certification to support industrial deployment and regulatory acceptance.

9.5 Bio-aerogels: Composition and Properties for Air Filtration

Bio-aerogels, engineered from natural biopolymers such as cellulose, chitosan, alginate, and proteins, possess a unique combination of ultralow density (0.01–0.5 g/cm^3), extensive hierarchical porosity (micro- to nano-scale), and exceptionally high specific surface area typically hundreds of m^2/g making them ideal nanostructured materials for advanced air filtration applications [79, 115]. The sol gel freeze-drying fabrication routes allow precise tuning of pore architectures and material composition: for instance, cellulose nanofibril (CNF)-based aerogels can be functionalized with phthalimide groups to achieve HEPA-level $PM_{2.5}$ capture efficiency (~99.95%) while retaining structural stability under humidity and temperature stress [118]. Similarly, composite aerogels such as silica-laden or super-amphiphobic cellulose hybrids maintain >98.5% $PM_{0.3}$ efficiency after multiple cycles, even under high humidity or corrosive environments, thanks to robust mechanical properties (Young's modulus ~220 kPa) and pore resilience [9, 149]. All-protein nanofiber/foam aerogels exhibit versatile pollutant trapping, effectively filtering both particulate and gaseous toxins while offering inherent biodegradability and regenerability [79]. The remarkable interplay between nanoscale pore networks, tailored surface chemistry, and biohybrid reinforcement thus enables bio-aerogels to rival conventional synthetic filters which surpassing or matching HEPA standards all while offering renewable sourcing, environmental compatibility, and potential for scalable, green fabrication. Recent studies have highlighted that the performance of bio-aerogels strongly depends on their chemical composition and functional additives, which can enhance mechanical strength, hydrophobicity, or even catalytic properties for simultaneous chemical degradation of pollutants. Table 9.3 summarizes recent (2020–2025) state-of-the-art bio-aerogel compositions used in air filtration research, highlighting their structural constituents and the corresponding application focus.

Table 9.3 highlights how tailoring the composition of bio-aerogels can optimize their performance in specific air filtration scenarios. Cellulose-based aerogels reinforced with bamboo charcoal demonstrate exceptional $PM_{2.5}$ removal efficiency

Table 9.3 Bio-aerogel compositions and air filtration applications

Composition	Functional additives	Key properties	Application in air filtration	References
Cellulose nanofiber (CNF)/poly(vinyl alcohol)	Bamboo charcoal	Hierarchical porosity, high $PM_{2.5}$ capture (>99.6%), reusable	Particulate matter ($PM_{2.5}$) filtration	[118]
Konjac glucomannan/ wheat straw fibers	None	High breathability, moderate hydrophobicity, ~93.5% $PM_{0.3}$ capture	Sustainable HVAC filters in humid conditions	[115]
Bacterial cellulose/ UiO-66-NH_2 (MOF)	MOF immobilization	Enhanced surface area, chemical adsorption, reusable	Combined $PM_{2.5}$ and VOC filtration	[30]
Protein nanofiber aerogel (e.g., zein)	None	Foam-nanofiber hybrid structure, biodegradable, $PM_{0.3}$ >95.5%	Versatile PM and gaseous pollutant capture	[80]
Polyaryl thioether sulfone/nanofiber aerogel	Grooved nanofibers	High-temperature resistance (330 °C), $PM_{0.3}$ ~99.7% at 17 pa ΔP	Industrial high-temperature flue gas filtration	[128]

with reusability, attributed to the hierarchical pore network and adsorption capability of charcoal. Konjac glucomannan-wheat straw composites prioritize breathability and moderate hydrophobicity, making them particularly suitable for humid environments where conventional filters tend to fail. The incorporation of metal-organic frameworks (MOFs) such as UiO-66-NH_2 within bacterial cellulose markedly enhances surface area and chemical adsorption sites, enabling dual removal of particulates and gaseous pollutants such as VOCs. Notably, protein-based nanofiber aerogels offer a biodegradable, eco-friendly solution while achieving >95% efficiency even for submicron particles. Finally, advanced synthetic-biohybrid aerogels like polyaryl thioether sulfone nanofibers excel in extreme industrial conditions by combining high-temperature stability with outstanding $PM_{0.3}$ capture and very low pressure drops. These findings collectively underscore the versatility of bio-aerogel platforms and demonstrate the feasibility of tailoring material chemistry to meet distinct operational and environmental requirements in air purification technologies.

Bio-aerogels as a novel class of nanostructured materials derived from renewable biopolymers, have gained considerable research attention for advanced air filtration applications due to their unique structural, functional, and environmental properties. Their ultralight density, highly porous architecture, and tunable surface chemistry make them excellent candidates for intercepting fine particulate matter and volatile organic compounds while ensuring low pressure drops and biodegradability. However, despite these advantages, several inherent limitations such as poor moisture stability, relatively low mechanical strength, and complex fabrication processes pose challenges for their large-scale implementation. A balanced assessment of these materials, considering both their benefits and drawbacks, is critical for guiding future optimization and industrial adoption. Table 9.4 summarizes the main advantages and limitations of bio-aerogels in air filtration.

Table 9.4 Advantages and limitations of bio-aerogels for air filtration

Advantages	Limitations	References
High filtration efficiency for $PM_{0.3}$-$PM_{2.5}$ (>99%) at low pressure drop	Susceptible to moisture and loss of performance in humid environments	[118]
Ultralight density and high porosity facilitate low airflow resistance	Mechanical fragility limits long-term durability under stress	[128]
Renewable, biodegradable, and environmentally sustainable materials	Fabrication is energy-intensive and time-consuming (e.g., freeze-drying)	[79]
Functionalizable surfaces allow integration of antimicrobial, hydrophobic, or catalytic properties	Lack of standardized testing and certification for industrial deployment	[115]
Potential to combine with inorganic nanomaterials (e.g., MOFs) for multifunctionality	Cost of raw materials and additives (e.g., MOFs) can be high	[30]

The data presented in Table 9.4 underscore the dual nature of bio-aerogels as advanced filtration materials: while their structural and environmental attributes position them as highly promising, significant technical and economic hurdles remain. Among their most notable strengths, the combination of high porosity and low density enables impressive particulate capture (often >99% for $PM_{0.3}$–$PM_{2.5}$) at low energy cost, a performance unattainable with conventional non-woven synthetic filters. Their biodegradability and renewable origin further align them with global sustainability goals. Additionally, the versatility of bio-aerogels allows functionalization with antimicrobial agents, hydrophobic coatings, or catalytic nanoparticles, which expands their applicability beyond particle removal to include bioaerosol inactivation and VOC adsorption. However, their intrinsic sensitivity to moisture due to the hydrophilic nature of most biopolymers that leads to diminished performance under high relative humidity, a common challenge in real-world conditions. Furthermore, the energy-intensive freeze-drying and solvent-exchange steps needed for aerogel fabrication present barriers to cost-effective scale-up, while mechanical fragility limits their durability during handling and operation. Finally, the lack of standardized industrial testing protocols for bio-aerogel filters hinders their certification and adoption in regulated markets. These limitations highlight the need for further material engineering, process innovation, and regulatory alignment.

Bio-aerogels represent a promising class of nanostructured air filtration materials that offer a unique balance of sustainability, performance, and compositional versatility. Their structure can be tailored with functional additives such as activated carbon, metal-organic frameworks (MOFs), or engineered nanofibers to enhance particulate removal efficiency, mechanical durability, and multifunctionality, addressing application-specific demands ranging from residential HVAC systems to industrial high-temperature environments. These advances, rooted in the integration of renewable biopolymers with cutting-edge nanomaterials, underscore the role of bio-aerogels as key enablers of next-generation, environmentally responsible air filtration technologies. However, fully realizing their potential will depend on addressing the technical and economic challenges that currently constrain their widespread adoption.

9.6 Types of Bio-aerogels for Air Filtration

The development of bio-aerogels as functional materials for air filtration has gained substantial momentum due to their exceptional physicochemical properties, such as high surface area, tunable porosity, low density, and biocompatibility [5]. Various sources of biopolymers and carbon-rich precursors have been utilized to fabricate aerogels with specific structural and chemical characteristics tailored for pollutant capture, antimicrobial activity, and particulate matter removal [45]. This section categorizes bio-aerogels according to their principal constituents, examining their synthesis methods, structural features, and filtration performance in airborne pollutant remediation. The four primary classes discussed include nanocellulose-based aerogels, chitosan and protein-derived aerogels, carbon-based aerogels, and hybrid aerogels containing metal or metal oxide components.

As research into sustainable filtration materials continues to expand, bio-aerogels have emerged as a leading candidate due to their versatility and compatibility with a range of bio-derived feedstocks [45]. The structural and functional diversity among different types of bio-aerogels allows for precise engineering of pore architecture, surface charge, and chemical affinity to suit specific air filtration needs [115]. In addition, recent advancements in material science have made it possible to incorporate hybrid components such as metal nanoparticles and carbon-based additives, which significantly improve the performance of bio-aerogels under a wide range of environmental conditions. Understanding the distinct properties and fabrication nuances of each class is essential for optimizing their deployment across various filtration platforms.

9.6.1 *Nanocellulose-Based Bio-aerogels*

Nanocellulose-based bio-aerogels have attracted significant attention in recent years due to their renewable origin, outstanding mechanical properties, and tunable nano-architecture [115]. These aerogels are typically produced from lignocellulosic biomass such as wood pulp, cotton, or agricultural residues, using methods like mechanical fibrillation, enzymatic treatment, or acid hydrolysis to generate cellulose nanocrystals (CNCs) or cellulose nanofibrils (CNFs) [10]. Bacterial nanocellulose (BNC), synthesized by microbial fermentation, has also been explored due to its high purity and ultrafine fiber network [101].

Production of nanocellulose aerogels begins with the preparation of stable aqueous suspensions, followed by physical or chemical gelation to form a three-dimensional network [10]. Techniques such as directional freezing (ice-templating) are commonly applied to manipulate pore alignment, resulting in aerogels with anisotropic structures favourable for air flow regulation [85]. To reinforce mechanical stability, chemical cross-linking agents like citric acid or polycarboxylic acids are often incorporated, forming covalent bonds that improve structural resilience, particularly under humid conditions [97]. Additionally, surface functionalization strategies such as carboxylation, phosphorylation, or the grafting of hydrophobic chains are commonly applied to modify the surface chemistry and meet specific performance requirements [21, 97].

Ongoing research has demonstrated that nanocellulose aerogels can be fabricated from a wide range of low-cost feedstocks, including agricultural waste streams, making them economically attractive for sustainable material development [115]. Nonetheless, scalability remains a challenge due to the energy-intensive drying processes and difficulty in handling delicate gel structures [4]. To address this, researchers are exploring the use of continuous processing methods, such as roll-to-roll freeze-casting and additive manufacturing of nanocellulose inks, which may enable industrial-scale fabrication [73]. There is also growing interest in integrating nanocellulose with conductive polymers or carbon-based materials to produce multifunctional aerogels capable of advanced filtration, sensing, or antimicrobial applications [101]. These innovations reflect the dynamic and expanding potential of nanocellulose-based aerogels in air filtration systems.

Beyond traditional aerogel forms, recent developments have focused on engineering nanocellulose into aerogel membranes and fibrous mats using electrospinning and 3D-printing-assisted deposition techniques [18]. These formats provide enhanced control over thickness, porosity gradients, and surface area-to-volume ratios, which are essential parameters for achieving tunable filtration properties. In parallel, efforts are being made to construct multilayered nanocellulose structures by sequentially combining them with other functional biopolymers or nanoscale additives [5, 115]. This approach enables the fabrication of stratified materials designed to selectively trap airborne particles, reinforce mechanical strength, and improve resistance to humidity. Moreover, studies have reported promising results in low-temperature, ambient-pressure drying methods, which aim to simplify aerogel production without compromising nano structural integrity [10]. Such innovations could significantly reduce production costs and energy inputs, advancing the feasibility of nanocellulose aerogels for widespread, scalable deployment in next-generation filtration technologies.

9.6.2 Chitosan and Protein-Derived Aerogels

Chitosan-based aerogels represent a versatile class of bio-aerogels, offering a combination of intrinsic bioactivity, high porosity, and environmental responsiveness. Derived from the deacetylation of chitin, chitosan possesses amino and hydroxyl functional groups that enable gelation through ionic interactions, pH-responsive behavior, or cross-linking reactions [1]. Its cationic nature in acidic media allows it to interact with a wide range of anionic polymers, proteins, and small molecules, which can be leveraged during the gelation phase to produce composite networks with enhanced stability. In aerogel formation, chitosan solutions are typically neutralized or ionically cross-linked, forming hydrogels that are later solvent-exchanged and dried to retain their internal structure [64]. While chitosan aerogels have been widely explored for biomedical, catalytic, and environmental remediation applications, recent efforts have also examined their suitability for air filtration [50, 72]. Studies suggest that their inherent antimicrobial properties, tunable surface chemistry, and ability to incorporate other functional additives may enable efficient capture of airborne pollutants and pathogens [73, 131]. However, compared to cellulose-based aerogels, the research in this area remains limited, and additional investigations are needed to establish their long-term performance, mechanical durability, and scalability in real-world filtration systems.

Similarly, protein-derived aerogels, synthesized from natural sources such as soy protein isolate, gelatin, casein, or albumin, have attracted interest due to their structural versatility, biocompatibility, and biodegradability [81]. These materials are typically formed through protein denaturation and gelation, induced by heating or enzymatic treatment, followed by supercritical carbon dioxide drying or freeze-drying to preserve the porous network. The resulting aerogels exhibit a broad range of porosities and surface chemistries, influenced by variables such as protein type, pH, ionic strength, and cross-linking strategy. Although applications have been extensively explored in biomedical and thermal insulation contexts, use in air filtration remains largely speculative [134, 141]. The combination of low density, tunable surface functionality, and highly interconnected pore structures suggests potential suitability for capturing airborne particulates or facilitating gas-phase interactions [79]. However, direct experimental evidence supporting effectiveness in filtration systems remains limited, highlighting the need for further research under realistic operational conditions.

Recent studies have focused on blending chitosan with other biopolymers, including gelatin, pectin, and alginate, to form hybrid aerogels with improved elasticity, reduced shrinkage, and enhanced water resistance [64, 131]. Innovations in ionic gelation, such as the use of tripolyphosphate or genipin, have enabled the formation of cross-linked networks without toxic reagents, supporting the development of environmentally friendly production routes [131]. Furthermore, co-assembly of chitosan with protein-based components has yielded aerogels with enhanced mechanical stability and functional surface chemistry, making them potential candidates for air filtration systems where selective adsorption, moisture tolerance, and mechanical resilience are essential [115].

In terms of scalability, recent efforts have focused on optimizing production techniques such as droplet-based gelation, bead formation, and electrospinning assisted gelation to fabricate chitosan- and protein-based aerogels in microscale or modular formats suitable for integration into industrial filtration systems [42, 137]. These methods facilitate uniform particle size distribution and reproducibility while allowing efficient incorporation into cartridges, membranes, or layered filter architectures. Although challenges persist, particularly regarding moisture sensitivity, mechanical fragility, and long-term stability under variable environmental conditions, advances in composite engineering and environmentally conscious cross-linking strategies are accelerating progress toward commercial deployment [42]. As formulations become more robust and production processes more efficient, these bio derived aerogels are increasingly positioned as practical options for future air filtration technologies.

9.6.3 Carbon-Based Aerogels

Carbon-based aerogels represent a prominent class of advanced porous materials, valued for its exceptional thermal stability, electrical conductivity, and chemical resistance [139]. These materials are generally synthesized through the pyrolysis of biomass-derived hydrogel precursors such as lignin, cellulose, starch, or tannin under an inert atmosphere [124, 138]. During this process, non-carbon elements are removed while the original gel network is preserved, yielding a lightweight, porous carbon matrix. The properties of the resulting carbon aerogel are influenced by several factors, including the type of precursor used, the pyrolysis temperature, and any post-treatment activation steps. Final structures can achieve large surface areas, with well-defined pore architectures that span the micro-, meso-, and macropore scales [20].

Biochar-based aerogels, synthesized from pyrolyzed agricultural residues such as rice husks, coconut shells, or sawdust, have attracted growing attention due to their low production cost, abundant feedstock availability, and environmental sustainability [56]. Typically, biochar is first produced via thermal decomposition of biomass under limited oxygen conditions and subsequently ground into fine particles. These particles are then re-dispersed in aqueous or polymeric systems to form gel precursors [20, 56]. Through physical blending, ionic cross-linking, or freeze-casting methods, the gel structure is stabilized and converted into a highly porous aerogel via freeze-drying or supercritical drying [20]. The resulting biochar-based aerogels combine the sorptive capacity of activated carbon with the structural lightness and tunability of bio-derived networks. Its high surface area, hierarchical porosity, and surface functional groups make them suitable for gas adsorption, volatile organic compound (VOC) capture, and particulate filtration in industrial settings [26].

Meanwhile, graphene-infused aerogels, are developed by incorporating graphene oxide (GO) sheets into biopolymer or nanocellulose matrices, followed by reduction to conductive graphene [45]. The dispersion of GO in aqueous gels enables uniform integration within the polymeric framework, contributing both mechanical reinforcement and electrical conductivity [23, 45]. Gelation is typically induced through physical cross-linking or chemical gelators, and after drying, the GO is reduced using agents such as ascorbic acid, hydrazine, or thermal/UV treatments [87]. These aerogels exhibit synergistic properties, such as, the graphene component imparts electrical conductivity, thermal stability, and tensile strength, while the biopolymer matrix provides flexibility and environmental compatibility [150]. Due to these combined features, graphene-infused aerogels are being explored for multifunctional air filtration applications, such as electrostatic particulate capture, antimicrobial enhancement, and integration into smart sensor-enabled filtration devices [99, 150].

Advances in fabrication have introduced new strategies to tailor the performance of carbon aerogels. Methods such as templated carbonization using silica or polystyrene spheres allow precise control over pore architecture [96]. Additionally, the use of microwave-assisted pyrolysis has shortened processing times and improved energy efficiency [7]. Besides, Recent developments also focus on heteroatom doping such as nitrogen, sulphur, or phosphorus, to modify surface chemistry and enhance adsorption capacity [38]. Despite its inherent brittleness, recent efforts have focused on enhancing mechanical durability through the use of flexible binders or dual-network cross-linking strategies [27].

Collectively, the evolution of carbon-based aerogels, from low-cost biochar variants to highly engineered graphene-infused hybrids, underscores the growing potential in next-generation air purification technologies. By tuning surface chemistry, porosity, and structural resilience, these materials can be customized for a range of functions, including pollutant adsorption, antimicrobial activity, and electrostatic filtration. As research continues to align fabrication strategies with scalable, eco-friendly processes, carbon aerogels are poised to bridge the gap between high-performance materials and practical, sustainable air filtration solutions.

9.6.4 Hybrid Aerogels Incorporating Metal or Metal Oxides

Hybrid bio-aerogels that integrate metal or metal oxide nanoparticles represent a rapidly growing area of research, aimed at creating multifunctional materials with enhanced structural and functional performance [85]. In these systems, metals such as silver, copper, or gold and metal oxides like titanium dioxide (TiO_2), zinc oxide (ZnO), and iron oxide (Fe_3O_4) are embedded within the porous matrix of biopolymer-derived aerogels. The incorporation is typically achieved through in situ synthesis, sol-gel deposition, or nanoparticle impregnation during or after gel formation [45, 97]. These hybrid structures benefit from the large surface area and interconnected porosity of the aerogel scaffold, combined with the catalytic, antimicrobial, or magnetic properties of the inorganic component [45].

Achieving a uniform dispersion of metal or metal oxide nanoparticles throughout the aerogel matrix is a central challenge in hybrid aerogel synthesis [59]. To overcome agglomeration and promote even distribution, surface functionalization of the biopolymer with carboxyl, hydroxyl, or amine groups is often applied [59, 157]. These functional groups serve as anchoring sites for metal precursors, facilitating their nucleation and stabilization within the gel network. The authors further discussed that the use of chelating agents or stabilizing ligands can further enhance the compatibility between organic and inorganic phases, resulting in more homogeneous hybrid structures.

Recent research has introduced green synthesis approaches that eliminate the need for toxic solvents or harsh reaction conditions [58]. For instance, plant extract-mediated reduction of metal salts is being used to produce nanoparticles in situ within chitosan or cellulose gels, offering a sustainable route to hybridization [123]. In other cases, metal alkoxides are hydrolysed within the polymer gel during sol-gel processing, simultaneously forming a metal oxide phase and reinforcing the biopolymer structure [157]. These methods enable the creation of monolithic or bead-shaped hybrid aerogels with tailored properties for specific performance goals.

On the technological side, hybrid aerogels present unique opportunities and hurdles for scale-up [115]. The complexity of integrating metal phases with biodegradable matrices requires tight process control to ensure reproducibility and safety [45,

157]. To address this, modular and continuous processing platforms are being developed, where gelation, nanoparticle formation, and drying occur in sequence under standardized conditions. Furthermore, advances in characterization techniques such as synchrotron-based tomography and in situ spectroscopy are offering deeper insights into the structure of hybrid interfaces and their performance behavior [133]. While commercial adoption is still in its early stages, hybrid aerogels are poised to play a central role in next-generation air filtration systems that demand multifunctional capabilities and sustainable performance.

9.7 Synthesis and Fabrication Techniques

The performance and structural integrity of bio-aerogels are fundamentally influenced by the synthesis and fabrication techniques used during their development. From sol-gel transitions to final drying methods, each processing step governs the aerogel's porosity, mechanical strength, surface functionality, and overall stability. Therefore, understanding the underlying mechanisms and process variables associated with each technique is essential not only for optimizing material properties but also for ensuring reproducibility and suitability across different biopolymer sources. Moreover, as aerogels move toward application in air filtration and beyond, selecting appropriate fabrication routes becomes crucial in balancing functional demands, production cost, and environmental impact. The fundamental workflow for the preparation of biopolymer-based aerogels is schematically illustrated in Fig. 9.1 (adapted from [115]).

This section explores the key processing approaches used in the fabrication of bio-aerogels, including the sol-gel formation process, various drying strategies such as freeze-drying and supercritical drying, and post-synthesis functionalization techniques that enhance aerogel performance. Additionally, emphasis is placed on green and scalable production strategies that support the sustainable manufacturing of bio-aerogels on an industrial scale. By examining the advantages and limitations of each method, this section aims to provide a comprehensive understanding of how processing decisions influence material design and scalability, ultimately forming the foundation for the practical implementation of bio-aerogels in advanced filtration systems and related applications.

9.7.1 Sol-Gel Process and Aerogel Formation

The sol-gel process represents a foundational technique in the synthesis of aerogels and is particularly well-suited for forming homogenous, highly porous networks from both inorganic and organic precursors. This method involves the transition of a colloidal suspension ("sol") into a continuous solid network ("gel") through a series of hydrolysis and polycondensation reactions [35]. In the context of bio-aerogels, the sol-gel method has been adapted to integrate biopolymers such as

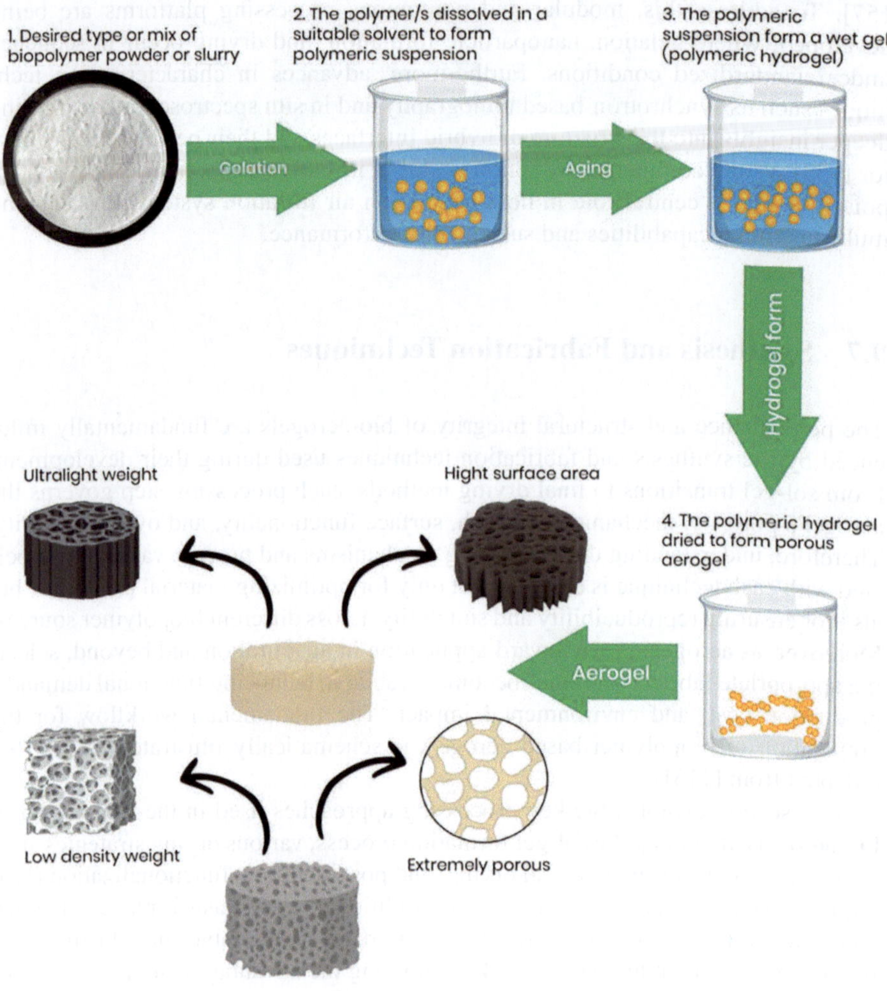

Fig. 9.1 Illustration of the basic principle of nanostructured bioaerogel fabrication. (Adapted from Saleh et al. [115])

cellulose, chitosan, gelatin, or lignin into three-dimensional networks with precisely controlled micro- and nano-scale porosity. The flexibility of this technique lies in its ability to operate at relatively mild temperatures and under aqueous or solvent-based systems, which is critical when working with temperature-sensitive biomaterials.

The formation of the gel network is typically initiated by altering pH, introducing cross-linking agents, or applying thermal or ionic triggers that facilitate polymer chain entanglement and solidification. In bio-aerogel systems, biopolymers are often blended with functional precursors like silanes (e.g., tetraethyl orthosilicate, TEOS) or metal alkoxides to generate hybrid networks [44]. These networks can be physically entangled or chemically cross-linked, depending on the desired material stiffness, porosity, or compatibility with subsequent drying and post-processing steps. For instance, TEOS-derived silica networks can be co-gelled with nanocellulose to enhance mechanical integrity and shape retention. This method also allows for fine control over the gelation kinetics, which directly impacts the final pore size distribution and network uniformity [120]. One of the major advantages of the sol-gel technique is its molecular-level mixing capability, enabling uniform dispersion of precursors and additives throughout the matrix. This feature is particularly valuable when fabricating hybrid or nanoparticle-embedded aerogels (as discussed further in Sect. 9.7.3), where the homogeneity of the final product is crucial for functional performance. Moreover, the sol-gel route enables the formation of aerogels with adjustable porosity, spanning from microporous to mesoporous and even macroporous structures, by modifying factors such as precursor concentration, catalyst strength, and aging conditions. However, one of the key limitations of this method is the relatively long gelation and aging times, often requiring hours to days depending on the system. Additionally, the gel networks formed via sol-gel processing are often fragile and prone to shrinkage during solvent removal, necessitating careful selection of drying techniques to preserve structural integrity.

Compared to alternative approaches such as freeze-drying or emulsion-templating, the sol-gel method provides superior control over the interconnectivity and pore architecture, but may be less suitable for raw materials with high crystallinity or low solubility, such as untreated lignocellulosic fibers [106]. In such cases, partial hydrolysis or functionalization of the biopolymer may be required prior to sol-gel processing. In contrast, materials such as gelatin, alginate, or starch, which naturally form gels in aqueous environments, are highly suited for sol-gel processing techniques. Additionally, the sol-gel method is highly adaptable for composite aerogels, where two or more polymers are co-gelled to achieve synergistic properties, such as combining chitosan's bioactivity with the mechanical strength of cellulose. While traditionally seen as more complex and time-consuming than physical gelation methods, the sol-gel process is increasingly being optimized for scalability, particularly through strategies like ambient drying, solvent exchange, and microwave-assisted gelation. Emerging efforts also focus on reducing environmental impact by replacing hazardous solvents (e.g., alcohols) with water or green solvents and by minimizing energy inputs. As a synthesis route, sol-gel processing remains one of the most versatile and tunable techniques for bio-aerogel fabrication, especially where precise control of structure-function relationships is critical for high-performance applications in filtration, sensing, or catalysis.

9.7.2 *Freeze-Drying and Supercritical Drying Methods*

Freeze-drying (also known as lyophilization) and supercritical drying are among the most widely used drying techniques in bio-aerogel fabrication, playing a crucial role in preserving the delicate porous network formed during the gelation stage. These methods differ in mechanism, energy input, scalability, and material compatibility, yet both aim to eliminate the solvent from the wet gel while minimizing structural collapse due to capillary forces [136]. Freeze-drying operates by first freezing the gel, typically at temperatures ranging from −20 °C to −80 °C, followed by sublimation of the frozen solvent under vacuum. This avoids the liquid-vapor phase transition that typically causes pore shrinkage in conventional drying. As water sublimes directly from solid to vapor, the aerogel retains much of its original porosity and structural integrity. The size of the ice crystals, which is controlled by the freezing rate, plays a crucial role in shaping the pore architecture. Rapid freezing leads to the formation of finer pores, whereas slower freezing generates larger and more open structures. This technique is especially compatible with water-based gels such as those derived from nanocellulose, gelatin, alginate, or starch, which naturally gel in aqueous systems. Additionally, freeze-drying is relatively safe, does not require toxic solvents or extreme pressures, and is easily adaptable for lab- and pilot-scale production.

Drying techniques profoundly influence the structure and performance of aerogels derived from biopolymer systems. Freeze-drying (lyophilization) is widely used for its relatively low cost, safety, and compatibility with biopolymers. Yet, it is energy-intensive, lengthy, and sensitive to gel thickness and composition. Thicker gels often experience poor heat transfer, leading to inhomogeneous drying, internal pore collapse, or microcrack formation due to the sublimation process and directional ice crystal growth [60]. These structural artifacts can compromise isotropic mechanical properties and transparency, making freeze-drying less suitable for optically demanding or thermosensitive applications. Nonetheless, it remains a preferred method when cost, safety, and equipment access are primary concerns [29]. By contrast, supercritical drying preserves the integrity of gel networks by replacing the solvent with a supercritical fluid, typically CO_2 and depressurizing above its critical point (31.1 °C, 7.38 MPa). This process avoids the liquid-gas phase boundary and the associated surface tension forces, thereby preventing pore collapse and shrinkage. Aerogels produced in this manner regularly exhibit ultrahigh porosity (>90%), low densities (0.003–0.2 g/cm^3), and large specific surface areas (>200 m^2/g) [67]. These characteristics make supercritical drying the method of choice for producing high-performance aerogels, particularly from hybrid sol-gel systems combining inorganic and biopolymer components. Despite its advantages, supercritical drying faces significant scalability and safety challenges. It requires specialized high-pressure equipment, rigorous solvent exchange protocols, and close control of temperature and pressure ramps, which collectively increase production cost and complexity. In addition, many biopolymer gels are naturally water-based and must undergo solvent exchange, typically to ethanol or acetone, before drying with CO_2. This added step can introduce complexities that potentially affect both biocompatibility and environmental safety. As a result, while supercritical drying excels in preserving fine nanostructures and minimizing shrinkage, it is less attractive for large-scale or eco-friendly production unless paired with green solvent systems or continuous-flow reactors.

In comparing the two, freeze-drying offers greater accessibility and is generally more suited for softer gels or highly hydrated systems with less demand for absolute pore fidelity. It accommodates a wider variety of biopolymer gels, including those with fragile or thermally sensitive components. Supercritical drying, by contrast, is often the preferred method when maintaining dimensional stability, achieving uniform pore size, and ensuring structural strength are critical, especially in hybrid or aerogels with high inorganic content. Both techniques play essential roles in the production of bio-aerogels, and the selection between them often involves weighing factors such as cost, scalability, safety, and specific material needs. Researchers are increasingly investigating hybrid drying strategies, such as combining freeze-drying with low-temperature thermal stabilization, as well as exploring alternative methods like ambient-pressure drying and microwave-assisted dehydration to harness the advantages of each approach.

9.7.3 Functionalization with Nanoparticles or Active Compounds

Post-synthesis functionalization of bio-aerogels with nanoparticles or active compounds represents a powerful strategy to enhance their functional performance and application specificity. While bio-aerogels inherently possess desirable properties such as high porosity, low density, and large surface area, their utility can be significantly expanded through the incorporation of antimicrobial agents, photothermal materials, catalytic nanoparticles, or adsorptive species. These functional additives alter the surface chemistry, electrical charge, and chemical reactivity of the aerogel matrix, enabling tailored interactions with target contaminants or stimuli. In air filtration applications, for example, passive mechanical sieving alone may be insufficient to capture ultrafine particulate matter, volatile organic compounds (VOCs), or airborne pathogens. Functionalization allows bio-aerogels to actively neutralize, adsorb, or degrade such pollutants through mechanisms including electrostatic interactions, chemical adsorption, catalysis, or photothermal inactivation. As a result, functionalized bio-aerogels offer a multifaceted approach to advanced environmental and biomedical applications, combining structural performance with selective, high-efficiency pollutant removal.

Functionalization can be performed either in situ, during the sol-gel or gelation process, or ex situ, following the formation and drying of the aerogel. In situ methods offer the advantage of better integration and dispersion of the functional agent throughout the matrix, as nanoparticles become embedded into the network before it solidifies. For instance, metal oxide nanoparticles such as TiO_2, ZnO, or Fe_3O_4 can be co-gelled with biopolymers like cellulose or chitosan to impart photocatalytic or magnetic properties. Similarly, silver or copper nanoparticles can be integrated to provide sustained antimicrobial activity [54]. Conversely, ex situ modification allows more precise control over surface loading and avoids potential interference with gelation kinetics, though it may result in less uniform distribution. Common ex situ techniques include immersion, spraying, or vapor-phase deposition of active agents onto dried aerogels, followed by thermal curing or chemical stabilization.

The choice of functionalization strategy is also influenced by the nature of the base material. For instance, cellulose-based aerogels, with their abundant hydroxyl groups, are well-suited for chemical grafting or ionic interaction with metal ions or organosilanes. Chitosan aerogels, on the other hand, offer amine groups that can chelate transition metals or interact with negatively charged compounds. Moreover, protein-derived aerogels contain carboxyl, amide, and thiol functionalities that expand the scope of surface chemistry modification [62]. Each biopolymer presents unique interaction sites that influence the efficiency, stability, and long-term activity of the functional agent. Furthermore, the porosity and surface morphology of the aerogel determine how well nanoparticles can be dispersed and retained, especially under flow or mechanical stress conditions typical in filtration systems [135].

Nevertheless, functionalization introduces several technical challenges that must be addressed to maintain the structural and functional integrity of the aerogel. Loading too much nanoparticle content can cause pore blockage, shrinkage, or mechanical brittleness, while insufficient dispersion may lead to hotspot formation or agglomeration. Moreover, chemical treatments, particularly those using strong acids, bases, or solvents, can damage the biopolymer networks or modify their surface charge, potentially affecting both biocompatibility and structural stability. To overcome these limitations, mild or green functionalization approaches are increasingly explored, such as using plant extracts for nanoparticle synthesis, employing water-based deposition, or applying enzymatic binding strategies. These not only preserve the sustainability profile of the bio-aerogels but also facilitate regulatory approval for environmental or biomedical applications.

In recent studies, multi-functional aerogels have been engineered by integrating combinations of active compounds, such as blending carbon nanomaterials with metal oxides or embedding enzymes alongside antimicrobial agents. These composite systems demonstrate synergistic effects, offering enhanced durability, reusability, and responsiveness to environmental stimuli like light, pH, or humidity. As research advances, functionalization is expected to become even more modular, enabling the design of aerogels with programmable surfaces or stimuli-responsive behavior [77]. This evolution complements the fundamental aerogel structure derived from prior synthesis steps, reinforcing the critical interplay between material composition, fabrication, and post-synthesis modification in shaping next-generation bio-aerogels for advanced applications.

9.7.4 Green and Scalable Production Approaches

As bio-aerogels transition from laboratory curiosities to practical filtration materials, the need for green and scalable production methods has become increasingly pressing. Conventional methods for fabricating aerogels, especially those that rely on organic solvents, high-energy drying processes, or intricate chemical precursors, present considerable environmental and economic challenges. To address these concerns, recent innovations have focused on replacing hazardous reagents with benign alternatives, minimizing energy input during synthesis, and drying, and adapting processes for continuous or batch-wise industrial throughput. The goal is to align aerogel manufacturing with the principles of green chemistry and circular bioeconomy, without compromising the structural or functional quality of the final product. One major advancement in this area is the adoption of water-based gelation systems, which eliminate the need for flammable or toxic solvents like ethanol, acetone, or supercritical fluids. Biopolymers such as alginate, cellulose, and starch are inherently compatible with aqueous environments, making them ideal candidates for green processing. Moreover, solvent exchange steps, which were previously required to make gels compatible with supercritical drying, can now be avoided thanks to advancements in ambient-pressure drying and low-temperature dehydration methods. These emerging methods reduce both the carbon footprint and operational cost of aerogel production while simplifying scale-up. Freeze-drying remains a widely used low-toxicity technique, but even it is being optimized with energy-efficient protocols, such as staged sublimation, pre-freezing alignment, and controlled humidity drying chambers.

Scalability is also being addressed through modular and continuous-flow systems, which allow consistent production of aerogels with minimal batch-to-batch variation. Techniques such as 3D printing, extrusion-based gel casting, and roll-to-roll coating are being repurposed to fabricate aerogel films, beads, or membranes at industrially relevant volumes. These techniques are particularly promising for air filtration applications, where consistent thickness, porosity, and surface area are critical for performance. Additionally, equipment design has evolved to support closed-loop systems for solvent recovery and recycling, thereby reducing waste and operational hazard. By integrating real-time monitoring and feedback control, such setups ensure reproducibility and reduce the margin for error in large-scale manufacturing. The suitability of green production approaches often depends on the biopolymer feedstock and the intended aerogel format. For example, nanocellulose and pectin are highly amenable to direct aqueous processing and ambient-pressure drying, whereas proteins or chitosan may require additional stabilization steps, such as ionic crosslinking or pH adjustment, to maintain their structural coherence. Furthermore, hybrid aerogels incorporating metal oxides or carbonaceous fillers can benefit from green sol-gel precursors, such as bio-derived silanes or bio-templated metal salts, which reduce toxicity and ecological burden. There is growing research interest in upcycling biomass waste, such as agricultural residues and by-products from food processing, to serve not only as sources of biopolymers but also as base materials for functional additives. This approach supports the development of fully sustainable material systems.

Ultimately, transitioning to greener and more scalable methods for aerogel fabrication represents more than just a technological improvement; it reflects a fundamental re-evaluation of the entire material lifecycle, encompassing raw material sourcing, processing, usage, and end-of-life considerations. Incorporating green chemistry principles at each step not only reduces environmental impact but also enhances commercial viability by aligning with regulatory standards, market demand for sustainable products, and the broader goals of climate resilience. As innovations continue to bridge the gap between laboratory protocols and industrial practices, green production methods are set to play a central role in enabling the mass adoption of bio-aerogels across diverse application sectors, especially those with strict environmental and safety requirements such as air purification, biomedicine, and food packaging.

9.8 Mechanisms of Air Filtration

Air filtration is a critical technology for mitigating airborne pollutants, including particulate matter, bioaerosols, and volatile organic compounds, to safeguard human health and environmental quality. The effectiveness of a filtration system fundamentally depends on the underlying mechanisms through which airborne contaminants are intercepted and retained within the filter medium. These mechanisms operate at the micro- and nanoscale, governed by particle dynamics, fluid flow behavior, and material-surface interactions. Classical filtration theory identifies interception, inertial impaction, diffusion, electrostatic attraction, and gravitational settling as the primary mechanisms responsible for particle capture, each dominating over distinct particle size ranges and flow regimes. In recent years, advances in nanostructured and functionalized materials, such as bio-aerogels, have introduced additional pathways such as enhanced surface adsorption and catalytic degradation that extend beyond conventional mechanical sieving. Understanding these mechanisms in detail is essential for rational design and optimization of high-performance, energy-efficient and environmentally sustainable filtration systems capable of meeting the increasingly stringent air quality standards of modern applications. This topic explores the scientific principles, relative contributions, and synergistic effects of the key filtration mechanisms, highlighting their relevance in the development of next-generation air filtration technologies.

9.8.1 *Filtration of Particulate Matter ($PM_{2.5}$ and PM_{10})*

The mechanisms governing the removal of airborne particulate matter ($PM_{2.5}$ and PM_{10}) are primarily categorized into mechanical and electrostatic forces, each contributing distinct yet complementary pathways to filtration efficiency. Mechanical capture is governed by size-dependent phenomena interception, inertial impaction, and diffusion which dominate for particles larger than the submicron range, while Brownian motion enhances the capture of ultrafine particles by diffusion. Interception occurs when particles following airflow streamlines come into contact with filter fibers due to their finite radius, a mechanism particularly effective for particles in the PM_{10} range [82]. Inertial impaction becomes significant when larger or denser particles deviate from the airflow due to their momentum, colliding directly with fibers, which is critical in high-velocity or turbulent flow conditions.

Diffusion, in contrast, arises from random Brownian motion and enhances the capture of $PM_{2.5}$ and smaller particles by increasing their likelihood of contacting filter surfaces. Complementing these mechanical processes, electrostatic attraction leverages the Coulombic forces between charged filter media and oppositely charged or polarizable particles, enabling enhanced capture without substantial increases in pressure drop. The integration of mechanical and electrostatic mechanisms in advanced filtration media, such as bio-aerogels and nanofiber composites, allows for highly efficient removal of $PM_{2.5}$ and PM_{10} across a broad size spectrum while maintaining low airflow resistance [43]. Table 9.5 shows the filtration mechanisms and processes for $PM_{2.5}$ and PM_{10} air filtration.

Table 9.5 presents a comparative overview of the primary mechanisms underlying particulate matter (PM) filtration, highlighting their respective operational principles, effective particle size ranges, and material innovations that enhance their performance. A critical assessment reveals that each mechanism plays a distinct yet complementary role in achieving efficient PM removal, and their relative contributions vary according to particle size, flow dynamics, and material properties. Interception is particularly effective for PM_{10} and larger particles, which maintain sufficient radius to cross streamlines and make contact with filter fibers. Recent developments in dense fibrous and hierarchically layered bio-aerogels, as demonstrated by Hamann et al. [49], have optimized interception by increasing surface area and creating tortuous pathways without substantially increasing airflow

Table 9.5 Filtration mechanisms and processes for $PM_{2.5}$ and PM_{10} air filtration

Filtration mechanism	Process description	Particle size range	Relevant material innovations	References
Interception	Particles following airflow streamlines come into contact with fibers when their radius exceeds the streamline offset	Typically PM_{10} and larger	Dense fibrous or layered bio-aerogels with hierarchical pores	[49]
Inertial impaction	Particles with sufficient mass deviate from flow due to inertia and collide with fibers, especially at high velocities	PM_{10} and coarse particles > 2.5 μm	Thick, high-strength biopolymer fiber mats	[36]
Diffusion	Ultrafine particles undergo random Brownian motion, increasing their chance of colliding with fibers	$PM_{2.5}$ and smaller (< 1 μm)	Nanofiber-reinforced aerogels with ultrafine pore networks	[61]
Electrostatic attraction	Oppositely charged or polarizable particles are drawn to charged fiber surfaces, enhancing capture	Broad spectrum (submicron to coarse)	Electrically charged cellulose or chitosan-based composites	[8]
Gravitational settling	Large particles (> 10 μm) settle onto surfaces under gravity in low-velocity regions	Coarse PM > 10 μm	Generally, not dominant in high-flow filters	[52]

resistance. However, interception alone is insufficient for submicron particles, where the probability of streamline deviation is minimal. Inertial impaction complements interception by targeting coarse particles (> 2.5 μm) with significant mass and momentum, causing them to deviate from streamlines and collide with filter fibers. Thick, high-strength biopolymer fiber mats, as reported by Dey et al. [36], have been effective in sustaining the structural integrity needed to withstand repeated impacts at high flow velocities. Nonetheless, inertial impaction becomes increasingly ineffective as particle size and mass decrease, leaving a gap in capture efficiency for $PM_{2.5}$ and below.

Diffusion driven by Brownian motion, dominates at the ultrafine scale ($PM_{2.5}$ and smaller), where random motion enhances the likelihood of particle-fiber collisions. Ji et al. [61] demonstrated that nanofiber-reinforced aerogels with ultrafine pore networks achieve high capture rates for nanoparticles, leveraging increased collision probability without compromising breathability. Yet, diffusion is inherently a low-velocity phenomenon, making it less efficient under high-flow conditions commonly encountered in industrial settings. Electrostatic attraction, as observed in the work of Aflaha et al. [8], offers the advantage of broad-spectrum particle removal by exploiting Coulombic and polarization effects. Functionalization of cellulose and chitosan composites to carry stable charges significantly enhances PM capture across a range of sizes while maintaining low pressure drops. However, this mechanism is sensitive to environmental factors, such as humidity, which can dissipate surface charges and reduce performance over time, highlighting the need for improved charge retention strategies. Finally, gravitational settling plays only a marginal role in high-velocity filtration contexts, as He et al. [52] noted, since its effect is largely limited to large particles (> 10 μm) and low-flow environments. While not a primary mechanism in engineered filter designs, it may still contribute passively in static or laminar flow zones. Critically, the interplay of these mechanisms suggests that high-performance filtration systems must be designed to maximize the synergistic effects of mechanical (interception, impaction, diffusion) and electrostatic forces, tailored to the intended application conditions and target PM size range. Material innovations, particularly those in bio-aerogel composites, have demonstrated considerable potential in achieving this balance by tuning fiber morphology, pore structure, and surface chemistry.

The evaluation of filtration mechanisms underscores the necessity of a multi-mechanism approach in air filtration material design. Mechanical processes effectively target coarse and fine particles through interception, impaction, and diffusion, while electrostatic forces provide broad-spectrum capture with minimal energy cost. Gravitational settling, though minor in engineered systems, remains relevant under specific conditions. Recent advances in bio-aerogel-based materials exemplify the potential of integrating these mechanisms into sustainable, high-efficiency filters [15]. To fully harness these benefits, future research must focus on optimizing structural parameters and surface functionality while mitigating environmental sensitivities, ensuring robust performance across diverse operational environments. Such integrated strategies are crucial for advancing next-generation air filtration technologies capable of meeting stringent air quality standards. Valuating both their strengths and weaknesses enables informed material selection and structural

Table 9.6 Advantages and limitations of filtration mechanisms for $PM_{2.5}$ and PM_{10}

Filtration mechanism	Advantages	Limitations
Interception	Effective for PM_{10} and larger particles. Following streamlines- does not significantly impede airflow	Ineffective for ultrafine particles ($PM_{2.5}$ and below) Dependent on fiber packing density
Inertial impaction	Highly efficient for coarse particles with significant momentum- robust at high flow velocities	Poor performance for fine and ultrafine particles May increase pressure drop if overdesigned
Diffusion	Dominates at $PM_{2.5}$ and smaller- enhanced by nanostructured, high-surface-area filters.	Less effective at high airflow rates- negligible impact on coarse particles.
Electrostatic attraction	Effective across a wide particle size range Low pressure drop due to non-mechanical capture.	Performance degrades in high humidity Charge stability over time remains a challenge
Gravitational settling	Passive mechanism requiring no energy- contributes to removal of very large particles	Negligible in high-velocity systems- limited to PM > 10 μm

tailoring, particularly when designing advanced sustainable filters such as bio-aerogels. Table 9.6 summarizes the principal advantages and limitations of the primary air filtration mechanisms, highlighting their relevance to particulate matter ($PM_{2.5}$ and PM_{10}) removal and providing a framework for future material innovations.

Table 9.6 provides a comprehensive evaluation of the main air filtration mechanisms, emphasizing their respective advantages and inherent limitations in the context of particulate matter ($PM_{2.5}$ and PM_{10}) removal. It is evident that each mechanism operates effectively within a defined particle size range and flow regime, underscoring the need for a multi-mechanism approach in advanced filtration material design. Interception is advantageous for removing PM_{10} and larger particles, as it relies on the direct contact of particles with filter fibers along airflow streamlines. Its minimal impact on airflow resistance makes it desirable in low-pressure systems. However, its dependence on fiber packing density and ineffectiveness against ultrafine particles limits its stand-alone utility in addressing $PM_{2.5}$ pollution. Similarly, inertial impaction demonstrates high efficiency for coarse particles and maintains robustness under high-velocity conditions. Yet, its contribution diminishes significantly for fine and ultrafine particles, and excessive reliance on this mechanism may lead to increased pressure drops if the filter structure is overdesigned. By contrast, diffusion dominates in capturing $PM_{2.5}$ and smaller particles through Brownian motion. This mechanism benefits from nanostructured, high-surface-area filter materials that increase collision probabilities. Nonetheless, diffusion becomes less effective at higher airflow rates, which can constrain its contribution in industrial applications. Electrostatic attraction, on the other hand, stands out for its ability to enhance capture efficiency across a wide particle size range without significantly increasing pressure drop. The use of charged fibers or functionalized biopolymers effectively supplements mechanical mechanisms, but electrostatic effects are often compromised under humid conditions and over extended operational times due to charge dissipation. On the other hand, gravitational settling, though energy-free and conceptually simple, has limited relevance in engineered filtration systems. Its contribution is negligible in high-velocity or turbulent airflow conditions and confined to larger, coarse particles (> 10 μm) that already pose less of a filtration challenge. These mechanisms highlights the trade-offs and synergies inherent in air filtration system design. Advanced materials, such as hierarchically structured bio-aerogels and charged nanofiber composites, exemplify the potential of integrating these mechanisms into a single medium to achieve high efficiency across a broad PM spectrum while minimizing drawbacks such as high pressure drop, poor humidity resistance, or material fragility. Moving forward, the strategic combination of these mechanisms through material innovation and structural optimization remains key to addressing stringent air quality standards in diverse application environments.

9.8.2 Adsorption of Volatile Organic Compounds (VOCs)

Volatile Organic Compounds (VOCs) represent a diverse class of carbon-based chemicals that readily evaporate at ambient conditions, contributing significantly to indoor and outdoor air pollution. Their sources range from industrial emissions, vehicular exhausts, and solvent use to natural biogenic activities. Prolonged exposure to VOCs such as benzene, toluene, formaldehyde, and xylene is associated with deleterious health effects including respiratory irritation, neurotoxicity, and carcinogenicity, as well as environmental impacts like photochemical smog and secondary organic aerosol formation. Given the increasing concern over air quality and public health, the removal of VOCs from air streams has become a critical area of research and technological development.

Among the various air purification strategies, adsorption-based filtration has emerged as a preferred approach due to its operational simplicity, high efficiency, and ability to remove VOCs at low concentrations. Adsorption, a surface phenomenon whereby VOC molecules accumulate on the surface of a solid adsorbent, offers distinct advantages over conventional thermal or catalytic oxidation, particularly in energy-constrained or low-temperature applications. Advances in materials science have led to the development of high-surface-area, porous adsorbents such as activated carbons, zeolites, and metal-organic frameworks (MOFs) which exhibit tunable surface chemistry and pore structures tailored for selective VOC capture [147]. This section elucidates the fundamental mechanisms underpinning VOC adsorption during air filtration, highlighting the physicochemical interactions involved and their dependence on both adsorbent properties and VOC characteristics. The initial stage in the adsorption of volatile organic compounds (VOCs) during air filtration is governed by the transport of these molecules from the bulk gas phase to the surface of the adsorbent. This stage is crucial because it determines the rate at which VOCs can access the available adsorption sites, and hence strongly influences the overall efficiency and dynamics of the filtration process. The transport of VOCs involves a multi-step mass transfer process, comprising convective transport, film diffusion, and intraparticle or pore diffusion, each of which is governed by distinct physical principles and system parameters [156].

In the first step of VOC adsorption within filtration systems, bulk airflow, driven by the pressure gradient of the ventilation system, transports VOC molecules toward the adsorbent surface, a process strongly correlated with airflow rate and pressure drop across the medium [88]. This convective transport sustains a high concentration of VOCs near the adsorbent, maintaining the concentration gradient that drives subsequent diffusion. However, if airflow velocity is too high, residence time near the adsorbent decreases, which can reduce adsorption efficiency. Upon nearing the adsorbent, VOC molecules encounter a stagnant gas layer, the boundary layeradjacent to the solid surface, where transport transitions from convection to diffusion [147]. Diffusion through this boundary layer is often the rate-limiting step, especially in laminar or poorly mixed flow conditions. Its thickness inversely correlates with flow turbulence and Reynolds number: more turbulent flows yield thinner boundary layers and enhanced mass transfer [68]. This stage can dominate the kinetics when the external surface area of adsorbent particles is large and interfacial mixing is limited. After VOC molecules diffuse across the external film, they penetrate into the porous structure of the adsorbent through intraparticle diffusion. This mechanism governs the migration of VOCs into the internal pore network, where the majority of the adsorption sites reside. Depending on the pore size and structure, diffusion can occur via:

- Knudsen diffusion [95], dominant in narrow micropores where molecule-wall collisions prevail over intermolecular collisions.
- Fickian diffusion, typical in larger mesopores and macropores where concentration gradients drive molecular migration [66].
- Surface diffusion, where adsorbed VOCs migrate along the internal surfaces to deeper sites [112].

The relative importance of these diffusion mechanisms depends on the pore size distribution, surface chemistry, and the molecular properties of the VOCs, including their size, polarity, and diffusivity. Several factors modulate the efficiency of VOC mass transfer to the adsorbent surface:

- Airflow velocity and pattern, which dictate convective delivery and boundary layer thickness.
- Temperature, which affects molecular kinetic energy, vapor pressure, and diffusivity.
- Relative humidity, which can compete for adsorption sites and modify the thickness of the adsorbed water layer, thereby altering effective diffusivity.

Adsorbent morphology including particle size, porosity, and surface area, which collectively determine the accessibility and distribution of adsorption sites [41]. Understanding and optimizing these parameters is fundamental in the design of high-performance air filtration systems. For instance, reducing particle size or enhancing porosity can shorten intraparticle diffusion paths, while increasing airflow turbulence can thin the boundary layer, thereby accelerating external mass transfer. Such optimizations improve the kinetics of VOC capture, particularly under dynamic conditions where breakthrough and regeneration behavior are critical. Once volatile organic compounds (VOCs) are transported through the mass transfer steps and reach the adsorbent surface, they undergo adsorption onto available active sites within the porous structure. This second stage of the adsorption process is governed by the nature of the interaction between the VOC molecules and the adsorbent, which can occur through several distinct mechanisms. The balance among these mechanisms depends on the physicochemical properties of the VOCs (e.g., polarity, molecular size, and vapor pressure), the surface chemistry and porosity of the adsorbent, and the surrounding environmental conditions.

The most fundamental mechanism of VOC capture is physisorption (physical adsorption), which arises from weak, non-specific van der Waals forces between the adsorbate molecules and the solid surface. Physisorption is a reversible process that does not involve the formation of new chemical bonds, making it particularly advantageous for applications where adsorbent regeneration is required. Its effectiveness is strongly enhanced by a high specific surface area and the presence of micropores, which provide a large number of adsorption sites and enable pore-filling effects [158]. The adsorption capacity by physisorption is also favored at lower temperatures, as the process is exothermic. For small, non-polar VOC molecules, microporous adsorbents such as activated carbons and zeolites are especially effective, as their narrow pore sizes promote capillary forces and enhance physical confinement of the molecules. In contrast, chemisorption (chemical adsorption) involves stronger and more specific interactions between the VOC molecules and reactive functional groups present on the adsorbent surface. This process is characterized by the formation of covalent or ionic bonds, often rendering the adsorption irreversible. Chemisorption is particularly relevant for polar or reactive VOCs, such as aldehydes and organic acids, which can interact with surface hydroxyl, carbonyl, or amine groups introduced through surface functionalization. While chemisorption generally provides higher selectivity and stronger binding, it limits the regenerability of the adsorbent and can reduce its usable lifetime. The deliberate modification of adsorbent surfaces with targeted functional groups has therefore become a key strategy in designing filters tailored to specific classes of VOCs.

A third, intermediate mechanism of VOC capture is capillary condensation, which occurs in mesoporous materials where pore diameters are comparable to the critical radius of the adsorbed vapor. Under conditions of elevated relative humidity, VOC vapors can condense into liquid-like phases within the mesopores, significantly increasing adsorption capacity. This mechanism is highly dependent on the pore size distribution and surface wettability of the adsorbent. While capillary condensation can enhance performance in some situations, it also poses challenges when water vapor competes for the same pore space, potentially displacing or blocking VOC adsorption [28]. The adsorption process proceeds until dynamic equilibrium is achieved between the concentration of VOCs in the gas phase and the amount adsorbed on the surface. At this point, the adsorbent is considered saturated and must be regenerated or replaced. The equilibrium relationship between adsorbed and free-phase VOCs can be described using models such as the Langmuir isotherm, which assumes monolayer adsorption on a homogeneous surface, or the Freundlich isotherm, which accounts for surface heterogeneity and multilayer adsorption. Similarly, the kinetics of adsorption, including the rate at which breakthrough occurs during dynamic filtration, are often modeled to predict filter lifetime and efficiency under varying operational conditions.

It is important to recognize that environmental factors, particularly temperature and relative humidity, exert significant influence over the adsorption process. Elevated temperatures can reduce adsorption capacity by increasing desorption rates, while high humidity levels can lead to competitive adsorption of water vapor on active sites, especially in hydrophilic adsorbents. Therefore, the development of advanced adsorbents with hydrophobic yet VOC-selective surfaces has emerged as a major research frontier. Such materials aim to mitigate moisture interference while maintaining strong affinity for organic vapors, thereby enabling robust and efficient VOC removal even under challenging environmental conditions. In summary, the interplay of physisorption, chemisorption, and capillary condensation governs the surface-level capture of VOCs, and understanding these mechanisms is essential for the rational design of high-performance adsorbents. The ongoing challenge lies in balancing adsorption capacity, selectivity, reversibility and resilience to environmental variability which continues to drive innovation in material science and air filtration technologies. Table 9.7 shows the adsorption mechanisms of volatile organic compounds (VOCs) in air filtration systems and Table 9.8 illustrated the advantages and disadvantages of VOC adsorption mechanisms in air filtration.

The adsorption of VOCs in air filtration systems is a complex but highly effective strategy that relies on a nuanced interplay of mass transport and surface interaction phenomena. A thorough understanding of the mechanisms physisorption, chemisorption, and capillary condensation enables the design of adsorbents with optimized pore structures, surface chemistries, and hydrophobic/hydrophilic balances to achieve efficient VOC capture under dynamic environmental conditions. While each mechanism presents unique strengths and trade-offs, integrating these principles through hybrid materials and advanced functionalization remains a promising direction for developing next-generation air filtration technologies. Future research should continue to focus on improving selectivity, capacity and regenerability while minimizing sensitivity to humidity and temperature fluctuations.

Table 9.7 Adsorption mechanisms of volatile organic compounds (VOCs) in air filtration systems

Mechanism	Method description	References
Physisorption	Removal of toluene and benzene using activated carbon via van der Waals forces and micropore filling. High surface area materials enhance adsorption at ambient temperature	[158]
Chemisorption	Formaldehyde removal using amine-functionalized mesoporous silica, were reactive amine groups chemically bond to aldehyde molecules irreversibly	[47]
Capillary condensation	Enhanced adsorption of VOCs at elevated humidity using mesoporous silica due to condensation of VOC vapors within mesopores smaller than critical radius	[152]

Table 9.8 Advantages and disadvantages of VOC adsorption mechanisms in air filtration

Mechanism	Advantages	Disadvantages
Physisorption	Reversible process; adsorbents can often be regenerated High adsorption capacity at ambient temperature Suitable for a wide range of non-polar and weakly polar VOCs Well-established, cost-effective materials (e.g., activated carbon)	Weak binding strength, sensitive to temperature and humidity fluctuations Non-selective, may adsorb undesired species (e.g., moisture) Limited effectiveness for highly polar or reactive VOCs
Chemisorption	Strong, selective binding to specific VOCs (e.g., formaldehyde) Effective at low concentrations Less affected by humidity compared to physisorption	Irreversible or difficult regeneration Finite reactive sites limit capacity Functionalization may increase cost and complexity of adsorbent fabrication
Capillary condensation	High capacity in mesoporous materials at high relative humidity Useful for vapors with high saturation pressures Exploits pore structure to enhance adsorption	Highly sensitive to humidity, water vapor may occupy pores and reduce VOC uptake Typically, less effective at low humidity May result in pore blockage or flooding

9.8.3 Antimicrobial Properties for Biological Contaminants

Biological contaminants, including bacteria, fungi, viruses, and their spores, represent a significant class of airborne pollutants that pose serious health risks, particularly in indoor and occupational environments. Exposure to bioaerosols has been linked to respiratory infections, allergic reactions, and even outbreaks of epidemic diseases. As global awareness of indoor air quality grows further heightened by pandemics such as COVID-19. There is increasing demand for advanced air filtration systems capable of not only capturing but also inactivating biological contaminants to prevent downstream proliferation or re-aerosolization. Conventional air filtration technologies, such as mechanical filters, are primarily designed to physically remove particles through interception, impaction, diffusion, and electrostatic attraction. While effective in trapping bioaerosols, these passive approaches can allow viable microorganisms to accumulate and colonize on filter surfaces, potentially leading to secondary contamination and reduced efficacy over time. To address these limitations, the integration of antimicrobial properties into filtration materials has emerged as a critical innovation. By embedding or coating filters with antimicrobial agents such as metal nanoparticles, biocidal polymers, or photocatalytic materials filters can achieve dual functionality which high-efficiency capture of biological contaminants and subsequent inactivation or destruction of the captured pathogens.

The antimicrobial function of air filters in mitigating biological contaminants involves two interdependent processes: the initial capture of biological particles and the subsequent inactivation of their viability on or near the filter surface. In the first stage, biological aerosols ranging from submicron viruses (~0.1 μm) to larger bacterial and fungal spores (~1–10 μm) are effectively transported and immobilized within the filter medium [53]. This capture is accomplished through a synergistic combination of classical air filtration mechanisms. Interception occurs when particles following airflow streamlines come into direct contact with and adhere to the filter fibers, a process particularly effective for particles comparable in size to the fiber separation. Larger and heavier particles are primarily removed via inertial impaction, whereby their momentum prevents them from following deflected airflow paths, causing collision with the fibers. For ultrafine biological particles, diffusion becomes dominant, as their random Brownian motion increases the likelihood of contact with fibers in low-velocity regions. Additionally, electrostatic attraction can enhance collection efficiency by drawing charged or polarizable microorganisms toward charged or polarized fibers, further increasing removal rates. Collectively, these mechanisms ensure the physical entrapment of viable microorganisms within the porous structure of the filter medium. However, in the absence of integrated antimicrobial functionalities, the retained microorganisms may remain viable, retain pathogenic potential, and under certain conditions proliferate on the filter surface. This microbial persistence can result in filter fouling, reduced airflow performance, and potential secondary bioaerosol release, underscoring the importance of combining physical capture with effective inactivation strategies for comprehensive bioaerosol control.

To ensure biological safety and maintain hygienic conditions within filtration systems, antimicrobial agents are strategically incorporated into filter materials to actively neutralize the viability of captured microorganisms. The mechanisms underlying microbial inactivation vary depending on the type of antimicrobial agent employed, but all aim to disrupt essential cellular structures and functions, rendering the pathogens non-viable. One widely adopted strategy involves the disruption of microbial cell membranes and walls, a process observed with metal-based nanoparticles such as silver (Ag), copper (Cu), and zinc oxide (ZnO). These materials release metal ions and/or generate reactive oxygen species (ROS), which compromise the structural integrity of microbial envelopes and lead to leakage of intracellular contents [98]. Similarly, oxidative stress induction is a hallmark of photocatalytic filters incorporating UV-activated titanium dioxide (TiO_2), which produces ROS that inflict damage on nucleic acids, proteins, and lipid membranes. Inactivation can also proceed via interference with metabolic and genetic processes, as seen with biocidal polymers such as quaternary ammonium compounds that penetrate cells, disrupt enzyme activity, and prevent replication. Moreover, some hydrophilic coatings exert antimicrobial effects through dehydration and osmotic imbalance, extracting water from microbial cells and inducing desiccation-induced death [113]. Importantly, these inactivation processes can occur in situ, either concomitant with or shortly after bioaerosol capture, significantly reducing the viability of retained microorganisms and mitigating risks of secondary contamination. Beyond enhancing biological safety, antimicrobial functionalities also contribute to prolonged filter service life by inhibiting biofilm formation, preserving airflow efficiency, and maintaining consistent filtration performance over extended operational

Table 9.9 Antimicrobial mechanisms in air filtration systems

Mechanism	Description	References
Membrane and wall disruption	Metal-based nanoparticles (e.g., Ag, Cu, ZnO) release metal ions or generate ROS that compromise microbial envelope integrity, leading to leakage and death	[94]
Oxidative stress induction	UV-activated photocatalysts (e.g., TiO_2) generate ROS that damage microbial DNA, proteins, and lipids, rendering cells non-viable	[34]
Metabolic and genetic interference	Biocidal polymers (e.g., quaternary ammonium compounds) disrupt cellular enzymes and DNA replication, preventing microbial proliferation	[11]
Dehydration and osmotic imbalance	Hydrophilic coatings extract water from microbial cells, leading to desiccation-induced death and structural collapse	[140]

periods. Together, these mechanisms underscore the critical role of antimicrobial integration in advancing air filtration technologies for effective bioaerosol control. Table 9.9 shows the antimicrobial mechanisms from the studies.

Table 9.9 highlights the principal mechanisms through which antimicrobial air filters achieve inactivation of captured microorganisms. Each mechanism targets a distinct vulnerability in microbial physiology, thereby ensuring effective and often synergistic inactivation. Membrane and wall disruption is a widely adopted approach that leverages the ion-releasing or ROS-generating capacity of metal nanoparticles such as silver, copper, and zinc oxide. This mechanism compromises the structural integrity of microbial envelopes, causing cell lysis. Similarly, oxidative stress induction, commonly achieved through UV-activated photocatalytic surfaces like titanium dioxide, generates reactive oxygen species that inflict widespread molecular damage to DNA, proteins, and lipids.

In contrast, metabolic and genetic interference directly disables essential biochemical pathways, exemplified by the action of biocidal polymers such as quaternary ammonium compounds, which disrupt enzymatic functions and DNA replication, effectively arresting microbial proliferation. Lastly, dehydration and osmotic imbalance exploit the dependence of microorganisms on water homeostasis. Hydrophilic filter coatings extract moisture from microbial cells, leading to desiccation and cell collapse. These mechanisms underscore the multidimensional strategies employed to ensure not only the capture but also the reliable inactivation of airborne biological contaminants. The selection of a particular mechanism or combination thereof depends on factors such as the target microorganisms, operational environment, and desired balance between efficacy, durability, and safety.

9.8.4 Catalytic Degradation of Air Pollutants

Airborne pollutants, comprising volatile organic compounds (VOCs), nitrogen oxides (NO_x), carbon monoxide (CO), ozone (O_3), and other hazardous gases, pose substantial risks to environmental and human health. These contaminants contribute to photochemical smog, climate change, respiratory illnesses, and even carcinogenesis. While conventional air filtration technologies such as mechanical filters and adsorption systems effectively remove particulate matter and trap gaseous pollutants, they often suffer from limitations including saturation, desorption, and secondary emissions of the retained pollutants [33]. To overcome these drawbacks, the integration of catalytic degradation mechanisms into air filtration systems has emerged as a promising strategy for achieving active, sustainable, and long-lasting pollutant abatement.

Catalytic air filtration systems rely on the principle of heterogeneous catalysis, whereby harmful gaseous species are chemically transformed into innocuous products such as carbon dioxide and water upon contact with a catalytically active surface. This approach offers several advantages over purely physical removal methods: continuous regeneration of the catalytic surface, elimination of secondary contamination risks, and the ability to target a broad spectrum of gaseous pollutants even at low concentrations [142]. Advances in nanomaterials and surface engineering have led to the development of high-surface-area catalysts such as noble metals, metal oxides, and photocatalysts that exhibit high activity, selectivity, and durability under ambient conditions. This section elucidates the fundamental processes and mechanisms underpinning catalytic degradation in air filtration systems, emphasizing the transport, adsorption, and reaction steps that govern overall performance.

The catalytic degradation of air pollutants in filtration systems proceeds through a series of interrelated steps, beginning with the transport of pollutants to the catalyst surface. In this initial stage, gaseous contaminants are conveyed from the bulk air stream to the catalyst interface through convective flow, diffusion across a stagnant boundary layer, and, in porous catalysts, intraparticle pore diffusion. The efficiency of transport is influenced by operational parameters such as airflow velocity, temperature, and the morphological characteristics of the catalyst material, including porosity and surface area. Once the pollutants arrive at the surface, they undergo adsorption and activation, a critical prerequisite for catalytic transformation. Adsorption can occur via physisorption or chemisorption, with chemisorption being more consequential for catalytic activity as it facilitates the electronic and structural activation of the reactant molecules. Simultaneously, molecular oxygen and in some cases, water vapor adsorb and dissociate on the catalytic surface, generating reactive oxygen species that drive subsequent oxidation reactions [2]. The core of the process, surface catalytic reactions, involves a sequence of elementary steps wherein adsorbed pollutants react with oxygen species or photocatalytically generated radicals to form benign products. For example, volatile organic compounds (VOCs) and carbon monoxide (CO) are oxidized into carbon dioxide and water, while nitrogen oxides (NO_x) are either reduced to nitrogen or oxidized to less harmful forms depending on the catalyst and environmental conditions.

On photocatalysts such as titanium dioxide (TiO_2), incident light promotes electron-hole pair generation, which reacts with oxygen and water to produce highly reactive radicals (OH, O_2^-) capable of mineralizing organic contaminants. These reactions are highly sensitive to the catalyst's composition, crystallinity, and the presence of promoters or dopants designed to enhance reactivity and stability. The final step, desorption of products, ensures the release of reaction products such as CO_2, H_2O, and N_2 back into the air stream, thereby regenerating active sites for continued catalytic cycles [12]. Maintaining efficient desorption is essential for sustained performance, as the accumulation of intermediates or byproducts can lead to catalyst poisoning and deactivation. This underscores the importance of advanced catalyst design and, when necessary, periodic regeneration to preserve long-term functionality. Together, these interconnected processes enable catalytic filtration systems to transform hazardous air pollutants into harmless end products in a continuous and energy-efficient manner. Table 9.10 shows the mechanisms of catalytic degradation of air pollutants in air filtration systems.

Catalytic degradation represents a transformative advancement in air filtration technology, addressing the limitations inherent in purely physical or adsorption-based approaches. As summarized in the table, multiple catalytic mechanisms are employed, each tailored to target specific classes of gaseous pollutants under diverse operational conditions. Oxidation on metal oxide catalysts exploits redox cycles and surface oxygen vacancies to efficiently oxidize VOCs and CO into harmless CO_2 and H_2O, providing a robust and thermally stable solution widely implemented in industrial settings. Photocatalytic oxidation, using UV-activated semiconductors such as TiO_2, expands this capability by leveraging light energy to generate reactive oxygen species (ROS) that mineralize organic contaminants. This approach has gained particular prominence for indoor and ambient-temperature applications due to its ability to operate effectively without external heating [31].

In the case of nitrogen oxides (NO_x), catalytic reduction and oxidation processes facilitated by noble metals and supported oxides exploit oxygen spillover phenomena to convert these highly toxic and reactive species into benign nitrogen or less harmful intermediates. The synergistic integration of plasma and catalysis further exemplifies innovation in this field, wherein non-thermal plasma produces highly reactive radicals and ozone that pre-activate pollutants for subsequent catalytic oxidation [71]. This hybrid approach improves overall efficiency, especially at low temperatures and low pollutant concentrations, while minimizing the energy footprint. The choice of catalytic mechanism and material is governed by factors including the target pollutant species, required selectivity, operating temperature and humidity, durability, and cost considerations. Although significant progress has been made in developing high-activity, high-surface-area catalysts with improved resistance to deactivation and poisoning, challenges remain in scaling these

Table 9.10 Mechanisms of catalytic degradation of air pollutants in air filtration systems

Mechanism	Description	References
Oxidation on metal oxides	Pollutants such as VOCs and CO are oxidized to CO_2 and H_2O on metal oxide surfaces (e.g., MnO_2, CeO_2, CuO) through redox cycles involving surface oxygen vacancies	[126]
Photocatalytic oxidation	UV-activated TiO_2 or doped photocatalysts generate electron-hole pairs that produce ROS (OH, O_2^-), mineralizing organic pollutants into CO_2 and H_2O	[111]
Oxygen spillover and reduction of NO_x	NO_x molecules adsorb and are reduced to N_2 or oxidized to less harmful NO_2 through oxygen spillover on catalysts such as Pt, Rh, or supported metal oxides.	[127]
Plasma-catalyst synergistic oxidation	Non-thermal plasma generates reactive species (e.g., O_3, radicals) that enhance catalytic oxidation of pollutants on downstream catalytic surfaces	[92]

technologies while maintaining long-term stability and avoiding secondary emissions. Notably, the sensitivity of some catalytic processes to ambient humidity and the potential generation of intermediate by-products highlights the need for further research into optimizing catalytic materials and operational strategies.

In conclusion, catalytic degradation mechanisms offer a paradigm shift in air filtration by not merely trapping but actively transforming airborne pollutants into innocuous products through surface-mediated chemical reactions. The sequence of pollutant transport, adsorption and activation, surface reactions, and product desorption underscores the intricacy of catalytic processes and their dependence on material design and environmental conditions. Among the various mechanisms, oxidation on metal oxides and photocatalytic oxidation stand out for their effectiveness and versatility, while innovations such as plasma-catalyst hybrid systems continue to push the boundaries of performance. The integration of catalytically active materials into filtration media combines high-efficiency capture of particulate and gaseous pollutants with sustainable, regenerable, and low-maintenance operation. Looking forward, continued research is needed to address challenges related to catalyst deactivation, selectivity, and energy efficiency, with an emphasis on developing cost-effective, scalable solutions that maintain performance under real-world conditions. Catalytic air filtration thus holds considerable promise as a cornerstone technology for improving air quality and safeguarding human and environmental health.

9.9 Applications of Bio-aerogels in Air Filtration

The distinctive physicochemical characteristics of bio-aerogels, including their extremely low density, high porosity, and adjustable surface functionality, position them as highly promising materials for air filtration applications. Unlike conventional synthetic filters, which often rely on petroleum-derived polymers and involve energy-intensive manufacturing processes, bio-aerogels are typically derived from renewable biopolymers and can be fabricated under milder, more environmentally friendly conditions [57, 67]. Their inherently porous structure enables efficient particle capture, while their large surface area supports adsorption of gaseous pollutants, volatile organic compounds (VOCs), and even pathogens when functionalized appropriately. This versatility allows bio-aerogels to serve as sustainable alternatives across a wide range of filtration contexts, from industrial emission control to personal protective equipment.

This section explores the practical applications of bio-aerogels in air filtration, organized across key domains where air quality management is critical. Each subtopic evaluates the suitability of bio-aerogels in replacing or complementing existing filtration technologies, considering factors such as material performance, biodegradability, production scalability, and regulatory viability. While the fundamental filtration mechanisms are rooted in physical entrapment and surface interaction, the actual deployment of bio-aerogels varies widely depending on the operating environment, pollutant type, and functional requirements. As such, this section outlines not only the current state of bio-aerogel use in each sector but also their future potential and associated challenges.

9.9.1 Industrial Air Filtration Systems

Industrial air filtration systems operate under demanding conditions, often requiring high thermal stability, chemical resistance, and long service lifespans to handle large volumes of contaminated air containing particulate matter, aerosols, and corrosive gases. Bio-aerogels, particularly those reinforced with nanocellulose, chitosan, or hybrid carbon-based components, have shown strong potential to meet these criteria while offering a more sustainable alternative to conventional fiberglass or synthetic polymer filters. Their ultrafine pore networks enable the efficient capture of PM 2.5 and PM 10 particles, while surface functionalization with metal oxides or activated carbon can enhance the adsorption of hazardous gaseous pollutants such as sulphur dioxide (SO_2), nitrogen oxides (NOx), and VOCs [149]. One of the major advantages of bio-aerogels in this sector is their lightweight and flexible nature, which facilitates integration into both rigid and modular filtration units without adding significant load to mechanical systems. Additionally, the customization of pore size, surface chemistry, and layer configuration enables the development of multifunctional filters capable of addressing diverse pollutant profiles. In contrast to synthetic filters that often deteriorate under extreme pH or temperature conditions, properly modified bio-aerogels, especially those enhanced with thermally stable fillers such as graphene or silica, can be designed to resist degradation. This improved durability extends their operational lifespan and reduces the frequency of replacement.

Nevertheless, the use of bio-aerogels in industrial settings is not without limitations. One key challenge is their mechanical fragility under high-flow or high-pressure conditions, which can compromise structural integrity and filtration performance. To address this, ongoing research focuses on creating composite aerogels with embedded fibrous scaffolds or crosslinked matrices to improve mechanical strength without sacrificing porosity. Moreover, although many bio-aerogels perform exceptionally well under controlled laboratory conditions, their long-term behavior in industrial settings, where they may be exposed to oils, reactive gases, and varying humidity levels, is still not well understood. These uncertainties currently limit their widespread adoption in critical infrastructure, though pilot studies and niche deployments have shown encouraging results. In terms of sustainability, bio-aerogels offer a distinct advantage over synthetic alternatives due to their origin from renewable or waste-derived biomaterials and their potential for biodegradability at the end of life. Their production can be aligned with green chemistry principles, particularly when using water-based systems and ambient drying methods [15]. As a result, bio-aerogels could support the shift toward greener industrial operations, especially in sectors aiming to reduce environmental impact and meet stricter air quality regulations. Potential sectors for deployment include cement and steel manufacturing, chemical processing, and waste incineration plants. These industries are facing growing pressure to reduce airborne emissions while maintaining production efficiency.

Looking ahead, the commercialization of bio-aerogels for industrial air filtration will depend on continued innovation in material robustness, cost-effective fabrication, and standardization for safety and regulatory compliance. Collaborative efforts between academia, material engineers, and industrial stakeholders will be crucial in scaling production, validating long-term performance, and developing modular designs tailored to specific industrial emissions. With these advancements, bio-aerogels could play a transformative role in modernizing industrial air filtration systems for a more sustainable and cleaner future.

9.9.2 Indoor Air Quality Improvement (HVAC Systems)

Indoor air quality (IAQ) has become a growing public health concern, especially in urban environments where people spend the majority of their time indoors. Heating, ventilation, and air conditioning (HVAC) systems are essential for maintaining comfortable and healthy indoor air, yet they also serve as potential vectors for circulating pollutants, allergens, and microorganisms. Bio-aerogels offer a novel solution for enhancing IAQ through integration into HVAC filters and standalone air purification units. Their ability to trap fine particulates, adsorb gaseous contaminants, and support antimicrobial activity makes them suitable candidates for next-generation indoor filtration technologies [115]. Unlike conventional HVAC filters, which typically rely on multilayer synthetic meshes or activated carbon inserts, bio-aerogels can be engineered with tailored porosity and surface functionality to optimize both airflow and pollutant capture. Their low density and high surface area make it possible to develop filters that operate efficiently without creating significant pressure drops, which is a critical factor in HVAC system design for maintaining energy efficiency. Furthermore, because HVAC filters are replaced regularly, using biodegradable or compostable materials such as cellulose, chitosan, or protein-based aerogels could significantly reduce plastic waste and environmental burden associated with frequent filter disposal.

In terms of performance, bio-aerogels demonstrate strong potential in removing fine particulate matter (PM 1, PM 2.5), formaldehyde, ammonia, and other volatile organic compounds that accumulate indoors from building materials, furniture, or human activity [79]. When modified with photocatalytic nanoparticles or antimicrobial agents, these materials can help suppress microbial growth and lower the concentration of airborne pathogens [110]. This functionality has become especially important in response to global health crises such as the COVID-19 pandemic. These functionalizations can be achieved using mild, green chemical treatments, making them safe for indoor use and compliant with health and safety standards.

However, the adoption of bio-aerogels in HVAC systems is currently limited by durability concerns, especially under fluctuating humidity and temperature conditions. While some aerogels are moisture-sensitive and may swell or degrade over time, this issue can be mitigated through surface hydrophobization or incorporation of moisture-stable components such as silica or carbon nanomaterials. Additionally, large-scale production and integration into existing HVAC infrastructure remain technical challenges, particularly in achieving uniform thickness and filter formats that fit standardized housing units. Nonetheless, modular and layered configurations, such as bio-aerogel mats or coatings applied to mesh substrates, are being actively explored as scalable and adaptable formats. Bio-aerogels represent a green and forward-looking alternative for IAQ improvement in homes, schools, hospitals, and office buildings. Their compatibility with sustainable fabrication processes, coupled with the potential for localized, small-scale manufacturing from regional biomass sources, aligns with broader goals of environmental sustainability and indoor health optimization. As research progresses and performance data under real-life HVAC conditions accumulate, bio-aerogels could redefine the standard for high-performance, eco-friendly air filtration in indoor environments.

9.9.3 Personal Protective Equipment (Face Masks and Filters)

The use of personal protective equipment (PPE), particularly face masks and respirators, has become increasingly widespread in both healthcare and public settings due to heightened awareness of airborne diseases, pollution, and occupational hazards. Bio-aerogels present a compelling alternative to traditional synthetic materials used in mask filters, offering lightweight, breathable, and highly porous structures that can efficiently capture airborne particles and pathogens without causing excessive airflow resistance. Their application in PPE not only meets filtration performance standards but also introduces the possibility of eco-friendly, biodegradable, and reusable protective gear.

Conventional mask filters often rely on melt-blown polypropylene layers that are non-biodegradable and energy-intensive to manufacture. In contrast, bio-aerogels derived from cellulose, chitosan, and protein-based biopolymers can be fabricated using water-based, solvent-free methods, significantly reducing environmental impact. These bio-based filters can be engineered with micro- to nano-scale porosity suitable for capturing fine particulate matter (PM 2.5 and smaller), bacterial aerosols, and viruses [72]. Moreover, they naturally support the integration of functional coatings, including silver nanoparticles, essential oils, or plant-based antimicrobial agents, which impart antiviral or antibacterial properties to the surface and further strengthen their protective capabilities.

An advantage of bio-aerogels in PPE is their breathability. Due to their open porous structure and ultralight weight, they offer high air permeability while maintaining efficient particulate filtration [115]. This makes them particularly suitable for prolonged wear, minimizing heat build-up and respiratory discomfort often reported with traditional synthetic masks [37]. Furthermore, some bio-aerogels possess moisture-regulating properties that help prevent condensation and inhibit microbial growth inside the mask, thereby enhancing both hygiene and user comfort. Despite these promising features, several limitations must be addressed before large-scale adoption of bio-aerogel-based PPE can be realized. The mechanical fragility of some aerogels poses a challenge for handling, packaging, and repeated use. Reinforcement through layered composites, nonwoven fabric backing, or crosslinking agents is being explored to overcome this issue. Additionally, the cost and complexity of current production methods remain a barrier to mass manufacturing, especially when compared to the low-cost extrusion processes used for synthetic filters [61]. Nevertheless, as fabrication techniques evolve and materials are optimized for mechanical resilience, the feasibility of producing disposable or semi-reusable bio-aerogel filters at scale continues to improve.

The potential impact of bio-aerogels in this sector spans various fields from medical-grade respirators and surgical masks to consumer-grade protective masks for urban dwellers, workers in dusty environments, and individuals in high-pollution zones. The compatibility with sustainability objectives, especially through the use of compostable or biodegradable mask components, makes bio-aerogels an appealing option in settings where environmental regulations are becoming more stringent and public demand for eco-friendly solutions is increasing. With continued research and investment, bio-aerogels could play a transformative role in developing the next generation of high-performance, eco-conscious PPE.

9.9.4 Automotive and Transportation Air Filtration

Air quality in automotive and transportation environments is an emerging health and engineering concern, particularly in densely populated urban regions where vehicles are both contributors to and victims of airborne pollution. In-vehicle air filtration systems are critical for maintaining cabin air quality and protecting occupants from harmful pollutants such as fine dust (PM 2.5), exhaust-derived VOCs, and allergens [44]. Bio-aerogels are gaining attention as next-generation filtration materials in this domain due to their superior porosity, tunable microstructure, and potential for sustainable production. Their use aligns with the automotive industry's broader transition toward greener, lighter, and more energy-efficient technologies.

Air quality within vehicles and transportation environments is an emerging health and engineering concern, particularly in densely populated urban regions where vehicles not only generate pollution but also expose occupants to airborne contaminants. Standard cabin filters typically employ nonwoven synthetic fibers and activated carbon layers to remove particulates and neutralize odors. However, these petroleum-derived filters are non-biodegradable and environmentally problematic. In contrast, bio-aerogels derived from renewable polymers such as cellulose, chitosan, or lignin demonstrate filtration efficiencies that match or surpass traditional materials while being lightweight, biodegradable, and customizable [115]. These nanostructured materials combine >99% particulate removal with low pressure drop, making them ideal for reducing vehicle weight, an important consideration for improving fuel efficiency and lowering emissions. Moreover, bio-aerogels can simultaneously filter particulate matter and adsorb volatile organic compounds, streamlining filter architecture by reducing multilayer components. Their flexibility enables integration into complex or compact HVAC geometries, and reinforcement with agents like nanoclays or carbon nanotubes improves their mechanical and thermal stability for use under vehicle operating conditions [151]. Nevertheless, transitioning to bio-aerogel-based filters in transportation systems is not without challenges. Durability under cyclic humidity, exposure to exhaust gases, and mechanical wear remain critical concerns. Although early-stage research has shown promise in producing composite or laminated bio-aerogel filters with enhanced robustness, large-scale deployment still requires rigorous testing under real-world driving conditions. In addition, cost competitiveness and ease of integration with current automotive assembly lines must be addressed. These factors are particularly important in mass-market vehicles, where manufacturing efficiency and part standardization are central to feasibility.

Bio-aerogels offer strong potential for use in automotive air filtration, not only in passenger vehicles but also in public transportation systems such as buses, trains, and aircraft cabins. These environments demand high air turnover and must meet strict filtration standards. With increasing consumer demand for greener vehicles and regulatory pressures to reduce environmental footprint, the integration of biodegradable and high-performance filter components like bio-aerogels could become a key differentiator in the transportation industry. As material performance continues to improve and fabrication processes become more streamlined, bio-aerogels are poised to play a central role in redefining clean air technology across all modes of transportation.

9.10 Performance Optimization of Bio-Aerogels

Bio-aerogels, an emerging class of ultralight and highly porous materials derived from biopolymers, have garnered significant attention in recent years as sustainable alternatives to conventional synthetic aerogels. Their unique combination of high specific surface area, open porosity, low thermal conductivity, and biodegradability makes them particularly attractive for a broad range of applications, including thermal insulation, air and water purification, biomedical devices, and energy storage. However, translating the inherent material potential of bio-aerogels into real-world performance remains a challenging task, as their mechanical fragility, processability, and long-term stability often fall short of application requirements.

Performance optimization of bio-aerogels is therefore critical to fully unlock their functional capabilities and meet specific application demands. Such optimization entails the rational design and fine-tuning of material properties such as pore structure, density, mechanical strength, thermal stability, and surface chemistry through control of the synthesis process and incorporation of functional additives. By systematically understanding and engineering the key factors that influence aerogel performance, it is possible to achieve an optimal balance between lightweight structure, functional properties, and durability, while retaining the intrinsic environmental benefits of biopolymer-based materials. This section elaborates on the processes involved in optimizing bio-aerogel performance, highlighting the interplay between material formulation, processing parameters, and targeted functional outcomes.

9.10.1 Enhancing Adsorption Capacity and Efficiency

Bio-aerogels, derived from renewable biopolymers such as cellulose, chitosan, alginate, and starch, have emerged as promising sustainable materials for adsorption-based applications, particularly in air and water purification. Their distinctive properties extremely high porosity, low density, and large specific surface area make them ideal candidates for removing a variety of contaminants, including heavy metals, dyes, oils, and volatile organic compounds (VOCs) [125]. However, despite these inherent advantages, the pristine forms of bio-aerogels often exhibit limited adsorption capacity, poor selectivity, and suboptimal mechanical stability under operational conditions. These drawbacks constrain their practical implementation in demanding environmental and industrial settings. Enhancing the adsorption capacity and efficiency of bio-aerogels is a critical component of their performance optimization, aimed at maximizing functionality without compromising their lightweight structure and environmental sustainability. This optimization is achieved through purposeful alterations to the aerogel's chemical composition, pore architecture, and surface functionalities, all designed to improve pollutant-adsorbent interactions, increase the density of active sites, and facilitate effective mass transport within the porous matrix. This section presents the principal strategies employed to enhance bio-aerogel adsorption performance, emphasizing how material design, surface chemistry, and structural engineering work synergistically to deliver high adsorption efficiency and selectivity for specific contaminants.

Improving the adsorption performance of bio-aerogels involves a multifaceted approach that integrates chemical functionalization, network reinforcement, and surface tailoring to optimize both the thermodynamic and kinetic aspects of pollutant capture. One of the most effective strategies is modifying the biopolymer matrix by introducing functional groups such as amine ($-NH_2$), carboxyl (-COOH), thiol (-SH), and sulfonic acid ($-SO_3H$). These groups enhance the aerogel's affinity for specific pollutants via electrostatic attraction, chelation, hydrogen bonding, and covalent interactions. Additionally, grafting or crosslinking the aerogel with synthetic polymers or blending it with polyelectrolytes increases surface charge density and mechanical resilience [15]. These modifications not only elevate the material's adsorption capacity and selectivity but also preserve its structural integrity under operational stress, ensuring long-term durability and effectiveness in demanding environmental remediation applications.

A critical approach to enhancing the adsorption performance of bio-aerogels is structural tuning and porosity control, which focuses on optimizing the internal architecture of the aerogel to maximize surface accessibility and facilitate efficient mass transfer. Designing a hierarchical pore structure composed of interconnected micro-, meso-, and macropores not only increases the specific surface area but also promotes rapid diffusion of contaminants into the interior of the aerogel matrix. This is particularly important when dealing with large molecular species or when operating at high pollutant concentrations, where diffusion limitations can otherwise hinder adsorption efficiency. Controlled sol-gel xprocessing and templating techniques are commonly employed to achieve such hierarchical porosity, ensuring a well-connected network that prevents pore blockage and sustains high adsorption kinetics.

In addition to structural optimization, the incorporation of active nanomaterials into the bio-aerogel matrix has emerged as an effective strategy for further improving adsorption capacity and functional performance. Embedding nanomaterials such as metal nanoparticles (e.g., Ag, Fe_3O_4, TiO_2), graphene oxide, or metal-organic frameworks (MOFs) introduce additional active sites, catalytic properties, or magnetic responsiveness that complement the inherent properties of the bio-aerogel [17]. These hybrid materials benefit from the synergy between the aerogel's lightweight, highly porous structure and the nanomaterials' high reactivity and specificity, resulting in enhanced adsorption capacity, improved selectivity toward targeted pollutants, and, in the case of magnetic nanoparticles, ease of recovery. This integration of nanomaterials represents a powerful avenue for tailoring bio-aerogels to meet the demands of specific environmental remediation challenges.

Another important aspect of enhancing the adsorption performance of bio-aerogels is surface area and hydrophobicity engineering, which involves tailoring the surface energy and wettability characteristics of the material to suit specific application requirements. For instance, in applications such as oil-water separation and the adsorption of volatile organic compounds (VOCs), it is advantageous to increase the hydrophobicity and oleophilicity of the aerogel surface to achieve selective uptake of non-polar contaminants while resisting water intrusion. This can be accomplished through surface modification with hydrophobic silanes (e.g., methyltrimethoxysilane, MTMS) or fluorinated agents, which lower the surface free energy and impart water-repellent properties without compromising the aerogel's high porosity and low density [107]. Conversely, for aqueous-phase adsorption of dissolved pollutants, such as heavy metals or dyes, a hydrophilic surface with appropriate charge density and wettability is essential to maximize the contact between the adsorbent and waterborne species. Strategies to retain or enhance hydrophilicity include preserving hydroxyl-rich surfaces or introducing charged functional groups that promote affinity for ionic or polar contaminants [6]. Therefore, enables the development of application-specific bio-aerogels with superior adsorption selectivity and capacity in diverse environments.

Enhancing the adsorption capacity and efficiency of bio-aerogels demands a holistic approach that integrates chemical, structural, and functional design principles. By leveraging advanced functionalization, porosity engineering, and hybridization techniques, bio-aerogels can be transformed into high-performance, application-specific adsorbents with superior kinetics, capacity, and selectivity. These optimized materials hold significant promise for sustainable pollutant remediation and resource recovery in line with circular economy principles.

9.10.2 Balancing Filtration Efficiency with Airflow Resistance

One of the critical challenges in the design and optimization of bio-aerogels for air filtration applications lies in achieving a balance between high filtration efficiency and low airflow resistance. Filtration efficiency is largely determined by the aerogel's ability to intercept, diffuse, and adsorb airborne particles and gaseous contaminants, which typically improves with higher density and smaller pore size. However, increasing the material density and reducing pore size inevitably increases pressure drop across the filter, leading to higher airflow resistance and energy consumption. Therefore, a careful design of the aerogel's microstructure is essential to maximize contaminant removal while maintaining acceptable breathability and low operational costs.

The strategy involves the control of pore size distribution and porosity to achieve a hierarchical structure that effectively captures both fine and ultrafine particles while preserving sufficient void space for unimpeded airflow. For example, Abidin et al. [6] demonstrated that cellulose nanofiber (CNF)-based aerogels with interconnected meso- and macropores can achieve over 99% filtration efficiency for $PM_{2.5}$ with a pressure drop below 120 Pa, meeting standards for indoor air quality applications. This was accomplished by fine-tuning the nanofiber concentration and freeze-drying parameters to adjust fiber packing density and pore connectivity. Additionally, the incorporation of functional nanomaterials and surface modifications can enhance particle capture efficiency without significantly increasing material density. For instance, loading CNF aerogels with electrospun nanofibers or surface charges can improve electrostatic attraction to fine particles, thereby increasing capture rates at lower material thicknesses [83]. Such hybrid approaches exploit synergies between mechanical sieving, electrostatic interactions, and adsorption to reduce reliance on purely dense microstructures.

Another effective approach is anisotropic structural engineering where directional freezing techniques create aligned pore channels that facilitate preferential airflow while maintaining tortuous pathways for particle interception. This anisotropic architecture has been shown to lower pressure drop significantly compared to isotropic aerogels of the same density while retaining comparable filtration efficiency [46]. Lastly, post-processing treatments such as compressive densification can be applied judiciously to adjust fiber contact and pore size at a minimal compromise to porosity, enabling designers to meet specific application standards such as HEPA or ULPA filtration performance levels.

In summary, balancing filtration efficiency with airflow resistance in bio-aerogels requires an integrative design strategy that combines microstructural control, material functionalization, and directional architecture optimization. These approaches allow for the development of advanced bio-aerogel filters that deliver high contaminant removal efficiency at low pressure drop, enabling energy-efficient and breathable filtration systems suitable for diverse environmental and occupational settings.

9.10.3 Durability and Reusability in Real-World Applications

Durability and reusability are critical parameters that dictate the practical viability of bio-aerogels in real-world applications, particularly in filtration, adsorption, and catalysis. While bio-aerogels offer compelling advantages such as low density, high porosity, and biodegradability, their inherent mechanical fragility and susceptibility to structural collapse, fouling, or degradation under repeated use and harsh operational conditions often limit their long-term performance. Therefore, optimizing the durability and reusability of bio-aerogels requires targeted strategies that reinforce mechanical stability, preserve functional properties after regeneration cycles, and minimize performance deterioration over time.

A primary approach involves crosslinking and chemical modification of the biopolymer matrix to enhance mechanical strength and water resistance. Covalent or ionic crosslinking agents, such as glutaraldehyde, genipin, or multivalent cations, can improve the rigidity of the aerogel's skeletal structure and reduce its tendency to shrink or disintegrate during drying, wetting, or cyclic loading. For example, Mendoza et al. [90] reported that genipin-crosslinked chitosan aerogels exhibited markedly improved compressive strength and retained over 90% of their adsorption capacity after five regeneration cycles. Similarly, the introduction of hydrophobic modifications, such as silanization, can render the surface more resistant to moisture-induced collapse and improve resistance to fouling in humid or aqueous environments.

Another effective strategy involves incorporation of reinforcing fillers or hybridization with mechanically robust nanomaterials. Embedding cellulose nanocrystals (CNCs), graphene oxide, carbon nanotubes, or inorganic nanoparticles within the aerogel matrix enhances load-bearing capacity, distributes stress, and mitigates structural fatigue during repeated compression or use cycles. Mokhena et al. [93], showed that cellulose nanofiber-graphene hybrid aerogels maintained structural integrity and functional performance over multiple adsorption-desorption cycles due to improved resilience and network stabilization.

In addition to mechanical reinforcement, ensuring efficient and mild regeneration processes is crucial for maintaining reusability. Thermal annealing, solvent washing, or mild desorption techniques such as low-temperature vacuum drying should be optimized to remove adsorbed contaminants effectively without damaging the aerogel's porous network or altering its surface chemistry. Overly aggressive regeneration, such as high-temperature calcination, can lead to pore collapse or loss of functional groups, thereby diminishing long-term performance.

Finally, designing antifouling and self-cleaning surfaces can further enhance durability by reducing irreversible fouling and clogging of pores. Coating the aerogel with photocatalytic or antimicrobial agents, such as TiO_2 or silver nanoparticles, has been shown to inhibit microbial growth and organic fouling, prolonging operational lifespan in environmental applications [39]. In summary, enhancing the durability and reusability of bio-aerogels requires an integrated approach that combines chemical crosslinking, nanomaterial reinforcement, optimized regeneration protocols, and antifouling surface engineering. These strategies collectively ensure that bio-aerogels maintain their structural integrity and functional performance across multiple operational and regeneration cycles, thereby advancing their readiness for deployment in real-world environmental and industrial settings.

9.11 Comparison with Traditional Air Filtration Materials

The evolution of air filtration technologies has been driven by increasing demands for higher efficiency, lower energy consumption, sustainability, and multifunctionality. Traditional air filtration materials, such as fiberglass mats, polymeric nonwovens, and cellulose-based papers, have long been used due to their availability, simplicity, and proven effectiveness in removing particulate matter. However, these conventional materials often suffer from drawbacks, including high pressure drop, limited selectivity, susceptibility to microbial growth, and lack of environmental compatibility.

Recent advances in material science have led to the development of modern air filtration materials, including nanofiber membranes, bio-aerogels, metal-organic frameworks (MOFs), and hybrid multifunctional composites. These emerging materials offer significant improvements in terms of surface area, porosity control, antimicrobial properties, and adaptability to specific pollutants such as VOCs, bioaerosols, and ultrafine particles. The following table provides a systematic comparison between traditional and modern air filtration materials, highlighting their respective advantages, limitations, and representative examples in the context of current technological trends. Table 9.11 shows the comparison of traditional and modern air filtration materials.

The comparison presented in the Table 9.11 underscores the substantial progress that has been achieved in the field of air filtration materials, transitioning from conventional fibrous mats and papers to advanced, multifunctional materials. Traditional materials, such as fiberglass mats and cellulose-based papers, remain widely used due to their cost-effectiveness, manufacturing scalability, and proven ability to capture coarse and fine particulate matter. However, their inherent limitations high pressure drop, low selectivity, environmental non-compatibility, and vulnerability to microbial colonization have motivated the exploration of more advanced alternatives, particularly in high-performance and environmentally sensitive applications.

Modern nanofiber membranes address some of these limitations by offering tunable pore structures, high specific surface areas, and improved filtration efficiency, particularly for ultrafine particles ($PM_{0.3}$ and below). Their electrospinning-based fabrication enables precise control of fiber diameter and porosity, but challenges such as mechanical fragility and relatively high production costs still impede widespread adoption. In parallel, bio-aerogels, derived from renewable biopolymers, present an attractive solution for sustainable filtration with high porosity, low density, and potential antimicrobial properties. Their biodegradability aligns well with circular economy principles, although issues of mechanical weakness and moisture sensitivity remain critical hurdles to overcome.

MOF-based filters represent a more specialized and chemically sophisticated class of materials, offering unparalleled selectivity toward specific contaminants, such as volatile organic compounds and hazardous gases, due to their highly tunable pore chemistry. Yet, their high cost, limited mechanical robustness, and fabrication

Table 9.11 Comparison of traditional and modern air filtration materials

Category	Material examples	Advantages	Limitations
Traditional materials	Fiberglass mats, cellulose papers, polyester/cotton blends	Cost-effective Established manufacturing processes Effective for coarse and fine particulates	High pressure drop Poor selectivity Non-biodegradable (synthetics) Prone to microbial growth
Modern nanofiber membranes	Electrospun polymer nanofibers (e.g., PAN, PVDF)	High surface area-to-volume ratio Efficient for ultrafine particles ($PM_{0.3}$ and below) Tunable pore size	Mechanical fragility Higher production cost
Bio-aerogels	Cellulose, chitosan, alginate-based aerogels	Renewable and biodegradable High porosity and low density Potential antimicrobial properties	Mechanical weakness Moisture sensitivity High fabrication complexity
Metal-organic framework (MOF)-based filters	MOF coatings or composites	High chemical selectivity (e.g., VOC adsorption) Tailorable pore chemistry High adsorption capacity	Expensive precursors Limited mechanical robustness
Hybrid multifunctional composites	Nanofiber-nanoparticle, aerogel-nanofiber hybrids	Synergistic properties (mechanical, chemical, antimicrobial) Customizable for specific pollutants Improved durability	Complex fabrication Scalability challenges

challenges restrict their current use to niche or research applications. Finally, hybrid multifunctional composites, combining the advantages of nanofibers, aerogels, and nanoparticles, exemplify the trend toward synergistic designs that address multiple performance metrics simultaneously mechanical strength, selectivity, antimicrobial activity, and durability. While promising, these hybrid systems remain hampered by fabrication complexity and scalability concerns.

The comparison highlights the ongoing trade-offs between cost, sustainability, performance, and manufacturability that characterize air filtration material development. Emerging technologies increasingly aim to reconcile these factors by engineering materials that are not only effective and selective but also environmentally responsible and economically viable. Future research directions will likely focus on optimizing hybrid systems and scalable fabrication methods, ensuring that the advantages of modern materials can be fully realized in real-world air quality management.

9.12 Summary

This chapter presents a comprehensive exploration of bio-aerogels as next-generation nanostructured materials for sustainable air filtration, addressing the limitations of conventional systems such as HEPA filters and activated carbon beds in removing ultrafine particulates and gaseous pollutants. Derived from renewable biopolymers like cellulose, chitosan, alginate, and proteins, bio-aerogels offer ultra-light density, hierarchical porosity, high surface area, and functional tunability, making them ideal for eco-friendly filtration technologies. The chapter highlights recent advances in composition and hybridization including cellulose-MOF composites and protein-based aerogels that achieve exceptional filtration efficiencies (>99% for $PM_{0.3}$) with low pressure drops. It further details the mechanisms by which these materials function interception, diffusion, electrostatic attraction, VOC adsorption, and antimicrobial inactivation and examines how tailored pore structures, surface chemistries, and embedded additives enhance performance. Four main bio-aerogel categories are classified and evaluated for their fabrication methods and filtration applications, spanning HVAC systems, industrial exhaust, and personal protective equipment. While acknowledging challenges such as moisture sensitivity and mechanical fragility, the chapter underscores ongoing innovations in crosslinking, surface engineering, and green processing as viable solutions. Ultimately, bio-aerogels are positioned as a transformative platform that unites high-efficiency air purification with biodegradability and sustainability, offering significant potential for future industrial and environmental implementation.

References

1. Abdelhamid HN (2025) Degradable biopolymer-based nanocomposites in water treatment and air purification: a review. ACS Appl Poly Mat 7(11):6576–6596. https://doi.org/10.1021/acsapm.5c00600
2. Abdelkader M, Assadi AA, Guiza M, Elfalleh W, Khezami L, Tahraoui H, Baaloudj O, Mouni L, Zhang J, Amrane A (2025) Photocatalytic degradation of pollutants in air streams using luminous textiles under ultraviolet light illumination: a pilot-scale remediation study. Catalysts 15(3). https://doi.org/10.3390/catal15030262
3. Abdul Khalil HPS, Adnan AS, Yahya EB, Olaiya NG, Safrida S, Hossain MS, Balakrishnan V, Gopakumar DA, Abdullah CK, Oyekanmi AA, Pasquini D (2020) A review on plant cellulose nanofibre-based aerogels for biomedical applications. Polymers 12(8):1759. https://doi.org/10.3390/polym12081759
4. Abdul Khalil HPS, Jha K, Yahya EB, Panchal S, Patel N, Garai A, Kumari S, Jameel M (2023) Insights into the potential of biopolymeric aerogels as an advanced soil-fertilizer delivery systems. Gels 9(8). https://doi.org/10.3390/gels9080666
5. Abdullah, Zou YC, Farooq S, Walayat N, Zhang H, Faieta M, Pittia P, Huang Q (2023) Bio-aerogels: fabrication, properties and food applications. Crit Rev Food Sci Nutr 63(24):6687–6709. https://doi.org/10.1080/10408398.2022.2037504
6. Abidin MNZ, Goh PS, Ismail AF, Said N, Othman MHD, Hasbullah H, Abdullah MS, Ng BC, Kadir SHSA, Kamal F (2018) Highly adsorptive oxidized starch nanoparticles for efficient urea removal. Carbohydr Polym 201:257–263. https://doi.org/10.1016/J.CARBPOL.2018.08.069

7. Adeola AO, Duarte MP, Naccache R (2023) Microwave-assisted synthesis of carbon-based nanomaterials from biobased resources for water treatment applications: emerging trends and prospects. Frontiers in Carbon 2(July):1–29. https://doi.org/10.3389/frcrb.2023.1220021
8. Aflaha R, Putri LA, Farrel A, Anzinger S, Rianjanu A, Yulianto N, Fueldner M, Roto R, Peiner E, Wasisto HS, Triyana K (2025) Crafting high-temperature stable and hydrophobic nanofiber membranes for particulate matter filtration. Commun Mat 6(1). https://doi.org/10.1038/s43246-025-00799-y
9. Ahmadi Heidari N, Fathi M, Hamdami N, Taheri H, Siqueira G, Nyström G (2023) Thermally insulating cellulose nanofiber aerogels from brewery residues. ACS Sustain Chem Eng 11(29):10698–10708. https://doi.org/10.1021/acssuschemeng.3c01113
10. Al Abdallah H, Tannous JH, Abu-Jdayil B (2024) Cellulose and nanocellulose aerogels, their preparation methods, and potential applications: a review. Cellulose 31(4):2001–2029. https://doi.org/10.1007/s10570-024-05743-w
11. Alkarri S, Bin Saad H, Soliman M (2024) On antimicrobial polymers: development, mechanism of action, international testing procedures, and applications. Polymers 16(6). https://doi.org/10.3390/polym16060771
12. Anucha CB, Altin I, Bacaksiz E, Stathopoulos VN (2022) Titanium dioxide (TiO_2)-based photocatalyst materials activity enhancement for contaminants of emerging concern (CECs) degradation: in the light of modification strategies. Chem Eng J Adv 10(February):100262. https://doi.org/10.1016/j.ceja.2022.100262
13. Atoufi Z, Cinar Ciftci G, Reid MS, Larsson PA, Wågberg L (2022) Green ambient-dried aerogels with a facile pH-tunable surface charge for adsorption of cationic and anionic contaminants with high selectivity. Biomacromolecules 23(11):4934–4947. https://doi.org/10.1021/acs.biomac.2c01142
14. Azfaralariff A, Razali N, Abdul Khalil HPS, Ahmad MI, Abdullah CK, Yahya EB, Iqbal M, Zain MSC (2025) Transforming agricultural waste into sustainable bioadsorbents and advanced bioaerogels from durian (Durio zibethinus) peel. Biofuels Bioprod Biorefin. https://doi.org/10.1002/BBB.2771;WGROUP:STRING:PUBLICATION
15. Bahadi SA, Hassan AH, Onaizi SA (2025) Recent advances in CO2 adsorption using aerogel materials: a review. Next Mat 7(September 2024):100388. https://doi.org/10.1016/j.nxmate.2024.100388
16. Bakhori NM, Ismail Z, Hassan MZ, Dolah R (2023) Emerging trends in nanotechnology: aerogel-based materials for biomedical applications. Nano 13(6):1–32. https://doi.org/10.3390/nano13061063
17. Barrios E, Fox D, Yee Y, Sip L, Catarata R, Calderon JE, Azim N, Afrin S, Zhang Z, Zhai L (2019) Aerogel composites: a review. Polymers 11(4):1–41
18. Barthwal S, Uniyal S, Barthwal S (2024) Nature-inspired superhydrophobic coating materials: drawing inspiration from nature for enhanced functionality. Micromachines 15(3). https://doi.org/10.3390/mi15030391
19. Bheekhun N, Abu Talib AR, Hassan MR (2013) Aerogels in aerospace: an overview. Adv Mater Sci Eng 2013. https://doi.org/10.1155/2013/406065
20. Bishnoi P, Siwal SS, Singh N, Thakur MK, Hart P, Thakur VK (2025) Advanced synthesis and fabrication strategies for 2D mesoporous carbon materials in energy storage and conversion. Adv Energy Sustain Res Mc. https://doi.org/10.1002/aesr.202500163
21. Budtova T, Aguilera DA, Beluns S, Berglund L, Chartier C, Espinosa E, Gaidukovs S, Klimek-kopyra A, Kmita A, Lachowicz D, Liebner F, Platnieks O, Rodríguez A, Navarro LKT, Zou F, Buwalda SJ (2020) Biorefinery approach for aerogels. Polymers 12(12):1–63. https://doi.org/10.3390/polym12122779
22. Buratti C, Moretti E, Belloni E, Agosti F (2014) Development of innovative aerogel based plasters: preliminary thermal and acoustic performance evaluation. Sustainability:5839–5852. https://doi.org/10.3390/su6095839
23. Carvalho AF, Kulyk B, Fernandes AJS, Fortunato E, Costa FM (2022) A review on the applications of graphene in mechanical transduction. Adv Mater 34(8):1–55. https://doi.org/10.1002/adma.202101326

24. Chang X, Cheng X, Zhang H, Li W, He L, Yin X, Liu X, Yu J, Liu YT, Ding B (2023) Superelastic carbon aerogels: an emerging material for advanced thermal protection in extreme environments. Adv Funct Mater 33(26):2215168. https://doi.org/10.1002/ADFM.202215168. WGROUP:STRING:PUBLICATION
25. Chen YX, Klima KM, Brouwers HJH, Yu Q (2022) Effect of silica aerogel on thermal insulation and acoustic absorption of geopolymer foam composites: the role of aerogel particle size. Compos Part B 242(June):110048. https://doi.org/10.1016/j.compositesb.2022.110048
26. Cheng X, Sun X, Zheng S, Zhou W, Yuan Z, Zhu L, Chen X (2024) Quaternized plant-based porous biochar for direct air capture of CO2 by moisture-swing adsorption. Ind Eng Chem Res 63(5):2320–2328. https://doi.org/10.1021/acs.iecr.3c03023
27. Cho MJ, Yiu J, Lin LT, Hua Q, Karaaslan MA, Renneckar S (2025) Lignin nanofiber flexible carbon aerogels for self-standing Supercapacitors. ChemSusChem 18(3):1–10. https://doi.org/10.1002/cssc.202400932
28. Choma J, Szczęśniak B, Kapusta A, Jaroniec M (2024) A concise review on porous adsorbents for benzene and other volatile organic compounds. Molecules 29(23):1–37. https://doi.org/10.3390/molecules29235677
29. Członka S, Bertino MF, Kośny J, Shukla N (2018) Freeze-drying method as a new approach to the synthesis of polyurea aerogels from isocyanate and water. J Sol-Gel Sci Technol 87(3):685–695. https://doi.org/10.1007/s10971-018-4769-9
30. Dai Q, Liu J, Zheng J, Fu B (2023) Immobilization of UiO-66-NH2 into bacterial cellulose aerogels for efficient particulate matter filtration. Sustainability (Switzerland) 15(18):1–11. https://doi.org/10.3390/su151813382
31. Das A, Peu SD, Hossain MS, Nahid MMA, Karim FRB, Chowdhury H, Porag MH, Argha DBP, Saha S, Islam ARMT, Salah MM, Shaker A (2023) Advancements in adsorption based carbon dioxide capture technologies- a comprehensive review. Heliyon 9(12):e22341. https://doi.org/10.1016/j.heliyon.2023.e22341
32. Dash R, Li Y, Ragauskas AJ (2012) Cellulose nanowhisker foams by freeze casting. Carbohydr Polym 88(2):789–792. https://doi.org/10.1016/J.CARBPOL.2011.12.035
33. David E, Niculescu VC (2021) Volatile organic compounds (Vocs) as environmental pollutants: occurrence and mitigation using nanomaterials. Int J Environ Res Public Health 18(24). https://doi.org/10.3390/ijerph182413147
34. de Jager TL, Cockrell AE, Du Plessis SS (2017) Ultraviolet light induced generation of reactive oxygen species. Adv Exp Med Biol 996:15–23. https://doi.org/10.1007/978-3-319-56017-5_2
35. Dervin S, Pillai SC (2017) An introduction to sol-gel processing for aerogels, pp 1–22. https://doi.org/10.1007/978-3-319-50144-4_1
36. Dey E, Choudhary U, Ghosh SK (2017) A review on surface modification of textile fibre by high efficiency particulate air (HEPA) filtration process. Am J Eng Res (AJER) 6:190–193. www.ajer.org
37. Ding C, Liu Y, Xie P, Lan J, Yu Y, Fu X, Yang X, Zhong WH (2021) A novel carbon aerogel enabling respiratory monitoring for bio-facial masks. J Mater Chem A 9(22):13143–13150. https://doi.org/10.1039/D1TA00794G
38. Fan Q, Song C, Fu P (2024) Advances in the improvement of the quality and efficiency of biomass-derived porous carbon: a comprehensive review on synthesis strategies and heteroatom doping effects. J Clean Prod 452(March). https://doi.org/10.1016/j.jclepro.2024.142169
39. Feng B, Xu K, Gu S, Zheng S, Zou Q, Xu Y, Yu L, Lou F, Yu F, Jin T, Li Y, Sheng J, Yen HL, Zhong Z, Wei J, Chen Y (2021) Multi-route transmission potential of SARS-CoV-2 in healthcare facilities. J Hazard Mater 402(June 2020):123771. https://doi.org/10.1016/j.jhazmat.2020.123771
40. Fischer F, Rigacci A, Pirard R, Berthon-Fabry S, Achard P (2006) Cellulose-based aerogels. Polymer 47(22):7636–7645. https://doi.org/10.1016/j.polymer.2006.09.004
41. Fouda-Mbanga BG, Onotu OP, Tywabi-Ngeva Z (2024) Advantages of the reuse of spent adsorbents and potential applications in environmental remediation: a review. Green Anal Chem 11(September):100156. https://doi.org/10.1016/j.greeac.2024.100156

42. Ganesan K, Budtova T, Ratke L, Gurikov P, Baudron V, Preibisch I, Niemeyer P, Smirnova I, Milow B (2018) Review on the production of polysaccharide aerogel particles. Materials 11(11):1–37. https://doi.org/10.3390/ma11112144
43. Gao Y, Tian E, Zhang Y, Mo J (2022) Utilizing electrostatic effect in fibrous filters for efficient airborne particles removal: principles, fabrication, and material properties. Appl Mater Today 26:101369. https://doi.org/10.1016/j.apmt.2022.101369
44. García-González CA, Budtova T, Durães L, Erkey C, Del Gaudio P, Gurikov P, Koebel M, Liebner F, Neagu M, Smirnova I (2019) An opinion paper on aerogels for biomedical and environmental applications. Molecules 24(9):1815. https://doi.org/10.3390/MOLECULES24091815
45. García-González CA, Blanco-Vales M, Barros J, Boccia AC, Budtova T, Durães L, Erkey C, Gallo M, Herman P, Kalmár J, Iglesias-Mejuto A, Malfait WJ, Zhao S, Manzocco L, Plazzotta S, Milovanovic S, Neagu M, Nita LE, Paraskevopoulou P et al (2025) Review and perspectives on the sustainability of organic aerogels. ACS Sustain Chem Eng 13(18):6469–6492. https://doi.org/10.1021/acssuschemeng.4c09747
46. Garemark J, Yang X, Sheng X, Cheung O, Sun L, Berglund LA, Li Y (2020) Top-down approach making anisotropic cellulose aerogels as universal substrates for multifunctionalization. ACS Nano 14(6):7111–7120. https://doi.org/10.1021/acsnano.0c01888
47. Girimonte R, Testa F, Turano M, Leone G, Gallo M, Golemme G (2022) Amine-functionalized mesoporous silica adsorbent for CO2 capture in confined-fluidized bed: study of the breakthrough adsorption curves as a function of several operating variables. PRO 10(2). https://doi.org/10.3390/pr10020422
48. Gordon SB, Bruce N, Grigg J, Hibberd P, Kurmi O, Lam K-B, Mortimer K, Asante KP, Balakrishnan K, Balmes J, Bar-Zeev N, Benka-Coker M, Berhanu N, Budzko L, Buser M, Chafe Z, Clifford M, Cogswell M, Collishaw N et al (2014) Respiratory risks from household air pollution in low and middle income countries. Lancet Respir Med 2(10):823–860. https://doi.org/10.1016/S2213-2600(14)70168-7
49. Hamann L, Foat T, Blanke A (2025) The diversity of biological models for bio-inspired aerosol filters. J R Soc Interface 22(227):1–15. https://doi.org/10.1098/rsif.2025.0221
50. Han X, Xue Y, Lou R, Ding S, Wang S (2023a) Facile and efficient chitosan-based hygroscopic aerogel for air dehumidification. Int J Biol Macromol 251(April):126191. https://doi.org/10.1016/j.ijbiomac.2023.126191
51. Han M, Zhang R, Wu Q, Wu N, Liu J (2023b) Biopolymer-based aerogels for electromagnetic wave shielding and absorbing. Chin J Chem 41(3):322–334. https://doi.org/10.1002/CJOC.202200552;PAGEGROUP:STRING:PUBLICATION
52. He, J., Chen, K., & Xu, J. (2017). Urban air pollution and control. Encycl Sustain Technol, 243-257. https://doi.org/10.1016/B978-0-12-409548-9.10182-4, Elsevier
53. Hoon D, Haeng Y, Piri A, An S, Hwang J (2020) Since January 2020 Elsevier has created a COVID-19 resource centre with free information in English and Mandarin on the novel coronavirus COVID- 19. The COVID-19 resource centre is hosted on Elsevier Connect, the company' s public news and information. January
54. Hou Y, Chen J, Pan D, Zhao L (2023) Directional-freezing-assisted in situ sol-gel strategy to synthesize high-strength, fire-resistant, and hydrophobic wood-based composite aerogels for thermal insulation. Gels 9(2). https://doi.org/10.3390/gels9020170
55. Hu BC, Zhang HR, Li SC, Chen WS, Wu ZY, Liang HW, Yu HP, Yu SH (2023) Robust carbonaceous nanofiber aerogels from all biomass precursors. Adv Funct Mater 33(1). https://doi.org/10.1002/ADFM.202207532
56. Hui B, Chen H, Zhou C, Cai L, Zhang K, Quan F, Yang D (2022) Biochar aerogel-based electrocatalyst towards efficient oxygen evolution in acidic media. Biochar 4(1). https://doi.org/10.1007/s42773-022-00163-0
57. Husain FM, Khan A, Khan RA, Siddique JA, Oves M, Parwaz Khan AA, Ansari MO, Cancar HD (2021) Bio-based aerogels and their environment applications: an overview. In: Adv Aerogel Comp Environ Remed, pp 347–356. https://doi.org/10.1016/B978-0-12-820732-1.00018-7

58. Ihsanullah I, Sajid M, Khan S, Bilal M (2022) Aerogel-based adsorbents as emerging materials for the removal of heavy metals from water: progress, challenges, and prospects. Sep Purif Technol 291(March):120923. https://doi.org/10.1016/j.seppur.2022.120923
59. Inonu Z, Keskin S, Erkey C (2018) An emerging family of hybrid nanomaterials: metal-organic framework/aerogel composites. ACS Appl Nano Mat 1(11):5959–5980. https://doi.org/10.1021/acsanm.8b01428
60. Jeong Y, Patel R, Patel M (2024) Biopolymer-based biomimetic aerogel for biomedical applications. Biomimetics 9(7). https://doi.org/10.3390/biomimetics9070397
61. Ji X, Huang J, Teng L, Li S, Li X, Cai W, Chen Z, Lai Y (2023) Advances in particulate matter filtration: materials, performance, and application. Green Energy Environ 8(3):673–697. https://doi.org/10.1016/j.gee.2022.03.012
62. Ji Q, Li Z, Su L, Boateng ID, Zhou C, Liu X (2025) Preparation, functional modification and industrial application of nanocellulose aerogels – a comprehensive review. Biomass Bioenergy 193(September 2024):107591. https://doi.org/10.1016/j.biombioe.2025.107591
63. Jia G, Guo J, Guo Y, Yang F, Ma Z (2024) CO_2 adsorption properties of aerogel and application prospects in low-carbon building materials: a review. Case Stud Constr Mater 20:e03171. https://doi.org/10.1016/j.cscm.2024.e03171
64. Jin T, Liu T, Lam E, Moores A (2021) Chitin and chitosan on the nanoscale. Nanoscale Hor 6(7):505–542. https://doi.org/10.1039/d0nh00696c
65. Karamikamkar S, Naguib HE, Park CB (2020) Advances in precursor system for silica-based aerogel production toward improved mechanical properties, customized morphology, and multifunctionality: a review. Adv Colloid Interf Sci 276:102101. https://doi.org/10.1016/j.cis.2020.102101
66. Kärger J, Ruthven DM (2016) Diffusion in nanoporous materials: fundamental principles, insights and challenges. New J Chem 40(5):4027–4048. https://doi.org/10.1039/c5nj02836a
67. Khan NR, Sharmin T, Bin Rashid A (2024) Exploring the versatility of aerogels: broad applications in biomedical engineering, astronautics, energy storage, biosensing, and current Progress. Heliyon 10(1):e23102. https://doi.org/10.1016/j.heliyon.2023.e23102
68. Khararoodi MG (2023, August) Systematic development and validation of predictive models for the removal of indoor gaseous pollutants using carbon-
69. Kim CH, Youn HJ, Lee HL (2015) Preparation of cross-linked cellulose nanofibril aerogel with water absorbency and shape recovery. Cellulose 22(6):3715–3724. https://doi.org/10.1007/s10570-015-0745-5
70. Kistler SS (1931) Coherent expanded aerogels and jellies. Nature 127(3211):741. https://doi.org/10.1038/127741a0
71. Ko LS, Pan KL, Chang MB (2024) Review on catalytic oxidation of NO by gas-phase oxidants. Aerosol Air Qual Res 24(12):240153. https://doi.org/10.4209/AAQR.240153
72. Lachowicz D, Kmita A, Wirecka R, Berent K, Szuwarzyński M, Zapotoczny S, Pajdak A, Cios G, Mazur-Panasiuk N, Pyrc K, Bernasik A (2023) Aerogels based on cationically modified chitosan and poly(vinyl alcohol) for efficient capturing of viruses. Carbohydr Polym 312. https://doi.org/10.1016/j.carbpol.2023.120756
73. Lázár I, Čelko L, Menelaou M (2023) Aerogel-based materials in bone and cartilage tissue engineering—a review with future implications. Gels 9(9). https://doi.org/10.3390/gels9090746
74. Lelieveld J, Evans J, Fnais M et al (2015) The contribution of outdoor air pollution sources to premature mortality on a global scale. Nature 525:367–371. https://doi.org/10.1038/nature15371
75. Leventis N, Chandrasekaran N, Sotiriou-Leventis C, Mumtaz A (2009) Smelting in the age of nano: iron aerogels. J Mater Chem 19(1):63–65. https://doi.org/10.1039/b815985h
76. Li C, Zhang G, Lin L, Wu T, Brunner S, Galmarini S, Bi J, Malfait WJ, Zhao S, Ostrikov KK (2023) Silica aerogels: from materials research to industrial applications. Int Mater Rev 68(8):1480–1515. https://doi.org/10.1080/09506608.2023.2167547

77. Li L, Jin G, Shen J, Guo M, Song J, Li Y, Xiong J (2025) Carbon aerogels: synthesis, modification, and multifunctional applications. Gels 11(7):1–37. https://doi.org/10.3390/gels11070548
78. Lin L, Li Z, Mao H, Li W, Wang C (2021) Optically active polyurethane/silica aerogel coated cotton fabrics for thermal protection. Front. Mater. 8:681678. https://doi.org/10.3389/fmats.2021.681678
79. Lin S, Fu X, Luo M, Zhong WH (2023) A protein aerogel with distinctive filtration capabilities for formaldehyde and particulate pollutants. Sep Purif Technol 310(January):123179. https://doi.org/10.1016/j.seppur.2023.123179
80. Lin S, Liu W, Fu X, Luo M, Liu H, Zhong WH (2023c) An all-protein aerogel with a nanofiber/foam structure for versatile air filtration. Mater Today Chem 34(June):101760. https://doi.org/10.1016/j.mtchem.2023.101760
81. Lin S, Liu W, Fu X, Luo M, Zhong WH (2024) Protein-based materials for sustainable, multifunctional air filtration. Sep Purif Technol 335(October 2023):126252. https://doi.org/10.1016/j.seppur.2023.126252
82. Liu W, Wang X, Zhao B, Lu J (2023) New insights into the synergistic effect on condensable particulate matter based on the formation, characteristics and removal. Aerosol Air Qual Res 23(12):230145. https://doi.org/10.4209/AAQR.230145
83. Lu X, Lin P, Huang Y, He X, Yi C, Sun J, Farid MU, An AK, Guo J (2025) Advances in electrospun Nanofibrous aerogels: pioneering methods, versatile applications, and future horizons. In: Advanced fiber materials (Issue 0123456789). Springer, Singapore. https://doi.org/10.1007/s42765-025-00552-7
84. Lv A, Wang M, Shi H, Lu S, Zhang J, Jiao S (2023) A carbon aerogel lightweight Al battery for fast storage of fluctuating energy. Adv Mater 35(35):2303943. https://doi.org/10.1002/ADMA.202303943;SUBPAGE:STRING:ACCESS
85. Mandić V, Bafti A, Panžić I, Radovanović-Perić F (2024) Bio-based aerogels in energy storage systems. Gels 10(7). https://doi.org/10.3390/gels10070438
86. Martín-Cruz Y, Bordón P, Pulido-Melián E, Saura-Cayuela T, Monzón M (2024) Development of cellulose air filters for capturing fine and ultrafine particles through the valorization of banana cultivation biomass waste. Environments MDPI 11(3). https://doi.org/10.3390/environments11030050
87. Maskim Vladimirovich G, Pavlovich Melnikov V (2018) Graphene oxide/reduced graphene oxide aerogels. In: Graphene oxide applications and opportunities, pp 39–55. https://doi.org/10.5772/intechopen.78987
88. Maury-Micolier A, Huang L, Jolliet O (2023) Coupled mass and heat transfer modelling in building envelopes to consistently assess human exposure and energy performance in indoor environments. J Build Perform Simul 16(6):734–748. https://doi.org/10.1080/19401493.2023.2200377
89. Mazrouei-Sebdani Z, Begum H, Schoenwald S, Horoshenkov KV, Malfait WJ (2021) A review on silica aerogel-based materials for acoustic applications. J Non-Cryst Solids 562(March):120770. https://doi.org/10.1016/j.jnoncrysol.2021.120770
90. Mendoza DJ, Browne C, Raghuwanshi VS, Simon GP, Garnier G (2019) One-shot TEMPO-periodate oxidation of native cellulose. Carbohydr Polym 226:115292. https://doi.org/10.1016/J.CARBPOL.2019.115292
91. Meo SA, Salih MA, Alkhalifah JM, Alsomali AH, Almushawah AA (2024) Effect of air pollutants particulate matter $PM_{2.5}$, PM_{10}, carbon monoxide (CO), nitrogen dioxide (NO_2), sulfur dioxide (SO_2), and ozone (O_3) on fractional exhaled nitric oxide (FeNO). Pak J Med Sci 40(8):1719–1723. https://doi.org/10.12669/pjms.40.8.9630
92. Mohamed RYA, Kumarachari RK, Bukke SPN, Neerugatti D, Mekasha YT, Bandarapalle K (2025) Plasma catalysis for sustainable industry: lab-scale studies and pathways to upscaling. Discover Appl Sci 7(4). https://doi.org/10.1007/s42452-025-06718-7. Springer
93. Mokhena, T. C., Mochane, M. J., Mtibe, A., Sigonya, S., Ntsendwana, B., Masibi, E. G., Sikhwivhilu, L., & Motsoeneng, T. S. (2024). Recent advances on nanocellulose-graphene oxide composites: a review. In Cellulose (Vol. 31, 12). Springer Dordrecht. https://doi.org/10.1007/s10570-024-06055-9

94. Mondal SK, Chakraborty S, Manna S, Mandal SM (2024) Antimicrobial nanoparticles: current landscape and future challenges. RSC Pharm 1(3):388–402. https://doi.org/10.1039/d4pm00032c
95. Mourkou M, Yu H, Baltussen S, Snead N, Kapil N, Coppens MO (2024) A novel ultra-high vacuum diffusion setup to study Knudsen diffusion. React Chem Eng 9(11):3047–3059. https://doi.org/10.1039/D4RE00267A
96. Mudassir MA, Kousar S, Ehsan M, Usama M, Sattar U, Aleem M, Naheed I, Saeed OB, Ahmad M, Akbar HF, Ud Din MA, Ansari TM, Zhang H, Hussain I (2023) Emulsion-derived porous carbon-based materials for energy and environmental applications. Renew Sust Energ Rev 185(July):113594. https://doi.org/10.1016/j.rser.2023.113594
97. Nita LE, Ghilan A, Rusu AG, Neamtu I, Chiriac AP (2020) New trends in bio-based aerogels. Pharmaceutics 12
98. Olawade DB, Jimoh MO, Afolabi AS (2024) Metal oxide nanomaterials in antimicrobial therapy: mechanisms, applications, and future perspectives. J Nanostructure Chem 14(1):45–67. https://doi.org/10.1007/s40097-023-00564-5
99. Omoseebi A (2025) Waste-to-resource innovation: harnessing hazardous wastes to combat carbon emissions with graphene. January. https://www.researchgate.net/publication/387933809
100. Pääkkö M, Vapaavuori J, Silvennoinen R, Kosonen H, Ankerfors M, Lindström T, Berglund LA, Ikkala O (2008) Long and entangled native cellulose I nanofibers allow flexible aerogels and hierarchically porous templates for functionalities. Soft Matter 4(12):2492–2499. https://doi.org/10.1039/B810371B
101. Panahi-Sarmad M, Alikarami N, Guo T, Haji M, Jiang F, Rojas OJ (2024) Aerogels based on bacterial nanocellulose and their applications. Small 20(44):1–31. https://doi.org/10.1002/smll.202403583
102. Payanda Konuk O, Alsuhile AAM, Yousefzadeh H, Ulker Z, Bozbag SE, García-González CA, Smirnova I, Erkey C (2023) The effect of synthesis conditions and process parameters on aerogel properties. Front Chem 11:1294520. https://doi.org/10.3389/fchem.2023.1294520
103. Pekala RW (1989) Organic aerogels from the polycondensation of resorcinol with formaldehyde. J Mater Sci 24(9):3221–3227. https://doi.org/10.1007/BF01139044
104. Pierre AC, Pajonk GM (2002) Chemistry of aerogels and their applications. Chem Rev 102(11):4243–4265. https://doi.org/10.1021/cr0101306
105. Portugal J, Bedia C, Amato F, Juárez-Facio AT, Stamatiou R, Lazou A, Campiglio CE, Elihn K, Piña B (2024) Toxicity of airborne nanoparticles: facts and challenges. Environ Int 190:108889. https://doi.org/10.1016/j.envint.2024.108889
106. Qian L, Zhang H (2011) Controlled freezing and freeze drying: A versatile route for porous and micro–/nano-structured materials. J Chem Technol Biotechnol 86(2):172–184. https://doi.org/10.1002/JCTB.2495;JOURNAL:JOURNAL:10974660;WGROUP:STRING:PUBLICATION
107. Qiao S, Tian T, Duan X, Zhou J, Cheng Y (2013) Novel single-stage autotrophic nitrogen removal via co-immobilizing partial nitrifying and anammox biomass. Chem Eng J 230:19–26. https://doi.org/10.1016/J.CEJ.2013.06.048
108. Qiao S, Yan J, Wang Z, Wang Y, Yu J, Hu Z (2023) Tough and lightweight polyimide/cellulose nanofiber aerogels with hierarchical porous structures as an efficient air purifier. Sep Purif Technol 325(June):124668. https://doi.org/10.1016/j.seppur.2023.124668
109. Radwan-Pragłowska J, Piatkowski M, Janus Ł, Bogdał D, Matysek D, Cablik V (2018) Microwaveassisted synthesis and characterization of antioxidant chitosan-based aerogels for biomedical applications. Int J Polym Anal Charact 23(8):721–729. https://doi.org/10.1080/1023666X.2018.1504471
110. Rai A, Tebbi SO, Ben Mahfoud C, Bourbala M, Boudraa R, Amrane A, Maaloul N, Rendueles M, Oulego P, Alharthi MN, Mouni L (2025) Innovative antibacterial air filters impregnated with photocatalytic MgFe2O4 nanoparticles for improved microbiological air quality. Catalysts 15(4). https://doi.org/10.3390/catal15040365

111. Ramesh N, Lai CW, Johan MRB, Mousavi SM, Badruddin IA, Kumar A, Sharma G, Gapsari F (2024) Progress in photocatalytic degradation of industrial organic dye by utilising the silver doped titanium dioxide nanocomposite. Heliyon 10(24):e40998. https://doi.org/10.1016/j.heliyon.2024.e40998
112. Rigby SP (2025) Modelling and simulation of surface diffusion in heterogeneous porous materials. Surfaces 8(2):31. https://doi.org/10.3390/surfaces8020031
113. Rokicka-Konieczna P, Morawski AW (2024) Photocatalytic bacterial destruction and mineralization by TiO2-based photocatalysts: a mini review. Molecules 29(10). https://doi.org/10.3390/molecules29102221
114. Roth TB, Anderson AM, Carroll MK (2008) Analysis of a rapid supercritical extraction aerogel fabrication process: prediction of thermodynamic conditions during processing. J Non-Cryst Solids 354(31):3685–3693. https://doi.org/10.1016/J.JNONCRYSOL.2008.04.003
115. Saleh WM, Ahmad MI, Yahya EB, Abdul AK (2023) Nanostructured bioaerogels as a potential solution for particulate matter pollution. Gels 9(7). https://doi.org/10.3390/gels9070575
116. Saleh WM, Yahya EB, Ahmad MI, Khalil HPSA (2024) Exploring the properties of gelatin/chitosan bioaerogel scaffolds: toward ultra-light and highly porous materials. Polym Eng Sci 64(8):3993–4001. https://doi.org/10.1002/pen.26827
117. Saleh WM, Yahya EB, Ahmad MI, Abdul Khalil HPS (2025) Hydrophobic nanostructured bioaerogel filters for efficient particulate matter removal. Express Polym Lett 19(5):457–469. https://doi.org/10.3144/expresspolymlett.2025.34
118. Sepahvand S, Jonoobi M, Ashori A, Rabie D, Gauvin F, Brouwers HJH, Yu Q, Mekonnen TH (2022) Modified cellulose nanofibers aerogels as a novel air filters; synthesis and performance evaluation. Int J Biol Macromol 203:601–609. https://doi.org/10.1016/j.ijbiomac.2022.01.156
119. Shang Z, Pu F, Zhang X, Jin H, Chen S, Ding Y, Hu A (2023) A hyper-cross-linked aerogel with rigid conjugated polymers as building blocks for efficient iodine capture. ACS Appl Poly Mat 5(5):3827–3834. https://doi.org/10.1021/acsapm.3c00411
120. Shange MG, Khumalo NL, Mohomane SM, Motaung TE (2024) Factors affecting silica/cellulose nanocomposite prepared via the sol-gel technique: a review. Materials 17(9):1–23. https://doi.org/10.3390/ma17091937
121. Shen J, Guan D, Wu X, Zhao K (2023) Preparation and application of carbon aerogels. Springer Handbooks, Part F1485, pp 921–940. https://doi.org/10.1007/978-3-030-27322-4_35
122. Shi L, Steenland K, Li H, Liu P, Zhang Y, Lyles RH, Requia WJ, Ilango SD, Chang HH, Wingo T, Weber RJ, Schwartz J (2021) A national cohort study (2000–2018) of long-term air pollution exposure and incident dementia in older adults in the United States. Nat Commun 12:6754. https://doi.org/10.1038/s41467-021-27049-2
123. Singh RP (2019) Potential of biogenic plant-mediated copper and copper oxide nanostructured nanoparticles and their utility. In: Plant nanobionics (Issue October), pp 115–177. https://doi.org/10.1007/978-3-030-16379-2
124. Soffian MS, Abdul Halim FZ, Aziz F, A.Rahman M, Mohamed Amin MA, Awang Chee DN (2022) Carbon-based material derived from biomass waste for wastewater treatment. Environ Adv 9(March):100259. https://doi.org/10.1016/j.envadv.2022.100259
125. Stark FW, Thue PS, Missio AL, Machado FM, Delucis R d A, Andreazza R (2025) Cellulose-based aerogels for environmentally sustainable applications: a review of the production, modification, and sorption of environmental contaminants. Polymers 17(2). https://doi.org/10.3390/polym17020236
126. Stoian M, Rogé V, Lazar L, Maurer T, Védrine JC, Marcu IC, Fechete I (2021) Total oxidation of methane on oxide and mixed oxide ceria-containing catalysts. Catalysts 11(4). https://doi.org/10.3390/catal11040427
127. Su YQ, Xia GJ, Qin Y, Ding S, Wang YG (2021) Lattice oxygen self-spillover on reducible oxide supported metal cluster: the water-gas shift reaction on Cu/CeO2catalyst. Chem Sci 12(23):8260–8267. https://doi.org/10.1039/d1sc01201k

128. Su Q, Zhu C, Gong A, Long S, Zhang G, Wei Z, Wang X, Yang J (2024) Novel multi-hierarchical nanofiber aerogel for high efficient filtration of high-temperature flue gas. Sep Purif Technol 347(February):127573. https://doi.org/10.1016/j.seppur.2024.127573
129. Sumiyoshi T, Adachi I, Enomoto R, Iijima T, Suda R, Leonidopoulos C, Marlow DR, Prebys E, Kawabata R, Kawai H, Ooba T, Nanao M, Suzuki K, Ogawa S, Murakami A, Khan MHR (1999) Silica aerogel Cherenkov counter for the KEK B-factory experiment. Nucl Instrum Methods Phys Res Sect A 433(1):385–391. https://doi.org/10.1016/S0168-9002(99)00460-X
130. Sun H, Chai C, Qian Y, Ma L, Hao J, Dong S (2024) Biobased aerogels fabricated by the ice crystal method for smoke purifying. ACS Appl Poly Mat 6(3):1708–1719. https://doi.org/10.1021/acsapm.3c02375
131. Takeshita S, Zhao S, Malfait WJ, Koebel MM (2021) Chemistry of chitosan aerogels: three-dimensional pore control for tailored applications. Angewandte Chemie Int Edn 60(18):9828–9851. https://doi.org/10.1002/anie.202003053
132. Tillotson TM, Hrubesh LW (1991) Transparent ultralow-density silica aerogels prepared by a two-step sol-gel process. J Non Crystal Sol 145:44–50
133. Valencia L, Handa R, Monti S, Jasso-Salcedo AB, Georgouvelas D, Magaña I, Díaz de León R, Velikov KP, Mathew AP, Kumar S (2022) On the mineralization of nanocellulose to produce functional hybrid materials. J Mater Chem A 10(17):9248–9276. https://doi.org/10.1039/d2ta00457g
134. Vrabič-Brodnjak U (2024) Hybrid materials of bio-based aerogels for sustainable packaging solutions. Gels 10(1). https://doi.org/10.3390/gels10010027
135. Wan C, Jiao Y, Wei S, Zhang L, Wu Y, Li J (2019) Functional nanocomposites from sustainable regenerated cellulose aerogels: a review. Chem Eng J 359(August 2018):459–475. https://doi.org/10.1016/j.cej.2018.11.115
136. Wang Z, Zhu W, Huang R, Zhang Y, Jia C, Zhao H, Chen W, Xue Y (2014) Fabrication and characterization of cellulose acetate nanofibers. Adv Mater Res 1030-1032:78–81. https://doi.org/10.4028/www.scientific.net/AMR.1030-1032.78
137. Wang L, Zhang C, Gao F, Pan G (2016) Needleless electrospinning for scaled-up production of ultrafine chitosan hybrid nanofibers used for air filtration. RSC Adv 6(107):105988–105995. https://doi.org/10.1039/c6ra24557a
138. Wang C, Eisenreich F, Tomović Ž (2023a) Closed-loop recyclable high-performance polyimine aerogels derived from bio-based resources. Adv Mater 35(8). https://doi.org/10.1002/adma.202209003
139. Wang X, Li H, Liu H, Feng Z, Zhang B, Wei D, Liao X, Han R (2023b) A novel carbon-based fiber aerogel with interfacial thermal resistance: temperature insulation, oxidation resistance, and mechanical performance. Ceram Int 49(9):13698–13707. https://doi.org/10.1016/j.ceramint.2022.12.247
140. Wang Y, Liu L, Hou S (2025) Surface engineering as a potential strategy to enhance desiccation tolerance of beneficial bacteria. Front Microbiol 16(April):1–8. https://doi.org/10.3389/fmicb.2025.1576511
141. Wani MS, Denzer B, Caggiano NJ, Prud'homme RK, Arnold CB (2025) Hierarchically porous graphitic aerogels via thermal morphogenesis of proteins for environmental remediation. ACS Appl Nano Mater 8(16):8464–8472. https://doi.org/10.1021/acsanm.5c01156
142. Whang HS, Lim J, Choi MS, Lee J, Lee H (2019) Heterogeneous catalysts for catalytic CO2 conversion into value-added chemicals. BMC Chem Eng 1(1):1–19. https://doi.org/10.1186/s42480-019-0007-7
143. World Health Organization (2023) WHO ambient air quality database, 2022 update: Status report. World Health Organization. https://iris.who.int/server/api/core/bitstreams/34c921bb-f3ce-4dcd-a7f8-e405ecaed126/content
144. Wu X, Fan M, Mclaughlin JF, Shen X, Tan G (2018) A novel low-cost method of silica aerogel fabrication using fly ash and trona ore with ambient pressure drying technique. Powder Technol 323:310–322. https://doi.org/10.1016/J.POWTEC.2017.10.022

145. Yadav S, Kapoor TS, Vernekar P, Phuleria HC (2023) Examining the chemical and optical properties of biomass-burning aerosols and their impact on oxidative potential. Aerosol Air Qual Res 23(11):230102. https://doi.org/10.4209/aaqr.230102
146. Yahya EB, Idris M, Idrus R, Khalid K, Ngadiman S, Tye YY (2020) A review on revolutionary natural biopolymer-based aerogels in antibacterial delivery. Mater Adv 1(11):3971–3984. https://doi.org/10.1039/D0MA00624D
147. Yang X, Zhong H, Zhang W, Liu Y, Sun N, Kuang R, Wang C, Zhan A, Zhang J, Tang Q, Li Z (2023) Progress in adsorptive removal of volatile organic compounds by zeolites. Aerosol Air Qual Res 23(5):220442. https://doi.org/10.4209/AAQR.220442
148. Yao Y, Zhang X, Guo Z, Liu W, Hu C, Ru Y, Zhang L, Jiang C, Qiao J (2023) Preparation and application of recyclable polymer aerogels from styrene-maleic anhydride alternating copolymers. Chem Eng J 455(November 2022):140363. https://doi.org/10.1016/j.cej.2022.140363
149. Yu Q, Mu Y, Li P, Zhou W, Zhang J, Li J, Wei Y, Wang S (2025) Durable and high-efficiency air filtration by superamphiphobic silica composite aerogel. Colloid Interf 9(3). https://doi.org/10.3390/colloids9030038
150. Zafar M, Muhammad Imran S, Iqbal I, Azeem M, Chaudhary S, Ahmad S, Kim WY (2024) Graphene-based polymer nanocomposites for energy applications: recent advancements and future prospects. Results Phys 60(March):107655. https://doi.org/10.1016/j.rinp.2024.107655
151. Zeng Z, Ma XYD, Zhang Y, Wang Z, Ng BF, Wan MP, Lu X (2019) Robust lignin-based aerogel filters: high-efficiency capture of ultrafine airborne particulates and the mechanism. ACS Sustain Chem Eng 7(7):6959–6968. https://doi.org/10.1021/acssuschemeng.8b06567
152. Zhang W, Liang F, Li C, Qiu LG, Yuan YP, Peng FM, Jiang X, Xie AJ, Shen YH, Zhu JF (2011) Microwave-enhanced synthesis of magnetic porous covalent triazine-based framework composites for fast separation of organic dye from aqueous solution. J Hazard Mater 186(2–3):984–990. https://doi.org/10.1016/J.JHAZMAT.2010.11.093
153. Zhang H, Li Y, Xu Y, Lu Z, Chen L, Huang L, Fan M (2016) Versatile fabrication of a superhydrophobic and ultralight cellulose-based aerogel for oil spillage clean-up. Phys Chem Chem Phys 18(40):28297–28306. https://doi.org/10.1039/C6CP04932J
154. Zhang Q, Zheng Y, Tong D, Shao M, Wang S, Zhang Y, Xu X, Wang J, He H, Liu W, Ding Y, Lei Y, Li J, Wang Z, Zhang X, Wang Y, Cheng J, Liu Y, Shi Q, Yan L et al (2019) Drivers of improved $PM_{2.5}$ air quality in China from 2013 to 2017. Proc Natl Acad Sci USA 116(49):24463–24469. https://doi.org/10.1073/pnas.1907956116
155. Zhang SN, Pang HQ, Fan TH, Ye Q, Cai QL, Wu X (2022) Thermal insulation performance of SiC-doped silica aerogels under large temperature and air pressure differences. Gels 8(5). https://doi.org/10.3390/gels8050320
156. Zheng J, Yang C, Xue M, Li X, Zhao X (2024) Mass transfer kinetics of volatile organic compound desorption from a novel adsorbent. Processes 12(9):2031. https://doi.org/10.3390/PR12092031
157. Zhi M, Tang H, Wu M, Ouyang C, Hong Z, Wu N (2022) Synthesis and Photocatalysis of metal oxide aerogels: a review. Energy Fuel 36(19):11359–11379. https://doi.org/10.1021/acs.energyfuels.2c01049
158. Zhou S, Yuan H, Ma X, Liu Y (2017) Hair chemical element contents and influence factors of reproductive-age women in the West Ujimqin Banner, Inner Mongolia, China. Chemosphere 166:528–539. https://doi.org/10.1016/J.CHEMOSPHERE.2016.09.126

Chapter 10
Challenges and Prospective in Bio-based Nanotechnology for Environmental and Energy Applications

10.1 Challenges in Bio-based Nanotechnology

Bio-based nanotechnology, which utilizes nanomaterials derived from biological sources such as plants and microbes, holds significant promise for useful applications such as environmental remediation, agriculture, and biomedicine. However, the development and implementation of these technologies face several challenges. One major hurdle is the difficulty in processing biopolymers like cellulose and chitin into nanofibers due to their limited solubility and strong hydrogen bonding, necessitating specialized solvent systems and fabrication strategies. Additionally, the biosynthesis of nanomaterials using microorganisms, while environmentally friendly, often results in variability in the size and shape of the nanoparticles, posing challenges for standardization and scalability [28]. Environmental and health concerns also present significant obstacles. Nanoparticles can enter the environment through various pathways, potentially interacting with existing contaminants and causing unforeseen biological effects. In agricultural settings, the use of nanomaterials may lead to phytotoxicity, affecting plant growth and development. Moreover, the potential for nanoparticles to cause oxidative stress and DNA damage raises concerns about their safety for both human health and ecosystems. Economic factors further complicate the adoption of bio-based nanotechnologies. High production costs and the need for complex pretreatment processes can limit the scalability of these technologies, particularly in developing economies. Addressing these challenges requires continued research into safer synthesis methods, comprehensive risk assessments, and the development of cost-effective production techniques to ensure the sustainable integration of bio-based nanotechnology into various industries.

M. I. Ahmad, A. K. H. P. S., *Environmental and Energy Technology*, Green Energy and Technology, https://doi.org/10.1007/978-3-032-10706-0_10

10.1.1 Synthesis and Scalability of Bio-based Nanomaterials

The synthesis of bio-based nanomaterials frequently depends on biological sources such as plants, bacteria, fungi, and algae, which act as natural reducing and stabilizing agents. However, a major challenge lies in the inherent variability of these biological systems. Genetic diversity, fluctuations in environmental factors like temperature, pH, light intensity, and nutrient availability, as well as differences in the metabolic states of organisms, can all significantly influence the synthesis process. This variability often leads to inconsistencies in nanoparticle characteristics such as size distribution, shape, surface charge, and functional properties, ultimately affecting the reproducibility and reliability of the final product. For example, Gupta et al. [26] reported that even minor changes in microbial culture conditions can result in noticeable batch-to-batch variation, posing obstacles for quality control in large-scale production. To mitigate these issues, researchers are exploring strategies such as optimizing growth conditions, standardizing extract preparations, and utilizing genetic engineering to develop more robust, consistent, and predictable biosynthetic pathways for nanoparticle production.

Green synthesis methods provide a sustainable and environmentally benign alternative to conventional chemical approaches for producing nanomaterials. These methods typically utilize biological agents such as plant extracts or microbial cultures to mediate nanoparticle formation under mild conditions, reducing the need for toxic solvents and high-energy inputs. However, scaling these processes from the laboratory to industrial levels poses considerable technical and economic challenges. One major difficulty is maintaining the distinct physicochemical properties of nanoparticles such as uniform size, morphology, and stability when moving to larger volumes, where factors like mixing efficiency and temperature control become harder to regulate. Additionally, preventing nanoparticle aggregation during synthesis and storage at scale requires careful optimization of reaction conditions and stabilization strategies. The costs associated with cultivating biological systems on a large scale, along with the need for sterile environments and skilled operators, can make industrial implementation less economically viable. Thus, innovations in bioprocess engineering and scalable bioreactor designs are essential to ensure reproducibility, efficiency, and cost-effectiveness in green nanomaterial production [30].

Although green synthesis of nanomaterials is widely praised for its sustainability and reduced environmental footprint, it is crucial to assess the potential environmental and health risks associated with bio-based nanomaterials. Once released into the environment whether through manufacturing, application, or disposal, nanoparticles can interact with various biotic and abiotic components, potentially altering ecological dynamics. These particles may bind to existing pollutants, change their bioavailability, or introduce new forms of toxicity. In agricultural applications, some nanoparticles have been found to induce phytotoxic effects, inhibiting seed germination, root elongation, or chlorophyll production [30]. Furthermore, their ability to penetrate biological membranes and accumulate in tissues raises concerns about cytotoxicity, oxidative stress, and genotoxicity in humans and animals. The long-term behavior of bio-based nanoparticles in soil, water, and air remains poorly understood, necessitating thorough environmental fate studies. Therefore, comprehensive life-cycle assessments and standardized toxicological evaluations are essential to ensure that the adoption of these nanomaterials does not come at the cost of ecosystem or public health.

Optimizing the conditions for biological synthesis of nanomaterials is essential for ensuring high yields and consistent product quality, especially in industrial-scale production. Parameters such as pH, temperature, and nutrient supply must be carefully controlled, as even small deviations can lead to significant variations in nanoparticle size, shape, and functional properties. For instance, a study by Alsaiari et al. [7] reported that microorganisms used in synthesis processes require stable environmental conditions to thrive, and fluctuations can negatively impact the efficiency of nanoparticle formation. Bioreactors play a crucial role in maintaining optimal growth conditions by providing controlled environments for microbial cultures. These systems are designed to regulate parameters like temperature, agitation, and gas exchange to facilitate maximum productivity. Moreover, incorporating biological synthesis processes into existing manufacturing infrastructure presents challenges, such as ensuring technical compatibility with current systems, optimizing space for bioreactor integration, and training personnel to handle and monitor the biological processes effectively. The complexity of these operations underscores the need for precise planning and expertise.

The commercialization of bio-based nanomaterials faces significant regulatory and market hurdles, which can delay their widespread adoption. One of the primary challenges is the adherence to stringent regulatory standards that ensure the safety of these materials for human use and environmental impact. Given the novel nature of bio-based nanomaterials, regulatory bodies often lack standardized testing protocols, making it difficult to assess their safety and efficacy in a consistent and reliable manner. This uncertainty may lead to delays in product approval or even regulatory rejection. Additionally, bio-based nanomaterials face stiff competition from conventional, synthetic counterparts that are already well-established in the market and often less costly to produce. Overcoming market resistance, educating consumers, and ensuring trust in the new materials are critical to boosting consumer acceptance. To address these challenges, there is a need for strong collaborative efforts between government agencies, industry stakeholders, and academic institutions to develop clear regulatory frameworks and effectively communicate the benefits and safety of bio-based nanomaterials to the public [41].

Economic factors significantly influence the adoption and scalability of bio-based nanotechnologies. One of the most pressing challenges is the high production costs associated with bio-based processes, which often require expensive raw materials, specialized equipment, and complex pretreatment methods to prepare biomass feedstocks. These expenses can make bio-based nanomaterials less competitive compared to conventional synthetic alternatives, particularly in developing economies where financial resources and infrastructure may be limited. Ensuring a consistent, reliable, and economically viable supply of biomass is crucial to maintaining cost-effectiveness in large-scale production. Additionally, integrating biological synthesis processes into existing manufacturing systems often requires significant capital investment in new equipment, as well as retraining personnel to manage and operate these processes efficiently. Addressing these economic challenges demands continued research to identify cost-effective production techniques, improve the efficiency of biomass processing, and explore alternative feedstocks [35]. Moreover, supportive government policies, including subsidies, tax incentives, and research grants, are essential to encourage investment in the bio-based nanotechnology sector and foster its growth.

10.1.2 Stability and Durability in Real-World Applications

Nanomaterials possess a high surface area-to-volume ratio, which imparts them with extraordinary physical, chemical, and biological properties that are not typically observed in their bulk counterparts. This unique trait is largely responsible for their widespread utility across a broad range of applications, including medicine, electronics, catalysis, and environmental technologies. However, the same feature that makes nanomaterials advantageous also introduces a significant challenge instability. The high surface energy of nanoparticles, arising from the large proportion of surface atoms, makes them thermodynamically unstable and more reactive. As a result, they are susceptible to structural and compositional changes over time, such as agglomeration, dissolution, or oxidation. These changes can severely impact their performance and reliability in real-world conditions. For example, silver nanoparticles tend to oxidize easily, producing silver ions that can reduce their antimicrobial efficiency. Therefore, maintaining the structural and chemical stability of nanomaterials is essential for ensuring their long-term functionality and safe integration into various applications [9].

The surface chemistry of nanomaterials is a crucial factor that significantly influences their overall stability, durability, and performance in various environments. Nanoparticles possess a high proportion of surface atoms compared to bulk materials, and these surface atoms have fewer neighboring atoms, making them chemically and physically more reactive. This increased reactivity often leads to spontaneous interactions with molecules or ions in the surrounding medium, which can result in surface modifications, agglomeration, or even the breakdown of the nanoparticles. In biological systems, for instance, proteins and other biomolecules can adsorb onto the surface of nanoparticles, forming what is known as a "protein corona" [48]. This corona can drastically alter the physicochemical characteristics of the nanoparticle, affecting its biodistribution, cellular uptake, and ultimately, its effectiveness particularly in applications like targeted drug delivery. To overcome these issues, researchers often employ surface modification techniques using polymers, surfactants, or biocompatible coatings to improve colloidal stability and preserve the desired functionality of the nanoparticles.

Environmental factors play a critical role in determining the stability and performance of nanomaterials across different applications. External conditions such as pH, temperature, and ionic strength can greatly influence the physical and chemical behavior of nanoparticles, particularly their dispersion, surface interactions, and tendency to aggregate. For example, changes in pH can modify the surface charge of nanoparticles, which affects their electrostatic repulsion. When this repulsion is weakened, nanoparticles are more likely to aggregate, resulting in a loss of their distinct nanoscale properties. Similarly, in environments with high ionic strength, such as saline or biological fluids, ions in the medium can neutralize surface charges and diminish colloidal stability, further promoting agglomeration [24]. Temperature fluctuations can also pose challenges by affecting the kinetic energy of particles, accelerating reactions or destabilizing surface coatings. Because nanomaterials are often employed in complex or sensitive environments, carefully controlling these external conditions is essential to preserving their structure, function, and long-term effectiveness.

In biomedical applications, ensuring the stability and durability of nanomaterials is absolutely critical for their successful performance and therapeutic effectiveness. Nanoparticles designed for drug delivery must remain physically and chemically stable in the dynamic and complex environment of the human body, particularly within the bloodstream. Instability can lead to premature degradation, aggregation, or rapid clearance by immune cells, preventing the nanoparticles from reaching their intended targets [38]. In industrial and environmental applications, the long-term durability and resilience of nanomaterials under harsh operational conditions are of utmost importance. Nanoparticles used in processes such as catalysis, wastewater treatment, and pollution control are often exposed to challenging environments that include strong oxidizing or reducing agents, fluctuating temperatures, high pressures, and extreme pH levels. These factors can induce chemical or structural degradation of the nanoparticles, which in turn diminishes their performance, shortens their operational life, and increases the need for frequent replacement or maintenance. Additionally, compromised nanomaterials may lose their functional properties, such as catalytic efficiency or adsorption capacity. A further concern is the potential environmental risk posed by the unintended release of degraded or transformed nanoparticles into ecosystems, where they might interact with living organisms or accumulate over time. To address these issues, researchers are focused on engineering nanomaterials with enhanced durability, using strategies such as core-shell structures, surface coatings, and hybrid materials to improve resistance against environmental and operational stressors [68].

Overcoming the stability and durability challenges associated with nanomaterials demands a comprehensive and interdisciplinary approach. Innovations in material synthesis methods are essential for producing nanoparticles with more uniform sizes, controlled shapes, and tailored surface chemistries that contribute to greater stability. Surface engineering techniques, such as functionalizing nanoparticles with stabilizing agents, biocompatible polymers, or inorganic coatings, can significantly enhance resistance to chemical degradation, aggregation, and environmental stressors. The development of core-shell nanostructures, where a stable outer shell protects a functional core, has also proven effective in extending nanoparticle lifespan under harsh conditions [25]. Equally important is the rigorous characterization and testing of nanomaterials under simulated real-world conditions to assess their long-term behavior and interactions with surrounding media. Achieving reliable and sustainable nanomaterial applications ultimately relies on collaborative research efforts that integrate knowledge from chemistry, physics, materials science, environmental science, and biology to design nanoparticles that maintain functionality, safety, and durability throughout their lifecycle.

10.1.3 Toxicity and Environmental Safety Concerns

The toxicity of nanomaterials in biological systems is a significant concern, given their ability to interact with cellular structures at the nanoscale. Due to their small size, nanoparticles can easily penetrate biological barriers, such as cell membranes, allowing them to directly interact with cellular components. Once inside the body, these particles may accumulate in vital organs like the liver, lungs, and kidneys, potentially leading to toxic effects [54]. Nanomaterials can induce oxidative stress, inflammation, and cell apoptosis, causing cellular damage. Their ability to modify gene expression or interact with DNA also raises concerns about mutagenicity and carcinogenicity. Some nanoparticles, particularly those composed of metals like silver or titanium, are known to release ions that further contribute to toxicity. In addition, a study by Zhao et al. [71] reported that the surface properties of nanoparticles, such as charge and coating materials, can influence their interaction with biological systems, making it essential to assess their toxicity in a range of biological contexts.

Nanomaterials pose significant environmental risks, particularly when they are released into ecosystems through industrial activities, wastewater, or agricultural processes. Due to their small size and high surface area, nanoparticles can easily travel through water, soil, and air, making them highly mobile in the environment. When released into aquatic environments, these particles can accumulate in the tissues of aquatic organisms, leading to bioaccumulation and potential toxic effects on marine life. Nanoparticles can also disrupt soil health by interacting with microorganisms, affecting biodiversity and ecosystem functions. Additionally, their ability to transport other pollutants, such as heavy metals or organic contaminants, can enhance the toxicity of these substances. As nanomaterials are not always biodegradable, their persistence in ecosystems may lead to long-term environmental damage. The potential for unintended ecological consequences emphasizes the need for further research to understand how these materials behave in the environment and how to mitigate their negative impacts [71].

The effects of nanomaterials on aquatic life are of growing concern, particularly due to their small size and reactivity. Nanoparticles can easily enter aquatic organisms, either through ingestion or dermal absorption, leading to potential toxic effects. In many cases, nanoparticles can induce oxidative stress in fish, algae, and other aquatic species, damaging cellular structures such as membranes, proteins, and DNA. In some instances, exposure to nanomaterials has been shown to impair reproduction, growth, and survival rates in aquatic organisms. Furthermore, the interaction between nanoparticles and other pollutants can amplify toxicity, potentially leading to more severe environmental consequences [59]. Nanoparticles can also alter the microbial communities in water, affecting nutrient cycles and food webs. The ability of nanoparticles to accumulate in the tissues of aquatic species raises concerns about bioaccumulation in the food chain, which could ultimately impact human health if contaminated fish or shellfish are consumed. Understanding these risks is critical for managing the safe use of nanomaterials.

One of the most concerning aspects of nanomaterial toxicity is their potential for bioaccumulation. Nanoparticles, due to their small size and surface reactivity, can enter organisms via the skin, lungs, or digestive system. Once inside, they can accumulate in various organs, including the liver, kidneys, and spleen, where they may exert long-term toxic effects. The bioaccumulation of nanomaterials can lead to persistent exposure in organisms, even after the initial exposure has ended, potentially resulting in chronic toxicity. This is particularly concerning in the food chain, as nanomaterials can accumulate in the tissues of animals, passing through trophic levels and eventually reaching humans. For instance, fish or livestock that ingest nanomaterials may transfer these particles to humans through consumption [67]. This bioaccumulation can result in prolonged exposure to harmful substances, raising concerns about long-term health risks and environmental impacts. Therefore, addressing this issue requires understanding the mechanisms of bioaccumulation and developing safer, biodegradable nanomaterials.

The toxicity of nanomaterials is not uniform; rather, it is influenced by several key factors, including their size, shape, surface charge, and chemical composition. Smaller nanoparticles, with their higher surface area, tend to be more reactive and may exhibit greater toxicity due to increased interactions with biological and environmental systems. For example, nanoparticles with a high surface energy or reactive surface groups are more likely to induce oxidative stress or disrupt cellular structures. The shape of the nanoparticles also plays a significant role, as elongated or rod-like particles may penetrate cells more readily than spherical particles. Surface charge can influence the particle's ability to interact with biomolecules, potentially enhancing or reducing toxicity. Additionally, the chemical composition of the nanoparticles determines their potential for ion release, which can contribute to toxic effects. As a result, understanding how these properties influence toxicity is crucial for designing safer nanomaterials with controlled environmental and biological impact [71].

To address the toxicity concerns associated with nanomaterials, researchers have developed several strategies aimed at enhancing their safety profiles. One approach is surface modification, where nanoparticles are coated with biocompatible materials such as polyethylene glycol (PEG) to reduce reactivity and improve stability in biological environments. PEGylation, for instance, can help reduce the immune response and prolong the circulation time of nanoparticles in the bloodstream. Another strategy is the development of biodegradable nanomaterials, which break down over time into non-toxic byproducts, reducing their potential for long-term environmental or biological accumulation. Additionally, researchers are working on designing nanomaterials that are less likely to accumulate in living organisms or the environment, further minimizing risks. Rigorous testing and evaluation of nanomaterials in relevant biological and environmental conditions are necessary to ensure their safety. Furthermore, establishing standardized safety guidelines and regulatory frameworks will be essential for the responsible use of nanomaterials in various industries [37]. Figure 10.1 explores Schematic representation of the interaction pathways of nanomaterials in the environment and biological systems. Nanomaterials can originate from or impact the atmosphere, terrestrial ecosystems, and aquatic systems. These materials may enter the food chain through animals and plants, ultimately reaching humans and potentially causing direct or indirect effects on human health.

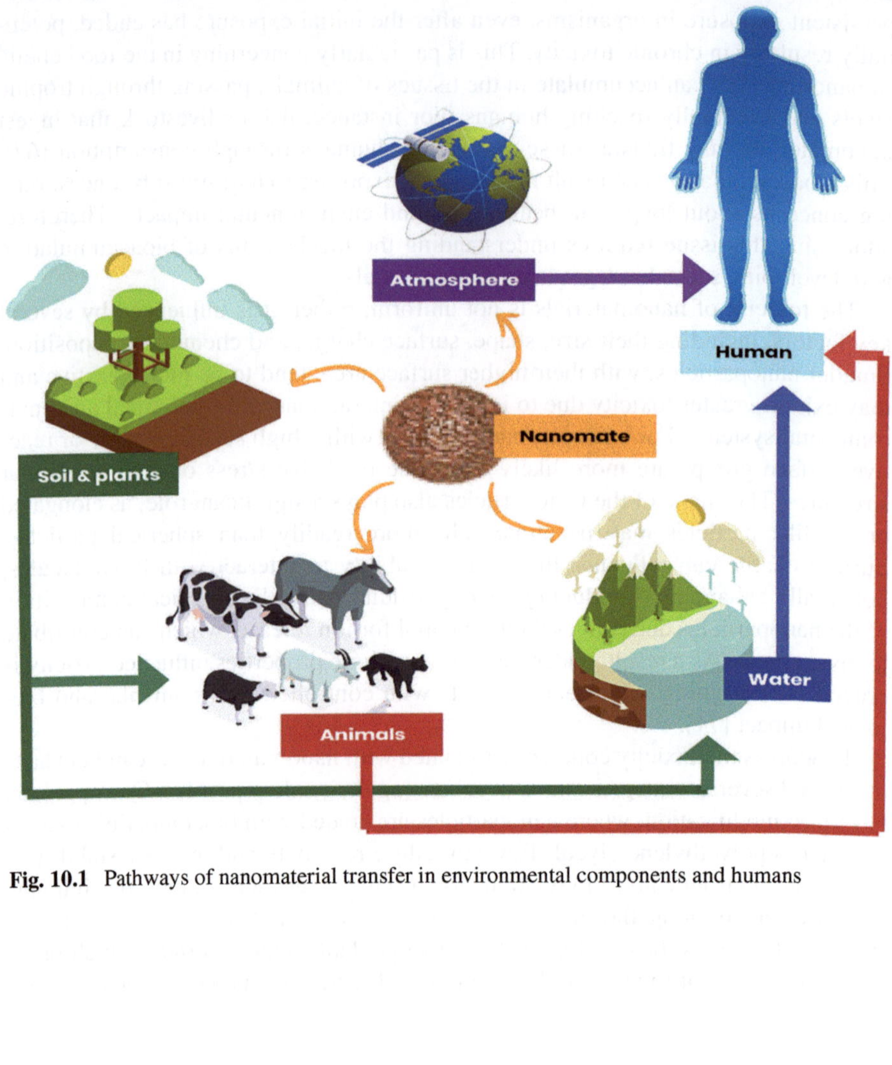

Fig. 10.1 Pathways of nanomaterial transfer in environmental components and humans

10.1.4 Economic Viability and Cost Barriers

The economic viability of nanomaterials is significantly limited by their high production costs. Nanomaterials often require advanced synthesis techniques, such as chemical vapor deposition, sol-gel processing, or atomic layer deposition, which are resource-intensive. These methods typically demand high energy input, clean room environments, and costly raw materials, thereby inflating the overall manufacturing expense. Moreover, specialized labor is essential for operating complex equipment and ensuring precision, further adding to operational costs [29]. Unlike conventional materials that benefit from decades of optimization and mass production, nanomaterials are still largely in a developmental stage, lacking the scale needed to reduce costs. This economic imbalance makes it difficult to justify their use in industries with tight profit margins, such as packaging, textiles, or automotive manufacturing. Without significant breakthroughs in low-cost, scalable production methods, the financial barriers to implementing nanomaterials will continue to restrict their broader industrial uptake and technological integration.

Integrating nanomaterials into existing industrial processes is not straightforward and can require substantial infrastructure changes. Most conventional manufacturing systems are not designed to handle the unique physical and chemical properties of nanomaterials, such as their tendency to agglomerate or their reactivity at nanoscale. As a result, industries must often invest in retrofitting existing equipment or purchasing entirely new systems tailored to nanomaterial handling. Additionally, the production and storage of nanomaterials may demand specialized containment to prevent contamination or degradation, especially in the case of reactive nanoparticles like silver or titanium dioxide. These adjustments significantly raise the initial investment costs, particularly for small- to medium-sized enterprises that may not have the capital flexibility to absorb such expenses. Without financial incentives, grants, or clearer cost-saving projections, many companies view these upgrades as prohibitively expensive. This presents a critical barrier to widespread adoption, as industries tend to favor technologies that align with existing infrastructure and workflows [29, 55].

The evolving regulatory landscape surrounding nanomaterials adds another layer of economic complexity. Because nanomaterials exhibit unique behaviors compared to their bulk counterparts, regulatory agencies often require additional safety assessments to evaluate their environmental and health impacts. These evaluations involve extensive testing, including toxicity studies, long-term environmental fate analyses, and compliance with emerging international standards. The cost of such testing can be enormous, especially when repeated across different jurisdictions [51] Companies must also invest in documentation, legal guidance, and staff training to ensure regulatory compliance—factors that can escalate operational expenditures. Furthermore, shifting policies and lack of harmonization between regulatory bodies worldwide create an unpredictable environment for investors and developers. Businesses may hesitate to commit to nanomaterials without assurance that future regulations won't render their products obsolete or force costly redesigns. These financial uncertainties discourage innovation and commercialization, particularly among startups that cannot afford the overhead of navigating complex and dynamic regulatory frameworks.

Economic viability is further challenged by issues related to public perception and market acceptance of nanomaterials. Despite the scientific enthusiasm, consumer trust in nanotechnology remains fragile, particularly in sectors such as food, cosmetics, and healthcare. Concerns over potential health risks and environmental contamination often dominate public discourse, regardless of scientific consensus. In response, companies may need to invest in public education campaigns, product labeling, and transparent communication strategies to build consumer confidence. These initiatives require significant marketing budgets and time, with no guaranteed returns. Moreover, the spread of misinformation or high-profile controversies can rapidly erode public trust and lead to regulatory or consumer backlash. The resulting drop in demand can hinder economies of scale, which are essential for driving down nanomaterial costs. Until widespread societal acceptance is achieved, nanomaterial producers must bear the cost of both mitigating consumer fears and operating within a cautious market environment, all of which impacts their bottom line [21].

Intellectual property plays a crucial role in determining the economic trajectory of nanomaterials, but it also presents notable barriers. Developing novel nanomaterials or innovative methods for their application involves intensive research and development, often over many years. Securing patents to protect these innovations is expensive and involves detailed legal processes. Additionally, navigating the complex landscape of existing patents is a challenge—companies must ensure they are not infringing upon prior claims, which may require hiring patent attorneys or conducting comprehensive intellectual property audits [52]. These procedures are financially and logistically burdensome, especially for startups and research-driven firms. Furthermore, intellectual property disputes can stall product rollouts and lead to costly litigation. The uncertainty surrounding freedom-to-operate and intellectual property exclusivity can dissuade potential investors and discourage collaborative efforts. In essence, while intellectual property protection is essential for ensuring return on innovation, the associated costs and complexities represent a formidable economic barrier to nanomaterials' advancement

A critical but often overlooked economic challenge in nanomaterial deployment is the absence of standardized testing and characterization methods. Without uniform benchmarks for properties like particle size distribution, surface area, or toxicity, it becomes difficult to compare products, ensure consistency, or meet regulatory and commercial requirements. This lack of standardization increases research and development costs, as each company must develop its own testing protocols or seek specialized services. It also complicates quality assurance and supply chain management, particularly when sourcing nanomaterials from third parties. Moreover, the absence of recognized industry standards can result in product rejections or recalls, further increasing financial risk. Smaller companies, in particular, may struggle to compete with larger firms that have in-house testing capabilities. Establishing global standards would streamline these processes and reduce duplication of effort, but until such frameworks are in place, the fragmented testing landscape will continue to impede cost-effective nanomaterial adoption and scale-up across industries [53].

10.1.5 Regulatory and Standardization Gaps

One of the primary challenges in regulating nanomaterials is the absence of universally accepted definitions and frameworks. Different countries and organizations have developed their own criteria for what constitutes a nanomaterial, leading to inconsistencies in regulation and oversight. For instance, the European Commission's recommendation defines nanomaterials based on particle size distribution, while other bodies may consider additional factors like surface area or reactivity. This lack of harmonization complicates international trade and collaboration, as manufacturers must navigate varying regulatory landscapes [6]. Furthermore, the rapid pace of nanotechnology development often outstrips the ability of regulatory bodies to adapt, resulting in outdated or inadequate guidelines. The inconsistency in definitions and frameworks not only hampers innovation but also poses risks to public health and the environment due to potential oversight gaps. Addressing this issue requires concerted efforts to develop standardized, globally recognized definitions and regulatory approaches for nanomaterials

The absence of standardized testing and characterization methods for nanomaterials presents significant challenges for their safe and effective application. Without uniform protocols, it becomes difficult to assess the toxicity, environmental impact, and overall safety of nanomaterials consistently. This variability can lead to conflicting results across studies, undermining the reliability of safety assessments. Moreover, the unique properties of nanomaterials, such as their high surface area-to-volume ratio and quantum effects, necessitate specialized testing methods that are not yet universally established. The lack of standardization also impedes regulatory approval processes, as authorities may question the validity of non-standardized test results. To mitigate these issues, international collaboration is essential to develop and adopt standardized testing and characterization protocols. Such standardization would enhance the comparability of data, facilitate regulatory approvals, and ultimately ensure the safe integration of nanomaterials into various industries [13].

Assessing and managing the risks associated with nanomaterials is complicated by their diverse and dynamic nature. Traditional risk assessment models often fall short in accounting for the unique behaviors of nanomaterials, such as their ability to penetrate biological membranes or their potential to accumulate in the environment [17]. Additionally, the long-term effects of exposure to nanomaterials are not well understood, making it challenging to predict their impact on human health and ecosystems. The lack of comprehensive risk assessment frameworks hinders the development of effective management strategies and regulatory policies. Furthermore, the uncertainty surrounding nanomaterial risks can lead to public apprehension and resistance to their adoption. To address these challenges, it is imperative to invest in research that elucidates the behavior and effects of nanomaterials and to develop risk assessment models tailored to their specific characteristics. Such efforts would inform evidence-based regulations and promote the responsible use of nanotechnology [31].

Effective regulation of nanomaterials is often hindered by limitations in oversight and enforcement capabilities. Regulatory agencies may lack the resources, expertise, or authority to monitor the production and use of nanomaterials adequately. This deficiency can result in insufficient surveillance of nanomaterial applications, increasing the risk of unintentional harm to health and the environment. Moreover, the rapid evolution of nanotechnology can outpace the development of appropriate regulatory tools and enforcement mechanisms. In some cases, existing regulations may not explicitly cover nanomaterials, leading to regulatory gaps and ambiguities [16]. Strengthening regulatory oversight requires enhancing the capacity of agencies through training, funding, and the development of specialized tools for monitoring nanomaterials. Additionally, establishing clear legal frameworks that specifically address nanomaterials will provide a more robust basis for enforcement and ensure that safety standards are upheld across the industry.

The integration of nanomaterials into consumer products and medical applications raises ethical considerations that are often overlooked in regulatory discussions. Issues such as informed consent, privacy, and the potential for unintended consequences necessitate careful ethical scrutiny. Furthermore, public engagement is crucial in shaping the acceptance and responsible development of nanotechnology. A lack of transparency and public involvement can lead to mistrust and resistance, as seen in past controversies surrounding genetically modified organisms. Regulatory frameworks must incorporate mechanisms for ethical review and public consultation to address societal concerns effectively. This approach ensures that the development and application of nanomaterials align with public values and ethical standards. By fostering open dialogue and incorporating ethical considerations into regulatory processes, stakeholders can build public trust and facilitate the responsible advancement of nanotechnology [22].

The global nature of nanotechnology development and commerce necessitates international harmonization of regulations and standards. Disparities in regulatory approaches across countries can create barriers to trade, hinder collaborative research, and lead to inconsistent safety standards. For instance, a nanomaterial approved for use in one country may be restricted or banned in another due to differing regulatory criteria. Such inconsistencies can impede the global deployment of beneficial nanotechnologies and create confusion among manufacturers and consumers [61]. International collaboration is essential to develop cohesive regulatory frameworks that facilitate the safe and equitable distribution of nanomaterials. Organizations like the Organisation for Economic Co-operation and Development (OECD) and the International Organization for Standardization (ISO) play pivotal roles in promoting harmonization efforts. By working together, countries can establish unified standards and regulations that support innovation while ensuring safety and public confidence in nanotechnology applications.

10.2 Sustainability and Life Cycle Considerations

Nanomaterials have revolutionized numerous industries, offering enhanced performance and novel functionalities. However, their production and application raise concerns regarding environmental and health impacts throughout their life cycle. Life Cycle Assessment (LCA) serves as a comprehensive tool to evaluate these impacts, encompassing stages from raw material extraction to end-of-life disposal. Despite its importance, many LCA studies on nanomaterials lack detailed inventory data and often overlook toxicity assessments, leading to incomplete evaluations. Addressing these gaps is crucial for developing sustainable nanotechnologies that minimize adverse effects while maximizing benefits. In this context, environmental impact, biodegradability, and circular economy will be discussed [44].

10.2.1 Environmental Impact of Bio-based Nanomaterials

Bio-based nanomaterials are increasingly gaining attention as sustainable and eco-friendly alternatives to traditional, synthetic nanomaterials. They are synthesized from renewable biological sources such as plants, fungi, algae, and bacteria, making them biodegradable and less likely to accumulate in the environment. Common examples include nanocellulose derived from plant fibers, chitosan obtained from crustacean shells, and lignin nanoparticles sourced from wood waste [65]. These materials are being explored for a variety of applications, ranging from agricultural enhancement and food packaging to water purification and environmental cleanup. One of their key environmental advantages lies in their production, which typically requires less energy and fewer harsh chemicals, resulting in lower greenhouse gas emissions and a reduced dependency on fossil resources. However, to fully understand the environmental impact of bio-based nanomaterials, thorough life cycle assessments are necessary. These should evaluate raw material sourcing, processing, usage, and end-of-life disposal to ensure these materials do not create unforeseen ecological or sustainability challenges.

In the agricultural sector, bio-based nanomaterials have emerged as promising tools for improving crop health, productivity, and resilience to environmental stressors. These nanomaterials, derived from renewable biological sources, interact with plants at the cellular level to boost nutrient absorption and stress resistance. For example, cellulose nanoparticles, obtained from plant biomass, can significantly enhance nutrient uptake by increasing root surface area and optimizing nutrient transport mechanisms within plant cells. Additionally, these nanoparticles have been found to trigger defense pathways, regulating gene expression related to stress tolerance and enabling plants to better endure harsh conditions such as drought, salinity, and temperature fluctuations. Chitosan nanoparticles, sourced from the deacetylation of chitin in crustacean shells, possess natural antimicrobial properties. They effectively inhibit the growth of pathogenic fungi and bacteria, reducing the risk of crop diseases. Overall, the use of bio-based nanomaterials in agriculture offers a sustainable approach to increasing yield while minimizing reliance on synthetic agrochemicals [63, 73].

Bio-based nanomaterials have demonstrated considerable potential in environmental remediation due to their unique physicochemical properties, particularly their high surface area, porosity, and abundance of reactive functional groups. These characteristics make them highly effective in adsorbing a wide range of environmental pollutants, including heavy metals, dyes, and organic compounds from contaminated water bodies. Nanocellulose, derived from plant fibers, and chitosan-based nanomaterials, obtained from natural biopolymers, have been successfully used to remove toxic substances from wastewater, offering an environmentally friendly, biodegradable alternative to conventional treatment materials [56]. Moreover, their hydrophobic nature allows certain bio-based nanomaterials to be employed in oil spill cleanup, where they can absorb and retain hydrocarbons efficiently. These capabilities support their use in restoring polluted ecosystems while reducing the ecological footprint associated with synthetic nanomaterials. Despite their advantages, further research is needed to assess the scalability, cost-effectiveness, and long-term environmental impacts of deploying bio-based nanomaterials in large-scale remediation projects.

Although bio-based nanomaterials are often promoted as environmentally friendly alternatives, their overall impact on the environment requires careful scrutiny. The cultivation of biomass to produce these nanomaterials can result in significant land use changes, potentially leading to habitat loss and deforestation. Additionally, large-scale production may demand considerable amounts of water and agricultural inputs, which could place pressure on natural resources and even compete with food production. Another concern lies in the environmental fate of bio-based nanomaterials once they are released into ecosystems. While they are generally regarded as biodegradable and less toxic than synthetic nanomaterials, their nano-scale dimensions and reactive surfaces might still present unforeseen risks to aquatic organisms and soil microbiota. Their degradation products, though organic, may have unknown ecological effects. As a result, comprehensive ecotoxicological assessments and long-term environmental monitoring are essential. These will ensure that bio-based nanomaterials are used safely and sustainably, minimizing unintended consequences to ecosystems and biodiversity [18].

The regulatory landscape surrounding bio-based nanomaterials is still in its formative stages, marked by a lack of clear, comprehensive guidelines specifically tailored to their unique properties and applications. Although these materials are derived from renewable sources and often considered more environmentally benign than their synthetic counterparts, their nano-scale features introduce complexities that are not yet fully addressed by existing regulatory frameworks. There is an urgent need for standardized testing protocols that evaluate their toxicity, environmental persistence, and behavior throughout their life cycle from production and usage to eventual degradation or disposal. This regulatory gap presents challenges for industries looking to adopt bio-based nanomaterials while ensuring safety for human health and the environment. Developing a robust, science-based regulatory framework requires an interdisciplinary approach that integrates insights from material science, toxicology, environmental chemistry, and public policy. Such frameworks should not only assess the risks but also promote innovation by offering clear guidance on safe production practices and allowable exposure limits. By doing so, regulatory agencies can help industries implement bio-based nanomaterials responsibly while minimizing potential adverse effects. Furthermore, comprehensive life cycle assessments and ecotoxicological evaluations must be embedded into regulatory planning to capture the full environmental footprint of bio-based nanomaterials. Collaborative efforts among researchers, industry leaders, and policymakers are essential to foster informed decision-making. As bio-based nanomaterials continue to play an increasing role in sustainable development, particularly in agriculture, water treatment, and packaging, the establishment of tailored regulations will be vital in ensuring their benefits are realized without compromising environmental or human safety [12, 32, 56]. Table 10.1 elaborates the positive and negative impacts of nanomaterials on the environment.

Table 10.1 Positive and negative environmental impacts of nanomaterials

Positive environmental impacts	Negative environmental impacts
Pollution remediation Nanomaterials can effectively remove pollutants from air, water, and soil, aiding in environmental cleanup efforts	Ecotoxicity Certain nanomaterials can be toxic to aquatic and terrestrial organisms, affecting biodiversity
Enhanced energy efficiency Incorporation of nanomaterials in solar cells and batteries improves energy conversion and storage, promoting renewable energy use	Bioaccumulation Nanoparticles can accumulate in organisms, leading to potential biomagnification through the food chain
Sustainable agriculture Nanopesticides and nanofertilizers can increase crop yields while reducing chemical runoff, minimizing environmental impact	Environmental persistence Some nanomaterials are resistant to degradation, leading to long-term environmental presence and potential harm
Waste management Nanotechnology facilitates advanced recycling processes, enabling the recovery of valuable materials and reducing landfill waste	Regulatory challenges The rapid development of nanotechnology outpaces regulatory frameworks, leading to potential environmental risks due to insufficient oversight
Climate change mitigation Nanomaterials enhance carbon capture and storage technologies, contributing to the reduction of greenhouse gas emissions	Unintended consequences The interaction of nanomaterials with ecosystems can lead to unforeseen and potentially harmful effects

10.2.2 Biodegradability and Recycling Challenges

Nanomaterials pose considerable challenges concerning their biodegradability and recyclability, which in turn affect overall environmental sustainability. Due to their unique characteristics such as their extremely small size, high surface area, and enhanced reactivity, they tend to persist in ecosystems, making their management and disposal particularly difficult. These properties not only complicate waste treatment processes but also raise concerns about their long-term environmental impact. The biodegradability of nanomaterials is governed by multiple interrelated factors, including their chemical structure, physical dimensions, and the conditions of the surrounding environment. For example, while some biodegradable polymers are designed to break down over time, they may not do so efficiently under natural environmental conditions especially in marine or agricultural settings, potentially resulting in the buildup of microplastics and causing unexpected ecological disruptions [66]. Additionally, environmental variables like temperature, pH, oxygen levels, and microbial activity significantly influence degradation rates. Hence, thoroughly understanding these interactions is crucial for evaluating ecological risks and improving the design of environmentally friendly nanomaterials.

Recycling nanomaterials presents a complex and pressing challenge due to their extremely small size, unique physicochemical properties, and widespread integration into a variety of commercial and industrial products. These materials are often embedded in composites, coatings, electronics, and packaging, making their detection and separation from bulk waste streams particularly difficult. Traditional recycling processes, which are designed for larger and more easily identifiable materials, are typically ineffective in isolating nanomaterials, resulting in their unintentional disposal in landfills or destruction through incineration. This contributes to potential environmental and health risks, as nanomaterials may persist in the ecosystem or release harmful byproducts during incineration. Compounding the issue is the absence of standardized protocols and regulatory frameworks for identifying, collecting, and recovering nanomaterials. To address these concerns, the development of innovative and scalable recycling technologies—such as magnetic separation, solvent extraction, or advanced filtration is essential. These methods could enable more sustainable management of nanomaterial waste and reduce ecological harm [19, 72].

The persistence of nanomaterials in the environment has sparked growing concern due to their potential toxicity, bioaccumulation, and long-term ecological impacts. Because of their nanoscale size and high reactivity, these materials can easily penetrate biological membranes, leading to adverse effects on living organisms. Research has demonstrated that certain nanomaterials, such as metal oxides and carbon-based nanoparticles, can induce oxidative stress, inflammation, and cytotoxicity in aquatic organisms. These toxic effects not only harm individual species but can also disrupt entire food chains, threatening biodiversity and ecosystem stability. Furthermore, the uncontrolled and often unmonitored release of nanoparticles into the environment whether through improper waste disposal, industrial runoff, or degradation of nanocomposites can result in cumulative and persistent environmental damage. To mitigate these risks, it is crucial to implement comprehensive environmental risk assessments and establish robust regulatory frameworks. These should guide the production, usage, and disposal of nanomaterials, ensuring they are managed responsibly and sustainably [11, 64].

As nanotechnology continues to advance rapidly, there is an increasing need for the development of robust and adaptive regulatory frameworks to manage the unique challenges associated with nanomaterials. Unlike conventional materials, nanomaterials exhibit distinct physical, chemical, and biological behaviors due to their nanoscale dimensions, which can pose unforeseen health and environmental risks. Despite their growing use in medicine, electronics, agriculture, and consumer products, there is currently a noticeable gap in comprehensive regulations specifically tailored to govern the production, application, and disposal of these materials. This regulatory shortfall creates uncertainties in risk management and public safety [60]. Therefore, establishing standardized testing protocols, toxicity evaluations, and safety assessment procedures is essential for guiding responsible innovation. Policymakers must work closely with scientists, environmental experts, and industry stakeholders to develop evidence-based guidelines that address the specific characteristics of nanomaterials. Such collaborative efforts will ensure their safe and sustainable integration into modern technologies while protecting human health and the environment.

To mitigate the environmental challenges posed by nanomaterials, scientists and researchers are increasingly turning to green synthesis methods that emphasize environmental safety and sustainability. These eco-friendly approaches involve the use of non-toxic solvents, biodegradable materials, and renewable feedstocks such as plant extracts, sugars, and natural polymers instead of hazardous chemicals typically used in conventional nanomaterial production. Green synthesis not only reduces the generation of harmful byproducts but also lowers energy consumption and waste, thereby minimizing the overall environmental footprint. Furthermore, integrating LCAs into the design, production, and post-use stages of nanomaterials is becoming an essential practice. LCAs allow researchers to evaluate the environmental impacts associated with each phase of a nanomaterial's lifespan, from raw material extraction to disposal [3]. This holistic evaluation helps identify hotspots of environmental concern and supports the development of safer, more sustainable nanotechnologies. Together, these strategies pave the way toward responsible innovation in the field of nanoscience.

Effectively addressing the challenges related to the biodegradability and recycling of nanomaterials demands a comprehensive, interdisciplinary approach that combines scientific innovation with policy development. These challenges are complex, involving not only technical limitations but also gaps in environmental understanding and regulatory oversight. Future research should prioritize the development of efficient, scalable degradation and recycling methods tailored to the unique properties of nanomaterials. In addition, it is critical to investigate the long-term environmental and ecological impacts of nanomaterials, including their behavior, transformation, and accumulation in natural systems. Establishing robust, science-based regulatory frameworks is equally important to ensure safe production, use, and disposal practices. Achieving these goals will require strong collaboration among researchers across disciplines, industry stakeholders committed to sustainable practices, and policymakers who can translate scientific findings into effective regulations. This cooperative effort is vital for advancing environmentally responsible nanotechnology and minimizing potential risks to ecosystems and public health [42]. Table 10.2 explores some biodegradable nanomaterials with their advantages and limitations.

10.2.3 Role in Circular Economy and Sustainable Development

Integrating nanomaterials into the circular economy is challenging due to their unique physicochemical properties and behaviors at the nanoscale. While these materials can significantly enhance product performance, their small size and high reactivity complicate recycling, reuse, and safe disposal. Unlike bulk materials, nanomaterials often exhibit unpredictable degradation patterns, which undermines traditional circular economy principles like longevity and recyclability. Their presence in complex composite products makes material separation and recovery inefficient or economically unfeasible. This situation is exacerbated by a lack of universal classification systems or standardized end-of-life protocols for nanomaterials, making it difficult to track, sort, and recycle them effectively. Furthermore, the infrastructure required to manage these materials sustainably is still underdeveloped. This gap hinders circular material flows and poses risks to both the environment and human health if nanomaterials persist or accumulate in ecosystems. Therefore, adapting circular economy frameworks to accommodate these materials requires innovative recycling strategies and targeted policy support [27, 36].

The environmental and health risks posed by nanomaterials across their lifecycle present major obstacles to their sustainable development. From synthesis to disposal, these materials can release harmful nanoparticles into ecosystems, often bypassing conventional filtration systems. In construction and manufacturing sectors, airborne nanoparticles may be inhaled or absorbed, posing toxicity risks such as oxidative stress or inflammatory responses. Once released into the environment, some nanoparticles persist and bioaccumulate, potentially disrupting aquatic food chains or altering soil microbiomes [8]. This raises serious concerns about their long-term ecological impact. These risks are further amplified by the limited biodegradability of most engineered nanomaterials. Moreover, current environmental risk assessments often lack the precision to evaluate nanoscale interactions, making regulatory oversight inconsistent. Without a thorough understanding of exposure pathways and bioavailability, it becomes difficult to establish safe thresholds for usage. Therefore, integrating robust environmental risk assessments into circular

Table 10.2 Pros and cons of some biodegradable nanomaterials

Biodegradable nanomaterials	Pros	Cons
Polyactic acid micelles	Hydrophobicity is helpful when transporting poorly soluble medications Physical and chemical characteristics that are readily adjustable to achieve desired pharmacokinetic and biodegradable results	Non-specific absorption by the systems of reticuloendothelium Low capacity to load drugs Low effectiveness of encapsulating Immunotoxicity that is depending on particle size Degradation rate can easily impact
Polylactic-co-glycolic acid micelles	Broad variety of erosion times Mechanical qualities that can be changed By varying the PLA:PGA ratio and their molecular weights, the rate of degradation can be altered	Some medications and bioactive compounds cannot be used with PLGA monomers due of their acidic nature Achieving the ideal medication release profile is challenging Dose-dependent, non-linear, and readily modified pharmacokinetics and biodistribution
Poly-e-caprolactone nanoparticles	Acidic byproducts are not produced during degradation Ideal for long-term implantation devices, slow degradation rate Adaptable mechanical characteristics	Its manufacturing is limited by its hydrophobicity
Chitosan nanoparticles	Good moisture retention, permeability, and absorbability Degradable with ease Minimal toxicity	Extremely susceptible to ambient temperature Environmental pH has an impact on the pace of degradation Inadequate stability throughout time
Dendrimers	High degree of branching and polyvalency, with very high surface-to-volume ratio, enabling high drug carrying efficiency Capable of carrying drugs with poor solubility High water solubility beneficial as an agent for MRI imaging	Cytotoxicity may occur during the interactions with cell membrane depending on the charge and modifications made on the surface of the dendrimer
Liposomes	Self-assembly, which makes drug loading simple High efficiency of loading Protect encapsulated drugs from early inactivation, degradation and dilution in the circulation can be prepared in a variety of ways for different administration routes Surface modification allows for easy functionalization.	Fast clearance rate Poor stability Iintricate production process

Adapted from Su and Kang [62])

economy models is vital for mitigating unintended consequences of nanomaterial use.

Recycling nanomaterials remains a critical bottleneck in achieving circularity. Their small size and tendency to embed into composite materials render conventional mechanical or thermal recycling methods ineffective. For example, electronic waste containing nanomaterial coatings or additives often ends up in landfills because separating nanoscale components without damaging them or releasing them into the environment is technically and economically challenging. Furthermore, there are currently no widely adopted technologies or industrial standards for identifying, sorting, or recovering nanomaterials from waste streams. Even advanced methods like solvent extraction, magnetic separation, or chemical leaching often struggle with scalability, cost, and potential secondary pollution. These limitations significantly reduce the recyclability rate of products containing nanomaterials and compromise the principles of a closed-loop economy. Overcoming these hurdles will require the development of novel recovery technologies tailored to nanoscale properties, as well as eco-design principles that consider recyclability from the outset of material and product development [45].

The regulatory landscape for nanomaterials is fragmented and underdeveloped, hindering their integration into circular economy strategies. Most current regulations classify materials based on bulk properties and fail to account for the unique behaviors of nanostructured materials. This gap leads to regulatory blind spots, particularly regarding their environmental persistence and health risks. Inconsistencies across national and international jurisdictions make it difficult for manufacturers and recyclers to adopt uniform sustainability practices. Furthermore, the absence of standardized testing protocols for nanomaterial toxicity, environmental impact, and recyclability prevents effective risk communication and hinders public trust. These regulatory deficiencies also delay innovation, as companies remain uncertain about compliance expectations. Collaborative efforts between scientists, policymakers, and industry stakeholders are needed to establish globally harmonized standards and lifecycle-based policies. Such frameworks should promote transparency in nanomaterial usage and disposal, enable data sharing, and incentivize sustainable design through certifications or extended producer responsibility models [23].

Sustainable development of nanomaterials must begin with greener synthesis methods that minimize environmental harm and resource depletion. Traditional nanomaterial production often involves toxic solvents, high energy consumption, and hazardous byproducts. In contrast, green synthesis utilizes biodegradable solvents, renewable feedstocks, and energy-efficient techniques to reduce environmental footprints. For instance, biosynthesis approaches using plant extracts or microbial systems offer promising alternatives that align with circular economy principles. However, these methods still face challenges such as scalability, consistency in particle quality, and economic feasibility [44]. Moreover, integrating sustainability at the design stage requires comprehensive LCAs that evaluate the full environmental cost of nanomaterials from raw material extraction to end-of-life disposal. LCAs can help identify hotspots for improvement and guide innovation toward more circular, eco-friendly alternatives. Research and development in this area must be supported by public and private investments to accelerate the transition toward truly sustainable nanotechnologies.

The complex challenges associated with the circular economy and sustainability of nanomaterials cannot be addressed in isolation. Interdisciplinary collaboration is essential, combining expertise from material science, environmental engineering, toxicology, economics, and policy-making. Researchers must work alongside industry leaders to design materials that are both high-performing and sustainable throughout their lifecycle. At the same time, policymakers must stay informed by scientific developments to craft adaptive regulations that encourage innovation without compromising safety. Education and public awareness also play a role, raising consumer understanding of nanomaterials impacts and promoting demand for greener alternatives. Moreover, integrating circular economy thinking into nanotechnology innovation hubs and funding programs can incentivize sustainable design from the earliest stages. Ultimately, the goal is to create a value chain where nanomaterials contribute to, rather than compromise, sustainability goals. Achieving this vision demands coordinated action, long-term commitment, and shared responsibility across disciplines and sectors [27, 36].

10.3 Future Prospects and Opportunities

Nanomaterials hold immense promise for the future, with opportunities spanning across various sectors including medicine, energy, electronics, and environmental protection. Their unique properties such as high surface area, enhanced strength, and improved conductivity, make them ideal for developing advanced drug delivery systems, efficient solar cells, and lightweight, durable materials. In medicine, nanomaterials offer targeted therapies with reduced side effects, while in energy, they contribute to better storage and conversion systems. As research advances, sustainable and large-scale production of nanomaterials is becoming more feasible, paving the way for their integration into everyday technologies and offering innovative solutions to global challenges [46]. In this section, multifunctional bio-nanomaterials, integration with AI and IoT, updates on energy systems, and collaborative research and policies will be discussed.

10.3.1 Smart and Multifunctional Bio-nanomaterials

Smart and multifunctional bio-nanomaterials represent a transformative frontier in biomedical science, offering versatile and adaptive capabilities that respond to specific physiological stimuli. These innovative materials are engineered to change their physical, chemical, or biological properties when exposed to triggers such as pH variations, temperature changes, enzymatic activity, or specific biomolecular interactions within the body. This responsiveness allows for highly targeted therapeutic interventions and accurate diagnostic procedures, improving overall treatment outcomes. For example, temperature-sensitive polymers can transition between states at normal body temperatures, enabling site-specific and controlled drug release. Likewise, pH-responsive nanocarriers take advantage of the acidic environment commonly found in tumors to selectively deliver chemotherapeutic drugs, thereby reducing damage to healthy tissues and minimizing systemic toxicity. Such smart delivery systems increase drug efficacy while lowering side effects. The integration of these stimuli-responsive mechanisms into nanomaterials significantly enhances their performance, positioning them as essential components in the design of next-generation diagnostic tools and personalized therapeutic systems [1].

In the realm of biosensing, smart bio-nanomaterials have significantly transformed the detection, identification, and monitoring of a wide range of biomolecules with high accuracy and efficiency. These advanced materials leverage the unique properties of nanostructures such as carbon nanotubes, quantum dots, and metal nanoparticles to enhance the performance of biosensors in terms of sensitivity, specificity, and responsiveness. The incorporation of such nanomaterials enables the detection of minute concentrations of analytes by amplifying signals generated during molecular interactions. This results in faster and more reliable signal transduction, which is crucial for real-time monitoring of vital physiological parameters [39]. A prominent example is the development of nanomaterial-based glucose sensors, which allow for continuous glucose monitoring in diabetic patients, greatly improving personalized disease management. Furthermore, the versatility of smart bio-nanomaterials supports the creation of multiplexed sensing platforms capable of simultaneously identifying multiple biomarkers. This innovation facilitates the advancement of comprehensive diagnostic tools for early disease detection and precision healthcare.

The application of smart bio-nanomaterials in targeted drug delivery systems has opened new avenues for improving the precision and effectiveness of medical treatments. Traditional drug delivery methods often suffer from limitations such as non-specific distribution, low bioavailability, and undesirable side effects. Smart nanocarriers, however, can overcome these challenges through strategic surface modifications that enable active targeting of diseased cells, such as cancerous tissues or inflamed areas. These modifications may involve the attachment of ligands, antibodies, or peptides that recognize specific cellular markers, ensuring that therapeutic agents are directed precisely to the affected site [20]. Moreover, these nanocarriers can be engineered to respond to specific intracellular conditions such as acidic pH, redox potential, or the presence of particular enzymes -common in pathological environments. Once inside the targeted cells, the nanocarriers release their drug payload in a controlled manner. This smart response not only increases the therapeutic efficacy of the drug but also significantly minimizes systemic toxicity and side effects.

Beyond therapeutic applications, smart bio-nanomaterials have emerged as powerful tools in the fields of tissue engineering and regenerative medicine, where they contribute to the restoration of damaged tissues and organs. These materials are designed to closely mimic the architecture and biochemical cues of the natural extracellular matrix, which is essential for supporting cell adhesion, migration, proliferation, and differentiation. Nanostructured scaffolds, particularly those made from electrospun nanofibers, offer a highly porous and biocompatible environment conducive to tissue growth. These scaffolds can also be functionalized to release bioactive molecules, such as growth factors or cytokines, in a controlled and sustained manner, thereby enhancing tissue repair and regeneration processes. Moreover, integrating conductive nanomaterials, such as graphene or gold nanoparticles, into scaffolds supports the electrical communication necessary for regenerating electrically active tissues like neurons and cardiac muscle. These multifunctional smart systems represent a critical advancement in regenerative therapies, enabling more effective and personalized tissue repair strategies [14].

In the context of environmental monitoring and food safety, smart bio-nanomaterials provide cutting-edge solutions for the rapid detection and effective control of contaminants. These materials can be engineered into highly sensitive nanosensors capable of detecting extremely low concentrations of harmful pathogens, chemical residues, or toxins in food, water, and air. Their high surface area, coupled with unique optical or electrical properties, enables quick and accurate responses, making them ideal for real-time, on-site analysis. This allows for early identification of contamination, supporting swift decision-making to prevent the spread of harmful agents and ensuring public safety. Additionally, certain nanomaterials, such as silver or zinc oxide nanoparticles, possess strong antimicrobial properties [2]. These can be integrated into food packaging materials to create active packaging systems that inhibit the growth of bacteria and fungi, thereby prolonging the shelf life of perishable products. These diverse capabilities highlight the potential of smart bio-nanomaterials in safeguarding environmental and food systems, extending far beyond biomedical applications.

Looking to the future, the integration of smart bio-nanomaterials with advanced digital technologies is expected to revolutionize healthcare by enabling personalized medicine and continuous health monitoring. Wearable and implantable devices embedded with nanomaterial-based sensors can monitor a range of physiological signals such as heart rate, glucose levels, body temperature, or biochemical markers in real time. This continuous data collection provides critical insights for early diagnosis, timely interventions, and improved disease management. Furthermore, the emergence of theranostic platforms, systems that combine therapeutic delivery with diagnostic capabilities which allows for precise, targeted treatment based on a patient's unique physiological condition. For example, a single nanodevice could detect the presence of a tumor, deliver medication directly to it, and monitor treatment progress simultaneously [40]. However, to bring these innovations to widespread clinical use, several challenges must be addressed. Key issues include ensuring the long-term biocompatibility of nanomaterials, developing cost-effective large-scale manufacturing methods, and navigating complex regulatory frameworks to ensure safety and efficacy.

Figure 10.2 illustrates various smart bio-nanomaterial applications in fabric textiles, highlighting their versatility across multiple sectors. At the center is the integration of bio-nanomaterials, which enhance textile performance. These applications

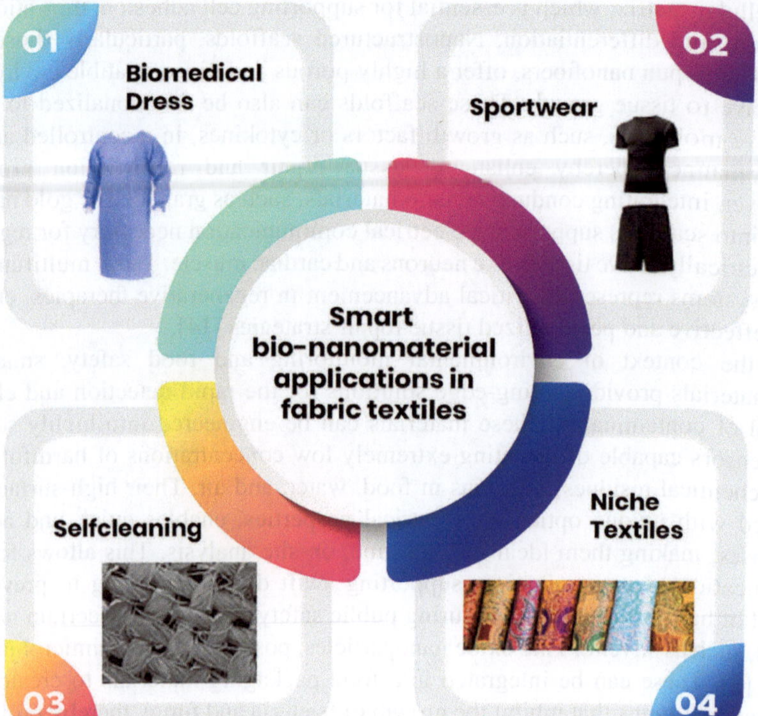

Fig. 10.2 Smart bio-nanomaterial applications as fabric textiles

include self-cleaning fabrics, which use nanotechnology to repel dirt and liquids, and biomedical dresses, designed for hygienic and protective use in healthcare settings. Sportswear benefits from improved breathability and antibacterial properties, while niche textiles showcase advanced aesthetics and functionality for specialized or luxury uses. Together, these innovations demonstrate how smart bio-nanomaterials are transforming traditional fabrics into multifunctional, high-performance textiles for diverse industries [58].

10.3.2 Integration with AI and IoT for Environmental Monitoring

The integration of nanomaterials with Artificial Intelligence (AI) and the Internet of Things (IoT) is revolutionizing environmental monitoring by enabling real-time detection and analysis of pollutants. Nanomaterials, due to their high surface area and unique physicochemical properties, serve as sensitive components in sensors that detect various environmental contaminants. When these sensors are connected via IoT networks, they facilitate continuous data collection across diverse locations.

AI algorithms process this vast amount of data to identify patterns, predict pollution trends, and provide actionable insights. For instance, AI-driven systems can analyze data from nanomaterial-based sensors to detect hazardous chemicals in air, water, and soil, allowing for timely interventions and mitigation strategies [49]. This synergy enhances the precision and responsiveness of environmental monitoring systems, contributing to more effective pollution management and protection of public health.

In air quality monitoring, the combination of nanomaterials, AI, and IoT technologies has led to the development of advanced sensor networks capable of detecting airborne pollutants with high sensitivity. Nanomaterial-based sensors can identify trace levels of harmful substances such as volatile organic compounds, nitrogen oxides, and particulate matter. These sensors transmit data in real-time through IoT networks to centralized systems where AI algorithms analyze the information to assess air quality, identify pollution sources, and forecast pollution levels. Such systems enable authorities to issue timely health advisories and implement pollution control measures. Moreover, the scalability of IoT networks allows for widespread deployment of these sensors, facilitating comprehensive air quality monitoring across urban and rural areas [70]. This integrated approach enhances our ability to understand and manage air pollution, ultimately contributing to healthier environments

Water quality monitoring has also benefited significantly from the integration of nanomaterials with AI and IoT technologies. Nanomaterial-enhanced sensors can detect contaminants such as heavy metals, pesticides, and pathogens at very low concentrations. These sensors, when connected through IoT networks, provide continuous monitoring of water bodies, transmitting data to centralized systems for analysis. AI algorithms process this data to identify contamination events, assess the severity of pollution, and predict future water quality trends. This real-time monitoring capability allows for prompt responses to contamination incidents, ensuring the safety of drinking water and the protection of aquatic ecosystems. Additionally, AI-driven analysis can optimize water treatment processes by adjusting parameters in response to changing water quality, enhancing the efficiency and effectiveness of water purification systems [49].

In soil pollution monitoring, the integration of nanomaterials, AI, and IoT technologies has enabled the development of sophisticated systems for detecting and managing soil contaminants. Nanomaterial-based sensors can identify pollutants such as heavy metals, pesticides, and hydrocarbons with high sensitivity. These sensors, deployed in agricultural fields and industrial sites, transmit data through IoT networks to centralized platforms where AI algorithms analyze the information to assess soil health and contamination levels. This analysis supports decision-making in agriculture by informing soil remediation strategies and guiding the application of fertilizers and pesticides [10]. Furthermore, AI-driven models can predict the movement and transformation of pollutants in the soil, aiding in the development of long-term land management plans. This integrated approach enhances our ability to maintain soil quality and ensure sustainable land use.

Despite the significant advancements, the integration of nanomaterials with AI and IoT for environmental monitoring presents several challenges. Ensuring the accuracy and reliability of sensor data is critical, as environmental conditions can affect sensor performance. Regular calibration and maintenance of sensors are necessary to maintain data quality. Data security and privacy are also major concerns, as the transmission and storage of environmental data must be protected against unauthorized access and cyber threats. Additionally, the interoperability of devices from different manufacturers poses challenges in creating cohesive monitoring networks [70]. Standardization of communication protocols and data formats is essential to address this issue. Furthermore, the cost of deploying and maintaining these advanced monitoring systems can be prohibitive, particularly in resource-limited settings. Addressing these challenges requires collaborative efforts among researchers, industry stakeholders, and policymakers to develop robust, secure, and cost-effective solutions.

Looking ahead, the continued integration of nanomaterials with AI and IoT technologies holds great promise for advancing environmental monitoring. Emerging developments such as the use of machine learning algorithms for predictive analytics, the deployment of autonomous sensor networks, and the integration of satellite data with ground-based sensors are expected to enhance the capabilities of monitoring systems. [69, 70]. These advancements will enable more comprehensive and proactive environmental management strategies. Moreover, the development of low-cost, energy-efficient sensors will facilitate the widespread adoption of these technologies, including in developing regions where environmental monitoring is often limited. As these technologies evolve, they will play a crucial role in addressing global environmental challenges, supporting sustainable development, and protecting public health.

10.3.3 Bio-based Nanotechnology in Emerging Energy Systems

Bio-based nanotechnology is increasingly recognized as a transformative force in the field of energy systems, offering a powerful convergence of sustainability and cutting-edge material science. At its core, this discipline leverages nanomaterials derived from renewable biological sources such as cellulose, lignin, chitin, and other biopolymers to replace or enhance traditional, often fossil-fuel-based materials used in energy applications. These bio-derived nanomaterials present an environmentally friendly alternative, not only because they originate from sustainable sources but also due to their biodegradability and lower carbon footprint throughout their lifecycle. What makes them particularly suitable for modern energy systems are their remarkable physicochemical properties. These include a high surface area, tunable porosity, and modifiable surface chemistry, which together enhance their ability to store and transfer energy efficiently [33]. Such characteristics make them ideal candidates for use in batteries, supercapacitors, fuel cells, and solar energy devices. Furthermore, the advancement of bio-based nanomaterials supports broader global initiatives aimed at mitigating climate change, reducing greenhouse gas emissions, and decreasing reliance on non-renewable fossil fuels. By integrating these green nanotechnologies into energy infrastructure, researchers and industries alike are paving the way for a more sustainable and resilient energy future.

Nanocellulose, a nanoscale material derived from the structural components of plant biomass, has emerged as one of the most promising bio-based nanomaterials for use in advanced energy systems. It is typically extracted from cellulose-rich sources like wood, agricultural residues, or even algae through chemical or mechanical processess [57]. What sets nanocellulose apart is its unique combination of high mechanical strength, renewability, biodegradability, and extraordinarily large surface area, all of which are key attributes for enhancing the efficiency and durability of energy devices. In energy storage technologies such as batteries and supercapacitors, nanocellulose has proven especially valuable as a component in electrodes and separators, where it improves ion transport and structural stability. These features translate to higher charge storage capacity, longer cycle life, and improved safety. Furthermore, when integrated into composite materials, nanocellulose facilitates the development of lightweight, flexible, and bendable energy storage systems, which are essential for the evolution of wearable electronics and portable technologies.

Carbon-based nanomaterials derived from biomass, particularly from lignocellulosic sources such as lignin, cellulose, and hemicellulose, are rapidly emerging as highly versatile and sustainable materials for energy applications. Through various carbonization and activation processes, these natural polymers can be transformed into nanostructured carbon materials with excellent electrical conductivity, large surface areas, and highly tunable pore architectures. These characteristics are critical for their effectiveness in energy storage and conversion technologies, such as batteries, fuel cells, and especially supercapacitors. Bio-derived carbon electrodes, in particular, have demonstrated high energy density and outstanding power capabilities, enabling rapid charge-discharge cycles and long-term electrochemical stability. Beyond their functional advantages, the use of biomass as a carbon precursor offers substantial environmental and economic benefits [15, 47]. It reduces reliance on fossil-based carbon sources while simultaneously valorizing agricultural and industrial byproducts. This approach supports the development of a circular bioeconomy, where waste materials are repurposed into high-value nanomaterials for clean energy solutions.

The integration of bio-based nanomaterials into energy systems represents a significant advancement toward achieving sustainable, high-performance, and environmentally responsible energy technologies. These nanomaterials, derived from renewable biological sources such as agricultural residues, algae, and forest byproducts, offer unique benefits including biodegradability, low toxicity, and minimal environmental impact. Because they originate from naturally abundant and renewable feedstocks, their production helps reduce dependence on finite fossil resources, ultimately lowering the carbon footprint associated with traditional energy storage and conversion devices. Additionally, they can often be sourced from waste materials, enhancing the economic feasibility of their use and contributing to circular economy models. Technological progress in nanotechnology now allows scientists to manipulate the structural, morphological, and chemical characteristics of these materials at the nanoscale. This degree of control enables the fine-tuning of properties such as conductivity, porosity, and thermal stability to meet the specific demands of diverse energy systems, from lithium-ion batteries to supercapacitors and fuel cells. As a result, bio-based nanomaterials can significantly improve energy density, charge-discharge rates, and long-term device stability. However, despite their promising potential, there are challenges that must be addressed to fully realize their commercialization. Inconsistencies in raw biomass sources and variations in processing methods can result in unpredictable material performance. To overcome this, researchers are working on developing standardized synthesis protocols, scalable manufacturing methods, and quality control mechanisms [4]. Collaborative efforts among academic institutions, industrial stakeholders, and governmental bodies are vital to accelerate progress and bring these innovations to market. Bio-based nanotechnology offers a compelling and sustainable path forward for the energy sector. With continued investment in research and development, these materials have the potential to revolutionize the way energy is stored and utilized, supporting global efforts to combat climate change and foster energy resilience for future generations [57].

10.3.4 Collaborative Research and Policy Development

Collaborative research and policy development are essential for the responsible and effective advancement of nanomaterials. Given the highly interdisciplinary nature of nanotechnology, successful innovation depends on the cooperation of researchers, engineers, industry leaders, and policymakers. This collaboration helps address complex scientific, ethical, environmental, and regulatory challenges that arise with the rapid development of nanomaterials. A leading example of this approach is the National Nanotechnology Initiative in the United States, which brings together more than 20 federal agencies under a unified strategy to support research, foster innovation, and ensure that nanotechnologies are developed with consideration for human health and environmental safety [50]. Through such collaborative frameworks, stakeholders can share scientific knowledge, technical resources, and regulatory expertise. This integrated effort not only accelerates technological progress but also ensures that new discoveries are guided by sound ethical and safety standards. Ultimately, these partnerships lay the foundation for more informed, transparent, and adaptive policy-making in the nanotechnology sector.

Public-private partnerships are fundamental in bridging the gap between laboratory research and real-world applications in the field of nanotechnology. These collaborations foster synergy between academic institutions, government agencies, and private sector companies, enabling the transformation of theoretical innovations into commercially viable technologies. The European Union actively promotes such partnerships through funding programs and strategic initiatives designed to accelerate the adoption of nanotechnology across key industries [34]. These joint efforts allow for the sharing of technical expertise, infrastructure, and financial resources, significantly speeding up the research-to-market pipeline. By fostering open communication between researchers and industrial stakeholders, public-private partnerships ensure that scientific advancements are tailored to meet practical demands in sectors such as healthcare, renewable energy, and advanced electronics. This alignment between innovation and market needs not only enhances the societal impact of nanotechnology but also improves its economic sustainability. Ultimately, these partnerships are vital for ensuring that nanotech solutions are both innovative and accessible.

International cooperation is increasingly vital in the establishment of standardized regulatory frameworks for nanomaterials, as nanotechnology transcends national boundaries through global research collaborations, trade, and industrial applications. The diversity of nanomaterial types and their widespread usage across sectors such as medicine, electronics, and energy makes regulatory harmonization a critical priority. Without consistent global standards, discrepancies in safety evaluations and labeling can hinder innovation and pose risks to public health and the environment. Organizations like the Organization for Economic Cooperation and Development (OECD) play a leading role in fostering international dialogue and collaboration. The OECD's efforts include developing comprehensive guidelines that take into account both the intrinsic physicochemical properties of nanomaterials and their behavior in specific environmental and biological systems [34]. These harmonized protocols aim to streamline regulatory approval processes, reduce redundancy in testing, and build public trust in nanotechnology.

Government agencies play a pivotal role in ensuring the responsible and sustainable development of nanotechnology by overseeing research, regulation, and safety assessments. Agencies such as the U.S. Environmental Protection Agency (EPA) are at the forefront of investigating the environmental fate, transport, and potential toxicity of nanomaterials. Through rigorous studies, the EPA evaluates how these materials behave in air, water, and soil, and assesses their long-term impacts on ecosystems and human health. This research is essential for developing evidence-based regulations that minimize potential environmental and public health risks ([43]. Likewise, the National Institute for Occupational Safety and Health (NIOSH) plays a critical role in protecting workers who manufacture or handle nanomaterials. NIOSH collaborates with international bodies to establish occupational exposure limits, implement safe handling procedures, and monitor emerging risks. Together, these agencies produce valuable data, safety guidelines, and risk assessment frameworks that shape national policies and support industry compliance with safety and environmental standards

Despite significant progress through collaborative research and policy initiatives, several challenges continue to hinder the effective regulation and societal acceptance of nanotechnology. One of the primary obstacles lies in the inherent complexity and diversity of nanomaterials. These materials vary widely in size, shape, composition, and behavior, making it difficult to establish standardized definitions and uniform regulatory frameworks. This complexity can delay the approval of new nanotech products and impede international trade due to inconsistent safety protocols across countries. Public perception also presents a notable challenge. Many individuals remain skeptical about the potential risks of nanomaterials, particularly concerning their long-term effects on health and the environment. This skepticism is often fueled by a lack of accessible information and limited public involvement in policy-making processes. Therefore, transparent communication and inclusive stakeholder engagement are essential. Educating the public, industry professionals, and policymakers about the benefits, risks, and regulatory efforts surrounding nanotechnology can help build trust and promote informed decision-making. Fostering interdisciplinary and cross-sector partnerships remains a cornerstone of sustainable nanotechnology development [5]. By working together, scientists, governments, industries, and communities can align technological progress with ethical standards, public interest, and safety expectations. Ongoing dialogue and cooperation are crucial to responsibly unlocking nanotechnology's transformative potential

10.4 Summary

Bio-based nanotechnology is emerging as a promising field that bridges sustainability and advanced material science, particularly in environmental and energy applications. Derived from renewable biological sources such as cellulose, lignin, and other plant-based materials, bio-based nanomaterials offer biodegradable, eco-friendly alternatives to conventional nanomaterials. Their unique structural features in terms of high surface area, tunable porosity, and biocompatibility, make them highly effective in energy storage systems, including batteries and supercapacitors, and in environmental remediation processes. Nanocellulose and bio-derived carbon nanomaterials have shown significant potential in improving energy efficiency, reducing reliance on fossil fuels, and promoting circular economy practices by utilizing agricultural and industrial waste. However, despite their promise, the widespread adoption of bio-based nanomaterials faces multiple challenges. These include difficulties in scaling up production, variability in raw material sources, and a lack of standardized characterization and regulatory frameworks. Public perception and risk awareness concerning nanomaterial safety also pose barriers to commercialization. Addressing these concerns requires robust, interdisciplinary collaboration among researchers, industry leaders, government agencies, and policymakers. Efforts such as international cooperation for harmonized regulation, public-private partnerships for innovation, and government-led safety assessments are instrumental in overcoming current limitations. Looking ahead, continued investment in research, transparent communication, integrations with AI and IoT, and inclusive policy development will be essential to fully realize the transformative potential of bio-based nanotechnology. With strategic collaboration and scientific advancement, bio-based nanomaterials can play a critical role in shaping a sustainable, low-carbon future.

References

1. Aflori M (2021) Smart nanomaterials for biomedical applications—a review. Nanomaterials 11(2):396. https://doi.org/10.3390/nano11020396
2. Aghababai Beni A, Jabbari H (2022) Nanomaterials for environmental applications. Res Eng 15:100467. https://doi.org/10.1016/j.rineng.2022.100467
3. Ahmed SF, Mofijur M, Rafa N, Chowdhury AT, Chowdhury S, Nahrin M, Islam ABMS, Ong HC (2022) Green approaches in synthesising nanomaterials for environmental nanobioremediation: technological advancements, applications, benefits and challenges. Environ Res 204:111967. https://doi.org/10.1016/j.envres.2021.111967
4. Al Rai A, Yanilmaz M (2021) High-performance nanostructured bio-based carbon electrodes for energy storage applications. Cellulose 28(9):5169–5218. https://doi.org/10.1007/s10570-021-03881-z
5. Albalawi F, Hussein MZ, Fakurazi S, Masarudin MJ (2021) Engineered nanomaterials: the challenges and opportunities for nanomedicines. Int J Nanomedicine Volume 16:161–184
6. Allan J, Belz S, Hoeveler A, Hugas M, Okuda H, Patri A, Rauscher H, Silva P, Slikker W, Sokull-Kluettgen B, Tong W, Anklam E (2021) Regulatory landscape of nanotechnology and nanoplastics from a global perspective. Regul Toxicol Pharmacol 122:104885. https://doi.org/10.1016/j.yrtph.2021.104885
7. Alsaiari NS, Alzahrani FM, Amari A, Osman H, Harharah HN, Elboughdiri N, Tahoon MA (2023) Plant and microbial approaches as green methods for the synthesis of nanomaterials: synthesis, applications, and future perspectives. Molecules 28(1):463. https://doi.org/10.3390/molecules28010463
8. Angrisano M, Fabbrocino F (2023) The relation between environmental risk analysis and the use of nanomaterials in the built environment sector: a circular economy perspective. Recent Progress in Materials 05(01):1–21. https://doi.org/10.21926/rpm.2301005
9. Baig N, Kammakakam I, Falath W (2021) Nanomaterials: a review of synthesis methods, properties, recent progress, and challenges. Mat Adv 2(6):1821–1871. https://doi.org/10.1039/D0MA00807A
10. Bala M, Khanna V (2025) Comparative analysis of nanomaterials and artificial intelligence for sustainable nutrient management in soil, pp 137–158. https://doi.org/10.1007/978-3-031-76953-5_6
11. Bystrzejewska-Piotrowska G, Golimowski J, Urban PL (2009) Nanoparticles: their potential toxicity, waste and environmental management. Waste Manag 29(9):2587–2595. https://doi.org/10.1016/j.wasman.2009.04.001
12. Catalán J, Norppa H (2017) Safety aspects of bio-based nanomaterials. Bioengineering 4(4):94. https://doi.org/10.3390/bioengineering4040094
13. Cerrillo C, Barandika G, Igartua A, Areitioaurtena O, Mendoza G (2017) Key challenges for nanotechnology: standardization of ecotoxicity testing. J Environ Sci Health C 35(2):104–126. https://doi.org/10.1080/10590501.2017.1298361
14. Choi S, Tripathi A, Singh D (2014) Smart nanomaterials for biomedics. J Biomed Nanotechnol 10(10):3162–3188. https://doi.org/10.1166/jbn.2014.1933
15. Chowdhury MS, Oliullah MS, Islam RT, Hurayra MA, Al Mahmud MZ, Hasan N, Rahman MK, Islam MS, Khan J, Hossain N (2025) Biomaterials for energy storage: synthesis, properties, and performance. Green Technologies and Sustainability 3(2):100152. https://doi.org/10.1016/j.grets.2024.100152
16. Creutzenberg O (2021) Nanoparticles and their regulation. In: Regulatory toxicology. Springer, Berlin/Heidelberg, pp 1–17. https://doi.org/10.1007/978-3-642-36206-4_122-1
17. Dyadyura K, Ivakhniuk T, Hrebenyk L, Ivakhniuk U, Sukhodub L (2021) Standardization issues of test methods for engineering nanomaterials, pp 797–805. https://doi.org/10.1007/978-3-030-68014-5_77
18. Elsaid K, Olabi AG, Wilberforce T, Abdelkareem MA, Sayed ET (2021) Environmental impacts of nanofluids: a review. Sci Total Environ 763:144202. https://doi.org/10.1016/j.scitotenv.2020.144202

19. Falinski MM, Turley RS, Kidd J, Lounsbury AW, Lanzarini-Lopes M, Backhaus A, Rudel HE, Lane MKM, Fausey CL, Barrios AC, Loyo-Rosales JE, Perreault F, Walker WS, Stadler LB, Elimelech M, Gardea-Torresdey JL, Westerhoff P, Zimmerman JB (2020) Doing nano-enabled water treatment right: sustainability considerations from design and research through development and implementation. Environ Sci Nano 7(11):3255–3278. https://doi.org/10.1039/D0EN00584C
20. Fatima H, Naz MY, Shukrullah S, Aslam H, Ullah S, Assiri MA (2022) A review of multi-function smart nanoparticle based drug delivery systems. Curr Pharm Des 28(36):2965–2983. https://doi.org/10.2174/1381612828666220422085702
21. Ferreira MT, Soldado E, Borsoi G, Mendes MP, Flores-Colen I (2023) Nanomaterials applied in the construction sector: environmental, human health, and economic indicators. Appl Sci 13(23):12896. https://doi.org/10.3390/app132312896
22. Festus-Ikhuoria IC, Obiuto NC, Adebayo RA, Olajigae OK (2023) Nanotechnology in consumer products: a review of applications and safety considerations. World J Adv Res Rev 21(3):2050–2059. https://doi.org/10.30574/wjarr.2024.21.3.0923
23. Foulkes R, Man E, Thind J, Yeung S, Joy A, Hoskins C (2020) The regulation of nanomaterials and nanomedicines for clinical application: current and future perspectives. Biomater Sci 8(17):4653–4664. https://doi.org/10.1039/D0BM00558D
24. García-Quintero A, Palencia M (2021) A critical analysis of environmental sustainability metrics applied to green synthesis of nanomaterials and the assessment of environmental risks associated with the nanotechnology. Sci Total Environ 793:148524. https://doi.org/10.1016/j.scitotenv.2021.148524
25. Gottardo S, Mech A, Drbohlavová J, Małyska A, Bøwadt S, Riego Sintes J, Rauscher H (2021) Towards safe and sustainable innovation in nanotechnology: state-of-play for smart nanomaterials. NanoImpact 21:100297. https://doi.org/10.1016/j.impact.2021.100297
26. Gupta D, Boora A, Thakur A, Gupta TK (2023) Green and sustainable synthesis of nanomaterials: recent advancements and limitations. Environ Res 231:116316. https://doi.org/10.1016/j.envres.2023.116316
27. Hansen SF, Arvidsson R, Nielsen MB, Hansen OFH, Clausen LPW, Baun A, Boldrin A (2022) Nanotechnology meets circular economy. Nat Nanotechnol 17(7):682–685. https://doi.org/10.1038/s41565-022-01157-6
28. Hunt N, Kestens V, Rasmussen K, Badetti E, Soeteman-Hernández LG, Oomen AG, Peijnenburg W, Hristozov D, Rauscher H (2025) Regulatory preparedness for multicomponent nanomaterials: current state, gaps and challenges of REACH. NanoImpact 37:100538. https://doi.org/10.1016/j.impact.2024.100538
29. Iavicoli I, Leso V, Ricciardi W, Hodson LL, Hoover MD (2014) Opportunities and challenges of nanotechnology in the green economy. Environ Health 13(1):78. https://doi.org/10.1186/1476-069X-13-78
30. Jafarzadeh S, Nooshkam M, Zargar M, Garavand F, Ghosh S, Hadidi M, Forough M (2024) Green synthesis of nanomaterials for smart biopolymer packaging: challenges and outlooks. J Nanostructure Chem 14(2):113–136. https://doi.org/10.1007/s40097-023-00527-3
31. Jillavenkatesa A (2017) Introduction: an overview of Nanotechnolgy and nanomaterial standardization and opportunities and challenges. In: Metrology and standardization of nanotechnology. Wiley, pp 1–14. https://doi.org/10.1002/9783527800308.ch1
32. Kabir E, Kumar V, Kim K-H, Yip ACK, Sohn JR (2018) Environmental impacts of nanomaterials. J Environ Manag 225:261–271. https://doi.org/10.1016/j.jenvman.2018.07.087
33. Kausar A (2024) Energy systems and green sourced nanomaterials—a today's outlook. Mat Technol Rep 2(2):1879. https://doi.org/10.59400/mtr1879
34. Khatoon UT, Velidandi A (2025) An overview on the role of government initiatives in nanotechnology innovation for sustainable economic development and research Progress. Sustainability 17(3):1250. https://doi.org/10.3390/su17031250

35. Kulkarni AG, De Britto S, Jogaiah S (2021) Economic considerations and limitations of green synthesis vs chemical synthesis of nanomaterials. In: Advances in nano-fertilizers and nano-pesticides in agriculture. Elsevier, pp 459–468. https://doi.org/10.1016/B978-0-12-820092-6.00018-5
36. Kumar A, Tyagi PK, Tyagi S, Ghorbanpour M (2024) Integrating green nanotechnology with sustainable development goals: a pathway to sustainable innovation. Discov Sustain 5(1):364. https://doi.org/10.1007/s43621-024-00610-x
37. Liu Y, Zhu S, Gu Z, Chen C, Zhao Y (2022) Toxicity of manufactured nanomaterials. Particuology 69:31–48. https://doi.org/10.1016/j.partic.2021.11.007
38. Mabrouk M, Das DB, Salem ZA, Beherei HH (2021) Nanomaterials for biomedical applications: production, characterisations, recent trends and difficulties. Molecules 26(4):1077. https://doi.org/10.3390/molecules26041077
39. Malik S, Singh J, Goyat R, Saharan Y, Chaudhry V, Umar A, Ibrahim AA, Akbar S, Ameen S, Baskoutas S (2023) Nanomaterials-based biosensor and their applications: a review. Heliyon 9(9):e19929. https://doi.org/10.1016/j.heliyon.2023.e19929
40. Manoharan AK, Batcha MIK, Mahalingam S, Raj B, Kim J (2024) Recent advances in two-dimensional nanomaterials for healthcare monitoring. ACS Sens 9(4):1706–1734. https://doi.org/10.1021/acssensors.4c00015
41. Nanomater JMS, Laura D (2024) Biocompatibility of nanomaterials: challenges and solutions. J Mat Sci Nanomat 8(2). https://doi.org/10.4172/jmsn.100124
42. Nath H, Subhranshu SS, Manohara SR (2021) Current guidelines and regulatory challenges, insight into the legal, societal, and ethical issues of nanomaterials. In: Nanotoxicology. CRC Press, pp 445–463
43. Neu-Baker NM, Eastlake A, Hodson L (2022) Results of the 2019 survey of engineered nanomaterial occupational health and safety practices. Int J Environ Res Public Health 19(13):7676. https://doi.org/10.3390/ijerph19137676
44. Nizam NUM, Hanafiah MM, Woon KS (2021) A content review of life cycle assessment of nanomaterials: current practices, challenges, and future prospects. Nanomaterials 11(12):3324. https://doi.org/10.3390/nano11123324
45. Omran BA, Baek K-H (2022) Valorization of agro-industrial biowaste to green nanomaterials for wastewater treatment: approaching green chemistry and circular economy principles. J Environ Manag 311:114806. https://doi.org/10.1016/j.jenvman.2022.114806
46. Ong W-J, Zheng N, Antonietti M (2021) Advanced nanomaterials for energy conversion and storage: current status and future opportunities. Nanoscale 13(22):9904–9907. https://doi.org/10.1039/D1NR90103F
47. Österberg M, Henn KA, Farooq M, Valle-Delgado JJ (2023) Biobased nanomaterials—the role of interfacial interactions for advanced materials. Chem Rev 123(5):2200–2241. https://doi.org/10.1021/acs.chemrev.2c00492
48. Phan HT, Haes AJ (2019) What does nanoparticle stability mean? J Phys Chem C 123(27):16495–16507. https://doi.org/10.1021/acs.jpcc.9b00913
49. Popescu SM, Mansoor S, Wani OA, Kumar SS, Sharma V, Sharma A, Arya VM, Kirkham MB, Hou D, Bolan N, Chung YS (2024) Artificial intelligence and IoT driven technologies for environmental pollution monitoring and management. Front Environ Sci 12. https://doi.org/10.3389/fenvs.2024.1336088
50. Powers CM, Grieger KD, Beaudrie C, Hendren CO, Michael Davis J, Wang A, Sayes CM, MacDonell M, Gift JS (2015) Data dialogues: critical connections for designing and implementing future nanomaterial research. Environ Syst Decis 35(1):76–87. https://doi.org/10.1007/s10669-014-9518-1
51. Qureshi F, Yusuf M, Tahir M, Haq M, Mohamed MMI, Kamyab H, Nguyen H-HT, Vo D-VN, Ibrahim H (2023) Renewable hydrogen production via biological and thermochemical routes: nanomaterials, economic analysis and challenges. Process Saf Environ Prot 179:68–88. https://doi.org/10.1016/j.psep.2023.07.075

52. Rachamalla H, Mukherjee A (2022) Nanotechnology application and intellectual property right. In: Cell culture: advanced technology and applications in medical and life sciences, vol 117
53. Radnik J, Kersting R, Hagenhoff B, Bennet F, Ciornii D, Nymark P, Grafström R, Hodoroaba V-D (2021) Reliable surface analysis data of nanomaterials in support of risk assessment based on minimum information requirements. Nanomaterials 11(3):639. https://doi.org/10.3390/nano11030639
54. Sajid M (2022) Nanomaterials: types, properties, recent advances, and toxicity concerns. Curr Opin Environ Sci Health 25:100319. https://doi.org/10.1016/j.coesh.2021.100319
55. Saleh TA (2024) Trends in nanomaterial types, synthesis methods, properties and uses: toxicity, environmental concerns and economic viability. Nano Struct Nano Obj 37:101109. https://doi.org/10.1016/j.nanoso.2024.101109
56. Saraswat P, Singh S, Prasad M, Misra R, Rajput VD, Ranjan R (2023) Applications of bio-based nanomaterials in environment and agriculture: a review on recent progresses. Hybrid Adv 4:100097. https://doi.org/10.1016/j.hybadv.2023.100097
57. Sezali NAA, Ong HL, Villagracia AR, Hoang T (2024) Bio-based nanomaterials for energy application: a review. Vietnam J Chem 62(1):1–12. https://doi.org/10.1002/vjch.202300158
58. Sherlin HSN, Tarangini K (2023) A review on augmentation of natural fabric materials with novel bio/nanomaterials and their multifunctional perspectives. Hybrid Advances 2:100020. https://doi.org/10.1016/j.hybadv.2023.100020
59. Shinde SA, More PR, Ingle AP (2023) Hazardous effects of nanomaterials on aquatic life. In: Nanotechnology in agriculture and agroecosystems. Elsevier, pp 423–450. https://doi.org/10.1016/B978-0-323-99446-0.00012-X
60. Slimani Y, Hannachi E, Yasin G (2022) Air pollution management by nanomaterials. In: Nanotechnology for environmental pollution decontamination. Apple Academic Press, pp 409–441
61. Soltani AM, Pouypouy H (2019) Standardization and regulations of nanotechnology and recent government policies across the world on nanomaterials. In: Advances in phytonanotechnology. Elsevier, pp 419–446. https://doi.org/10.1016/B978-0-12-815322-2.00020-1
62. Su S, Kang PM (2020) Systemic review of biodegradable nanomaterials in nanomedicine. Nanomaterials 10(4):656. https://doi.org/10.3390/nano10040656
63. Verma SK, Kumar P, Mishra A, Khare R, Singh D (2024) Green nanotechnology: illuminating the effects of bio-based nanoparticles on plant physiology. Biotechnol Sustain Mat 1(1):1. https://doi.org/10.1186/s44316-024-00001-2
64. Vineeth Kumar CM, Karthick V, Kumar VG, Inbakandan D, Rene ER, Suganya KSU, Embrandiri A, Dhas TS, Ravi M, Sowmiya P (2022) The impact of engineered nanomaterials on the environment: release mechanism, toxicity, transformation, and remediation. Environ Res 212:113202. https://doi.org/10.1016/j.envres.2022.113202
65. Wiederrecht GP, Bachelot R, Xiong H, Termentzidis K, Nominé A, Huang J, Kamat PV, Rozhkova EA, Sumant A, Ostraat M, Jain PK, Heckle C, Li J, Pupek KZ (2023) Nanomaterials and Sustainability. ACS Energy Lett 8(8):3443–3449. https://doi.org/10.1021/acsenergylett.3c01303
66. Yaseen AH, Ghazi A, Awad AM, Al-Dabagh MZN, Bodnar N, Najem NR, Ahmed SS, Salman AS (2024) A comprehensive review on the environmental impacts and innovations in nanostructured biodegradable polymers. J Nanostruct 14(4):1107–1121
67. Zeng H, Lv Z, Sun X, Tong Y, Wu W, Dong S, Mao L (2024) Predicting bioaccumulation of nanomaterials: modeling approaches with challenges. Environ Health 2(4):189–201. https://doi.org/10.1021/envhealth.3c00138
68. Zhang L (2020) Applications, challenges and development of nanomaterials and nanotechnology. J Chem Soc Pak 42(5)
69. Zhang P, Guo Z, Ullah S, Melagraki G, Afantitis A, Lynch I (2021) Nanotechnology and artificial intelligence to enable sustainable and precision agriculture. Nat Plants 7(7):864–876. https://doi.org/10.1038/s41477-021-00946-6

70. Zhang S, Wei S, Liu Z, Li T, Li C, Huang XL, Wang C, Xie Z, Al-Hartomy OA, Al-Ghamdi AA, Wageh S, Gao J, Tang Y, Wang H, Wang Q, Zhang H (2022) The rise of AI optoelectronic sensors: from nanomaterial synthesis, device design to practical application. Mat Today Phys 27:100812. https://doi.org/10.1016/j.mtphys.2022.100812
71. Zhao J, Lin M, Wang Z, Cao X, Xing B (2021) Engineered nanomaterials in the environment: are they safe? Crit Rev Environ Sci Technol 51(14):1443–1478. https://doi.org/10.1080/10643389.2020.1764279
72. Živković D, Balanović L, Mitovski A, Talijan N, Štrbac N, Sokić M, Manasijević D, Minić D, Ćosović V (2014) Nanomaterials environmental risks and recycling: actual issues. Reciklaza i Odrzivi Razvoj 7(1):1–8. https://doi.org/10.5937/ror1401001Z
73. Żukowska G, Durczyńska Z, Roszkowski S, Myszura-Dymek M, Bik-Małodzińska M (2024) Possibilities of using bio-based nanomaterials in sustainable agriculture. J Ecol Eng 25(3):313–322. https://doi.org/10.12911/22998993/182966

Index

M. I. Ahmad, A. K. H. P. S., *Environmental and Energy Technology*, Green Energy and Technology, https://doi.org/10.1007/978-3-032-10706-0

MIX
Papier aus verantwortungsvollen Quellen
Paper from responsible sources
FSC® C105338

If you have any concerns about our products,
you can contact us on
ProductSafety@springernature.com

In case Publisher is established outside the EU,
the EU authorized representative is:
Springer Nature Customer Service Center GmbH
Europaplatz 3, 69115 Heidelberg, Germany

Printed by Libri Plureos GmbH
in Hamburg, Germany